Y0-BSD-947

The Gravitational Constant: Generalized Gravitational Theories and Experiments

NATO Science Series

A Series presenting the results of scientific meetings supported under the NATO Science Programme.

The Series is published by IOS Press, Amsterdam, and Kluwer Academic Publishers in conjunction with the NATO Scientific Affairs Division

Sub-Series

I. Life and Behavioural Sciences	IOS Press
II. Mathematics, Physics and Chemistry	Kluwer Academic Publishers
III. Computer and Systems Science	IOS Press
IV. Earth and Environmental Sciences	Kluwer Academic Publishers
V. Science and Technology Policy	IOS Press

The NATO Science Series continues the series of books published formerly as the NATO ASI Series.

The NATO Science Programme offers support for collaboration in civil science between scientists of countries of the Euro-Atlantic Partnership Council. The types of scientific meeting generally supported are "Advanced Study Institutes" and "Advanced Research Workshops", although other types of meeting are supported from time to time. The NATO Science Series collects together the results of these meetings. The meetings are co-organized bij scientists from NATO countries and scientists from NATO's Partner countries – countries of the CIS and Central and Eastern Europe.

Advanced Study Institutes are high-level tutorial courses offering in-depth study of latest advances in a field.
Advanced Research Workshops are expert meetings aimed at critical assessment of a field, and identification of directions for future action.

As a consequence of the restructuring of the NATO Science Programme in 1999, the NATO Science Series has been re-organised and there are currently Five Sub-series as noted above. Please consult the following web sites for information on previous volumes published in the Series, as well as details of earlier Sub-series.

http://www.nato.int/science
http://www.wkap.nl
http://www.iospress.nl
http://www.wtv-books.de/nato-pco.htm

Series II: Mathematics, Physics and Chemistry – Vol. 141

The Gravitational Constant: Generalized Gravitational Theories and Experiments

edited by

Venzo de Sabbata
University of Bologna, Italy

George T. Gillies
University of Virginia, U.S.A.

and

Vitaly N. Melnikov
Moscow University, Russia

Kluwer Academic Publishers

Dordrecht / Boston / London

Published in cooperation with NATO Scientific Affairs Division

Proceedings of the NATO Advanced Study Institute on
The Gravitational Constant: Generalized Gravitational Theories and Experiments
Erice, Italy
April 30–May 10, 2003

A C.I.P. Catalogue record for this book is available from the Library of Congress.

ISBN 1-4020-1955-6
ISBN 1-4020-2242-5 (e-book)

Published by Kluwer Academic Publishers,
P.O. Box 17, 3300 AA Dordrecht, The Netherlands.

Sold and distributed in North, Central and South America
by Kluwer Academic Publishers,
101 Philip Drive, Norwell, MA 02061, U.S.A.

In all other countries, sold and distributed
by Kluwer Academic Publishers,
P.O. Box 322, 3300 AH Dordrecht, The Netherlands.

Printed on acid-free paper

All Rights Reserved
© 2004 Kluwer Academic Publishers
No part of this work may be reproduced, stored in a retrieval system, or transmitted
in any form or by any means, electronic, mechanical, photocopying, microfilming,
recording or otherwise, without written permission from the Publisher, with the exception
of any material supplied specifically for the purpose of being entered
and executed on a computer system, for exclusive use by the purchaser of the work.

Printed in the Netherlands.

CONTENTS

PREFACE

This XVIII Course of the International School of Gravitation and Cosmology held in Erice from April 30 to May 10, 2003, has been very successful and has provided an up-to-date understanding of the progress and current problems on the Gravitational constant both on "Generalized Gravitational Theories and Experiments" either in Laboratory with Casimir force measurements, or in space at solar system distances and in Cosmological observations.

We have had a qualified participation of more than fifty ASI research-workers on different fields (theoretical and experimental physicists, astrophysicists, astronomers, cosmologists from a variety of backgrounds, mathematicians) coming from various NATO and partner NATO countries (as, for instance, Belarius, Bulgaria, France, Germany, Israel, Italy, Kyrgyzistan, Latvia, Norway, Poland, Russia, Turkey, U.K., USA).

We had lectures on different aspects of the state and prediction of unified theories of the physical interactions including gravitation as a cardinal link, on the role of experimental gravitation and observational cosmology in discriminating between them, on the problem of precise measurements and stability of fundamental physical constants in space and time, and of the gravitational constant in particular.

Thus we have had timely lectures in recent advances in unified and scalar-tensor theories, theories in diverse dimensions and their observational windows, gravitational experiments in space, rotational and torsional effects in gravity, basic problems in cosmology, early universe as an arena for testing unified models; basis for Quantum Gravity, brane-inspired models in Gravitation and Cosmology.

Moreover we have considered cosmological variations of G, absorption and screening in prerelativistic and relativistic theories, constraints of Non-Newtonian Gravity from the recent Casimir Force measurements. Search for scalar-tensor Gravity with Lunar laser ranging. Special attention has been paid to the increasing role of fundamental gravitational experiments in space, the role of basic standards and determination of cosmological parameters, prediction of 5-dimensional projective unified field theory on cosmological, astrophysical and quantum effects, long range spin interaction, space-based determination of G, interface of quantum mechanics and gravity, Quaternion program.

Besides the theoretical aspect, a large part of the Course was devoted to the examination of laboratory experiments and space experiments for the measurement of big G and tests of the equivalence principle including the measurement of big G using a superconductor gravimeter.

We wish to thank all the lecturers and seminar speakers who did some much time to make the School successful, and all participants for contributing to the very scientific and human atmosphere.

Venzo de Sabbata

Venzo de Sabbata
(Director of the School)

Erice, 1 May 2003

WELCOME

Only some points before the Gillies lecture

It is a pleasure to welcome you all at this eighteenth Course of the International School of Cosmology and Gravitation.

The Director of Ettore Majorana Center for Scientific Culture Professor Antonino Zichichi, who cannot be present at this moment, (maybe he will come some days) has entrusted me with the welcome address on his behalf.

I thank all lecturers, who have accepted our invitations to come to Erice to give lectures (whitout any renumeration, as is usual in this School), from their different points of view, on the problem of Gravitational constant.

As you know from the poster, we will present different aspects of the state and predictions of unified theories of physical interactions including gravitation as a cardinal link, the role of experimental gravitation and observational cosmology in discriminating between them, the problem of precise measurement and stability of fundamental physical constants in space and time, the gravitational constant in particular, the basis for non Newtonian gravity, the constraints on non Newtonian gravity from the recent Casimir force measurements, gravity and non locality, neutrinos and gravitation, absorption and screening in relativistic theories, temporal and spatial dependence of empirical gravitational constant, charges and fundamental constants in unified theory, torsional effects in gravity, observational windows of extra dimensions and unified models, searching for scalar-tensor gravity with lunar laser ranging, motions of a precise torsion pendulum, early universe as an arena for testing unified models, 5-dimensional projective unified field theory for cosmological astrophysical and quantum effects, cosmic microwave background anysotropy as test of variability of fundamental constants and so on.

I thank all lecturers also, because they accepted as far as they were able, to be present the whole time and to participate in all discussions, which certainly will arise among all participants.

In fact this is one of the important tasks of these Courses. As perhaps many participants already know, because they were present in previous Courses, the atmosphere of the School is very friendly, and exchanges of ideas can occur quite freely among all participants including lecturers.

So we will have a very broad spectrum of subjects, of which at first sight each has sufficient material to constitute a Course by itself. We have considered it more important to exhibit the various theories that attempt to give some answer to the many problem that we have in our hands and to exchange information on the latest developments discussing future prospects.

I really hope that this Course with all these different topics serves to make some further progress on this fascinating subject. Before asking Professor Gillies to address the meeting, some few points regarding the organization of the lectures:
as you can see from the tentative time-table, we will have five lectures each day, three in the morning and two in the afternoon (with a break, both in the morning and in the afternoon, for coffee). Every lecture will last no more than fifty minutes, and after every lecture there will be a few minutes of discussion. For lunch and dinner you can choose any of the restaurants approwed for the School (you will see the list near the entrance), signing a list marked "Cosmology" after every meal and either "lunch" or "dinner". You have to pay only for beverages. Now, before that Professor Gillies will address the meeting, we like to commemorate Peter G.Bergmann.

Peter G.Bergmann
(24 March 1915 - 19 October 2002)

Erice, 1 May 2003

Memorial day

Peter was native of Germany: he left the Germany of Hitler in 1933 and completed his doctorate (theoretical Physics) in Prague 1936. Worked with Einstein in Princeton 1936-1941 on unitary field theories. After world war II became a professor at Syracuse University in 1947. He was also connected with New York University

Co-founder of International Committee on General Relativity and Gravitation (1959), and of the International Society on General Relativity and Gravitation (1973). Author of several books and of numerous articles. Many lectures, summer Schools, meetings throughout the world.

In Italy: Sestriere 1958 Erice 1975, 1977, 1979, 1982, 1985, Napoli 1983, 1984, Bologna 1984.

This is a bare curriculum that he give to me in 1985, hand written. You can have this manuscript if you like.

Now every people know how deep were and are the works of Bergmann on general relativity and in developing Einstein theory of gravitation starting from the remote 1936 when he went from Prague to the Princeton Institute for Advanced Study to work with Einstein. In fact Peter come to Prague in 1933 as a refugee from Hitler's Germany where as a Jew he could not have completed his academic training. He had read some scientific publications of Einstein and suddenly he realizes that just this kind of theoretical research he would like to do. So he wrote to Einstein asking to accept him as a collaborator (but he wrote at a wrong address). As after one month there was no reply, he wrote again enclosing a copy of a dissertation. This time he received an answer in which Einstein wrote that he would be glad to talk with him. At that time Bergmann was unaware that Einstein in the meantime had written to Prof.Philipp Frank (the Director of the Institute of Theoretical Physics in Prague) asking him about Bergmann. So when Bergmann arrived in Princeton he began to work with Einstein. He in fact worked with Einstein from 1936 to 1941 at the Institute for Advanced Study in Princeton in an attempt to provide a geometrical unified field theory of gravitation and electromagnetism. Within this period he wrote two article together: A.Einstein and P.G.Bergmann in Ann.Math. 39, 65 (1938); and A.Einstein, V.Bargmann and P.G.Bergmann in Th. von Karman anniversary volume 212 (1941)

In 1942, Bergmann published the book "Introduction to the Theory of Relativity" which included a foreword by Albert Einstein. This book is a reference for the subject, either as a textbook for classroom use or for individual study. Einstein said in his foreword: "Bergmann's book seems to me to satisfy a definite need. much effort has gone into making this book logically and pedagogically satisfactory and Bergmann has spent many hours with me which were devoted to this end". Einstein said also in this foreword that if general relativity has played a rather modest role in the correlation of empirical facts and has contributed very little to understanding of quantum phenomena (was the year 1942) it is quite possible that some of the results of the general relativity such as general covariance of the laws of nature and their non-linearity may help to overcome the difficulties encountered at present in atomic and nuclear processes.

Many of the important works of Peter Bergmann are in fact concerned with the quantization of field theories which are covariant with respect to general coordinate transformations. As to these points, Bergmann's article "General Relativity" in the volume V of the "Encyclopaedia of Physics" was a pionier work.

Of course we cannot go through all the Bergmann's works but, as we have said, we like to mention the great contribution that he has made to the School of Cosmology and Gravitation in Erice from its inception in 1974: being ever present and always discussing every argument in great depth, always coming to the point and clarifying every aspects of the problem under discussion. For instance in the sixth Course of the School (1979) he discussed "the fading world point" which deals with the nature of space-time and of its elements, the world points. Other important discussions were on the "unitary field theories", during the eighth Course in 1983 where he discussed various unitary theories such as Kaluza-Klein, scalar-tensor theories and projective theories. During the ninth Course (1985) devoted to "Topological properties of space-time" and during the tenth Course of School in 1987 devoted to "Gravitational Measurements" Bergmann discussed 'gravitation at spatial infinity' and the 'observables in general relativity' where he showed in a very elegant way the profound difference of the notion of observable in general relativity from the corresponding concept in special relativity (or in Newtonian physics).

So Bergmann really has been the centre, the fulcrum of all discussions in all the Courses of our International School of Cosmology and Gravitation. We like to remember also that from the early fifties onwards, one of the main goals and challenging tasks for Bergmann was the quantization of the gravitational field. The difficulties which he faced and solved in dealing with the full non-linear properties of Einstein'stheory and with non-perturbative approaches were enormous. Bergmann and his collaborators investigated the canonical (i.e. Hamiltonian) formalism of classical and quantum covariant field theories.

These covariant field theories of general type of the theory of relativity are brought into the canonical form and then quantized. These works are of the year 1948 and 1949 [Phys.Rev. 75, 680 (1949) and (with Johanna H.M.Brunings) Rev.Mod.Phys.21, 480 (1949)], and constitute the basis of the modern researches in this field. They are followed by other important works [Phys.Rev.83, 1018 (1951) with James L.Anderson] on the constraints in covariant field theories and the application to the cases of gauge and coordinate invariance specializing the problem by assuming a quadratic lagrangian in the differentiated quantities; moreover a beautiful work on 'Spin and Angular Momentum in G.R.'[Phys.Rev.89, 400 (1953)] in presence of a semiclassical Dirac field.The problem of origin of constraints is considered also in another important work with Irwin Goldberg [Phys.Rev. 98, 531 (1954)] where there is analyzed the group-theoretical significance of the Dirac bracket in order to prepare the ground for the utilization of the Dirac bracket for the quantization of generally covariant theories.

Once again we cannot describe all of Bergmann work here but we would like to stress the fact that Bergmann clarified and made profound contributions in the different quantization programs based on the canonical formalism. As Bergmann himself said, "the resulting theory would give us answers to such questions as the nature of a fully quantized geometry of space-time , the role of world points in this geometry, the 'softening-up of the light cone', and the effect of this not only on the divergences associated with the gravitational field but with all other fields as well".

Really Peter Gabriel Bergmann is admirable both as scientist and as a man of great culture and humanity.

All people in Erice remember for ever his extraordinary presence together with his wife Margot.

One of the most precious aspect of Peter was his entire kindness as a human being. He was always courteous and gentle, always, in every, every, circunstance.

As professor Komar said (Art Komar and his wife Alice were very close to him and they had liked to be present during that ceremony but were prevented for some health disease) Peter must be remembered not only as a brillant physicist but also for his great and total kindness.

But now I like to give a good new: through Professor Komar I know that just before to the death he knew that the American

Physical Society awarded their first medal ever for gravitational reseaech. This medal was awarded jointly to Peter G.Bergmann and John A.Wheeler. So Peter was told about the joint award and knew that he had been awarded this honour he so well deserved, just a short while before the bad fall that he took from which he never regained consciousness.
As a little homage of our School to Peter, we propose to dedicate from now all the future Courses to Bergmann calling our School

THE INTERNATIONAL SCHOOL OF COSMOLOGY AND GRAVITATION
"PETER G.bERGMANN"

Now I invite all of you to a minute of silence.

Now let me to give a last comment: Peter was very sensible to the problem of peace: in fact in the years '75 and '77, (at that time, if I remember well, he was the President of the International Committee of General Relativity and Gravitation) Peter gave the best, during the Course of our School with Mercier (the secretary of the International Society of General Relativity and Gravitation) Hehl, Schmutzer, Rosen, Ivanenko, Sokolov, to reunite, to unify again the International Committee that was broken for some political reason, and he was successful. May be that Professor Schmutzer likes to give some words on that argument.

Venzo de Sabbata

Words in commemoration of Peter G. Bergmann

Much was said and has still to be said to pay tribute to Peter G. Bergmann's scientific work and his influence on younger generations of physicists. Allow me to add here some short, very personal sentences, also on behalf of my colleagues forming till 1991 the Potsdam-Babelsberg group of relativists founded by Hans-Jürgen Treder.

Peter Bergmann had a great influence on my colleagues' and my education and later scientific work. As to his books and papers that had the greatest fascination for us, I want to mention the text book "Introduction to the Theory of Relativity", the article "General Relativity" in volume V of the "Encyclopedia of Physics" and the pioneering works on the quantization of generally covariant field theories.

Aside from Peter Bergmann's substantial impact on theoretical physics and his great role as a teacher, there is still another aspect that was most important to us. Peter Bergmann will be remembered by us for his lasting friendly support of our Potsdam-Babelsberg group which helped us to join the international community of relativists and to take part in its work over decades.

Before 1989, Peter Bergmann and his wife Margot were several times guests of our institute, the Einstein Laboratory for Theoretical Physics. Vividly I remember the stimulating discussions on physics, culture and politics in the Einstein house in Caputh (at that time a part of our institute), at which Peter Bergmann and his wife stayed, and later discussions during the Courses on Cosmology and Gravitation here in Erice. With sincere gratitude, we also recall that, after the unification of Germany in 1990, when – to say it in Peter's own words – the situation throughout the academic institutions of the former GDR was grim, he together with Venzo de Sabbata and Antonio Zichichi became very active to reach that we could continue our scientific work.

Although Peter Bergmann was compelled by the Nazis to leave Germany, he was ready to help many colleagues in Germany after the Second World War, in West and East. I think that by Peter's attitude and friendship we were strengthened in our efforts to do our best to prevent that the outrageous German history repeats.

Those who had the privilege of knowing Peter personally will always remember his kindness, modesty and integrity. We shall miss him.

Horst-Heino v. Borzeszkowski
Technical University Berlin, Institute of Theoretical Physics

Some History of the Career of Prof. Peter G. Bergmann

After coming to the United States in 1936, Dr. Bergmann joined Albert Einstein at the Institute for Advanced Studies in Princeton, New Jersey. His contemporaries who also worked with Einstein included Leopold Infeld, Valentine Bargman, and others. Among the papers Bergmann published during this time was "On a Generalization of Kaluza's Theory of Electricity," Annals of Mathematics, Vol. 39, No. 3, p. 683 (July 1938), with Einstein as the co-author.

In 1941, following his time at the Institute, Bergmann joined the faculty of Black Mountain College, near Ashville, North Carolina. Although the college no longer exists, it is remembered as an interesting experiment in higher education in the United States. It was founded in 1933 and included among its faculty several artists, writers and thinkers who were prominent in their fields at the time. Einstein was a member of the College's Board of Advisors. During his one year term there, Bergmann was the only faculty member specializing in physics. The college had a very informal class structure and it sought to introduce the students to many kinds of innovations, both scholastically and architecturally, including the first construction and use of Buckminster Fuller's Geodesic Dome.

Immediately after his time at Black Mountain, Bergmann moved to Lehigh University in Pennsylvania, where he joined the faculty of the Department of Physics. He was with them full time for a period of two years and then began to spend an increasing fraction of his time on war-related research during the period from 1944 to 1946. While at Lehigh, he made a lasting impression on several of the undergraduates attending college there, including Lee Iacocca, who is familiar to many for his terms as President of Ford Motor Co. and Chrysler Motors, Inc. in the U.S. In fact, in his autobiography, Iacocca mentions some anecdotes about his various interactions with Bergmann.

It was also during this time that Bergmann's famous book, "Introduction to the Theory of Relativity," was published by Prentice-Hall, Inc., New York (1946). Einstein wrote a brief *Foreword* for this book, and in it he included the comment, "This book gives an exhaustive treatment of the main features of the theory of relativity which is not only systematic and logically complete, but also presents adequately its empirical basis." It is hard to imagine how one could receive any greater praise for his efforts than these words offered by Einstein himself.

The bulk of Bergmann's career, from 1947 to 1982, was spent at Syracuse University in New York. He founded the Relativity Group at Syracuse and worked with many faculty and students there over the years. Much of his effort there was focused on attempts to reconcile quantum mechanics with gravity via the introduction of a quantized theory of gravity. He and many others hoped that such a step would help lead towards the creation of a unified field theory, which of course was the great quest already begun during his time with Einstein. Characteristic of his publications in those days was his paper with I. Goldberg, "Dirac Bracket Transformations in Phase Space," published in the Physical Review, Vol. 98, No. 2, p. 531 (April 15, 1955).

During his years at Syracuse, and thereafter during his retirement appointment at New York University, Bergmann maintained a heavy involvement in gravitational physics and especially in the international aspects of promoting research in it. He did this by lending his support and credibility to many different scientific conferences on general relativity, including the Texas Symposia on Relativistic Astrophysics, the International Conferences on General Relativity and Gravitation ("GRG" Conferences), and of course the International Schools of Cosmology and Gravitation held at the Ettore Majorana Centre in Erice, Italy. He was a tireless traveler to such conferences, and was often accompanied by his wife Margot.

It is extremely fitting that just prior to his death at the age of 87, he and John Wheeler were jointly awarded the inaugural Einstein Prize in Gravitational Physics, granted under the auspices of the American Physical Society. Indeed, as the citation of the award says, he has been an "...inspiration to generations of researchers in general relativity."

George T. Gillies

Dear colleagues,

We heard a lot already here about the outstanding scientist and a man of wonderful human features and great culture, about Peter Gabriel Bergmann. That is why I will dwell upon only on my personal feelings from meetings and contacts with Peter's scientific works.

My first personal meeting with Prof. Peter Bergmann was here in Erice in 1987 at the 10th course of the International School of Cosmology and Gravitation "Gravitational Measurements, Fundamental Interactions and Constants", of which I was a director. After that course the new area appeared – gravitational relativistic metrology, which influenced very much a further development of precise space-time measurements, gravitational wave detectors networks, studies of fundamental physical constants, role of gravity in unified models etc. Peter presented a very nice and enlightening talk on "Observables in General Relativity", which was well received by the audience.

Of course, my knowledge of his works dates to more earlier times, when we studied relativistic theory of gravitation and problems of interrelations between general relativity and quantum fenomena in particular. First of all, I had to mention his book "Introduction to the Theory of Relativity" (with the foreword of Albert Einstein) as a good book used by many relativists, but what was even more important to us in the end of sixties and later, was his contributions to quantization of covariant field theories, canonical formalism of classical and quantum field theories, Dirac brackets etc.

It was a time when these problems were intensively studied in Russia due to pioneering works of K.P. Staniukovich, D.D. Ivanenko, M.A. Markov and later of Ya.B. Zeldovich, A.D. Sakharov and their colleagues and students, which became famous afterwards also.

In our group created by Prof. Staniukovich in the Russian (USSR) State Committee for Standards we started these investigations in 1967, inspired by ideas of unification of micro and macro-world fenomena in such important fields as particle-like exact solutions of self-consistent systems of fields including gravitational one, quantum cosmology with fields and the cosmological constant, quantization of fields in a given gravitational background, self-consistent treatment of quantum effects in cosmology, quantization of the gravitational field itself etc. Many fundamental works were done in this group by such well known now scientists as Profs. Yu.N. Barabanenkov, V.A. Belinski, K.A.Bronnikov, M.B. Mensky, G.A.Vilkovysky, V.D.Ivashchuk, V.R.Gavrilov etc. And in all these studies we used many results of Peter Bergmann, J. Wheeler, Bryce Dewitt and other outstanding scientists.

After 1987 I was here in Erice at practically all courses as a lecturer and this year as one of directors. And Peter was here at all courses except maybe one. I was always impressed by his encyclopedic and profound knowledge, his clear and up-to-date lectures, his friendly attitude, soft and gallant manners.

Usually we say in Russian that he was a very intelligent man. I am not sure that it is a proper word in English, but those who know Russian or, at any rate are familiar with the great Russian literature of the 19th century (I mean Tolstoi, Dostoevsky, Chekhov, Turgenev and others) may understand me properly.

Peter is not with us now, it is a very unusual situation here after many years of the working of this school, but I think the fact, that from now on the Erice school led by Prof. Venzo de Sabbata will be called as "Peter Bergmann School of Cosmology and Gravitation", is a good commemoration for him.

I wish a big success to this course and many-many courses in future!

Thank you.

Prof. Vitaly N. Melnikov,
Co-director of the 18th course,
President of the Russian Gravitational Society.

Erice, May 9, 2003.

In commemoration of Peter Gabriel Bergmann – famous scientist and good friend

Born in 1915 in Berlin, he started his university studies already in the age of 16 at the Technische Hochschule in Dresden, where he registered both for chemistry and physics laboratory exercises. Two semesters later he moved to Freiburg (Breisgau) to attend the lectures of Gustav Mie who then did research on Einstein´s General Relativity Theory. After having escaped Nazi terror he continued his studies in physics at the Charles University in Prague, where he finished his doctorate in 1936.

In that same year he moved to the Institute for Advanced Study in Princeton, collaborating with Albert Einstein on the attempt to develop a geometrical unified field theory of gravitation and electromagnetism – the well-known aim of research of the late Einstein. During the following ten years he published two common papers in this field: Einstein and Bergmann (1938), Einstein, Bargmann and Bergmann (1942). About 20 years later, when Peter Bergmann and I met first in 1962 at the Jablonna Conference on General Relativity and Gravitation (near Warsaw), we had a discussion in context with my own research on projective relativity theory. Bergmann told me that then in Princeton he investigated an attempt to generalize this type of theory, as P. Jordan did it later in 1945: In a discussion with Einstein he asked him for a common publication, but Einstein refused. At least at that time Einstein did not believe in a 5-dimensional structure of such a new theory intended..

Already at Jablonna we both Peter and I spontaneously became friends: He an expelled German and I a German of the younger generation totally condemning Nazism. I was strongly impressed by this great human attitude of Peter, in particular also insofar as I knew of the Nazi crimes that happened to the family of Peter's wife Margot.

After the International Conference on General Relativity and Gravitation in Copenhagen in July 1971, when Peter became president of the corresponding society and I a member of the International Committee of this society, we both met for several committee meetings etc., particularly also in Erice/Sicily in connection with the Schools on General Relativity and Gravitation organized by V. de Sabbata. At the GR8-conference at Waterloo in 1977 it was decided by this committee that the next GR9-conference should take place in Jena (then in the German Democratic Republic) under my leadership. In context with the politically very complicated situation under the conditions of the Cold War it was necessary for me to have a permanent contact with Peter, since he was the president of the society. Peter helped me always in many consultation always, when the emerging problems seemed to be unsolvable.

Peter' s idea of having the GR9-conference to Jena based on different reasons: One important point of performing this conference on German soil was the fact that Einstein developed his General Relativity Theory in Berlin 1915. Another argument resulted from the fact of the existence of the Iron Curtain preventing the getting-together of scientist from the east and the west who had contact by mail but never met personally. Last but not least, in Jena existed a rather strong relativity group that I had founded in 1957. I had to promise to the International Committee to do all I could making sure that participants from all countries, particularly from Chile, Israel, South Africa, Taiwan and South Korea obtained entrance visas without political interference. Further, after the assassination of the Israel athletes during the Olympic Summer Games in Munich in 1972, I had to take care of the security of the participants, particularly from Israel. Through very difficult negotiations with my superior authorities I could meet these very strong conditions posed to me by the committee.

I succeeded and the conference with 830 participants from 51 countries took place successfully in July 1980, running in a good scientific and humane climate. I was strongly supported by the Secretary of the Local Organisational Committee, R. Collier. Peter Bergmann's personality gave me much energy and strong mental help. He was not only an outstanding scientist but also a man with great political foresight – on behalf of the international relativity community in the terrible political situation of the Cold War. I have to thank him much for ever!

Ernst Schmutzer, Jena

The group of the participants at the 2003 Course in Erice

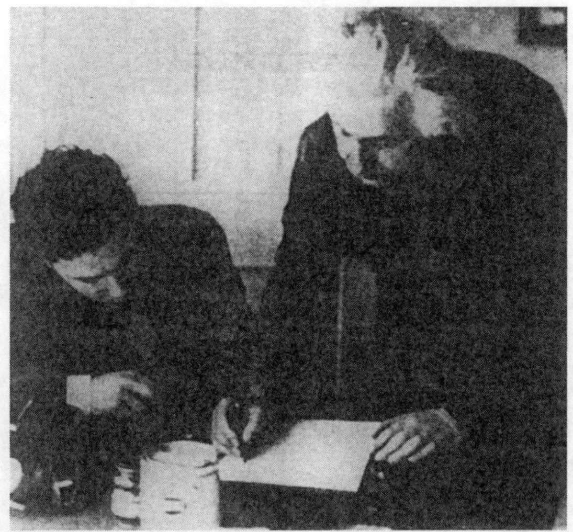

Bergmann with Einstein

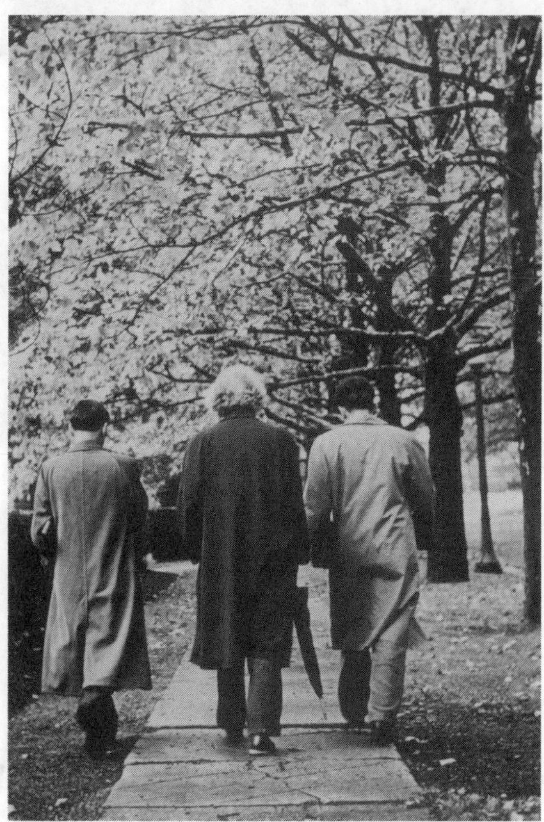

Bergmann with Einstein and Bargmann (from the left: Bargmann, Einstein, Bergmann) at Princeton University

During the memorial day in Erice: Gillies, Schmutzer, de Sabbata, Borzeszkowski and Melnikov

de Sabbata, Bergmann and Datta

Bergmann and de Sabbata

de Sabbata and Bergmann in Erice

SCREENING AND ABSORPTION OF GRAVITATION IN PRE-RELATIVISTIC AND RELATIVISTIC THEORIES

H.-H. VON BORZESZKOWSKI AND T. CHROBOK
Institut für Theoretische Physik, Technische Universität Berlin, Hardenbergstr. 36, D-10623 Berlin, Germany

AND

H.-J. TREDER
Rosa-Luxemburg-Str. 17a, D-14482 Potsdam, Germany

Abstract. After commenting on the early search for a mechanism explaining the Newtonian action-at-a-distance gravitational law we review non-Newtonian effects occurring in certain ansatzes for shielding, screening and absorption effects in pre-relativistic theories of gravity. Mainly under the aspect of absorption and suppression (or amplification), we then consider some implications of these ansatzes for relativistic theories of gravity and discuss successes and problems in establishing a general framework for a comparison of alternative relativistic theories of gravity. We examine relativistic representatives of theories with absorption and suppression (or amplification) effects, such as fourth-order theories, tetrad theories and the Einstein-Cartan-Kibble-Sciama theory.

1. Introduction

All deviations from the gravitational theories of Newton and Einstein touch fundamental problems of present-day physics and should be examined experimentally. In particular, such examination provides further tests of Einstein's general theory of relativity (GRT), which contains Newton's theory as an approximate case. Therefore, it makes sense to systematically analyze all effects that differ from the well-known Newtonian and post-Newtonian ones occurring in GRT (let us call them non-Einsteinian effects). If these effects can be excluded experimentally, then this would provide further support for GRT; otherwise one would have to change basic postulates of

V. de Sabbata et al. (eds.),
The Gravitational Constant: Generalized Gravitational Theories and Experiments, 1–37.
© 2004 *Kluwer Academic Publishers. Printed in the Netherlands.*

present-day physics. Of all possible non-Einsteinian effects, we shall focus in this paper on the effects of shielding, absorption and suppression of gravitation.

The experience made with GRT shows that it is not probable to find any non-Einsteinian solar-system effects. Nevertheless, it is useful to search for them on this scale, too, for it can establish further null experiments in support of GRT. However, a strong modification of GRT and corresponding effects can be expected on the microscopic and, possibly, cosmological scale. As to the microscopic scale, this conjecture is insofar suggesting itself as there is the problem of quantum gravitation unsolved by GRT. The cosmological part of this conjecture concerns the standpoint with respect to the Mach principle. Of course, there is no forcing physical argument for a realization of this principle, and GRT did not leave unanswered the question as to it. But one can be unsatisfied with this answer and look for a generalization of GRT satisfying this principle more rigorously than GRT. If this is the case, one has a further argument for a modification of GRT predicting also non-Einsteinian effects. (In [12], it was shown that the tetrad theories discussed in Sec. 3.2 could open up new perspectives for solving both problems simultaneously.)

We begin in Sec. 2 with remarks on the early search for a mechanism explaining the Newtonian action-at-a-distance gravitational law and then discuss non-Newtonian effects occurring in certain ansatzes for shielding, screening and absorption effects in pre-relativistic theories of gravity. In Sec. 3, mainly under the aspect of absorption and suppression (or amplification), we consider successes and problems in establishing a general framework for a comparison of alternative relativistic theories of gravity. We discuss then the following relativistic representatives of theories with such effects, such as: in Sec. 3.1 fourth-order derivative theories of the Weyl-Lanczos type formulated in Riemann space-time, in Sec. 3.2 tetrad theories formulated in teleparallelized Riemann space-time, and in Sec. 3.3 an effective GRT resulting from the Einstein-Cartan-Kibble-Sciama theory based on Riemann-Cartan geometry. Finally, in Sec. 4 we summarize our results and compare them with GRT. The resulting effects are interpreted with respect to their meaning for testing the Einstein Equivalence Principle and the Strong Equivalence Principle.

2. Shielding, absorption and suppression in pre-relativistic theories of gravity

To analyze non-Einsteinian effects of gravitation it is useful to remember pre-relativistic ansatzes of the nineteenth century, sometimes even going back to pre-gravitational conceptions. One motivation for these ansatzes was to find a mechanical model that could explain Newton's gravitational inverse-square-law by something (possibly atomic) that might exist between the attracting bodies. In our context, such attempts are interesting to discuss, because they mostly imply deviations from Newton's law. Another reason for considering such rivals of Newton's law was that there were several anomalous geodesic, geophysical, and astronomical effects which could not be explained by Newtonian gravitational theory. Furthermore, after the foundation of GRT, some authors of the early twentieth century believed that there remained anomalies which also could not be explained by GRT. As a result, pre-relativistic assumptions continued to be considered and relativistic theories competing with GRT were established.

One influential early author of this story was G.-L. Le Sage [79, 80]. In the eighteenth century, he proposed a mechanical theory of gravity that was to come under close examination in the nineteenth century (for details on Le Sage's theory, see also P. Prevost [105] and S. Aronson [2]). According to this theory, space is filled with small atomic moving particles which due to their masses and velocities exert a force on all bodies on which they impinge. A single isolated body is struck on all sides equally by these atoms and does not feel any net force. But two bodies placed next to each other lie in the respective shadows they cast upon each other. Each body screens off some of the atoms and thus feels a net force impelling it toward the other body.

Under the influence of the kinetic theory of gases founded in the 1870's Le Sage's theory was revived by Lord Kelvin [118][1], S. T. Preston [103, 104], C. Isenkrahe [64] and P. Drude [33], bringing Le Sage's hypothesis up to the standard of a closed theory. However, this approach to gravitation was rejected by C. Maxwell [90] with arguments grounded in thermodynamics and the kinetic theory of gases. On the basis of these arguments, it was discussed critically by Poincaré and others (for the English and French part of this early history, cf. Aronson [2]).

The search for a mechanistic explanation of gravitation and the idea of a shielding of certain fluxes that intermediate gravitational interaction were closely related to the question of the accuracy of Newton's gravitational law. In fact, this law containing only the masses of the attracting bodies and their mutual distances can only be exactly valid when neither the intervening space nor the matter itself absorb the gravitational force or potential. Therefore, it is not surprising that the possibility of an absorption of gravitation had already been considered by Newton in his debate with E. Halley and N. Fatio de Duillier[2], as is documented in some of the Queries to Newton's "Opticks." The first research program looking for an experi-

[1]Later he reconsidered it from the view of radioactivity [119].

[2]Le Sage himself stated that his speculations go also back to the work of Fatio.

mental answer to this question was formulated by M.W. Lomonosov [83] in a letter to L. Euler in 1748. The program was however only realized 150 years later by R. von Eötvös and Q. Majorana, without explicit reference to Lomonosov. At about the same time, Euler discussed with Clairaut, a prominent specialist in celestial mechanics, the possibility of detecting deviations from Newton's gravitational law by analyzing the lunar motion. Clairaut believed for a time that he had found a fluctuation of the lunar motion testifying to an absorption of gravitation by matter, in this case, by the earth.

For a long time, the lunar motion has been the strongest criterion for the validity of the Newtonian and, later, the Einsteinian theory of gravity (today one would study the motions of artificial satellites). This is due to the fact that the gravitational influence of the sun on the moon exceeds the influence of the earth by the factor 9/4. This solar action varies in dependence on the distance of the system 'earth-moon' from the sun. Regarding this effect and, additionally, the action of the other planets on the lunar motion, a reaming fluctuation of the motion of the moon could possibly be due to an absorption of solar gravity when the earth stands between the sun and the moon. This early idea of Euler was later revived by von Seeliger, and just as Clairaut had analyzed the lunar motion in order to corroborate it, later Bottlinger [21, 22] did the same in order to find support for the hypothesis of his teacher von Seeliger [114].

The first ansatz for an exact mathematical description of absorption in the sense of Euler and Lomonosov was made by Laplace [76] in the last volume of his *Mécanique Céleste.* He assumed that the absorption $d\vec{F}$ of the flow $\vec{F}$ of the gravitational force is proportional to the flow $\vec{F}$ itself, the density ρ, and the thickness dr of the material penetrated by the gravitational flow, $d\vec{F} = k\rho\vec{F}dr$. Accordingly a mass element dm_1 exerts on another element dm_2 the force

$$|d\vec{F}| = \frac{Gdm_1dm_2}{r^2}e^{(-k\rho r)}, \tag{1}$$

where k is a universal constant of the dimension $(mass)^{-1}(length)^2$ and G is Newtons gravitational constant.

In the early twentieth century, when Newton's gravitational theory was replaced by GRT, the two aforementioned attempts by Bottlinger and Majorana were made to furnish observational and experimental proof of absorption effects in the sense of Euler and von Seeliger. Such effects do not exist in GRT and so evidence for them would have been a blow against the theory.

Using H. von Seeliger's hypothesis of 1909 (von Seeliger [114]), F.E. Bottlinger [21, 22] tried to explain short-period fluctuations of the motion of the moon (later it became clear that this explanation was not correct), while Majorana attempted to detect such absorption effects by laboratory experiments from 1918 till 1930. Being aware of previous experiments performed to detect an absorption of gravitation by matter, Majorana turned to this problem in 1918. He speculated that gravitation was due to a flow of gravitational energy from all bodies to the surrounding space which is attenuated on passing through matter. The attenuation would depend exponentially on the thickness of the matter and its density. Based on a theoretical estimation of the order of magnitude of this effect he carried out experiments the results of which seemed to confirm the occurrence of gravitational absorption. According to present knowledge, they must have been erroneous (for details of the history of these experiments, see, e.g., Crowley et al. [27], Gillies [55], Martins [1]).

Another conception competing with absorption or shielding of gravitation by matter also goes back to papers by von Seeliger [113]. In these papers, now for cosmological reasons, he considered a modification of Newton's law by an exponential factor. Similar ideas were proposed by C. Neumann [96]. At first sight, it would appear to be the same modification as in the former cases. This is true, however, only insofar as the form of the gravitational potential or force is concerned. For the differential equations to which the respective potentials are solutions, there is a great difference. In the first case, instead of the potential equation of Newtonian theory,

$$\Delta\phi = 4\pi G\rho, \tag{2}$$

one has equations with a so-called potential-like coupling,

$$\Delta\phi = 4\pi G\rho\phi, \tag{3}$$

while in the second case one arrives at an equation with an additional vacuum term,

$$\Delta\phi - k^2\phi = 4\pi G\rho. \tag{4}$$

The latter equation requires the introduction of a new fundamental constant k corresponding to Einstein's cosmological constant Λ. As will be seen in Sec. 3.2, in relativistic theory a combination of vacuum and matter effects can occur, too.

In [122] it is shown that, regarding cosmological consequences, both approaches are equivalent. Indeed, if one replaces in (3) ϕ by $\phi + c^2$ and considers a static cosmos with an average matter density $\bar{\rho} = const$ then the corresponding average potential $\bar{\phi}$ satisfies the equation

$$\Delta\bar{\phi} - k^2\bar{\phi} = 4\pi G\bar{\rho} \tag{5}$$

where

$$k^2 = 4\pi G\bar{\rho}c^{-2} \tag{6}$$

which is identical to the averaged equation (4).

Equation (3) shows that the potential-like coupling of matter modelling the conception of an absorption of the gravitational flow by the material penetrated can also be interpreted as a dependence of the gravitational number on the gravitational potential and thus on space and time. Therefore, to some extent it models Dirac's hypothesis within the framework of a pre-relativistic theory, and a relativistic theory realizing Dirac's idea could have equation (3) as a non-relativistic approximation. Another possible interpretation of (3) is that of a suppression of gravitation (self-absorption) by the dependence of the active gravitational mass on the gravitational potential. Indeed, the product of the matter density and the gravitational potential can be interpreted as the active gravitational mass.

Not so much for historical reasons but for later discussion, it is interesting to confront the potential equations (2) and (4) with the bi-potential equations [18],

$$\triangle\triangle\phi = 4\pi G\rho \tag{7}$$

$$\triangle(\triangle - k^2)\phi = 4\pi G\rho \tag{8}$$

or, for a point-like distribution of matter described by the Dirac delta-function δ,

$$\triangle\triangle\phi = 4\pi a\delta(\vec{r}) \tag{9}$$

$$\triangle(\triangle - k^2)\phi = 4\pi ak^2\delta(\vec{r}). \tag{10}$$

Eqs. (9) and (10) show that the elementary solutions are given by the Green functions,

$$\phi = \frac{ar}{2} \tag{11}$$

$$\phi = \frac{a}{r} - \frac{a}{r}e^{-kr} \tag{12}$$

indicating that there is a long-range (Newtonian) and a short-range (self-absorption) part of gravitational interaction.

In order to select these physically meaningful solutions from a greater manifold of solutions for the fourth-order equations one had to impose yet an additional condition on the structure of sources: The distribution of matter must be expressed by a monopole density. This becomes obvious when one considers the general spherical-symmetric vacuum solutions of (2, 10) which are given by

$$\phi = \frac{ar}{2} + b + \frac{c}{r} + \frac{d}{r^2} \tag{13}$$

$$\phi = \frac{A}{r} + B + C\frac{e^{-kr}}{r} + D\frac{e^{kr}}{r} \tag{14}$$

respectively, satisfying the equations

$$\triangle \triangle \phi = -4\pi a\delta(\vec{r}) - 4\pi c \triangle \delta(\vec{r}) \tag{15}$$

$$\triangle(\triangle - k^2)\phi = -4\pi(A + C) \triangle \delta(\vec{r}) - 4\pi k^2 C\delta(\vec{r}). \tag{16}$$

Originally, such equations where discussed in the framework of the Bopp-Podolsky electrodynamics. This electrodynamics states that, for $A = -C$, there exist two kinds of photons, namely massless and massive photons. They satisfy equations which, in the static spherical-symmetric case, reduce to Eq. (8). It was shown in [65] that its solution (12) is everywhere regular and $\phi(0) = ak$ ($a = e$ is the charge of the electron). The same is true for gravitons: A theory with massless and massive gravitons requires fourth order equations (see Sec. 3.1).[3]

The atomic hypotheses assuming shielding effects lead in the static case to a modification of Newton's gravitational law that is approximately given by the potential introduced by von Seeliger and Majorana. Instead of the r^2-dependence of the force between the attracting bodies given by the Newtonian fundamental law,

$$\vec{F}_{12} = -G \int \frac{\rho(\vec{r}_1)\rho(\vec{r}_2)}{r_{12}^3} \vec{r}_{12} d^3x_1 d^3x_2 \tag{17}$$

where G is the Newtonian gravitational constant, one finds,

$$\vec{F}_{12} = -G \int \frac{\rho(\vec{r}_1)\rho(\vec{r}_2)}{r_{12}^3} \vec{r}_{12} e^{(-k\int \rho dr_{12})} d^3x_1 d^3x_2. \tag{18}$$

Here the exponent $-k \int \rho dr_{12}$ means an absorption of the flow of force $\vec{F}$ by the atomic masses between the two gravitating point masses. Since, for observational reasons, one has to assume that the absorption exponent is much smaller than 1, as a first approximation, (18) may be replaced by Laplace's expression,

$$\vec{F}_{12}^* = -G\frac{M_1 M_2}{r_{12}^3} \vec{r}_{12} \exp/(-k\rho dr) \approx -G\frac{M_1 M_2}{r_{12}^3} \vec{r}_{12}(1 - k\rho\Delta r). \tag{19}$$

[3]In [120] it was shown that it is erroneous to assume that massive gravitons occur in Einsteins GRT with a cosmological term.

Equation (19) contains a new fundamental constant, namely Majorana's "absorption coefficient of the gravitational flow"

$$k \geq 0, [k] = cm^2 g^{-1}. \tag{20}$$

This value can be tested by the Eötvös experiment, where one can probe whether the ratio of the gravitational and the inertial mass of a body depends on its physical properties. In the case of absorption of gravitation the value of this ratio would depend on the density of the test body. Gravimetric measurements of the gravitational constant carried out by Eötvös by means of a torsion pendulum and gravitational compensators showed that k has to be smaller than

$$k < 4\times 10^{-13} cm^2 g^{-1}. \tag{21}$$

By comparison Majorana [86, 87, 88] obtained in his first experiments the value

$$k \approx 6.7\times 10^{-12} cm^2 g^{-1}, \tag{22}$$

which was compatible with his theoretical analysis. (Later, after some corrections, he arrived at about half this value [89]).

A more precise estimation of k can be derived from celestial-mechanical observations. As mentioned above, Bottlinger hypothesized that certain (saros-periodic) fluctuations of the motion of the moon are due to an absorption of solar gravity by Earth when it stands between Sun and Moon[4]. If we assume this hypothesis, then, following Crowley et al. [27], the amplitude λ of these fluctuations is related to the absorption coefficient k via

$$\lambda \approx 2ka\rho, \tag{23}$$

where ρ denotes the mean density and a the radius of the earth. If one assumes that

$$k \approx 6.3\times 10^{-15} cm^2 g^{-1}, \tag{24}$$

then the value of λ is in accordance with the so-called great empirical term of the moon theory. This, however, also shows that, if the fluctuations of the motion of moon here under consideration had indeed been explained by von Seeliger's absorption hypothesis, then greater values than the one given by (24) are not admissible as they would not be compatible with the motion theory. That there is this celestial-mechanical estimation of an

[4] A. Einstein commented Bottlinger's theory in [36, 38, 39] (see also [123]).

upper limit for k had already been mentioned by Russell [107] in his critique of Majorana's estimation (22).

A better estimate of k has been reached by measurements of the tidal forces. According to Newton's expression, the tidal force acting upon earth by a mass M at a distance R is

$$Z = -2\frac{GM}{R^2}\frac{a}{R} \tag{25}$$

where a denotes the radius of the earth. However von Seeliger's and Majorana's ansatz (18) provided

$$Z^* \approx -2\frac{GM}{R^2}\frac{a}{R} - \frac{\lambda GM}{2R^2} \approx -2\frac{GMa}{R^3}(1 + \frac{\lambda R}{4a}) \tag{26}$$

with the absorption coefficient of the earth body

$$\lambda \approx 2ka\rho \approx 6.6{\times}10^9 cm^{-2} g \times k. \tag{27}$$

Considering now the ratio of the tidal forces due to the sun and moon, Z_s and Z_m, one finds in the Newtonian case

$$\frac{Z_s}{Z_m} \approx \frac{M_s}{M_m}\left(\frac{R_m}{R_s}\right)^3 = \frac{5}{11} \tag{28}$$

and in the von Seeliger-Majorana case

$$\frac{Z_s^*}{Z_m^*} \approx \frac{M_s}{M_m}\left(\frac{R_m}{R_s}\right)^3 (1 + \lambda\frac{R_s}{a}) = \frac{5}{11}(1 + 4k \times 10^{13} cm^{-2} g). \tag{29}$$

Measurements carried out with a horizontal pendulum by Hecker [60] gave the result

$$\frac{Z_s^*}{Z_m^*} \leq 1, \quad \text{that is,} \quad k < 2{\times}10^{-14} g^{-1} cm^2. \tag{30}$$

This result is still compatible with Bottlinger's absorption coefficient, but not with Majorana's value (22), which provided a sun flood much greater than the moon flood (Russell [107]).

Later, Hecker's estimation was confirmed by Michelson and Gale [91], who by using a "level" obtained

$$\frac{Z_s^*}{Z_m^*} = 0.69 \pm 0.004, \quad \text{i. e.,} \quad k < 1.3{\times}10^{-14} g^{-1} cm^2. \tag{31}$$

(The real precision of these measurements, however, was not quite clear.)

Bottlinger [21, 22] had also proposed to search for jolting anomalies in gravimeter measurements occurring during solar eclipses due to a screening of the gravitational flow of the sun by the moon. In an analysis performed by Slichter et al. [115], however, this effect could not be found and those authors concluded that k has the upper limit $3\times10^{-15} cm^2 g^{-1}$. However, as argued earlier [11], measurements of this effect provide by necessity null results due to the equivalence of inertial and passive gravitational masses verified by Eötvös.

The latest observational limits on the size of the absorption coefficient is $k < 10^{-21} cm^2 g^{-1}$. It was established by a reanalysis of lunar laser ranging data (Eckardt [34], cf. also Gillies [56]). This would rule out the existence of this phenomenon, at least in the way that it was originally envisioned. (For an estimation of that part which, from the viewpoint of measurement, is possibly due to shielding effects, cf. [139].) The actual terrestrial experimental limit provides $k < 2 \times 10^{-17} cm^2 g^{-1}$ [130].

About the same estimate follows from astrophysics [116, 123]. Indeed, astrophysical arguments suggest that the value for k has to be much smaller than $10^{-21} cm^2 g^{-1}$. This can be seen by considering objects of large mass and density like neutron stars. In their case the total absorption can no longer be described by the Seeliger-Majorana expression. However, one can utilize a method developed by Dubois-Reymond (see Drude [33]) providing the upper limit $k = 3/R\rho$, where R is the radius and ρ the density of the star. Assuming an object with the radius $10^6 cm$ and a mass equal to $10^{34} g$ one is led to $k = 10^{-22} cm^2 g^{-1}$.

As in the aforementioned experiments, these values for k exclude an absorption of gravitation in accordance with the Seeliger-Majorana model. But it does not rule out absorption effects as described by relativistic theories of gravity like the tetrad theory, where the matter source is coupled potential-like to gravitation [11]. The same is true for other theories of gravity competing with GRT that were systematically investigated as to their experimental consequences in Will [138]. For instance, in the tetrad theory, the relativistic field theory of gravity is constructed such that, in the static non-relativistic limit, one has (for details see Sec 3.2)

$$\triangle\phi = 4\pi G\rho \left(1 - \frac{\alpha|\phi|}{c^2 r_{12}}\right), \quad \alpha = const. \tag{32}$$

From this equation it follows that there is a suppression of gravitation by another mass or by its own mass. In the case of two point masses, the mutual gravitational interaction is given by

$$m_1 \ddot{\vec{r}}_1 = -\frac{Gm_1 m_2}{r_{12}^3}\vec{r}_{12}\left(1 - \frac{2Gm_1}{c^2 r_{12}}\right),$$

$$m_2 \ddot{\vec{r}}_2 = -\frac{Gm_1 m_2}{r_{12}^3}\vec{r}_{12}\left(1 - \frac{2Gm_2}{c^2 r_{12}}\right). \tag{33}$$

Thus the effective active gravitational mass m is diminished by the suppression factor $(1 - 2Gm/c^2 r_{12})$. Such effects can also be found in gravitational theories with a variable 'gravitational constant' (Dirac [32], Jordan [66, 67], Brans and Dicke [23]). Furthermore, in the case of an extended body one finds a self-absorption effect. The effective active gravitational mass $\bar{M}$ of a body with Newtonian mass M and radius r is diminished by the body's self-field,

$$\bar{M} = GM\left(1 - \frac{4\pi G}{3c^2}\rho r^2\right) \tag{34}$$

where exact calculations show that the upper limit of this mass is approximately given by the quantity $c^2/\sqrt{G\rho}$.

The modifications of the Newtonian law mentioned above result from modifications of the Laplace equation. In their relativistic generalization, these potential equations lead to theories of gravity competing with GRT. On one hand GRT provides, in the non-relativistic static approximation, the Laplace equation and thus the Newtonian potential and, in higher-order approximations, relativistic corrections. On the other hand, the competing relativistic theories lead, in the first-order approximation, to the above mentioned modifications of the Laplace equation and thus, besides the higher-order relativistic corrections, to additional non-Newtonian variations. All these relativistic theories of gravity (including GRT) represent attempts to extend Faraday's principle of the local nature of all interactions to gravitation. Indeed, in GRT the geometrical interpretation of the equivalence principle realizes this principle insofar as it locally reduces gravitation to inertia and identifies it with the local world metric; this metric replaces the non-relativistic potential and Einstein's field equations replace the Poisson's potential equation. Other local theories introduce additional space-time functions which together with the metric describe the gravitational field. In some of the relativistic rivals of GRT, these functions are of a non-geometric nature. An example is the Jordan or Brans-Dicke scalar field, which, in accordance with Dirac's hypothesis, can be interpreted as a variable gravitational 'constant' G. (For a review of theories involving absorption and suppression of gravitation, see [11].) In other theories of gravity, these additional functions are essentials of the geometric framework, as for instance tetrad theories working in teleparallel Riemannian space and the metric-affine theories working in Riemann-Cartan space-times that are characterized by non-vanishing curvature and torsion (see Secs 3.2, 3.3).

3. On absorption, self-absorption and suppression of gravitation in relativistic theories

The original idea of shielding gravitation assumed that the insertion of some kind of matter between the source of gravitation and a test body could reduce the gravitational interaction. This would be a genuine non-Einsteinian effect. However, in the present paper, our main concern will be relativistic theories. Those theories satisfy the Einstein Equivalence Principle (EEP) [11, 138] stating the following:

i) The trajectory of an uncharged test body depends only on the initial conditions, but not on its internal structure and composition, and
ii) the outcome of any non-gravitational experiment is independent of the velocity of the freely falling apparatus and of the space-time position at which it is performed (local Lorentz invariance and local position invariance).

The point is that, as long as one confines oneself to theories presupposing the EEP, shielding effects will not occur. For, this principle implies that there is only one sign for the gravitational charge such that one finds quite another situation as in electrodynamics where shielding effects appear ("Faraday screen"). But other empirical possibilities for (non-GRT) relativistic effects like absorption and self-absorption are not excluded by the EEP.

In the case of absorption the intervening matter would not only effect on the test body by its own gravitational field but also by influencing the gravitational field of the source. If the latter field is weakened one can speak of absorption. (For a discussion of this effect, see also [129]). In contrast to absorption, self-absorption describes the backreaction of the gravitational field on its own source so that the effective active gravitational mass of a body (or a system of bodies) is smaller than its uneffected active gravitational mass. Of course, both absorption and self-absorption effects can occur simultaneously, and generally both violate the Strong Equivalence Principle (SEP) [11, 138] stating:

i) The trajectories of uncharged test bodies as well as of self-gravitating bodies depend only on the initial conditions, but not on their internal structure and composition, and
ii) the outcome of any local test experiment is independent of the velocity of the freely falling apparatus and of the space-time position at which it is performed.

GRT fulfils SEP. The same is true for fourth-order metric theories of gravity, although there one has a suppression (or amplification) of the matter source by its own gravitational field (see Sec. 3.1).

To empirically test gravitational theories, in particular GRT, it is helpful to compare them to other gravitational theories and their empirical predictions. For this purpose, it was most useful to have a general parameterized framework encompassing a wide class of gravitational theories where the parameters, whose different values characterize different theories, can be tested by experiments and observations. The most successful scheme of this type is the Parameterized Post-Newtonian (PPN) formalism [138] confronting the class of metric theories of gravity with solar system tests.

This formalism rendered it possible to calculate effects based on a possible difference between the inertial, passive and active gravitational masses. It particularly was shown that some theories of gravity predict a violation of the first requirement of the SEP that is due to a violation of the equivalence between inertial and passive gravitational masses. This effect was predicted by Dicke [31] and calculated by Nordtvedt [97]. It does not violate the EEP. Other effects result from a possible violation of the equivalence of passive and active gravitational masses. For instance, Cavendish experiments could show that the locally measured gravitational constant changes in dependence on the position of the measurement apparatus. This can also be interpreted as a position-dependent active gravitational mass, i.e., as absorption.

The PPN formalism is based on the four assumptions:

- **i)** slow motion of the considered matter,
- **ii)** weak gravitational fields,
- **iii)** perfect fluid as matter source, and
- **iv)** gravitation is described by so-called metric theories (for the definition of metric theories, see below).

As noticed in [138], due to the first three assumptions, this approach is not adequate to compare the class of gravitational theories described in (iv) with respect to compact objects and cosmology. And, of course, in virtue of assumption (iv), the efficiency of this approach is limited by the fact that it is dealing with metric theories satisfying EEP.

During the last decades great efforts were made to generalize this framework to nonmetric gravitational theories and to reach also a theory-independent parameterization of effects which have no counterpart in the classical domain. (For a survey, see e. g. [73, 57].) The generalization to a nonmetric framework enables one to test the validity of the EEP, too, and the analysis of "genuine" quantum effects leads to new test possibilities in gravitational physics, where these effects are especially appropriate to test the coupling of quantum matter to gravitation.

First of all, this generalized framework is appropriate to analyze terrestrial, solar system, and certain astrophysical experiments and observations. Of course, it would be desirable to have also a general framework for a systematic study of gravitational systems like compact objects or cosmological models in alternative theories of gravity. But one does not have it so that one is forced to confine oneself to case studies what, to discuss relativistic absorption effects, in the next section will be done.

Our considerations focus on theories that use the genuinely geometrical structures metric, tetrads, Weitzenböck torsion, and Cartan torsion for describing the gravitational field. Accordingly, the following case studies start with theories formulated in a Riemannian space, go then over to theories in a teleparallelized space, which is followed by a simple example of a theory established in a Riemann-Cartan space[5].

But before turning to this consideration, a remark concerning the relation between EEP and metric theories of gravity should be made. This remark is motivated by the fact that it is often argued that the EEP necessarily leads to the class of metric theories satisfying the following postulates:

i) Spacetime is endowed with a metric $\mathbf{g}$,
ii) the world lines of test bodies are geodesics of that metric, and
iii) in local freely falling frames, called local Lorentz frames, the non-gravitational physics are those of special relativity.

(Therefore, presupposing the validity of EEP, in [138] only such theories are compared.)

The arguments in favor of this thesis are mainly based on the observation (see, e.g., [138], p.22 f.) that one of the aspects of EEP, namely the local Lorentz invariance of non-gravitational physics in a freely falling local frame, demands that there exist one or more second-rank tensor fields which reduce in that frame to fields that are proportional to the Minkowski met-

[5] Of course, there are other alternatives to GRT which are also interesting to be considered with respect to non-Einsteinian effects, such as scalar-tensor theories of Brans-Dicke type and effective scalar-tensor theories resulting from higher-dimensional theories of gravity by projection, compactification or other procedures (for such theories, see also the corresponding contributions in this volume). At first the latter idea was considered by Einstein and Pauli [44, 52]. They showed that, if a Riemannian V_5 with signature (-3) possesses a Killing vector which is orthogonal to the hypersurface $x^4 = const$, V_5 reduces to a V_4 with a scalar field because then one has $(A, B = 0, \ldots, 4)$: $g_{AB,4} = 0, g_{i4} = 0, g_{44} = (g^{44})^{-1} = g_{44}(x^l)$.
It should be mentioned that the fourth-order theories discussed in Sec. 3.1 and the scalar-tensor theories are interrelated (cf., the references to equivalence theorems given in Sec.3.1). Furthermore, there is also an interrelation between fourth-order theories, scalar-tensor theories, and theories based on metric-affine geometry which generalizes Riemann-Cartan geometry by admitting a non-vanishing non-metricity. First this became obvious in Weyl's theory [133] extending the notion of general relativity by the requirement of conformal invariance and in Bach's interpretation of Weyl's "unified" theory as a scalar-tensor theory of gravity [3] (see also, [17]).

ric η. Another aspect of EEP, local position invariance then shows that the scalar factor of the Minkowski metric is a universal constant from which the existence of a unique, symmetric, second-rank tensor field **g** follows. However, the point we want to make here is that this does not automatically mean that the gravitational equations have to be differential equations for the metric (and, possibly, additional fields, as for instance a scalar field which do not couple to gravitation directly, as, e.g. realized in the Brans-Dicke theory [23]). This conclusion can only be drawn if one assumes that the metric is a primary quantity. Of course, in physically interpretable theory, one needs a metric and from EEP it follows the above-said. But one can also consider gravitational theories, wherein other quantities are primary. Then one has field equations for these potentials, while the metric describing, in accordance with EEP, the influence of gravitation on non-gravitational matter, is a secondary quantity somehow derived from the primary potentials. For instance, such theories are tetrad (or teleparallelized) theories [93, 94, 101] (see also [11, 12, 15]) or affine theories of gravity [41, 42, 43, 50, 110] (see also [16, 17]).

3.1. FOURTH-ORDER THEORIES

Fourth-order derivative equations of gravitation have received great attention since several authors have proved that the fourth-order terms R^2, $R_{ik}R^{ik}$, and $R_{iklm}R^{iklm}$ can be introduced as counter-terms of the Einstein-Hilbert Lagrangian R to make GRT, quantized in the framework of covariant perturbation theory, one-loop renormalizable. In this context it was clarified that higher-derivative terms naturally appear in the quantum effective action describing the vacuum polarization of the gravitational field by gravitons and other particles [29, 30, 28, 25]. Such equations revive early suggestions by Bach [3], Weyl [134, 135], Einstein [40], and Eddington [35] (see also Pauli [99] and Lanczos [74, 75]).

While early papers treated such equations formulated in non-Riemannian spacetime in order to unify gravitational and electromagnetic fields[6], later they were also considered as classical theories. From the viewpoint of classical gravitational equations with phenomenological matter modifying the Einstein gravitation at small distances, the discussion of such field equations was opened by Buchdahl [24] and Pechlaner and Sexl [100]. Later this discussion was anew stimulated by the argument [121, 124] that such equations should be considered in analogy to the fourth-order electromagnetic theory of Bopp and Podolsky [20, 102] in order to solve the singularity and collapse problems of GRT. Especially, it was shown [121, 124]

[6]Higher-derivative equations were also studied within the framework of quantum field theory by Pais and Uhlenbeck [98].

that, for physical reasons, the Einstein-Hilbert part of the Lagrangian must be necessarily included (for this point, see also [58]) and that one has to impose supplementary conditions on the matter source term. These considerations were continued in [18, 19] (see also [8, 9, 10]). The consequences for singularities of massive bodies and cosmological collapse were also analyzed in [117, 54]. (In the following, we shall call fourth-order equations supplemented by the second-order Einstein term *mixed fourth-order equations.*)

The meaning of such higher order terms in early cosmology was considered in [5, 78]. Especially the influence of pure R^2-terms in cosmological scenarios can be interpreted as a pure gravitational source for the inflationary process [92, 6, 84]. Moreover it was shown in [137] that the transformation $\tilde{g}_{ab} = (1 + 2\epsilon R)g_{ab}$ leads back to GRT described by $\tilde{g}_{ab}$ with an additional scalar field R acting as a source. (For more general equivalence theorems, see [85] and the corresponding references cited therein.)

In the case of stellar objects in the linearized, static, weak field limit for perfect fluid the corresponding Lane-Emden equation provides a relative change of the radius of the object compared to Newtons theory depending on the value of the fundamental coupling constants [26]. This may decrease or increase the size of the object.

Mixed fourth-order equations may only be considered as gravitational field equations, if they furnish approximately at least the Newton-Einstein vacuum for large distances. To this end, it is not generally sufficient to demand that the static spherical-symmetric solutions contain the Schwarzschild solution the Newtonian potential in the linear approximation. One has to demand that the corresponding exact solution can be fitted to an interior equation for physically significant equations of state. In the case of the linearized field equations, this reduces to the requirement that there exist solutions with suitable boundary conditions which, for point-like particles, satisfy a generalized potential equation with possessing a delta-like source.

The most general Lagrangian containing the Hilbert-Einstein invariant as well as the quadratic scalars reads as follows[7] (the square of the curvature tensor need not be regarded since it can be expressed by the two other terms, up to a divergence):

$$L = \sqrt{-g}R + \sqrt{-g}(\alpha R_{ab}R^{ab} + \beta R^2)l^2 + 2\kappa L_M \tag{35}$$

where α and β are numerical constants, l is a constant having the dimension of length. Variation of the action integral $I = \int L d^4x$ results in the field

[7]In this section, we mainly follow [18, 19].

equations of the fourth order, viz.,

$$l^2 H_{ab} + E_{ab} := \tag{36}$$

$$= l^2\big(\alpha\Box R_{ab} + (\frac{\alpha}{2} + 2\beta)g_{ab}\Box R - (\alpha + 2\beta)R_{;ab} + 2\beta RR_{ab} -$$

$$-\frac{\beta}{2}g_{ab}R^2 + 2\alpha R_{acdb}R^{cd} - \frac{\alpha}{2}g_{ab}R_{cd}R^{cd}\big) + \left(R_{ab} - \frac{1}{2}Rg_{ab}\right) = -\kappa T_{ab}.$$

(Here $\Box$ denotes the covariant wave operator.) These equations consist of two parts, namely, the fourth-order terms $\propto l^2$ stemming from the above quadratic scalars and the usual Einstein tensor, where

$$T_{ab} := \frac{1}{\sqrt{-g}}\frac{\delta L_M}{\delta g^{ab}} \quad (\text{with} \quad T^a{}_{b;a} = 0). \tag{37}$$

For $\alpha = -2\beta = 1$ (assumed by Eddington in the case of pure fourth-order equations), in the linear approximation the vacuum equations corresponding to Eqs. (36) reduce to

$$l^2\Box_\eta E^1_{ab} + E^1_{ab} = 0 \tag{38}$$

and, for $\alpha = -3\beta = 1$ (assumed by Bach and Weyl in the case of pure fourth-order equations), to

$$l^2\Box_\eta R^1_{ab} + R^1_{ab} = 0 \tag{39}$$

(the latter follows because in this case the trace of the vacuum equations provides $R = 0$). The discussion of these equations performed in [18] shows that both cases are characterized in that they possess, besides massless gravitons, only *one* kind of gravitons with non-vanishing restmass. (These results were corroborated by subsequent considerations [8, 10].)

In the linear approximation for static fields with the Hilbert coordinate condition

$$g^{ik}{}_{,k} = \frac{1}{2}g^k{}_{k,i} \tag{40}$$

in the above-considered two cases, the pure and mixed fourth order field equations rewritten in a compact form as

$$l^2 H_{ik} = -\kappa T_{ik}, \tag{41}$$

$$l^2 H_{ik} + E_{ik} = -\kappa T_{ik}, \tag{42}$$

produce the simple potential equations (9) and (10) for all components of the metric tensor, if $\alpha = -3\beta$ or $\alpha = -2\beta$, where $k \propto l^{-1}$ is a reciprocal length, and a is the mass of a point-like particle which is given by the Dirac delta-function δ. As shown in Sec. 2, it is this condition imposed on the source that leads to the Green functions (11) and (12). From the expressions (11) and (13) it is evident that, in the case of pure field equations of fourth order, this condition will reduce the manifold of solutions to a functions that do not satisfy the correct boundary conditions at large distances, i.e., those field equations do not mirror the long-range Newtonian interaction. They are consequently at most field equations describing free fields, in other words, equations of a unified field theory in the sense of Weyl's [134] and Eddington's [35] ansatz.

The requirement of general coordinate covariance furnishes in the fourth-order case a variety of field equations depending upon the parameters α and β, whereas the Einstein equations are determined by this symmetry group up to the cosmological term. If one postulates that the field equations are invariant with respect to conform transformations, one obtains just the pure fourth-order equations of Bach and Weyl, where $\alpha = -3\beta$. However, if one wants to couple gravitation to matter one has to go a step further since, due to the conform invariance, the trace of H_{ik} vanishes. Accordingly, the trace of the matter must also vanish. That means that in this case there are supportive reasons for the replacement of pure by mixed field equations. Then the conform invariance is broken by the term E_{ik} being the source of massless gravitons [18].

The meaning of the short-range part of the gravitational potential of this theory becomes more evident when one has a glance at the stabilizing effect it has in the classical field-theoretical particle model (for more details, see [19]). Within GRT, such models first were tried to construct by Einstein [37], Einstein and Rosen [53], and Einstein and Pauli [52]. Later this work was continued in the frame of the so-called geon program by Wheeler and co-workers. Already in the first paper by Einstein dealing with this program it was clear that, in the Einstein-Maxwell theory, it was difficult to reach a stable model. In particular, it was shown that a stable model requires to assume a cut-off length and the equality of mass and charge. The fact that the situation in regard to the particle problem becomes more tractable under these two conditions is an indication that GRT should be modified, if one wants to realize this program. As far as very dense stars are concerned, such a modification should already become significant if the distances are of the order of magnitude of the gravitational radii.

Indeed, such a modification is given by the mixed field equations. In the linear approximation given by Eq. (10), one is led to the Podolsky type solution (Greek indices run from $1 \ldots 3$),

$$g_{00} = 1 + 2\phi, \quad g_{0\alpha} = 0, \quad g_{\alpha\beta} = -\delta_{\alpha\beta}(1 - 2\phi) \tag{43}$$

where

$$\phi = \frac{GM}{rc^2}(1 - e^{-kr}). \tag{44}$$

By calculating the affine energy-momentum tensor $t^i{}_k$ of the gravitational field with this solution and integrating $t^i{}_k$ over a spatial volume, one obtains the result that the Laue criterion [77] for a stable particle with finite mass yields:

$$\int t^0{}_0 d^3x = \frac{GM^2}{2}c^2 k, \quad \int t^0{}_\alpha d^3x = \int t^\alpha{}_0 d^3x = 0. \tag{45}$$

This character of the Podolsky potential can also be seen if one introduces it in the Einstein model [49] of a stellar object (see [7]).

The left-hand pure vacuum part of the field equations (36) with arbitrary α and β are constructed according to the same scheme as Einstein's second-order equations. They are determined by the requirement that they are not to contain derivatives higher than fourth order and fulfil the differential identity $H^k{}_{i;k} = 0$. By virtue of this identity and the contracted Bianchi identity for E_{ik}, $E^k{}_{i;k} = 0$, we get, as in the case of Einstein's equations, the dynamic equations

$$T^k{}_{i;k} = 0, \tag{46}$$

for the pure and mixed equations of fourth order. Consequently, the EEP is also satisfied for those equations.

As to the SEP, the fourth-order metric theory satisfies all conditions of this principle formulated above. But this does not exclude a certain back-reaction of the gravitational field on its matter source leading to the fact that the effective active gravitational mass differs from the inertial mass. This can made plausible by the following argument [9]: Regarding that for vacuum solutions of Eqs. (36) satisfying the asymptotic condition

$$g_{00} = 1 - \frac{2a}{r} \quad \text{for} \quad r \to \infty \tag{47}$$

the active gravitational mass is given by (see [52])

$$a = \int\int\int (E^0{}_0 - E^\alpha{}_\alpha)\sqrt{-g}d^3x \tag{48}$$

we obtain

$$a = \kappa \int\int\int (T^0{}_0 - T^\alpha{}_\alpha)\sqrt{-g}d^3x - l^2 \int\int\int (H^0{}_0 - H^\alpha{}_\alpha)\sqrt{-g}d^3x. \tag{49}$$

This relation replaces the GRT equation

$$a = \kappa \int\int\int (T^0{}_0 - T^\alpha{}_\alpha)\sqrt{-g}d^3x. \tag{50}$$

It represents a modification of the Einstein-Newtonian equivalence of active gravitational mass a and inertial mass m stemming from the hidden-matter term $\propto l^2$. It behaves like hidden (or "dark") matter since, due to the relation $H^{ik}{}_{;k} = 0$, H^{ik} decouples from the visible-matter term T^{ik}. (In the case $\alpha = -3\beta$, this term may be interpreted as a second kind of gravitons). In dependence on the sign of this term, it suppresses or amplifies the active gravitational mass.

3.2. TETRAD THEORIES

That class of gravitational theories which leads to a potential-like coupling is given by tetrad theories formulated in Riemannian space-time with teleparallelism. To unify electromagnetism and gravitation it was introduced by Einstein and elaborated in a series of papers, partly in cooperation with Mayer (for the first papers of this series, see [45, 48, 46, 47, 51]). From another standpoint, later this idea was revived by Møller [93, 94] and Pellegrini & Plebański [101], where the latter constructed a general Lagrangian based on Weitzenböck's invariants [81]. These authors regard all 16 components of the tetrad field as gravitational potentials which are to be determined by corresponding field equations. The presence of a non-trivial tetrad field can be used to construct, beside the Levi-Civita connection defining a Riemannian structure with non-vanishing curvature and vanishing torsion, a teleparallel connection with vanishing curvature and non-vanishing torsion. This enables one to build the Weitzenböck invariants usable as Lagrangians. (Among them, there is also the tetrad equivalent of Einstein-Hilbert Lagrangian, as was shown by Møller.)

A tetrad theory of gravity has the advantage to provide a satisfactory energy-momentum complex [93, 101]. Later Møller [94] considered a version of this theory that can free macroscopic matter configurations of singularities.

From the gauge point of view, in [63] tetrad theory was regarded as translational limit of the Poincaré gauge field theory, and in [95] such a theory was presented as a constrained Poincaré gauge field theory[8]. Another approach [59] considers the translational part of the Poincaré group as gauge group, where in contrast to Poincaré gauge field theory this theory is assumed to be valid on microscopic scales, too. A choice of the Lagrangian leading to a more predictable behavior of torsion than in the

[8]For Poincaré gauge field theory, see Sec. 3.3.

above-mentioned versions of the tetrad theory is discussed in [70] - from the point of Mach's principle, tetrad theory was considered in [12].

The general Lagrange density which is invariant under global Lorentz transformations and provides differential equations of second order is given by

$$L^* = \sqrt{-g}(R + aF_{Bik}F^{Bik} + b\Phi_A\Phi^A) + 2\kappa L_M \tag{51}$$

where a, b are constants, $R = g^{ik}R_{ik}$ is the Ricci scalar, L_M denotes the matter Lagrange density and

$$F_{Aik} := h_{Ak,i} - h_{Ai,k} = h_A{}^l(\gamma_{lki} - \gamma_{lik}), \quad \Phi_A := h_A{}^i\gamma^m{}_{im}. \tag{52}$$

Here $h_A{}^l$ denote the tetrad field and $\gamma_{lki} = h^A{}_l h_{Ak;i}$ the Ricci rotation coefficients ($A, B, \ldots$ are tetrad (anholonomic) indices, $i, k, \ldots$ are space-time (holonomic) indices).

To consider absorption and self-absorption mechanisms we confine ourselves to the case $b = 0$, such that the Lagrange density takes the form

$$L = \sqrt{-g}(R + aF_{Bik}F^{Bik}) + 2\kappa L_M. \tag{53}$$

(Einstein [45] and Levi-Civita [81] discussed the corresponding vacuum solution as a candidate for a unified gravito-electrodynamic theory.) Introducing the tensors

$$T_{ik} := \frac{1}{\sqrt{-g}}\frac{\delta L_M}{\delta h^i{}_A}h^A{}_k \tag{54}$$

$$G_{ik} := R_{ik} - \frac{1}{2}g_{ik}R + \kappa T_{ik} + 2a(\frac{1}{4}g_{ik}F_{Bmn}F^{Bmn} - F_{Bim}F^B{}_k{}^m) \tag{55}$$

and the 4-vector densities

$$S_A{}^i = \sqrt{-g}h_A{}^k G^i{}_k \tag{56}$$

and varying the Lagrangian (53) with respect to the tetrad field, it follow the gravitational field equations in the "Maxwell" form

$$F_A{}^{ik}{}_{;k} = \frac{1}{2a}S_A{}^i. \tag{57}$$

These equations can also be rewritten in an "Einstein" form. This form stems from a Lagrangian which differs from (53) by a divergence term [126]. Since, in the following paragraphs of this section, we assume a symmetric energy-momentum tensor this form be given for this special case:

$$R_{ik} - \frac{1}{2}g_{ik}R = -\kappa T_{ik} - \Theta_{(ik)} \tag{58}$$

$$\Theta_{ik} - \Theta_{ki} = 0 \tag{59}$$

where

$$\Theta_{ik} = a(\frac{1}{2}g_{ik}F_{Bmn}F^{Bmn} - F_{Bim}F^{B}{}_{k}{}^{m} + 2h^{A}{}_{i}F_{Ak}{}^{l}{}_{;l}). \tag{60}$$

The latter form of the field equations allows for an interesting interpretation, because the purely geometric term (60) can be regarded as matter source term[9]. Furthermore, for a symmetric tensor T_{ik}, due to the dynamical equations and Bianchi's identities, one has [12]

$$\Theta_{i}{}^{k}{}_{;k} = 0 \tag{61}$$

so that Θ_{ik} does not couple to visible matter described by T_{ik}. Again it behaves like hidden (or "dark") matter.

The Maxwell form (57) shows clearly that this theory can be considered as a general-relativistic generalization of Eq. (3). Interestingly, in this form the gravitational (suppressing or amplifying) effect appears as a combination of an absorption effect given by the potential-like coupling of usual matter and a hidden-matter effect. Of course, both versions of the theory must lead to the same empirical results.

To discuss absorption effects following from the potential-like coupling in more detail, let us consider the absorption of the active gravitational mass of Earth by the gravitational field of Sun, i.e., calculate the change of the spherical-symmetric part of the Earth field in dependence on the position in the Sun field. To do this, we shall follow a method used in [82] for equations similar to (57) (see also [11]). This effect gives an impression of the order of magnitude of the effects here under consideration.

To this end, we consider the field of Earth, $\underset{1}{h}{}^{A}{}_{i}$, as a perturbation of the field of Sun, $\underset{0}{h}{}^{A}{}_{i}$:

$$h^{A}{}_{i} = \underset{0}{h}{}^{A}{}_{i} + \underset{1}{h}{}^{A}{}_{i} \quad \text{with} \quad |\underset{1}{h}{}^{A}{}_{i}| \ll |\underset{0}{h}{}^{A}{}_{i}|. \tag{62}$$

Rewriting Eq. (57) in the form

$$\Box h^{A}{}_{i} - h^{Ak}{}_{,ik} = \frac{h_{,k}}{h}F^{A}{}_{i}{}^{k} + \frac{1}{2a}h^{A}{}_{l}E^{l}{}_{i} + \frac{\kappa}{2a}h^{A}{}_{l}T^{l}{}_{i} + \frac{1}{4}h^{A}{}_{i}F_{Bmn}F^{Bmn} - h^{Al}F_{Blm}F^{B}{}_{i}{}^{m} \tag{63}$$

[9]If one considers the vacuum version of (58) in the context of unified geometric field theory this interpretation is even forcing (see Schrödinger [110]).

(where $h := \sqrt{-g} = det|h^A{}_i|$ and $E_{ik} = R_{ik} - \frac{1}{2}g_{ik}R$) and inserting ansatz (62) one obtains in the first order approximation

$$\begin{aligned} \Box \underset{0}{h}{}^A{}_i + \Box \underset{1}{h}{}^A{}_i - \underset{0}{h}{}^{Ak}{}_{,ik} - \underset{1}{h}{}^{Ak}{}_{,ik} = \frac{\underset{0}{h}{}_{,k}}{\underset{0}{h}} \underset{0}{F}{}^A{}_i{}^k + \frac{\underset{0}{h}{}_{,k}}{\underset{0}{h}} \underset{1}{h}\underset{0}{F}{}^A{}_i{}^k + \underset{1}{h}{}_{,k}\underset{0}{F}{}^A{}_i{}^k + \quad (64) \\ + \frac{\underset{0}{h}{}_{,k}}{\underset{0}{h}} \underset{1}{F}{}^A{}_i{}^k + \frac{1}{2a}\underset{0}{h}{}^A{}_l \underset{1}{E}{}^l{}_i + \frac{1}{2a}\underset{1}{h}{}^A{}_l \underset{0}{E}{}^l{}_i + \frac{1}{2a}\underset{0}{h}{}^A{}_l \underset{0}{E}{}^l{}_i + \frac{1}{4}\underset{0}{h}{}^A{}_l \underset{0}{F}{}_{Bmn}\underset{0}{F}{}^{Bmn} + \\ + \frac{1}{4}(h^A{}_i F_{Bmn}F^{Bmn})_1 - \underset{0}{h}{}^{Al}\underset{0}{F}{}_{Blm}\underset{0}{F}{}^B{}_i{}^m - (h^{Al}F_{Blm}F^B{}_i{}^m)_1 + \\ + \frac{\kappa}{2a}\left(\underset{0}{h}{}^A{}_l \underset{E}{T}{}^l{}_i + \underset{1}{h}{}^A{}_l \underset{S}{T}{}^l{}_i + \underset{1}{h}{}^A{}_l \underset{E}{T}{}^l{}_i + \underset{0}{h}{}^A{}_l \underset{S}{T}{}^l{}_i\right), \end{aligned}$$

where $\underset{E}{T}{}^l{}_i$ and $\underset{S}{T}{}^l{}_i$ denote the energy-momentum tensor of Earth and Sun, respectively, and $\sqrt{-g} := h = \underset{0}{h} + \underset{1}{h}$. Regarding that the solar field $\underset{0}{h}{}^A{}_i$ is a solution of that equation which is given by the zero-order terms in Eq. (64) one gets as first-order equation for $\underset{1}{h}{}^A{}_i$

$$\begin{aligned} \Box \underset{1}{h}{}^A{}_i - \underset{1}{h}{}^{Ak}{}_{,ik} = \frac{\kappa}{2a}\left(\underset{0}{h}{}^A{}_l \underset{E}{T}{}^l{}_i + \underset{1}{h}{}^A{}_l \underset{S}{T}{}^l{}_i + \underset{1}{h}{}^A{}_l \underset{E}{T}{}^l{}_i\right) + \quad (65) \\ + \frac{\underset{0}{h}{}_{,k}}{\underset{0}{h}} \underset{1}{h}\underset{0}{F}{}^A{}_i{}^k + \underset{1}{h}{}_{,k}\underset{0}{F}{}^A{}_i{}^k + \frac{\underset{0}{h}{}_{,k}}{\underset{0}{h}} \underset{1}{F}{}^A{}_i{}^k + \frac{1}{2a}\underset{0}{h}{}^A{}_l \underset{1}{E}{}^l{}_i + \\ + \frac{1}{2a}\underset{1}{h}{}^A{}_l \underset{0}{E}{}^l{}_i + \frac{1}{4}(h^A{}_i F_{Bmn}F^{Bmn})_1 - (h^{Al}F_{Blm}F^B{}_i{}^m)_1. \end{aligned}$$

Now one can make the following assumptions

i) The terms in the second and third lines all contain cross-terms in $\underset{0}{h}{}^A{}_l$ and $\underset{1}{h}{}^A{}_l$. They describe, in the first-order approximation, the above-mentioned hidden matter correction of the active gravitational mass. In the following we assume that the constant a is so small that the usual matter terms in the first line of (65) are dominating.

ii) The term $\underset{1}{h}{}^A{}_l \underset{S}{T}{}^l{}_i$ is neglected. It describes the influence of Earth potential on the source of the solar field. Near the Earth, it causes a correction having no spherical-symmetric component with respect to the Earth field.

iii) The energy-momentum tensor is assumed to have the form $\underset{E}{T}{}^{ik} = \rho u^i u^k$ (with $\rho = const$). A more complicated ansatz regarding the internal structure of the Earth in more detail leads to higher-order corrections.

iv) The term $\underset{1}{h}{}^A{}_l \underset{E}{T}{}^l{}_i$ is neglected, too. It describes the influence of the Earth field on its own source leading to higher-order (self-absorption) effects.

As a consequence of these assumptions, Eqs. (65) reduce to the field equations

$$\Box \underset{1}{h}{}^A{}_i - \underset{1}{h}{}^{Ak}{}_{,ik} = \frac{\kappa}{2a} \underset{0}{h}{}^A{}_l \underset{E}{T}{}^l{}_i \tag{66}$$

Assuming coordinates, where the spherical-symmetric solution $\underset{1}{h}{}^A{}_i$ has the form $(\mu, \nu = 1, 2, 3)$,

$$\underset{1}{h}{}^{\hat{0}}{}_0 = \alpha, \quad \underset{1}{h}{}^{\hat{\mu}}{}_\nu = \delta^{\hat{\mu}}_\nu \beta, \quad \underset{1}{h}{}^{\hat{0}}{}_\nu = \underset{1}{h}{}^{\hat{\nu}}{}_0 = 0, \tag{67}$$

Eqs. (66) lead to the equations

$$\Delta\alpha = -\frac{\kappa}{2a} \underset{0}{h}{}^0{}_l T^l{}_0 \tag{68}$$
$$\Delta\beta = -\frac{\kappa}{4a} \underset{0}{h}{}^\mu{}_\nu T^\nu{}_\mu.$$

Therefore, up to a factor, the calculation of the absorption effect provides the same result as given in [125]. Thus a gravimeter would register an annual period in the active gravitational mass of the Earth depending on the distance Earth-Sun. The mass difference measured by a gravimeter at aphelion and perihelion results as

$$\frac{\Delta M}{M} = \frac{4GM_\odot}{c^2 R} \epsilon \frac{1}{2a} \tag{69}$$

($M_\odot =$ is the mass of the Sun, $R =$ the average distance of the planet from the Sun, and $\epsilon =$ the eccentricity of the planetary orbit). For Earth, this provides the value 6.6×10^{-10} and, for Mercury 2.06×10^{-8} if $2a = 1$.

It should be mentioned that the above-made estimation is performed under the assumption that the planet rests in the solar field (what enables one to assume (67)). However, one has to regard that the square of the velocity of the Earth with respect to the Sun is of the same order of magnitude as the gravitational field of the Sun. Thus, the velocity effect could compensate the absorption effect. As was shown in [82], the velocity effect is of the same order of magnitude, but generally it differs from the absorption effect by a factor of the order of magnitude 1.

3.3. EINSTEIN-CARTAN-KIBBLE-SCIAMA THEORY

The Einstein-Cartan-Kibble-Sciàma theory [111, 112, 68] (see also, for an early ansatz [131] and for a later detailed elaboration [61]) is the simplest example of a Poincaré gauge field theory of gravity, aiming at a quantum theory of gravity. Moreover, the transition to gravitational theories that are formulated in Riemann-Cartan space is mainly motivated by the fact that then the spin-density of matter (or spin current) can couple to the post-Riemannian structure.[10] Therefore, the full content of this theory becomes only obvious when one considers the coupling of gravitation to spinorial matter.

As far as the non-quantized version of the theory is concerned, there exist two elaborated descriptions of spinning matter in such a theory, the classical Weyssenhoff model [136, 71] and a classical approximation of the second quantized Dirac equation [4]. Moreover there is a number of interesting cosmological solutions which show that in such a theory it is possible to prevent the cosmological singularity [69, 72]. But, unfortunately this is not a general property of Einstein-Cartan-Kibble-Sciama cosmological solutions [71].

In the following we shall describe the situation of an effective Einstein-Cartan-Kibble-Sciama theory with Dirac-matter as source term. This theory is formulated in the Riemann-Cartan space U_4 [109], where the non-metricity Q_{ijk} is vanishing, $Q_{ijk} := -\nabla_i g_{jk} = 0$, such that the connection is given as:

$$\Gamma^k_{ij} = \left\{ {k \atop ij} \right\} - K_{ij}{}^k \tag{70}$$

with the Christoffel symbols $\left\{ {k \atop ij} \right\}$, the torsion $S_{ij}{}^k$, and the contorsion $K_{ij}{}^k$, which in the holonomic representation read,

$$\left\{ {k \atop ij} \right\} := \frac{1}{2} g^{kl}(g_{li,j} + g_{jl,i} - g_{ij,l}), \tag{71}$$

$$S_{ij}{}^k := \frac{1}{2}(\Gamma^k_{ij} - \Gamma^k_{ji}), \quad K_{ij}{}^k := -S_{ij}{}^k + S_j{}^k{}_i - S^k{}_{ij} = -K_i{}^k{}_j.$$

In the anholonomic representation, where all quantities are referred to an orthonormal tetrad of vectors $\vec{e}_A = e^i_A \partial_i$ (capital and small Latin indices run again from 0 to 3)[11], the pure metric part of the anholonomic connection $\tilde{\Gamma}$ is given by the Ricci rotation coefficients $\gamma_{AB}{}^i$.

The field equations are deduced by varying the action integral corresponding to a Lagrange density that consists of a pure gravitational and a

[10] For modern motivation and representation of metric-affine theories of gravity, see [62].

[11] In contrast to Sec. 3.2, we denote the tetrad with e instead of h since here they are coordinates in the Riemann-Cartan space but not a fixed tetrad field in Riemann space with teleparallelism.

matter part; in the holonomic version one has

$$L = L_G(g, \partial g, \Gamma, \partial\Gamma) + 2\kappa L_M(g, \partial g, \Gamma, \phi, \partial\phi) \tag{72}$$

and, in the anholonomic version,

$$L = L_G(e, \partial e, \tilde{\Gamma}, \partial\tilde{\Gamma}) + 2\kappa L_M(e, \partial e, \tilde{\Gamma}, \phi, \partial\phi). \tag{73}$$

In the holonomic representation, one has to vary (72) with respect to

g_{ij} and Γ^k_{ij} or, for $\Gamma = \Gamma(K, g)$, g_{ij} and $K_{ij}{}^k$
or, for $K = K(S, g)$, g_{ij} and $S_{ij}{}^k$

and, in the anholonomic representation, (73) with respect to

e^i_A and $\tilde{\Gamma}_{AB}{}^i$ or, for $\tilde{\Gamma} = \tilde{\Gamma}(K, \gamma)$, e^i_A and $K_{AB}{}^i$.

If matter with half-integer spin is to be coupled to gravitation, then one has to transit to the spinorial version of the latter, where one has to vary

$\gamma^i = e^i_A \gamma^A$ and $\omega_{AB}{}^i$ (or $S_{AB}{}^i$)

(γ^A denote the Dirac matrices defined by $\{\gamma^A, \gamma^B\} = 2\eta^{AB}$ and $\omega_{AB}{}^i$ the spinor connection and $S_{AB}{}^i$ the spinor torsion).

In the case of the Kibble-Sciama theory the pure gravitational Lagrangian is assumed to be the Ricci scalar such that $L_G = eR$ (with $e := \det(e^A_i)$, e^A_i is dual to e^i_A). Then the variation of (73) by e^i_A and ω_{iAB} provides the gravitational equations[12]:

$$R^A{}_i - \frac{1}{2} e^A{}_i R = -\kappa T^A{}_i, \tag{74}$$

$$T_{ikl} = -\frac{\kappa}{2} s_{ikl} \tag{75}$$

where

$$R^A{}_i - \frac{1}{2} e^A{}_i R := e^k{}_B R_{ik}{}^{AB} - \frac{1}{2} e^A{}_i e^l{}_C e^m{}_D R_{lm}{}^{CD} \tag{76}$$

$$T_{ikl} := S_{ikl} - 2\delta^l_{[i} S_{k]m}{}^m, \tag{77}$$

$$T_i{}^A := -\frac{1}{e} \frac{\delta L_M}{\delta e^i{}_A}, \tag{78}$$

$$s^{ikl} := -\frac{1}{e} \frac{\delta L_M}{\delta \omega_{iAB}} e^k{}_A e^l{}_B. \tag{79}$$

[12]Here we follow the (slightly modified) notations used in [4]. They differ from those ones in [61] by sign in R_{ik} and T_{ikl}, and spin density defined in [4] is twice that one in [61] (Therefore, it appear in [109] and our formulas (74) and (75) the minus signs and in Eq. (75) the factor 1/2).

Assuming the case of a Dirac matter field, where L_M is given by the expression,

$$L_D = \hbar c(\frac{i}{2}\overline{\psi}(\gamma^i\nabla_i - \overleftarrow{\nabla}_i\gamma^i)\psi - m\overline{\psi}\psi), \tag{80}$$

the source terms read (see, e.g., [61] and [4]),

$$T^A{}_i = \frac{\hbar c}{2}\overline{\psi}(i\gamma^A\nabla_i - \overleftarrow{\nabla}_i i\gamma^A)\psi, \tag{81}$$

$$s^{ikl} = s^{[ikl]} = \frac{\hbar c}{2}\overline{\psi}\gamma^{[i}\gamma^k\gamma^{l]}\psi = \frac{\hbar c}{2}\epsilon^{ikls}\overline{\psi}\gamma_s\gamma_5\psi, \tag{82}$$

where ϵ^{ikls} is the Levi-Civita symbol and $\gamma_5 := i\gamma^0\gamma^1\gamma^2\gamma^3$.

Due to the total anti-symmetry of the spin density and the field equations (77), the modified torsion T_{ikl} and the torsion S_{ikl} itself are also completely anti-symmetric. Therefore, because of (71), one has $\Gamma^k_{(ij)} = \left\{ {k \atop ij} \right\}$. (It should be mentioned that, for the total antisymmetry of S_{ikl}, the Dirac equations following by varying the Lagrangian (80) with respect to ψ and $\overline{\psi}$ couple only to metric but not to torsion (see [4]).)

Because of the second field equation (77) one can substitute the spin density for the torsion and this way arrive at effective Einstein equations. For this purpose, we split the Ricci tensor into the Riemannian and the non-Riemannian parts (^{0} denotes Riemannian quantities and $_;$ the covariant derivative with respect to $\{\}$),

$$\begin{aligned} R_{Bi} = e^l_A R^A{}_{Bli} = \\ = e^l_A(\overset{0}{R}{}^A{}_{Bli} - S^A{}_{Bl;i} + S^A{}_{Bi;l} - S^C{}_{Bl}S^A{}_{Ci} + S^C{}_{Bi}S^A{}_{Cl}) = \\ = \overset{0}{R}_{Bi} + S^l{}_{Bi;l}. \end{aligned} \tag{83}$$

Regarding that the Ricci scalar is given as

$$R = e^l_A e^{Bi} R^A{}_{Bli} = \overset{0}{R} + S_{CAD}S^{ACD} \tag{84}$$

we obtain Einstein tensor

$$R_{Bi} - \frac{1}{2}e_{Bi}R = \overset{0}{R}_{Bi} - \frac{1}{2}e_{Bi}\overset{0}{R} + S^l{}_{Bi;l} - \frac{1}{2}e_{Bi}S_{CAD}S^{ACD} \tag{85}$$

such that the effective Einstein equations have the form

$$\begin{aligned} \overset{0}{R}_{Bi} - \frac{1}{2}e_{Bi}\overset{0}{R} - \frac{\kappa^2}{2}e_{Bi}s_{klm}s^{klm} = \\ = -\kappa\left(\overset{0}{T}{}^D_{Bi} - \kappa\frac{i\hbar c}{8}\overline{\psi}\gamma^k\sigma^{lm}s_{klm}\psi e_{Bi} - \frac{1}{2}s^l{}_{Bi;l}\right) \end{aligned} \tag{86}$$

($\overset{0}{T}{}^{D}_{Bi}$ is the energy-momentum tensor of the Dirac field in the Riemannian space and $\sigma^{lm} = [\gamma^l, \gamma^m]$.) These equations differ by 3 terms from the Einstein equations:

i) On the left-hand side, it appears a cosmological term, where the effective cosmological constant Λ in the epoch under consideration, is given by the product: $1/2\times$ square of gravitational constant $\times$ square of the spin density of matter in this epoch. Due to the structure of s_{klm} given by Eq. (82), this term can be rewritten as follows[13]

$$\begin{aligned} -\frac{\kappa^2}{2} s_{klm} s^{klm} &= -\frac{\kappa^2}{8}(\epsilon_{klms}\overline{\psi}\gamma^s\gamma_5\psi)(\epsilon^{klmt}\overline{\psi}\gamma_t\gamma_5\psi) = \\ &= \frac{3\kappa^2}{8}\delta^t_s\psi^+\gamma_5\gamma^s\gamma_5\psi\psi^+\gamma_5\gamma_t\gamma_5\psi = \\ &= \frac{3\kappa^2}{8}\psi^+\gamma^s\gamma_s\psi|\psi|^2 = \frac{3\kappa^2}{2}|\psi|^4 \end{aligned} \tag{87}$$

where $|\psi|^2 = \psi^+\psi$. In other words, for a finite fermion number, the effective cosmological term is proportional to the square of the fermion density. Thus, a cosmological term can be dynamically induced or, if such a term is assumed to exist at the very beginning in the equations, tuned away in an effective Einstein-Cartan-Kibble-Sciama theory. (For other approaches, there is a vast number of papers that one has to regard. For the approach of effective scalar-tensor theories, see, e. g., [106] and the literature cited therein.)

ii) On the right-hand side, it appears an additional source term given by the spin density s_{klm} and a divergence term. Originally, the latter is also a pure geometric term, namely the divergence of the torsion $-S^l{}_{Bi;l}$. It has the same property as the dark-matter terms in fourth-order metric theory (Sec. 3.1) and the tetrad theory (Sec. 3.2):

$$S^l{}_{Bi;l}{}^{i} = -g_{si}S_B{}^{ls}{}_{;l}{}^{i} = -S_B{}^{ls}{}_{;ls} = 0. \tag{88}$$

But, in contrast to the above-discussed cases, in virtue of the second field equation (77), it can be rewritten in a visible-matter term.

There is still another way to go over from the Einstein-Cartan-Kibble-Sciama theory to an effective GRT. Instead of combining the field equations (74) and (75), this way performs the substitution $S_{ikl} \to \omega_{ikl}$ in the Lagrangian

$$\begin{aligned} &L(e^A_i, \partial e^a_i, S_{iAB}, \partial S_{iAB}, \psi, \partial\psi, \overline{\psi}, \partial\overline{\psi}) = \\ &= eR(e^A_i, \partial e^A_i, S_{iAB}, \partial S_{iAB}) + L_D(e^A_i, \partial e^A_i, S_{iAB}, \psi, \partial\psi, \overline{\psi}, \partial\overline{\psi}) \end{aligned} \tag{89}$$

[13] $\psi^+ = \overline{\psi}^*$ (complex-conjugate and transposed ψ)

where L_D is given by Eq. (80). This provides the effective Lagrangian

$$L^*(e_i^A, \partial e_i^A, \psi, \partial\psi, \overline{\psi}, \partial\overline{\psi}) = eR^*(e_i^A, \partial e_i^A) + \quad (90)$$
$$+2\kappa L_D^*(e_i^A, \partial e_i^A, S_{iAB}, \psi, \partial\psi, \overline{\psi}, \partial\overline{\psi})$$

$$R^*(e_i^A, \partial e_i^A) = \overset{0}{R} - \omega_{ikl}\omega^{ikl} = \quad (91)$$
$$= \overset{0}{R} - S_{ikl}S^{ikl} = \overset{0}{R} - \frac{\kappa^2}{4} s_{ikl}s^{ikl} = \overset{0}{R} + \frac{3\kappa^2}{2}|\psi|^4$$

$$L_D^*(e_i^A, \partial e_i^A, S_{iAB}, \psi, \partial\psi, \overline{\psi}, \partial\overline{\psi}) = \overset{0}{L}_D + \frac{1}{4} i\hbar ce(\overline{\psi}\gamma^i\sigma^{AB}\omega_{iAB}\psi) = \quad (92)$$
$$= \overset{0}{L}_D + \frac{1}{4} i\hbar ce(\overline{\psi}\gamma^i\sigma^{AB}S_{iAB}\psi) =$$
$$= \overset{0}{L}_D - \frac{1}{8} i\hbar ce(\overline{\psi}\gamma^i\sigma^{AB}s_{iAB}\psi).$$

The variation of L^* with respect to e_i^A provides the equations,

$$\overset{0}{R}_{Bi} - \frac{1}{2}e_{Bi}\overset{0}{R} - \frac{3\kappa^2}{2}e_{Bi}|\psi|^4 = -2\kappa\frac{1}{e}\frac{\delta L_D^*}{\delta e_k^B}g_{ki} \quad (93)$$

where

$$\frac{1}{e}\frac{\delta L_D^*}{\delta e_i^A} = (\overset{0}{T}{}^D)^i_A - \frac{1}{8}\frac{\kappa i\hbar c}{e}\frac{\delta e}{\delta e_i^A}(\overline{\psi}\gamma^c\sigma^{AB}s_{cAB}\psi) = \quad (94)$$
$$= (\overset{0}{T}{}^D)^i_A - \frac{1}{8} i\kappa\hbar c(\overline{\psi}\gamma^c\sigma^{AB}s_{cAB}\psi)e_A^i$$

such that , finally, one has:

$$\overset{0}{R}_{Bi} - \frac{1}{2}e_{Bi}\overset{0}{R} - \kappa^2 e_{Bi}\left(\frac{3}{2}|\psi|^4 - \frac{1}{4}i\hbar c(\overline{\psi}\gamma^c\sigma^{AD}s_{cAD}\psi)\right) = -\kappa\overset{0}{T}{}^D_{Bi}. \quad (95)$$

In contrast to Eq. (86), in Eq. (95) the term proportional to s_{klm} is absorbed in the effective cosmological term, while the "wattless" term is missing. But, generally, one meets again the above-described situation. To some extent, the effective theory represents a general-relativistic generalization of v. Seeliger's ansatz (4).

4. Conclusion

As introductorily mentioned, the idea of going beyond Riemannian geometry can be motivated differently. In dependence on the chosen approach, the question as to the matter coupled to gravitation has be answered in a different manner. As long as one wants to reach a unified geometric theory, there is no room left at all for any matter sources. Since all matter was to be described geometrically such sources are out of place. However, if one interprets the additional geometric quantities as describing gravitation, then one has to introduce matter sources. If one continues to assume the validity of SEP in this case, one has to confine oneself to the consideration of those purely metric theories which were reviewed in Sec. 3.1. In particular, this means that the source of gravitation is assumed to be the metric energy-momentum tensor. If one requires that the theory is derivable from a variation principle this follows automatically. This mathematical automatism has a deep physical meaning.

To make the latter point evident let us return to the discussion of SEP. The formulation of SEP given above focuses its attention on the coupling of gravitation to matter sources, without saying anything on the structure of the matter sources. However, reminding the starting point of the principle of equivalence, namely the Newtonian equivalence between inertial, passive gravitational and active gravitational masses, the SEP implies also a condition on the matter source. Indeed, while the EEP is the relativistic generalization of the equality of inertial and passive gravitational masses, the SEP generalizes the equality of all three masses. Therefore, the special-relativistic equivalent of the Newtonian inertial mass, i.e., the symmetrized special-relativistic energy-momentum tensor, has to be lifted into the curved space and chosen as relativistic equivalent of the active gravitational mass, that means, as source term in the gravitational equations. Of course, this condition is automatically satisfied if one starts from a Lagrangian in a Riemannian space. Therefore, assuming that EEP necessarily leads to a Riemannian space, in [138] this aspect of EEP was not mentioned explicitly. But if one considers theories based on non-Riemannian geometry one should keep in mind this condition required by SEP for the source term.

As was shown in Sec. 3.2, the latter limitation on matter can also be imposed on tetrad theories of gravity. However, it loses its meaning if one goes over to more general geometries and theories of gravity, respectively. In the case that the Lagrangian is given by the Ricci scalar this even leads to a trivialization of the geometric generalization [127]. Therefore, it is more consequent to consider non-Riemannian geometry and its perspectives for a generalization of GRT from the standpoint of the matter sources, as it was done in [111, 112, 68, 131, 61, 62].

As the examples given above demonstrate, in post-Einsteinian theories of gravity there are typical non-Einsteinian effects of gravitation which correspond to certain pre-relativistic ansatzes. One meets absorption (or/and self-absorption) and suppression (and/or amplification) effects.

In case that the additions (Θ_{ik}, H_{ik}, and S_{Bki}, respectively) to the matter source, have a suitable sign there exists an amplification of the matter source by its own gravitational field. Above we called these terms "hidden- or dark-matter terms". To some extent, this is justified by the following argument [13].

The fact that on large scales there is a discrepancy in the mass-to-light ratios can be explained in two alternative ways. Either one presupposes the validity of the theory of GRT and thus, in the classical approximation, i.e., for weak fields (where $GM/rc^2 \ll 1$) and low velocities (where $v \ll c$), the validity of Newtonian gravitational theory on large scales. Then one has to assume that there exist halos of dark matter which are responsible for this discrepancy. Or one takes a modification of GRT into consideration. Then, the classical approximation of the modified theory should represent a gravitational mechanics that is different from the Newtonian one. For, since the dynamical determination of masses of astrophysical systems is always performed in the classical approximation, this approximate mechanics has to work with a gravitational potential deviating from the usual $-GM/r$ form at large distances. From this view larger mass parameters seem to be responsible for the observed motions in astrophysical systems[14].

Interestingly, the above-discussed pure gravitation-field additions have one essential property of dark matter. They can be interpreted as source terms whose divergence vanishes. Accordingly, they do not couple to the optically visible matter and are themselves optically invisible. Of course, to answer the question whether they can really explain the astrophysical data one has to study the corresponding solutions of the respective gravitational equations.

To summarize, when one presupposes the framework of GRT in order to interpret the results of measurements or observations, then one finds the following situation: In all three examples there are dark-matter effects. In the case of tetrad theory, as an implication of the potential-like coupling, additionally one finds a variable active gravitational mass and a variable gravitational number G, respectively. In the case of the Einstein-Cartan-Kibble-Sciama model one finds a Λ-term simulated by spinorial matter. Thus, if one measures a variable G number one has an argument in favor of a theory with potential-like coupling. Moreover, dark-matter or Λ-effects do not necessarily mean that one has to search for exotic particles or that one has got an information about the value of Λ in GRT. These effects could also signal the need for a transition to an alternative theory of gravity.

[14]Another approach to a modified Newtonian law is given by the MOND model. (For this see [108] and the contribution of V. DeSabbata in this volume.)

References

1. de Andrade Martins, R. (1999) The Search for Gravitational Absorption in the Early Twentieth Century, in H. Goenner, J. Renn, J. Ritter and T. Sauer (eds.), *The Expanding Worlds of General Relativity*, Birkhäuser, Boston etc., pp. 3-44.
2. Aronson, S. (1964) The Gravitational Theory of Georges-Louis Le Sage, *The Natural Philosopher* **3**, 51-74.
3. Bach, R. (1921) Zur Weylschen Relativitätstheorie und der Weylschen Erweiterung des Krümmungsbegriffes, *Math. Z.* **9**, 110-123.
4. Bäuerle, G. G. A. and Haneveld, Chr. J. (1983) Spin and Torsion in the Very Early Universe, *Physica* **121A**, 541-551.
5. Barrow, J. D. and Ottewill, A. C. (1983) The Stability of General Relativistic Cosmological Theory, *J. Phys.* **A 16**, 2757-2776.
6. Berkin, A. L. (1990) R^2 Inflation in Anisotropic Universes, *Phys. Rev.* **D42**, 1017-1022.
7. v. Borzeszkowski, H.-H. (1976) On the Gravitational Collapse in Relativistic Theories of Gravitation, *Ann. Phys. (Leipzig)* **33**, 45-54.
8. v. Borzeszkowski, H.-H. (1981) High-frequency Waves in Gravitational Theories with Fourth-order Derivative equations, *Ann. Phys. (Leipzig)* **38**, 231-238.
9. v. Borzeszkowski, H.-H. (1981) On Singularities of General Relativity and Gravitational Equations of Fourth Order, *Ann. Phys. (Leipzig)* **38**, 239-248.
10. v. Borzeszkowski, H.-H. and Frolov, V. P. (1980) Massive Shell Models in the Gravitational Theories with Higher Derivatives, *Ann. Phys. (Leipzig)* **37**, 285-293.
11. v. Borzeszkowski, H.-H., Kasper, U., Kreisel, E., Liebscher, D.-E., and Treder, H.-J. (1971) *Gravitationstheorie und Äquivalenzprinzip*, ed. by H.-J. Treder, Akademie-Verlag, Berlin.
12. v. Borzeszkowski, H.-H. and Treder, H.-J. (1996) Mach-Einstein Doctrine and General Relativity, *Found. Phys.* **26**, 929-942.
13. v. Borzeszkowski, H.-H. and Treder, H.-J. (1998) Dark Matter versus Mach's Principle, *Found. Phys.* **28**, 273-290.
14. v. Borzeszkowski, H.-H. and Treder, H.-J. (1998) Mach's Principle Could Save the Gravitons, *Apeiron* **5**, 143.
15. v. Borzeszkowski, H.-H. and Treder, H.-J. (2000) Bohr's and Mach's Conceptions of Non-Locality in Gravity, in P. G. Bergmann, V. de Sabbata and J. N. Goldberg (eds.), *Classical and Quantum Nonlocality: Proceeding of 16th Course of the International School of Cosmology and Gravitation*, World Scientific, Singapore etc., pp. 1-20.
16. v. Borzeszkowski, H.-H. and Treder, H.-J. (2001) Spinorial Matter in Affine Theory of Gravity and the Space Problem, *Gen. Rel. Grav.* **33**, 1351-1369.
17. v. Borzeszkowski, H.-H. and Treder, H.-J. (2002) On Matter and Metric in Affine Theory of Gravity, *Gen. Rel. Grav.* **34**, 1909-1918.
18. v. Borzeszkowski, H.-H., Treder, H.-J., and Yourgrau, W. (1978) Gravitational Field Equations of Fourth Order and Supersymmetry, *Ann. Phys. (Leipzig)* **35**, 471-480.
19. v. Borzeszkowski, H.-H., Treder, H.-J., and Yourgrau, W. (1979) On Singularities in Electrodynamics and Gravitational Theory, *Astron. Nachr.* **300**, 57.
20. Bopp, F. (1940), *Ann. Phys. (Leipzig)* **38**, 354.
21. Bottlinger, K. F. (1912) Die Erklärung der empirischen Glieder der Mondbewegung durch die Annahme einer Extinktion der Gravitation im Erdinnern, *Astronomische Nachrichten* **191**, 147.
22. Bottlinger, K. F. (1912) *Die Gravitationstheorie und die Bewegung des Mondes*, C. Trömer, Freiburg.
23. Brans, C., and Dicke, R. H. (1961) Mach's Principle and a relativistic theory of gravitation, *Physical Review* **124**, 925-935.
24. Buchdahl, H. A. (1964), *Nuovo Cimento* **123**, 141.

25. Capper, D. M. and Duff, I. (1974) The One-loop Neutrino Contribution to the Graviton Propagator, *Nucl. Phys.* **B82**, 147-154.
26. Chen, Yihan, and Shao, Changgui (2001) Linearized Higher-order Gravity and Stellar Structure, *Gen. Rel. Grav.* **33**, 1267-1279.
27. Crowley, R. H., Woodward, J. F., and Yourgrau, W. (1974) Gravitational Attenuation and the Internal Heating of Planetary Bodies, *Astronomische Nachrichten* **295**, 203.
28. Deser, S. and van Nieuwenhuizen, P. (1974) One-loop Divergences of Quantized Einstein-Maxwell Fields, *Phys. Rev.* **D10**, 401-410; Nonrenormalizability of the Quantized Dirac-Einstein-System, *Phys. Rev.* **D10**, 411-420.
29. DeWitt, B. S. (1965) *Dynamical Theory of Groups and Fields*, Gordon and Breach, London.
30. DeWitt, B. S. (1975) Quantum Field Theory in Curved Spacetime, *Phys. Rep.* **19 C**, 295-357.
31. Dicke, R. H. (1964) Remarks on the Observational Basis of General Relativity, in H.-Y. Chiu and W. F. Hoffman (eds.), *Gravitation and Relativity*, Benjamin, New York, pp. 1-16.
32. Dirac, P. A. M. (1937) The Cosmological Constants, *Nature* **139**, 323.
33. Drude, P. (1897) Über Fernwirkungen, Beilage zu *Annalen der Physik* **62**.
34. Eckardt, D. H. (1990) Gravitational Shielding, *Physical Review* **D42**, 2144-2145.
35. Eddington, A. S. (1953) *Fundamental Theory*, Cambridge U. P., Cambridge.
36. Einstein, A. (1917) *Über spezielle und allgemeine Relativitätstheorie*, F. Vieweg, Braunschweig.
37. Einstein, A. (1919), *Sitzungsberichte der Preussischen Akademie der Wissenschaften zu Berlin*, p.349.
38. Einstein, A. (1919) Bemerkungen über periodische Schwankungen der Mondlänge, welche bisher nach der Newtonschen Mechanik nicht erklärbar schienen, *Sitzungsberichte der Preussischen Akademie der Wissenschaften zu Berlin*, 433-436.
39. Einstein, A. (1919) Bemerkungen zur vorstehenden Notiz, *Sitzungsberichte der Preussischen Akademie der Wissenschaften zu Berlin*, 711.
40. Einstein, A. (1921), *Sitzungsberichte der Preussischen Akademie der Wissenschaften zu Berlin*, p.261.
41. Einstein, A. (1923) Zur allgemeinen Relativitätstheorie, *Sitzungsberichte der Preussischen Akademie der Wissenschaften zu Berlin*, 32-38.
42. Einstein, A. (1923) Bemerkung zu meiner Arbeit 'Zur allgemeinen Relativitätstheorie', *Sitzungsberichte der Preussischen Akademie der Wissenschaften zu Berlin*, 76-77.
43. Einstein, A. (1923) Zur affinen Theorie, *Sitzungsberichte der Preussischen Akademie der Wissenschaften zu Berlin*, 137-140.
44. Einstein, A. (1927) Zu Kaluzas Theorie des Zusammenhanges von Gravitation und Elektrizität I, II, *Sitzungsberichte der Preussischen Akademie der Wissenschaften zu Berlin*, 23-30.
45. Einstein, A. (1928) Riemann-Geometrie mit Aufrechterhaltung des Begriffs des Fernparallelismus, *Sitzungsberichte der Preussischen Akademie der Wissenschaften zu Berlin*, 217-221.
46. Einstein, A. (1928) Neue Möglichkeiten für eine einheitliche Feldtheorie von Gravitation und Elektrizität, *Sitzungsberichte der Preussischen Akademie der Wissenschaften zu Berlin*, 224-227.
47. Einstein, A. (1929) Zur einheitlichen Feldtheorie, *Sitzungsberichte der Preussischen Akademie der Wissenschaften zu Berlin*, 2-7.
48. Einstein, A. (1929), *Sitzungsberichte der Preussischen Akademie der Wissenschaften zu Berlin*, p.124.
49. Einstein, A. (1940), *Ann. Math.* **40**, 922.

50. Einstein, A. (1950 and 1955) *The Meaning of Relativity*, 4th and 5th edn., Princeton U. P., Princeton.
51. Einstein, A. and Mayer, W. (1931) Systematische Untersuchung über kompatible Feldgleichungen, welche in einem Riemannschen Raum mit Fernparallelismus gesetzt werden können, *Sitzungsberichte der Preussischen Akademie der Wissenschaften zu Berlin*, 257-265.
52. Einstein, A. and Pauli, W. (1943) Non-Existence of Regular Stationary Solutions of Relativistic Field Equations, *Ann. Math.* **44**, 131-137.
53. Einstein, A. and Rosen, N. (1935) The Particle Problem in the General Theory of Relativity, *Phys. Rev.* **48**, 73-77.
54. Giesswein, M., Sexl, R., and Sreeruwitz (1976), in A. A. Sokolov (ed.), *Aktualnye problemi teoretitsheskoy fiziki*, Moscow U. P., Moscow.
55. Gillies, G. T. (1987) The Newtonian Gravitational Constant, *Metrologia-International Journal of Scientific Metrology* **24** (Supplement), 1.
56. Gillies, G. T. (1998) Modern Perspectives on Newtonian Gravity, in P. G. Bergmann, V. de Sabbata, G. T. Gillies, and P. I. Pronin (eds.), *Spin in Gravity. Is their a Possibility to Give an Experimental Basis to Torsion?*, World Scientific, Singapore etc., pp. .
57. Haugan, M. P. and Lämmerzahl, C. (2001) Principles of Equivalence: Their Role in Gravitation Physics and Experiments that Test them, in C. Lämmerzahl, C. F. W. Everitt, and F. W. Hehl (eds.), *Gyros, Clocks, Interferometers ...: Testing Relativistic Gravity in Space*, Springer, Berlin etc., pp. 195-213.
58. Havas, P. (1977), *Gen. Rel. Grav.* **8**, 631.
59. Hayashi, K. and Shirafuji, T. (1979) New General Relativity, *Phys. Rev.* **D19**, 3524-3553.
60. Hecker, O. (1907) Beobachtungen mit dem Horizontalpendel, *Veröffentlichungen des Geodätischen Instituts Potsdam*, No. 32.
61. Hehl, F. W., van der Heyde, P., and Kerlick, G. D. (1976) General Relativity with Spin and Torsion: Foundations and Prospects, *Rev. Mod. Phys.* **48**, 393-416.
62. Hehl, F. W., McCrea, J. D., Mielke, E. W., and Ne'eman, Y. (1995) Metric-affine Gauge Theory of Gravity: Field Equations, Noether Identities, World Spinors, and Breaking of Dilaton Invariance, *Phys. Rep.* **258**, 1-171.
63. Hennig, J. and Nitsch, J. (1981), *Gen. Rel. Grav.* **13**, 1947; Müller-Hoissen, F. and Nitsch, J. (1982) report, (unpublished)
64. Isenkrahe, C. (1879) *Das Rätsel von der Schwerkraft*, F. Vieweg, Braunschweig.
65. Ivanenko, D. D. and Sokolov, A. A. (1953) *Klassische Feldtheorie*, Akademie-Verlag, Berlin.
66. Jordan, P. (1947) *Die Herkunft der Sterne*, F. Vieweg, Braunschweig.
67. Jordan, P. (1952) *Schwerkraft und Weltall*, F. Vieweg, Braunschweig.
68. Kibble, T. W. B. (1961) Lorentz Invariance and the Gravitational Field, *J. Math, Phys.* **2**, 212-221.
69. Kopczyński, W. (1972; 1973) A Non-singular Universe with Torsion. *Phys. Lett.* **39A**, 219-220.; An Anisotropic Universe with Torsion. *Phys. Lett.* **43A**, 63-64.
70. Kopczyński, W. (1982) Metric-teleparallel Theories of Gravitation, *J. Phys.* **A15**, 493-506.
71. Korotkii, V.A. and Obukhov, Yu.N. (1987) The Weyssenhoff Fluid in Einstein-Cartan Theory, *Class. Quant. Grav.* **4**, 1633-1657.
72. Kuchowicz, B. (1975; 1976) The Einstein-Cartan Equations in Astrophysically Intersting Situations. I. The Case of Spherical Symmetry. *Acta Phys. Pol.* **B 6**, 555-575.; The Einstein-Cartan Equations in Astrophysically Intersting Situations. II. Homogeneous Cosmological Models of Axial Symmetry. *Acta Phys. Pol.* **B 7**, 81-97.
73. Lämmerzahl, C. (1998) Quantum Tests of Space-time Structure, in P. G. Bergmann, V. de Sabbata, G. T. Gillies and P. I. Pronin (eds.), *Spin in Gravity. Is their a Possibility to Give an Experimental Basis to Torsion?*, World Scientific,

Singapore etc., pp. 91-117.
74. Lanczos, C. (1938), *Ann. Math.* **39**, 842.
75. Lanczos, C. (1975) Gravitation and Riemannian Space, *Found. Phys.* **5**, 9-18.
76. de Laplace, P. S. (1825) *Mécanique Céleste*, Vol. 5, Bachelier, Paris; cf. also (1882) *Oeuvres*, Vol. 5, Bachelier, Paris.
77. v. Laue, M. (1953) *Die Relativitätstheorie, Vol. II*, 3rd ed. Vieweg, Braunschweig.
78. Le Denmat, G. and Sirousse, Zia H. (1987) Equivalence Between Fourth-order Theories of Gravity and General Relativity: Possible Implications for the Cosmological Singularity, *Phys. Rev.* **D35**, 480-482.
79. Le Sage, G.-L. (1749) *Essai sur l'origine des forces mortes*, MS, Univ. of Geneva Library.
80. Le Sage, G.-L. (1784) Lucréce Newtonien, *Memoires de l' Académie Royale des Science et Belles-Lettres de Berlin*, Berlin, pp. 1-28.
81. Levi-Civita, T. (1929 and 1931), *Sitzungsberichte der Preussischen Akademie der Wissenschaften zu Berlin*, p.2 and p.3257.
82. Liebscher, D.-E. (1969) On the Effect of Absorption of Gravitation, *Int. J. Theor. Phys.* **2**, 89-100.
83. Lomonosov, M. W. (1748) Letter no. 195 to Euler written in July 1748 (Correspondence between Euler and Lomonosov led in German and Latin), in: *Wegbereiter der deutsch-slawischen Wechselseitigkeit* , Vol. III, ed by A. P. Juskevic and E. Winter, Akademie-Verlag, Berlin, 1976; see also the other letters no. 193-202.
84. Maeda, Kei-ichi (1988) Inflation as a Transient Attractor in R^2 Cosmology, *Phys. Rev.* **D37**, 858-862.
85. Magnano, G. and Sokolowski, L. M. (1994) Physical Equivalence between Nonlinear Gravity Theories and a Genneral-relativistic Self-gravitating Scalar Field, *Phys. Rev.* **D50**, 5039-5059.
86. Majorana, Q. (1919) Sur la gravitation, *Comptes Rendus des Sèances de l'Académie de Sciences de Paris* **169**, 646-649.
87. Majorana, Q. (1919) Expériences sur la gravitation, *Comptes Rendus des Sèances de l'Académie de Sciences de Paris* **169**, 719.
88. Majorana, Q. (1920) On Gravitation. Theoretical and Experimental Researches, *Philosophical Magazine* **39** (Sixth series), 488-504.
89. Majorana, Q. (1930) Quelques recherches sur l'absorption de la gravitation par la matiére,*Journal de de Physique et le Radium* **1** (Seventh series), 314.
90. Maxwell, J. C. (1875) Atom, *Encyclopaedia Britannica*, 9th edn., Vol.3, Edinburgh.
91. Michelson, A. A. and Gale, H. L. (1919) The Rigidity of the Earth, *Astrophysical Journal* **50**, 342.
92. Mijic, M. B., Morris, M. S., and Suen, W. M. (1986) The R^2 Cosmology: Inflation without a Phase Transition, *Phys. Rev.* **D34**, 2934-2946.
93. Møller, C. (1961) Conservation Laws and Absolute Parallelism in General Relativity, *Math.-Fys. Skr. Dan. Vid. Selskab.* **1**, No.10.
94. Møller, C. (1978) On the Crisis in the Theory of Gravitation and a Possible Solution, *Math.-Fys. Skr. Dan. Vid. Selskab.* **39**, No.13.
95. Müller-Hoissen, F. and Nitsch, J. (1983) Teleparallelism - A Viable Theory of Gravity?, *Phys. Rev.* **28**, 718.
96. Neumann, C. (1870) *Allgemeine Untersuchungen über die Newtonsche Theorie der Fernwirkung*, B. G. Teubner, Leipzig.
97. Nordtvedt, K. (1968) Equivalence Principle for Massive Bodies. I. Phenomenology, *Phys. Rev.* **69**, 1014-1016; Equivalence Principle for Massive Bodies. II. Theory, *Phys. Rev.* **69**, 1017-1025.
98. Pais, A. and Uhlenbeck, E. (1950) Field Theories with Non-localized Action, *Phys. Rev.* **79**, 145-165.
99. Pauli, W. (1921), in *Enzyklopädie der mathematischen Wissenschaften*, vol.

V/2, B. G. Teubner, Leipzig.

100. Pechlaner, E. and Sexl, R. (1966) On Quadratic Lagrangians in General Relativity, *Comm. Math. Phys.* **2**, 165-175.

101. Pellegrini, C. and Plebański, J. (1962; 1963) A Theory of Gravitation, *Math.-Fys. Skr. Dan. Vid. Selskab.* **2**, No.2; Tetrad Fields and Gravitational Fields, *Math.-Fys. Skr. Dan. Vid. Selskab.* **2**, No.4.

102. Podolsky, B. (1941) A Generalized Electrodynamics, *Phys. Rev.* **62**, 68-71.

103. Preston, S. T. and Tolver, S. (1877) On Some Dynamical Conditions Applicable to Le Sage's Theory of Gravitation, *Philosophical Magazine* **4** (Fifth series), 365.

104. Preston, S. T. (1881) On the Importannce of Experiments in Relation to the Theory of Gravitation, *Philosophical Magazine* **11** (Fifth series), 391.

105. P. Prevost (1805), *Notice de la Vie et des Ecrits de Georges-Louis Le Sage*, J. J. Paschoud, Geneva.

106. Rooprai, N., Lohiya, D. (2001) Dynamically Tuning away the Cosmological Constant in Effective Scalar Tensor Theories, astro-ph/0101280.

107. Russell, H. N. (1921) On Majorana's Theory of Gravitation, *Astrophysical Journal* **54**, 334.

108. Sanders, R. H. and McGaugh, S. S. (2002) Modified Newtonian Dynamics as an Alternative to Dark Matter, astro-ph/0204521; to be published in: vol. **40** of Annual Reviews of Astronomy & Astrophysics.

109. Schouten, J. A. (1954) *Ricci-Calculus*, Springer, Berlin etc..

110. Schrödinger, E. (1950) *Space-Time-Structure*, Cambridge U. P., Cambridge.

111. Sciama, D. W. (1962), in *Recent Developments in General Relativity*, Pergamon, Oxford and PWN, Warsaw.

112. Sciama, D. W. (1964) The Physical Structure of General Relativity, *Rev. Mod. Phys.* **36**, 463-469 and 1103.

113. v. Seeliger, H. (1895) Über das Newton'sche Gravitationsgesetz, *Astronomische Nachrichten* **137**, 129.

114. v. Seeliger, H. (1909) Über die Anwendung der Naturgesetze auf das Universum, *Sitzungsberichte der Königlichen Bayrischen Akademie der Wissenschaften zu München, Mathem.-phys. Klasse* **39** (4. Abhandlung).

115. Slichter, L. B., Caputo, M., and Hager, C. L. (1965) TITEL???, *Journal of Geophysical Research* **70**, 1541.

116. Steenbeck, M. and Treder, H.-J. (1984) *Möglichkeiten der experimentellen Schwerkraftforschung*, Akademie-Verlag, Berlin.

117. Stelle, K. S. (1977), *Gen. Rel. Grav.* **9**, 353.

118. Thomson, W. (Lord Kelvin) (1873) On the Ultramundane Corpuscles of Le Sage; *Philosophical Magazine* **45** (Fourth series), 328.

119. Thomson, W. (Lord Kelvin) (1903) Discussions in: *Report of the Meeting of the British Association for the Advancement of Science* **73**, pp. 535-537.

120. Treder, H.-J. (1968) On the Question of a Cosmological Rest-mass of Gravitons, *Int. J. Theor. Phys.* **1**, 167-169.

121. Treder, H.-J. (1975) Zur unitarisierten Gravitationstheorie mit lang- und kurzreichweitigen Termen (mit ruhmasselosen und schweren Gravitonen), *Ann. Phys. (Leipzig)* **32**, 383-400.

122. Treder, H.-J. (1975) *Elementare Kosmologie*, Akademie-Verlag, Berlin.

123. Treder, H.-J. (1976) Bottlingers und Majoranas Absorption der Gravitationskraft und der Gezeitenkräfte, *Gerlands Beiträge der Geophysik* **85**, 513.

124. Treder, H.-J. (1977), in: *75 Jahre Quantentheorie*, Akademie-Verlag, Berlin.

125. Treder, H.-J. (1978) Teilchen-Modelle und effektive Radien in der Allgemeinen Relativitätstheorie, *Ann. Phys. (Leipzig)* **35**, 137-144.

126. Treder, H.-J. (1978) Einsteins Feldtheorie mit Fernparallelismus und Diracs Elektrodynamik, *Ann. Phys. (Leipzig)* **35**, 377-388.

127. Treder, H.-J. (1988) Zu den Einstein-Schrödingerschen Feldgleichungen mit

Materie, *Ann. Phys. (Leipzig)* **45**, 47-52.

128. Treder, H.-J., von Borzeszkowski, H.-H., van der Merwe, A., and Yourgrau, W. (1980) *Fundamental Principles of General Relativity Theories. Local and Global Aspects of Gravitation and Cosmology*, Plenum Press, New York and London.
129. Unnikrishnan, C. S. and Gillies, G. T. (2002) Constraints of Gravitational Shielding, in M. R. Edwards (ed.), *Pushing Gravity*, Apeiron, Montreal, pp. 259-266.
130. Unnikrishnan, S., Mohapatra, A. K. and Gillies, G. T. (2001) Anomalus Gravity Data During the 1997 Total Eclipse do not Support the Hypothesis of Gravitational Shielding, *Phys. Rev.* **D63**, 062002.
131. Utiyama, R. (1956) Invariant Theoretical Interpretation of Interaction, *Phys. Rev.* **101**, 1597-1607 .
132. Weitzenböck, R. (1928) Differentialinvarianten in der Einsteinschen Theorie des Fernparallelismus, *Sitzungsberichte der Preussischen Akademie der Wissenschaften zu Berlin*, p.466.
133. Weyl, H. (1918) Reine Infintesimal-Geometrie, *Math. Z.* **2**, 384-411; see also [135].
134. Weyl, H. (1919) Eine neue Erweiterung der Relativitätstheorie, *Ann. Phys. (Leipzig)* **59**, 101.
135. Weyl, H. (1921) *Raum, Zeit, Materie*, Springer, Berlin.
136. Weyssenhoff, J. and Raabe, A. (1947) Relativistic Dynamics of Spin-fluids and Spin-particles. *Acta Phys. Pol.* **9**, 7-18.
137. Whitt, B. (1984) Fourth-order Gravity as General Relativity plus Matter, *Phys. Lett.* **145B**, 176-178.
138. Will, C. M. (1993) *Theory and Experiment in Gravitational Physics. Revised Edition*, Cambridge U. P., Cambridge.
139. Woodward, J. F. (1972) *The Search for a Mechanism: Action-at-a-distance in Gravitational Theory*, Ph.D. thesis, University of Denver.

CONFORMAL FRAMES AND D-DIMENSIONAL GRAVITY

K.A. BRONNIKOV AND V.N. MELNIKOV
Centre for Gravitation and Fundamental Metrology, VNIIMS
3-1 M. Ulyanovoy St., Moscow 119313, Russia, and
Institute of Gravitation and Cosmology, PFUR
6 Miklukho-Maklaya St., Moscow 117198, Russia

Abstract. We review some results concerning the properties of static, spherically symmetric solutions of multidimensional theories of gravity: various scalar-tensor theories and a generalized string-motivated model with multiple scalar fields and fields of antisymmetric forms associated with p-branes. A Kaluza-Klein type framework is used: there is no dependence on internal coordinates but multiple internal factor spaces are admitted. We discuss the causal structure and the existence of black holes, wormholes and particle-like configurations in the case of scalar vacuum with arbitrary potentials as well as some observational predictions for exactly solvable systems with p-branes: post-Newtonian coefficients, Coulomb law violation and black hole temperatures. Particular attention is paid to conformal frames in which the theory is initially formulated and which are used for its comparison with observations; it is stressed that, in general, these two kinds of frames do not coincide.

1. Introduction

The known gravitational phenomena are rather well described in the framework of conventional general relativity (GR). However, in a more general context of theoretical physics, whose basic aims are to construct a "theory of everything" and to explain why our Universe looks as it looks and not otherwise, most of the recent advances are connected with models in dimensions greater than four: Kaluza-Klein type theories, 10-dimensional superstring theories, M-theory and their further generalizations. Even if such theories (or some of them) successfully explain the whole wealth of particle and astrophysical phenomenology, there remains a fundamental question

V. de Sabbata et al. (eds.),
The Gravitational Constant: Generalized Gravitational Theories and Experiments, 39–64.
© 2004 *Kluwer Academic Publishers. Printed in the Netherlands.*

of finding direct observational evidence of extra dimensions, which is of utmost importance for the whole human world outlook.

Observational "windows" to extra dimensions are discussed for many years [55, 11, 56]. Among the well-known predictions are variations of the fundamental physical constants on the cosmological time scale [1]–[6]. Such constants are, e.g., the effective gravitational constant G and the fine structure constant α. There exist certain observational data on G stability on the level of $\Delta G/G \sim 10^{-11} \div 10^{-12}\ \mathrm{y}^{-1}$ [1, 2, 7], which restrict the range of viable cosmological models. Some evidence on the variability of α has also appeared from quasar absorption spectra: $\Delta\alpha/\alpha \sim -0.72{\cdot}10^{-5}$ over the redshift range $0.5 < z < 3.5$ [8] (the minus means that α was smaller in the past).

Other possible manifestations of extra dimensions include excitations in compactified factor spaces [9], which can behave as particles with a large variety of masses and contribute to dark matter or to cross-sections of usual particle interactions; monopole modes in gravitational waves; various predictions for standard cosmological tests and generation of the cosmological constant [10], and numerous effects connected with local field sources, including, in particular, deviations from the Newton and Coulomb laws [4, 5, 11, 57, 12] and the properties of black holes, especially in the actively discussed brane-world framework [13].

In this paper we discuss solutions of multidimensional theories of gravity of Kaluza-Klein type, i.e., under the condition that neither the metric nor other fields depend on the additional (internal) coordinates [55, 11, 56]. The 4D metric is generally specified in such theories up to multiplying by a conformal factor depending on scalar fields and extra-dimension scale factors. This is the well-known problem of choice of a physical conformal frame (CF). Mathematically, a transition from one CF to another is nothing else but a substitution in the field equations, which can be solved using any variables. However, physical predictions about the behaviour of matter (except massless particles) are CF-dependent.

Among possible CFs one is distinguished: the so-called Einstein frame, in which the metric field Lagrangian contains the scalar curvature $\mathcal{R}$ with a constant coefficient. In other, so-called Jordan frames, $\mathcal{R}$ appears with field-dependent factors.

The choice of a physical CF in non-Einsteinian theories of gravity is rather widely discussed, but mostly in four dimensions in the context of scalar-tensor theories (STT) and in higher-order theories with curvature-nonlinear gravitational Lagrangians — see, e.g., [14, 15, 16] and numerous references therein. The review [15] classified the authors of published papers by their attitude to the problem: those (i) neglecting the issue; (ii) supporting the view that all frames are equivalent; (iii) recognizing the problem but giving no conclusive arguments; (iv) claiming that a Jordan frame physical; (v) asserting that the Einstein is physical. Each group included tens of names, and some names even got into more than one group.

Refs. [14] and [15] have presented arguments in favour of the Einstein frame, and the most important ones, applicable to STT and higher-order theories (and multiscalar-tensor theories obtainable from multidimensional gravity) are connected with the positivity of scalar field energy and the existence of a classically stable ground state.

In our view [2, 17, 18] [which turns out to be outside the groups (i)–(v)], the above arguments could be convincing if we dealt with an "absolute", or "ultimate" theory of gravity. If, however, the gravitational action is obtained in a certain limit of a more fundamental unified theory, theoretical requirements like the existence of a stable ground state should be addressed to this underlying theory rather than its visible manifestation. In the latter, the notion of a physical CF should be only related to the properties of instruments used for measuring masses, lengths and time intervals. Moreover, different sets of instruments (different measurement systems [2]) are described, in general, by different CFs. We thus suppose that there can be at least two different physical CFs: the fundamental one, in which the underlying field theory (or a field limit of a more fundamental theory) is specified, and the observational one, corresponding to a given set of instruments. One can say that the first CF describes what is happening "as a matter of fact", the second one — what we see.

The set of references used in the present observations is connected with atomic units, and the corresponding observational CF for any underlying theory is therefore the CF that provides geodesic motion for ordinary massive (fermionic) matter in 4 dimensions.

In what follows, we will first discuss the CF dependence of the properties of space-time using, as an example, static, spherically symmetric scalar field configurations in STT (Sec. 2) and in multidimensional theories of gravity with multiple factor spaces (Sec. 3). We shall see that some general theorems, valid in one CF, may be violated in another, and there are such conformal mappings that the space-times of different frames are even not in a one-to-one correspondence (the so-called conformal continuation [19]). Then, in Sec. 4, we will discuss the CF dependence of some observable quantities for a class of solutions of a generalized field model [20]–[23], containing multiple scalar fields and antisymmetric forms, associated with charged p-branes. This choice is motivated by the bosonic sector of the low-energy field approximation of superstring theories, M-theory and their generalizations [24]–[28]. The model is, however, not restricted to known theories since it assumes arbitrary dimensions of factor spaces, arbitrary ranks of antisymmetric forms and an arbitrary number of scalar fields.

Among the quantities to be discussed are (1) the post-Newtonian (PN) coefficients describing the weak field behaviour of the solutions, (2) for black hole solutions, the Hawking temperature $T_{\rm H}$ which is obviously important for small (e.g., primordial) black holes rather than those of stellar or galactic mass range and (3) the parameters of Coulomb law violation for the 4D components of the antisymmetric forms which behave as an electromagnetic field. We do not fix the underlying fundamental theory and thus have no reason to prescribe a particular CF, therefore the results are formulated in an arbitrary frame.

2. Scalar-vacuum configurations in STT

2.1. STT IN JORDAN AND EINSTEIN FRAMES

Consider a general (Bergmann-Wagoner-Nordtvedt) STT, in a D-dimensional manifold $\mathbb{M}_{\rm J}[g]$ with the metric $g_{\mu\nu}$ (to be called the Jordan conformal frame), for which the gravitational field action is written of the form

$$S_{\rm STT} = \int d^D x \sqrt{g}[f(\phi)\mathcal{R} + h(\phi)(\partial\phi)^2 - 2U(\phi)], \tag{1}$$

where $g = |\det(g_{\mu\nu})|$, $(d\phi)^2 = g^{\mu\nu}\,\partial_\mu\phi\,\partial_\nu\phi$ and f, h, U are arbitrary functions of the scalar field ϕ.

The action (1) can be simplified by the well-known conformal mapping which generalizes Wagoner's [29] 4-dimensional transformation,

$$g_{\mu\nu} = F(\phi)\overline{g}_{\mu\nu}, \qquad F(\phi) := |f(\phi)|^{-2/(D-2)}, \tag{2}$$

$$\frac{d\psi}{d\phi} = \pm\frac{\sqrt{|l(\phi)|}}{f(\phi)}, \qquad l(\phi) := fh + \frac{D-1}{D-2}\Big(\frac{df}{d\phi}\Big)^2, \tag{3}$$

removing the nonminimal scalar-tensor coupling expressed in the factor $f(\phi)$ before $\mathcal{R}$. The action (1) is now specified in the new manifold $\mathbb{M}_{\rm E}[\overline{g}]$ with the metric $\overline{g}_{\mu\nu}$ (the Einstein frame) and the new scalar field ψ:

$$S_{\rm E} = \int d^D x \sqrt{\overline{g}}\Big\{\operatorname{sign} f[\overline{\mathcal{R}} + (\operatorname{sign} l)(\partial\psi)^2] - 2V(\psi)\Big\}, \tag{4}$$

where the determinant $\overline{g}$, the scalar curvature $\overline{\mathcal{R}}$ and $(\partial\psi)^2$ are calculated using $\overline{g}_{\mu\nu}$, and

$$V(\psi) = |f|^{-D/(D-2)}(\psi)\,U(\phi). \tag{5}$$

The action (4) is similar to that of GR with a minimally coupled scalar field ψ but, in addition to arbitrary D, contains two sign factors. The usual sign of gravitational coupling corresponds to $f > 0$. On the other hand, theories with $l(\phi) < 0$ lead to an anomalous sign of the kinetic term of the ψ field in (4) — a "ghost" scalar field as it is sometimes called. Such fields violate all standard energy conditions and therefore easily lead to unusual solutions like wormholes [30, 31]. We will adhere to theories with $l > 0$. However, $f < 0$ in some regions of $\mathbb{M}_J$ will appear due to continuations to be discussed further.

Among the three functions of ϕ entering into (1) only two are independent since there is a freedom of transformations $\phi = \phi(\phi_{\rm new})$. We assume $h \geq 0$ and use this freedom, choosing in what follows $h(\phi) \equiv 1$.

2.2. NO-GO THEOREMS FOR THE EINSTEIN FRAME

Let us discuss some general properties of static, spherically symmetric scalar-vacuum configurations. We begin with the Einstein frame $\mathbb{M}_E(\overline{g})$ with the action (4) and put $\operatorname{sign} f = \operatorname{sign} l = 1$, thus obtaining D-dimensional GR with a minimally coupled scalar field ψ.

Choosing the radial coordinate ρ corresponding to the gauge condition $\overline{g}_{tt}\,\overline{g}_{\rho\rho} = -1$, we can write an arbitrary static, spherically symmetric metric in the form

$$ds_{\rm E}^2 = A(\rho)\,dt^2 - \frac{du^2}{A(\rho)} - r^2(\rho)\,d\Omega_{d_0}^2 \tag{6}$$

where $d_0 = D-2$ and $d\Omega_{d_0}^2$ is the linear element on the sphere $\mathbb{S}^{d_0}$ of unit radius. This gauge is preferable for considering Killing horizons, described as zeros of the function $A(\rho)$. The reason is that near a horizon ρ varies (up to a positive constant factor) like manifestly well-behaved Kruskal-like coordinates used for an analytic continuation of the metric. Thus, using this coordinate, which may be called *quasiglobal* [32], one can "cross the horizons" preserving the formally static expression for the metric.

Three independent field equations due to (4) for the unknowns $A(\rho)$, $r(\rho)$ and $\psi(\rho)$ may be written as follows:

$$(A' r^{d_0})' = -(4/d_0) r^{d_0} V; \tag{7}$$

$$d_0 r''/r = -\psi'^2; \tag{8}$$

$$A(r^2)'' - r^2 A'' = (d_0 - 2) r'(A'r - 2Ar') + 2(d_0 - 1), \tag{9}$$

where the prime denotes $d/d\rho$. These are three combinations of the Einstein equations; the scalar field equation $(Ar^{d_0}\psi')' = r^{d_0} dV/\psi$ can be obtained as their consequence.

Eqs. (7)–(9) cannot be exactly solved for a given arbitrary potential $V(\psi)$ but make it possible to prove some important theorems telling us what can and what cannot be expected from such a system:

A. The no-hair theorem [33, 34] claiming that asymptotically flat black holes cannot have nontrivial external scalar fields with nonnegative $V(\varphi)$. In other words, in case $V \geq 0$, the only asymptotically flat black hole solution is characterized outside the horizon by $V \equiv 0$, $\psi = \text{const}$ and the Schwarzschild (or Tangherlini in case $d_0 > 2$) metric, i.e., in (6) $r \equiv \rho$ and $A = A(r) = 1 - 2m/r^{d_0-1}$, $m = \text{const}$.

B. The generalized Rosen theorem [35, 36] asserting that particle-like solutions (i.e., asymptotically flat solutions with a regular centre) do not exist in case $V \geq 0$.

C. The nonexistence theorem for regular configurations without a centre (wormholes, horns, flux tubes with $\psi \neq \text{const}$) [32].

D. The causal structure theorem [32], asserting that the list of possible types of global causal structures (described by Carter-Penrose diagrams) for configurations with any potentials $V(\varphi)$ and any spatial asymptotics is the same as the one for $\varphi = \text{const}$, namely: Minkowski (or AdS), Schwarzschild, de Sitter and Schwarzschild–de Sitter.

These results will be referred to as Statements A, B, C, D, respectively.

Some comments are in order. Statement A is proved [33, 34] (see also [36] for $D > 4$) by finding an integral relation whose two parts have different signs unless the scalar field is trivial. There also exist no-hair theorems for black holes with de Sitter and anti-de Sitter asymptotics [37].

Statement B is proved in its most general form [36] by comparing two expressions for the mass: one written as an integral of the energy density and another given by the Tolman formula.

In Statement C, a *wormhole* is, by definition, a configuration with two asymptotics at which $r(\rho) \to \infty$, hence $r(\rho)$ must have at least one regular minimum. A *flux tube* is characterized by $r = \text{const} > 0$, i.e., it is a static (d_0+1)-dimensional cylinder. A *horn* is a configuration that tends to a flux tube at one of its asymptotics, i.e., $r(\rho) \to \text{const} > 0$ at one of the ends of the range of ρ. The statement is proved [32] using Eq. (8): e.g., it leads to $r'' \leq 0$, which is incompatible with a regular minimum of $r(\rho)$.

A proof of Statement D [32, 38] rests on Eq. (9) which implies that the function $A(\rho)/r^2$ cannot have a regular minimum, therefore $A(\rho)$ can have at most two simple zeros around a static (R) region with $A > 0$ or one double zero separating two nonstatic (T) regions ($A < 0$).

It should be stressed that the validity of Statements C and D is independent of any assumptions on the shape and even sign of the potential $V(\psi)$ and on the particular form of the spatial asymptotic.

In cases admitted by the above theorems, black hole and particlelike solutions can be obtained, as is confirmed by known explicit examples. Thus, there exist: (1) black holes possessing nontrivial scalar fields (scalar hair), with $V \geq 0$, but with non-flat and non-de Sitter asymptotics [39];

(2) black holes with scalar hair and flat asymptotics, but partly negative potentials [38]; (3) configurations with a regular centre, a flat asymptotic and positive mass, but also with partly negative potentials [38].

2.3. NO-GO THEOREMS FOR GENERIC SCALAR-TENSOR SOLUTIONS

In this section we discuss the possible validity of Statements A–D for STT solutions in a Jordan frame.

One can notice that when a space-time manifold $\mathbb{M}_\mathrm{E}[\overline{g}]$ (the Einstein frame) with the metric (6) is conformally mapped into another manifold $\mathbb{M}_\mathrm{J}[g]$ (the Jordan frame) equipped with the same coordinates according to the law (2), then a horizon $\rho = h$ in $\mathbb{M}_\mathrm{E}$ passes into a horizon of the same order in $\mathbb{M}_\mathrm{J}$, a centre ($r = 0$) and an asymptotic ($r \to \infty$) in $\mathbb{M}_\mathrm{E}$ pass into a centre and an asymptotic, respectively, in $\mathbb{M}_\mathrm{J}$ if the conformal factor $F = F(\rho)$ is regular (i.e., finite, at least C^2-smooth and positive) at the corresponding values of ρ. A regular centre passes to a regular centre and a flat asymptotic to a flat asymptotic under evident additional requirements.

The validity of Statements A–D in the Jordan frame thus depends on the nature of the conformal mapping (2) that connects $\mathbb{M}_\mathrm{J}[g]$ with $\mathbb{M}_\mathrm{E}[\overline{g}]$). Thus, if F vanishes or blows up at an intermediate value of ρ, there is no one-to-one correspondence between $\mathbb{M}_\mathrm{J}$ and $\mathbb{M}_\mathrm{E}$. In particular, if a singularity in $\mathbb{M}_\mathrm{E}$ is mapped to a regular sphere $\mathbb{S}_\mathrm{trans}$ in $\mathbb{M}_\mathrm{J}$, then $\mathbb{M}_\mathrm{J}$ should be continued beyond this sphere, and we obtain, by definition, a *conformal continuation* (CC) from $\mathbb{M}_\mathrm{E}$ into $\mathbb{M}_\mathrm{J}$ [40, 19].

Such continuations can only occur for special solutions: to be removed by a conformal factor, the singularity should be, in a sense, isotropic. Moreover, the factor F should behave precisely as is needed to remove it.

In more generic situations, for given $\mathbb{M}_\mathrm{E}$, there is either a one-to-one correspondence between the two manifolds, or the factor F "spoils" the geometry and creates a singularity in $\mathbb{M}_\mathrm{J}$, that is, in a sense, $\mathbb{M}_\mathrm{J}$ is "smaller" than $\mathbb{M}_\mathrm{E}$. In these cases Statement D is obviously valid in $\mathbb{M}_\mathrm{J}$. This is manifestly true for STT with $f(\phi) > 0$.

Statement C cannot be directly transferred to $\mathbb{M}_\mathrm{J}$ in any nontrivial case $F \neq \mathrm{const}$. In particular, minima of $g_{\theta\theta}$ (wormhole throats) can appear. Though, wormholes as global entities are impossible in $\mathbb{M}_\mathrm{J}$ if the conformal factor F is finite in the whole range of ρ, including the boundary values. Indeed, assuming that there is such a wormhole, we shall immediately obtain two large r asymptotics and a minimum of $r(\rho)$ between them even in $\mathbb{M}_\mathrm{E}$, in contrast to Statement C valid there.

Statements A and B can also be extended to $\mathbb{M}_\mathrm{J}$ for generic STT solutions, but here we will not concentrate on the details and refer to the papers [41] (see also Sec. 3).

Conformal continuations, if any, can in principle lead to new, maybe more complex structures.

2.4. CONFORMAL CONTINUATIONS

A CC from $\mathbb{M}_E$ into $\mathbb{M}_J$ can occur at such values of the scalar field ϕ that the conformal factor F in the mapping (2) is singular while the functions f, h and U in the action (1) are regular. This means that at $\phi = \phi_0$, corresponding to a possible transition surface $\mathbb{S}_{\rm trans}$, the function $f(\phi)$ has a zero of a certain order n. Then, in the transformation (3) near $\phi = \phi_0$ in the leading order of magnitude

$$f(\phi) \sim \Delta\phi^n, \qquad n = 1, 2, \ldots, \qquad \Delta\phi \equiv \phi - \phi_0. \tag{10}$$

One can notice, however, that $n > 1$ leads to $l(\phi_0) = 0$ (recall that by our convention $h(\phi) \equiv 1$). This generically leads to a curvature singularity in $\mathbb{M}_J$, as can be seen from the trace of the metric field equation due to (1) [19]. We therefore assume $l > 0$ at $\mathbb{S}_{\rm trans}$. Therefore, according to (3), we have near $\mathbb{S}_{\rm trans}$ ($\phi = \phi_0$):

$$f(\phi) \sim \Delta\phi \sim {\rm e}^{-\psi\sqrt{d_0/(d_0+1)}}, \tag{11}$$

where without loss of generality we choose the sign of ψ so that $\psi \to \infty$ as $\Delta\phi \to 0$.

In the CC case, the metric $\overline{g}_{\mu\nu}$ is singular on $\mathbb{S}_{\rm trans}$ while $g_{\mu\nu} = F(\phi)\,\overline{g}_{\mu\nu}$ is regular. There are two opportunities. The first one, to be called CC-I for short, is that $\mathbb{S}_{\rm trans}$ is an *ordinary regular surface* in $\mathbb{M}_J$, where both $g_{tt} = \mathcal{A} = FA$ and $-g_{\theta\theta} = R^2 = Fr^2$ (squared radius of $\mathbb{S}_{\rm trans}$) are finite. (Here θ is one of the angles that parametrize the sphere $\mathbb{S}^{d_0}$.) The second variant, to be called CC-II, is that $\mathbb{S}_{\rm trans}$ is a *horizon* in $\mathbb{M}_J$. In the latter case only $g_{\theta\theta}$ is finite, while $g_{tt} = 0$.

Given a metric $\overline{g}_{\mu\nu}$ of the form (6) in $\mathbb{M}_E$, a CC-I can occur if

$$F(\psi) = |f|^{-2/d_0} \sim 1/r^2 \sim 1/A \tag{12}$$

as $\psi \to \infty$, while the behaviour of f is specified by (11). In $\mathbb{M}_E$, the surface $\mathbb{S}_{\rm trans}$ ($r^2 \sim A \to 0$) is either a singular centre, if the continuation occurs in an R-region, or a cosmological singularity in the case of a T-region.

It has been shown [19] that necessary and sufficient conditions for the existence of CC-I are that $f(\phi)$ has a simple zero at some $\phi = \phi_0$, and $|U(\phi_0)| < \infty$. Then there is a solution in $\mathbb{M}_J$, smooth in a neighbourhood of the surface $\mathbb{S}_{\rm trans}$ ($\phi = \phi_0$), and in this solution the ranges of ϕ are different on different sides of $\mathbb{S}_{\rm trans}$.

Thus any STT with $h \equiv 1$ admitting a simple zero of $f(\phi)$ admits a CC-I. The smooth solution in $\mathbb{M}_J$ corresponds to two solutions on different sides of $\mathbb{S}_{trans}$ in two different Einstein frames. These solutions are special, being restricted by Eq. (12).

It is of interest that, under the CC-I conditions, any finite potential $V(\psi)$ is inessential near $\mathbb{S}_{trans}$: the solution is close to Fisher's scalar-vacuum solution [42] for $D = 4$ or its modification in other dimensions. For $U(\phi)$, the CC-I conditions do not lead to other restrictions than regularity at $\phi = \phi_0$.

In case $D = 3$, as follows from Eq. (9), a necessary condition for CC-I is $A/r^2 = \text{const}$.

A CC-II requires more special conditions [19], namely, there should be $D \geq 4$, $U(\phi) = 0$ and $dU/d\phi \neq 0$ at $\phi = \phi_0$. It then follows that $\mathbb{S}_{trans}$ is a second-order horizon, connecting two T-regions in $\mathbb{M}_J$. Thus the only kind of STT configurations admitting CC-II is a $D \geq 4$ Kantowski-Sachs cosmology consisting of two T-regions (in fact, epochs, since ρ is then a temporal coordinate), separated by a double horizon.

2.5. GLOBAL PROPERTIES OF CONTINUED SOLUTIONS

A solution to the STT equations may *a priori* undergo a number of CCs, so that each region of $\mathbb{M}_J$ between adjacent surfaces $\mathbb{S}_{trans}$ is conformally equivalent to some $\mathbb{M}_E$. However, the global properties of $\mathbb{M}_J$ with CCs are not so diverse as one might expect. In particular, Statement D, restricting possible causal structures, holds in $\mathbb{M}_J$ in the same form as in $\mathbb{M}_E$.

A key point for proving this is the observation that the quantity $B = A/r^2$ is insensitive to conformal mappings (both its numerator and denominator are multiplied by F) and is thus common to $\mathbb{M}_J$ and $\mathbb{M}_E$ which is equivalent to a given part of $\mathbb{M}_J$. Therefore zeros and extrema of B inside $\mathbb{M}_E$ preserve their meaning in the corresponding part of $\mathbb{M}_J$. Statement D rests on the fact that $B(\rho)$ cannot have a regular minimum in $\mathbb{M}_E$; the same is true in a region of $\mathbb{M}_J$ equivalent to some $\mathbb{M}_E$, and a minimum can only take place on a transition surface $\mathbb{S}_{trans}$ between such regions. A direct inspection shows [19] that this is not the case. Therefore Statement D is valid in $\mathbb{M}_J$ despite any number of CC's.

As for Statements A–C, the situation is more involved. To our knowledge, full analogues of Statements A and B (probably with additional restrictions) for a sufficiently general STT are yet to be obtained (see, however, [41]). Statement C is evidently violated due to CCs since wormholes are a generic product of such continuations [19].

Indeed, a generic behaviour of $\mathbb{M}_E$ is that r varies from zero to infinity. Let there be a family of such static solutions and let $f(\phi)$ have a simple zero. Then there is a subfamily of solutions admitting CC-I. A particular solution from this subfamily can come across a singularity beyond $\mathbb{S}_{\rm trans}$ [due to $f(\phi) \to \infty$ or $l(\phi) \to 0$], but if "everything is quiet", it will, in general, arrive at another spatial asymptotic and will then describe a wormhole.

It can be shown [19] that, under our assumption $l > 0$, there cannot be more than two values of ϕ where CCs are possible, i.e., where $f = 0$ and $df/d\phi \neq 0$. This does not mean, however, that an STT solution cannot contain more than two CCs. The point is that ϕ as a function of the radial coordinate is not necessarily monotonic, so there can be two or more CCs corresponding to the same value of ϕ. A transition surface $\mathbb{S}_{\rm trans} \in \mathbb{M}_{\rm J}$ corresponds to $r = 0$ in $\mathbb{M}_{\rm E}$, therefore an Einstein-frame manifold $\mathbb{M}_{\rm E}$, describing a region between two transitions, should contain two centres, more precisely, two values of the radial coordinate (say, ρ) at which $r = 0$. This property, resembling that of a closed cosmological model, is quite generic due to $r'' \leq 0$ in Eq. (8), but a special feature is that the conditions (12) should hold at both centres.

Well-known particular examples of CC-I are connected with massless nonminimally coupled scalar fields in GR, which may be described as STT with $f(\phi) = 1 - \xi\phi^2$, $h(\phi) = 1$, $U(\phi) = 0$. One such example is a black hole with a conformally coupled field ($\xi = 1/6$)) [44, 33], such that $\phi = \infty$ but the energy-momentum tensor is finite on the horizon. Other examples are wormholes supported by conformal [30] and nonconformal [46] fields. Ref. [19] contains an example of a configuration with an infinite number of CCs, built using a conformally coupled scalar field with a nonzero potential in three dimensions.

3. Theories with multiple factor spaces

3.1. REDUCTION

In Sec. 2 we have been concerned with STT solutions in D-dimensional space-times with the metrics $\overline{g}_{\mu\nu}$ given by (6) and $g_{\mu\nu} = F(\phi)\,\overline{g}_{\mu\nu}$. Let us now pass to space-times $\mathbb{M}^D$ with a more general structure

$$\mathbb{M}^D = \mathbb{R}_u \times \mathbb{M}_0 \times \mathbb{M}_1 \times \mathbb{M}_2 \times \cdots \times \mathbb{M}_n \tag{13}$$

where $\mathbb{M}_{\rm ext} = \mathbb{R}_u \times \mathbb{M}_0 \times \mathbb{M}_1$ is the "external" manifold, $\mathbb{R}_u \subseteq \mathbb{R}$ is the range of the radial coordinate u, $\mathbb{M}_1$ is the time axis, $\mathbb{M}_0 = \mathbb{S}^{d_0}$. Furthermore, $\mathbb{M}_2, \ldots, \mathbb{M}_n$ are "internal" factor spaces of arbitrary dimensions d_i, $i = 2, \ldots, n$, and, according to this notation, we also have $\dim \mathbb{M}_0 = d_0$ and

$\dim \mathbb{M}_1 = d_1 = 1$. The metric is taken in the form

$$ds_D^2 = -\,e^{2\alpha^0}du^2 - e^{2\beta^0}d\Omega_{d_0}^2 + e^{2\beta^1}dt^2 - \sum_{i=2}^{n} e^{2\beta^i}ds_i^2, \tag{14}$$

where ds_i^2 $(i = 2,\ldots,n)$ are metrics of Einstein spaces of arbitrary dimensions d_i and signatures while α^0 and all β^i are functions of the radial coordinate u.

Consider in $\mathbb{M}^D$ a field theory with the action

$$S = \int d^D x\,\sqrt{|g_D|}\Big[\mathcal{R}_D h_{ab}(\overline{\phi})g^{MN}\partial_M\phi^a,\,\partial_N\phi^b - 2V_D(\overline{\phi})\Big], \tag{15}$$

where $\mathcal{R}_D$ is the D-dimensional scalar curvature and the scalar field Lagrangian has a σ-model form. We assume that ϕ^a are functions of the external space coordinates x^μ $(\mu = 0,1,...,d_0+1)$, so that actually in (15) $g^{MN}\partial_M\phi^a,\,\partial_N\phi^b = g^{\mu\nu}\partial_\mu\phi^a\partial_\nu\phi^b$, where the metric $g_{\mu\nu}$ is formed by the first three terms in (14). The metric h_{ab} of the N'-dimensional target space $\mathbb{T}_\phi$, parametrized by ϕ^a, and the potential V are functions of $\overline{\phi} = \{\phi^a\} \in \mathbb{T}_\phi$.

The action (15) represents in a general form the scalar-vacuum sector of diverse supergravities and low-energy limits of string and p-brane theories [24, 25]. In many papers devoted to exact solutions of such low-energy theories (see Sec. 4) all internal factor spaces are assumed to be Ricci-flat, and nonzero potentials $V_D(\overline{\phi})$ are not introduced due to technical difficulties of solving the equations. Meanwhile, the inclusion of a potential not only makes it possible to treat massive and/or nonlinear and interacting scalar fields, but is also necessary for describing, e.g., the symmetry breaking and Casimir effects. (On the use of effective potentials for describing the Casimir effect in compact extra dimensions, see, e.g., [9] and references therein.)

Let us perform a dimensional reduction to the external space-time $\mathbb{M}_{\rm ext}$ with the metric $g_{\mu\nu}$. Eq. (15) is converted to

$$S = \int d^{d_0+2}x\sqrt{|g_{d_0+2}|}\,e^{\sigma_2}\Big\{\mathcal{R}_{d_0+2} + \sum_{i=2}^{n} d_i(d_i-1)K_i\,e^{-2\beta^i}$$
$$+\,2\nabla^\mu\nabla_\mu\sigma_2 + \sum_{i,k=2}^{n}(d_id_k + d_i\delta_{ik})(\partial\beta^i,\partial\beta^k) + L_{\rm sc}\Big\}, \tag{16}$$

where all quantities, including the scalar $\mathcal{R}_{d_0+2}$, are calculated with the aid of $g_{\mu\nu}$, and $\sigma_2 := \sum_{d=2}^{n} d_i\beta^i$, so that e^{σ_2} is the volume factor of extra dimensions.

It is helpful to pass in the action (1), just as in the STT (1), from the Jordan-frame metric $g_{\mu\nu}$ in $\mathbb{M}_{\rm ext}$ to the Einstein-frame metric

$$\overline{g}_{\mu\nu} = {\rm e}^{2\sigma_2/d_0} g_{\mu\nu}. \tag{17}$$

Then, omitting a total divergence, one obtains the action (15) in terms of $\overline{g}_{\mu\nu}$:

$$S = \int d^{d_0+2}x \sqrt{|\overline{g}|}\Big[\overline{\mathcal{R}} + H_{KL}(\partial\varphi^K, \partial\varphi^L) - 2V(\vec{\varphi})\Big]. \tag{18}$$

Here the set of fields $\{\varphi^K\} = \{\beta^i, \phi^a\}$, combining the scalar fields from (15) and the moduli fields β^i, is treated as a vector in the extended $N = (n-1+N')$-dimensional target space $\mathbb{T}_\varphi$ with the metric

$$(H_{KL}) = \begin{pmatrix} d_i d_k/d_0 + d_i\delta_{ik} & 0 \\ 0 & h_{ab} \end{pmatrix}, \tag{19}$$

while the potential $V(\varphi)$ is expressed in terms of $V_D(\overline{\phi})$ and β^i:

$$V(\vec{\varphi}) = {\rm e}^{-2\sigma_2/d_0}\left[V_D(\overline{\phi}) - \frac{1}{2}\sum_{i=2}^{n} K_i d_i(d_i-1)\,{\rm e}^{-2\beta^i}\right]. \tag{20}$$

3.2. EXTENDED NO-GO THEOREMS

The action (18) brings the theory (15) to a form quite similar to (4) ($f > 0$), but a single field ψ is now replaced by a σ model with the target space metric H_{KL}. It can be easily shown [36] that Statements A–D are entirly extended to the theory (18) under the condition that the metric H_{KL} is positive-definite, which is always the case as long as h_{ab} is positive-definite.

Let us now discuss the properties of the D-dimensional metric g_{MN} given by (14). Its "external" part $g_{\mu\nu}$ is connected with $\overline{g}_{\mu\nu}$ by the conformal transformation (17). Since the action (15) corresponds to GR in D dimensions, this frame may be called the D-dimensional Einstein frame, and we will now designate the manifold $\mathbb{M}^D$ endowed with the metric g_{MN} as $\mathbb{M}^D_{\rm E}$.

The nonminimal coupling coefficient in the action (1), being connected with the extra-dimension volume factor ${\rm e}^{\sigma_2}$, is nonnegative by definition, and the solution terminates where ${\rm e}^{\sigma_2}$ vanishes or blows up. Thus, in contrast to the situation in STT, conformal continuations are here impossible: one cannot cross a surface, if any, where ${\rm e}^{\sigma_2}$ vanishes. Roughly speaking, the Jordan-frame manifold $\mathbb{M}_{\rm ext}[g]$ can be smaller but cannot be larger than $\mathbb{M}_{\rm ext}[\overline{g}]$. If $\sigma_2 \to \pm\infty$ at an intermediate value of the radial coordinate, then the transformation (17) maps $\mathbb{M}_{\rm ext}[g]$ to only a part of $\mathbb{M}_{\rm ext}[\overline{g}]$.

Asymptotic flatness of the metric g_{MN} in $\mathbb{M}_{\rm E}^D$ implies an asymptotically flat Einstein-frame metric $\overline{g}_{\mu\nu}$ in $\mathbb{M}_{\rm ext}$ and finite limits of the moduli fields β^i, $i \geq 2$, at large radii. A similar picture is observed with the regular centre conditions: a regular centre in $\mathbb{M}_{\rm E}^D$ is only possible if there is a regular centre in $\mathbb{M}_{\rm ext}[\overline{g}]$ and β^i, $i \geq 2$ sufficiently rapidly tend to constant values. A horizon in $\mathbb{M}_{\rm E}^D$ always corresponds to a horizon in $\mathbb{M}_{\rm ext}[\overline{g}]$. (The opposite assertions are not always true, e.g., a regular centre in $\mathbb{M}_{\rm ext}[\overline{g}]$ may be "spoiled" when passing to g_{MN} by an improper behaviour of the moduli fields β^i.)

So the global properties of $\mathbb{M}_{\rm ext}[\overline{g}]$ and $\mathbb{M}_{\rm ext}[g]$ (and hence $\mathbb{M}_{\rm E}^D$), associated with Statements A–D, are closely related but not entirely coincide.

A. The **no-hair theorem** can be formulated for $\mathbb{M}_{\rm E}^D$ as follows:

Given the action (15) with h_{ab} positive-definite and a nonnegative potential (20) in the space-time $\mathbb{M}_{\rm E}^D$ with the metric (14), the only static, asymptotically flat black hole solution to the field equations is characterized in the region of outer communication by $\phi^a = \mathrm{const}$, $\beta^i = \mathrm{const}$ $(i = \overline{2,n})$, $V(\vec{\varphi}) \equiv 0$ and the Tangherlini metric $g_{\mu\nu}$.

In other words, the only asymptotically flat black hole solution is given by the Tangherlini metric in $\mathbb{M}_{\rm ext}$, constant scalar fields ϕ^a and constant moduli fields β^i outside the event horizon. Note that in this solution the metrics $g_{\mu\nu}$ and $\overline{g}_{\mu\nu}$ in $\mathbb{M}_{\rm ext}$ are connected by simple scaling with a constant conformal factor since $\sigma_2 = \mathrm{const}$.

Another feature of interest is that it is the potential (20) that vanishes in the black hole solution rather than the original potential $V_D(\overline{\phi})$ from Eq. (15). Theorem 5 generalizes the previously known property of black holes with the metric (14) when the internal spaces are Ricci-flat and the source is a massless, minimally coupled scalar field without a potential [43].

B. Particle-like solutions: Statement B is valid in $\mathbb{M}_{\rm E}^D$ in the same formulation as previously in $\mathbb{M}_{\rm E}$, but the condition $V \geq 0$ now also applies to the potential (20) rather than $V_D(\overline{\phi})$ from (15).

C. Wormholes and even wormhole throats are impossible with the metric $\overline{g}_{\mu\nu}$. The conformal factor $\mathrm{e}^{2\sigma_2/d_0}$ in (17) removes the prohibition of throats since for $g_{\mu\nu}$ a condition like $r'' \leq 0$ is no longer valid. However, a wormhole as a global entity with two large r asymptotics cannot appear in $\mathbb{M}_{\rm J} = \mathbb{M}_{\rm ext}[g_{\mu\nu}]$ for the same reason as in Sec. 2.3.

Flux-tube solutions with nontrivial scalar and/or moduli fields are absent, as before, but horns are not ruled out since the behaviour of the metric coefficient $g_{\theta\theta}$ is modified by conformal transformations.

It should be emphasized that all the restrictions mentioned in items A-C are invalid if the target space metric h_{ab} is not positive-definite.

D. The **global causal structure** of any Jordan frame cannot be more complex than that of the Einstein frame even in STT, where conformal continuations are allowed — see Sec. 2.5. The corresponding reasoning of [19] entirely applies to $\mathbb{M}_{\rm ext}[g]$ and hence to $\mathbb{M}^D_{\rm E}$. The list of possible global structures is again the same as that for the Tangherlini-de Sitter metric. This restriction does not depend (i) on the choice and even sign of scalar field potentials, (ii) on the nature of asymptotic conditions and (iii) on the algebraic properties of the target space metric. It is therefore the most universal property of spherically symmetric configurations with scalar fields in various theories of gravity.

A theory in $\mathbb{M}^D$ may, however, be initially formulated in another conformal frame than in (15), i.e., with a nonminimal coupling factor $f(\overline{\phi})$ before $\mathcal{R}_D$. Let us designate M^D in this case as $\mathbb{M}^D_{\rm J}$, a D-dimensional Jordan-frame manifold. (An example of such a construction is the so-called string metric in string theories [24, 25] where f depends on a dilaton field related to string coupling.) Applying a transformation like (2), we can recover the Einstein-frame action (15) in $\mathbb{M}^D_{\rm E}$, then by dimensional reduction pass to $\mathbb{M}_{\rm ext}[g]$ and after one more conformal mapping (17) arrive at the (d_0+2) Einstein frame $\mathbb{M}_{\rm ext}[\overline{g}]$. Addition of the first step in this sequence of reductions weakens our conclusions to a certain extent. The main point is that we cannot *a priori* require $f(\overline{\phi}) > 0$ in the whole range of $\vec{\varphi}$, therefore conformal continuations (CCs) through surfaces where $f = 0$ are not excluded.

Meanwhile, the properties of CCs have only been studied [19] for a single scalar field in $\mathbb{M}_{\rm ext}$ (in the present notation). In our more complex case of multiple scalar fields and factor spaces, such a continuation through the surface $f(\overline{\phi}) = 0$ in the multidimensional target space $\mathbb{T}_\phi$ can have yet unknown properties.

One can only say for sure that the no-hair and no-wormhole theorems fail if CCs are admitted. This follows from the simplest example of CCs in the solutions with a conformal scalar field in GR, leading to black holes [44, 45] and wormholes [30, 46] and known since the 70s although the term "conformal continuation" was introduced only recently [40]. A wormhole was shown to be one of the generic structures appearing as a result of CCs in STT ([19], see sect 2.4 of the present paper).

If we require that the function $f(\overline{\phi})$ should be finite and nonzero in the whole range $\mathbb{R}_u$ of the radial coordinate, including its ends, then all the above no-go theorems are equally valid in $\mathbb{M}^D_{\rm E}$ and $\mathbb{M}^D_{\rm J}$. One should only bear in mind that the transformation (2) from $\mathbb{M}^D_{\rm E}$ to $\mathbb{M}^D_{\rm J}$ modifies the potential $V_D(\overline{\phi})$ multiplying it by $f^{-D/(D-2)}$, which in turn affects the explicit form of the condition $V \geq 0$, essential for Statements A and B.

Statement D on possible horizon dispositions and global causal structures will be unaffected if we even admit an infinite growth or vanishing of $f(\overline{\phi})$ at the extremes of the range $\mathbb{R}_u$. However, Statement C will not survive: such a behaviour of f may create a wormhole or horn in $\mathbb{M}_J^D$. A simple example of this kind is a "horned particle" in the string metric in dilaton gravity of string origin, studied by Banks et al. [47].

4. p-branes and observable effects

4.1. THE MODEL AND THE TARGET SPACE $\mathbb{V}$

Let us now consider a model which can be associated with p-branes as sources of antisymmetric form fields. Namely, in the space-time (13) with the metric (14), we take, as in Refs. [20]–[23], [48, 49], the model action for D-dimensional gravity with several scalar dilatonic fields φ^a and antisymmetric n_s-forms F_s:

$$S = \frac{1}{2\kappa^2}\int d^D z\sqrt{|g|}\Big\{\mathcal{R} + \delta_{ab}g^{MN}\partial_M\varphi^a\partial_N\varphi^b - \sum_{s\in\mathcal{S}}\frac{1}{n_s!}\,\mathrm{e}^{2\lambda_{sa}\varphi^a}F_s^2\Big\}, \quad (21)$$

where $F_s^2 = F_{s,\,M_1\ldots M_{n_s}}F_s^{M_1\ldots M_{n_s}}$; λ_{sa} are coupling constants; $s\in\mathcal{S}$, $a\in\mathcal{A}$, where $\mathcal{S}$ and $\mathcal{A}$ are some finite sets. Essential differences from (15) are, besides the inclusion of the term with F_s^2, that (i) the potential U is omitted (i.e., φ^a are not self-coupled but coupled to F_s), (ii) the "extra" spaces $\mathbb{M}_i$ are assumed to be Ricci-flat and (for simplicity) spacelike and (iii) the target space metric h_{ab} is Euclidean, $h_{ab}=\delta_{ab}$. The "scale factors" e^{β^i} and the scalars φ^a are again assumed to depend on u only.

The F-forms (or, more precisely, their particular nonzero components, fixed up to permutation of indices, to be labelled with the subscript s) should also be compatible with spherical symmetry. They are naturally classified as *electric* ($F_{\mathrm{e}I}$) and *magnetic* ($F_{\mathrm{m}I}$) forms, and each of these forms is associated with a certain subset $I=\{i_1,\ldots,i_k\}$ ($i_1<\ldots<i_k$) of the set of numbers labelling the factor spaces: $\{i\}=I_0=\{0,\ldots,n\}$. By definition, an electric form $F_{\mathrm{e}I}$ carries the coordinate indices u and those of the subspaces $\mathbb{M}_i$, $i\in I$, whereas a magnetic form $F_{\mathrm{m}I}$ is built as a form dual to a possible electric one associated with I. Thus nonzero components of $F_{\mathrm{m}I}$ carry coordinate indices of the subspaces $\mathbb{M}_i$, $i\in\overline{I}:=I_0\setminus I$. One can write:

$$n_{\mathrm{e}I} = \operatorname{rank}F_{\mathrm{e}I} = d(I)+1, \qquad n_{\mathrm{m}I} = \operatorname{rank}F_{\mathrm{m}I} = D - n_{\mathrm{e}I} = d(\overline{I}) \quad (22)$$

where $d(I)=\sum_{i\in I}d_i = \dim\mathbb{M}_I$, $\mathbb{M}_I := \mathbb{M}_{i_1}\times\ldots\times\mathbb{M}_{i_k}$. The index s jointly describes the two types of forms.

If the time axis $\mathbb{R}_t$ belongs to $\mathbb{M}_I$, we are dealing with a true electric or magnetic form, directly generalizing the Maxwell field in $\mathbb{M}_{\rm ext}$; otherwise the F-form behaves in $\mathbb{M}_{\rm ext}$ as an effective scalar or pseudoscalar. Such F-forms will be called *quasiscalar*.

The forms F_s are associated with p-branes as extended sources of the spherically symmetric field distributions, where the brane dimension is $p = d(I_s) - 1$, and $d(I_s)$ is the brane world volume dimension. A natural assumption is that the branes only "live" in extra dimensions, i.e., $0 \not\in I_s, \ \forall s$.

The classification of F-forms can be illustrated using as an example $D = 11$ supergravity, representing the low-energy limit of M-theory [25]. The action (21) for the bosonic sector of this theory (truncated by omitting the Chern-Simons term) does not contain scalar fields, and the only F-form is of rank 4, whose various nontrivial components F_s (elementary F-forms, called simply F-forms according to the above convention) are associated with electric 2-branes [for which $d(I_s) = 3$] and magnetic 5-branes [such that $d(I_s) = 6$, see (22)].

Let us put $d_0 = 2$ and ascribe to the external space-time coordinates the indices $M = t,\ u,\ \theta,\ \phi$ (θ and ϕ are the spherical angles), and let the numbers $i = 2, \ldots, 8$ refer to the extra dimensions, each associated with an extra factor spaces $\mathbb{M}_i$ with the same number ($\mathbb{M}_i$ are thus assumed to be one-dimensional). The number $i = 1$ refers to the time axis, $\mathbb{M}_1 = \mathbb{R}_t$, as stated previously. Here are examples of different kinds of forms:

F_{ut23} is a true electric form, $I = \{123\}$; $\overline{I} = \{045678\}$.
$F_{\theta\phi 23}$ is a true magnetic form, $I = \{145678\}$; $\overline{I} = \{023\}$.
F_{u234} is an electric quasiscalar form, $I = \{234\}$; $\overline{I} = \{015678\}$.
$F_{\theta\phi t2}$ is a magnetic quasiscalar form, $I = \{345678\}$; $\overline{I} = \{012\}$.

Under the above assumptions, it is helpful to describe the system in the so-called σ model representation [22]). Namely (see more general and detailed descriptions in [22, 23]), let us choose the harmonic u coordinate in $\mathbb{M}$ ($\nabla^M \nabla_M u = 0$), such that

$$\alpha^0(u) = \sum_{i=0}^{n} d_i \beta^i \equiv d_0 \beta^0 + \sigma_1(u). \tag{23}$$

We use the notations

$$\sigma_i = \sum_{j=i}^{n} d_j \beta^j(u), \qquad \sigma(I) = \sum_{i \in I} d_i \beta^i(u). \tag{24}$$

Then the combination $\binom{u}{u} + \binom{\theta}{\theta}$ of the Einstein equations, where θ is one of the angular coordinates on $\mathbb{S}^{d_0}$, has a Liouville form, $\ddot\alpha - \ddot\beta^0 =$

$(d_0-1)^2\,\mathrm{e}^{2\alpha-2\beta^0}$ (an overdot means d/du), and is integrated giving

$$\mathrm{e}^{\beta^0-\alpha^0} = (d_0-1)s(k,u), \qquad s(k,u) := \begin{cases} k^{-1}\sinh ku, & k>0, \\ u, & k=0, \\ k^{-1}\sin ku, & k<0. \end{cases} \tag{25}$$

where k is an integration constant. Another integration constant is suppressed by properly choosing the origin of u. With (25) the D-dimensional line element may be written in the form

$$ds_D^2 = \frac{\mathrm{e}^{-2\sigma_1/\overline{d}}}{[\overline{d}s(k,u)]^{2/\overline{d}}}\left\{\frac{du^2}{[\overline{d}s(k,u)]^2} + d\Omega^2\right\} - \mathrm{e}^{2\beta^1}dt^2 + \sum_{i=2}^{n}\mathrm{e}^{2\beta^i}ds_i^2, \tag{26}$$

$\overline{d} := d_0 - 1$. The range of the u coordinate is $0 < u < u_{\max}$ where $u=0$ corresponds to spatial infinity while $u_{\max}$ may be finite or infinite depending on the form of a particular solution.

The Maxwell-like equations for F_s are integrated in a general form, giving the respective charges $Q_s = \text{const}$. The remaining set of unknowns $\beta^i(u)$, $\varphi^a(u)$ $(i=1,\ldots,n,\ a\in\mathcal{A})$ can be treated as a real-valued vector function $x^A(u)$ (so that $\{A\} = \{1,\ldots,n\}\cup\mathcal{A}$) in an $(n+|\mathcal{A}|)$-dimensional vector space $\mathbb{V}$ (target space). The field equations for x^A can be derived from the Toda-like Lagrangian

$$L = G_{AB}\dot{x}^A\dot{x}^B - V_Q(y), \qquad V_Q(y) = -\sum_s \epsilon_s Q_s^2\,\mathrm{e}^{2y_s} \tag{27}$$

(where $\epsilon_s = 1$ for true electric and magnetic form and $\epsilon_s = -1$ for quasi-scalar forms), with the "energy" constraint

$$E = G_{AB}\dot{x}^A\dot{x}^B + V_Q(y) = \frac{d_0}{d_0-1}k^2\,\mathrm{sign}\,k, \tag{28}$$

The nondegenerate symmetric matrix

$$(G_{AB}) = \begin{pmatrix} d_id_j/\overline{d} + d_i\delta_{ij} & 0 \\ 0 & \delta_{ab} \end{pmatrix} \tag{29}$$

specifies a positive-definite metric in $\mathbb{V}$; the functions $y_s(u)$ are defined as scalar products:

$$y_s = \sigma(I_s) - \chi_s\overline{\lambda}_s\overline{\varphi} \equiv Y_{s,A}x^A, \qquad (Y_{s,A}) = \Big(d_i\delta_{iI_s},\ \ -\chi_s\lambda_{sa}\Big), \tag{30}$$

where $\delta_{iI} = 1$ if $i\in I$ and $\delta_{iI} = 0$ otherwise; χ_s distinguish electric and magnetic forms: $\chi_{\mathrm{e}I} = 1$, $\chi_{\mathrm{m}I} = -1$. The contravariant components and scalar products of $\vec{Y}_s$ are found using the matrix G^{AB} inverse to G_{AB}:

$$(G^{AB}) = \begin{pmatrix} \delta^{ij}/d_i - 1/(D-2) & 0 \\ 0 & \delta^{ab} \end{pmatrix},$$

$$(Y_s{}^A) = \left(\delta_{iI} - \frac{d(I)}{D-2}, \quad -\chi_s\lambda_{sa}\right); \tag{31}$$

$$Y_{s,A}Y_{s'}{}^A \equiv \vec{Y}_s\vec{Y}_{s'} = d(I_s \cap I_{s'}) - \frac{d(I_s)d(I_{s'})}{D-2} + \chi_s\chi_{s'}\overline{\lambda}_s\overline{\lambda}_{s'}. \tag{32}$$

The equations of motion in terms of $\vec{Y}_s$ read

$$\ddot{x}^A = \sum_s q_s Y_s{}^A \,\mathrm{e}^{2y_s}, \qquad q_s := \epsilon_s Q_s^2. \tag{33}$$

One can notice that the metric (29) is quite similar to (19), the metric of $\mathbb{T}_\varphi$, especially if $h_{ab} = \delta_{ab}$. The difference is that in G_{AB} given by (29) we have $i, j = \overline{1, n}$ since $\mathbb{V}$ includes as a coordinate the metric function β^1, whereas in H_{KL} we have $i, j = \overline{2, n}$, so that $\mathbb{T}_\varphi$ is a subspace of $\mathbb{V}$.

4.2. SOME EXACT SOLUTIONS. BLACK HOLES

The integrability of the Toda-like system (27) depends on the set of vectors $\vec{Y}_s$. Each $\vec{Y}_s$ consists of input parameters of the problem and represents an F-form F_s with a nonzero charge Q_s, i.e., one of charged p-branes.

The simplest case of integrability takes place when $\vec{Y}_s$ are mutually orthogonal in $\mathbb{V}$ [23], that is,

$$\vec{Y}_s\vec{Y}_{s'} = \delta_{ss'}Y_s^2, \qquad Y_s^2 = d(I)[1 - d(I)/(D-2)] + \overline{\lambda}_s^2 > 0 \tag{34}$$

where $\overline{\lambda}_s^2 = \sum_a \lambda_{sa}^2$. Then the functions $y_s(u)$ obey the decoupled Liouville equations $\ddot{y}_s = \epsilon_s Q_s^2 Y_s^2 \,\mathrm{e}^{2y_s}$, whence

$$\mathrm{e}^{-2y_s(u)} = \begin{cases} Q_s^2 Y_s^2 \, s^2(h_s, \ u + u_s), & \epsilon_s = 1, \\ Q_s^2 Y_s^2 h_s^{-2} \cosh^2[h_s(u+u_s)], & \epsilon_s = -1, \quad h_s > 0, \end{cases} \tag{35}$$

where h_s and u_s are integration constants and the function $s(.,.)$ has been defined in (25). For the sought-for functions $x^A(u)$ and the “conserved energy” E we then obtain:

$$x^A(u) = \sum_s \frac{Y_s{}^A}{Y_s^2} y_s(u) + c^A u + \underline{c}^A, \tag{36}$$

$$E = \sum_s \frac{h_s^2 \operatorname{sign} h_s}{Y_s^2} + \vec{c}^{\,2} = \frac{d_0}{d_0 - 1}\, k^2 \operatorname{sign} k, \tag{37}$$

where the vectors of integration constants $\vec{c}$ and $\underline{\vec{c}}$ are orthogonal to each $\vec{Y}_s$: $c^A Y_{s,A} = \underline{c}^A Y_{s,A} = 0$.

Although many other solutions are known [21, 22, 48], their physical properties turn out to be quite similar to those of the present solutions for orthogonal systems (OS) of vectors $\vec{Y}_s$ [48, 49].

Black holes (BHs) are distinguished among other spherically symmetric solutions by the requirement that there should be horizons rather than singularities in $\mathbb{M}_{\rm ext}$ at $u = u_{\max}$. This leads to constraints upon the input and integration constants. The above OS solutions describe BHs if

$$h_s = k > 0, \qquad \forall\ s; \qquad\qquad c^A = k \sum_s Y_s^{-2} Y_s{}^A - k\delta_1^A, \tag{38}$$

where $A = 1$ corresponds to $i = 1$ (time). The constraint (37) then holds automatically. The value $u = u_{\max} = \infty$ corresponds to the horizon. The no-hair theorem of Ref. [49] states that BHs are incompatible with quasi-scalar F-forms, so that all $\epsilon_s = 1$.

Under the asymptotic conditions $\varphi^a \to 0$, $\beta^i \to 0$ as $u \to 0$, after the transformation

$$\mathrm{e}^{-2ku} = 1 - \frac{2k}{\overline{d} r^{\overline{d}}}, \qquad \overline{d} := d_0 - 1 \tag{39}$$

the metric (26) for BHs and the corresponding scalar fields may be written as

$$\begin{aligned} ds_D^2 = \Big(\prod_s H_s^{A_s}\Big)\bigg[& -dt^2 \Big(1 - \frac{2k}{\overline{d} r^{\overline{d}}}\Big) \prod_s H_s^{-2/Y_s^2} \\ & + \Big(\frac{dr^2}{1 - 2k/(\overline{d} r^{\overline{d}})} + r^2 d\Omega^2\Big) + \sum_{i=2}^n ds_i^2 \prod_s H_s^{A_s^i}\bigg]; \end{aligned}$$

$$A_s := \frac{2}{Y_s^2}\frac{d(I_s)}{D-2}; \qquad A_s^i := -\frac{2}{Y_s^2}\delta_{iI_s}; \tag{40}$$

$$\varphi^a = -\sum_s \frac{\lambda_{sa}}{Y_s^2} \ln H_s, \tag{41}$$

where H_s are harmonic functions in $\mathbb{R}_+ \times \mathbb{S}^{d_0}$:

$$H_s(r) = 1 + P_s/(\overline{d} r^{\overline{d}}), \qquad P_s := \sqrt{k^2 + Q_s^2 Y_s^2} - k. \tag{42}$$

The subfamily (38), (40)–(42) exhausts all OS BH solutions with $k > 0$ (non-extremal BHs). Extremal BHs, corresponding to minimum mass for given charges (the so-called BPS limit), are obtained either in the limit $k \to 0$, or directly from (35)–(37) under the conditions $h_s = k = c^A = 0$. The only independent integration constants in the BH solutions are k, related to the observed mass (see below), and the brane charges Q_s.

Other families of solutions, mentioned previously, also contain BH subfamilies. The most general BH solutions are considered in Ref. [50].

4.3. POST-NEWTONIAN PARAMETERS AND OTHER OBSERVABLES

One cannot exclude that real astrophysical objects (stars, galaxies, quasars, black holes) are essentially multidimensional objects, whose structure is affected by charged p-branes. (It is then unnecessary to assume that the antisymmetric form fields are directly observable, though one of them may manifest itself as the electromagnetic field.)

The **post-Newtonian** (PN) (weak gravity, slow motion) approximation of multidimensional solutions then determines the predictions of the classical gravitational effects: gravitational redshift, light deflection, perihelion advance and time delay (see [1, 7]).

For spherically symmetric configurations, a standard form of the PN metric uses the Eddington parameters β and γ in isotropic coordinates, in which the spatial part is conformally flat [1]:

$$ds^2_{\rm PN} = -(1 - 2V + 2\beta V^2)dt^2 + (1 + 2\gamma V)(d\rho^2 + \rho^2 d\Omega^2) \qquad (43)$$

where $d\Omega^2$ is the metric on $\mathbb{S}^2$, $V = GM/\rho$ the Newtonian potential, G the Newtonian gravitational constant and M the active gravitating mass.

Observations in the Solar system lead to tight constraints on the Eddington parameters [7]:

$$\gamma = 0.99984 \pm 0.0003, \qquad \beta = 0.9998 \pm 0.0006. \qquad (44)$$

The first restriction results from over VLBI observations [51], the second one from the γ data and an analysis of lunar laser ranging data [52, 53].

For a theory under consideration, the metric (43) should be identified with the asymptotics of the 4D metric in the observational CF. Preserving its choice yet undetermined, we can write according to (26) with $d_0 = 2$:

$$ds_4^* = {\rm e}^{2f(u)}\biggl\{ -{\rm e}^{2\beta^1}dt^2 + \frac{{\rm e}^{-2\sigma_1}}{s^2(k,u)}\biggl[\frac{du^2}{s^2(k,u)} + d\Omega^2\biggr]\biggr\} \qquad (45)$$

where $f(u)$ is an arbitrary function of u, normalized for convenience to $f(0) = 0$ (not to be confused with $f(\phi)$ that appeared in Sec. 2 and 3). Recall that by our notations $\sigma_1 = \beta^1 + \sigma_2$, the function $s(k,u)$ is defined in Eq. (25), and spatial infinity takes place at $u = 0$.

Passing to isotropic coordinates in (45), one finds that $u = 1/\rho$ up to cubic terms in $1/\rho$, and the decomposition in $1/\rho$ up to $O(\rho^{-2})$, needed for comparison with (43), precisely coincides with the u-decomposition near $u = 0$.

Using this circumstance, it is easy to obtain for the mass and the Eddington parameters corresponding to (45):

$$GM = -\beta^{1\prime} - f'; \qquad \beta = 1 + \frac{1}{2}\frac{\beta^{1\prime\prime} + f''}{(GM)^2}, \qquad \gamma = 1 + \frac{2f' - \sigma_2'}{GM}, \quad (46)$$

where $f' = df/du\Big|_{u=0}$ and similarly for other functions. The expressions (46) are quite general and are applicable to asymptotically flat, static, spherically symmetric solution of any theory where the energy-momentum tensor has the property $T^u_u + T^\theta_\theta = 0$, which leads to the metric (26) and, in particular, to the above solutions of the theory (21).

Two special choices of $f(u)$ can be distinguished. First, if, for some reasons, the 4D Einstein frame is chosen as the observational one, then, according to Eq. (17) with $d_0 = 2$, we have

$$f = f^{\rm E} = \sigma_2/2. \quad (47)$$

Second, let us try to add the matter action in (21) simply as const $\cdot \int L_m \sqrt{{}^D g}\, d^D x$, i.e., like the fermionic terms in the effective action in the field limit of string theory ([24], Eq. (13.1.49)]. In the observational frame with the metric $g^*_{\mu\nu}$ the matter action should read simply $\int d^4x \sqrt{g^*}\, L_m$. Identifying them, we obtain [17, 18] $g^*_{\mu\nu} = {\rm e}^{\sigma_2/2} g_{\mu\nu}$, whence

$$f = f^* = \sigma_2/4. \quad (48)$$

The parameter β can be calculated using (46) *directly from the equations of motion (33), without solving them*. This is true for any function f of the form $f = \vec{F}\vec{x}$ where $\vec{F} \in \mathbb{V}$ is a constant vector (i.e., f is a linear combination of β^i and the scalar fields φ^a):

$$\beta - 1 = \frac{1}{2(GM)^2} \sum_s \epsilon_s Q_s^2 (Y_s^1 + \vec{F}\vec{Y}_s)\, {\rm e}^{2y_s(0)}. \quad (49)$$

Explicit expressions for M and γ require knowledge of the solutions' asymptotic form. However, there is an exception: due to (47), in the 4D Einstein frame $\gamma = 1$, precisely as in GR, for all p-brane solutions in the general model (21).

Let us also present the quantities $\beta^{1\prime}$ and σ_2', needed for finding GM and γ, for OS BH solutions:

$$\beta^{1\prime} = -k - \sum_s P_s \frac{1-b_s}{Y_s^2}, \qquad \sigma_2' = -\sum_s \frac{1-2b_s}{Y_s^2} \tag{50}$$

where $b_s = d(I_s)/(D-2)$.

Some general observations can be made from the above relations [18]:

- The expressions for β depend on the input constants D, $d(I_s)$ (hence on p-brane dimensions), on the mass M and on the charges Q_s. For given M, they are independent of other integration constants, emerging in the solution of the Toda system (33), and also of p-brane intersection dimensions, since they are obtained directly from Eqs. (33) [18]. This means, in particular, that β is the same for BH and non-BH configurations with the same set of input parameters, mass and charges.
- All p-branes give positive contributions to β in both frames (47) and (48), which leads to a general restriction on the charges Q_s for given mass and input parameters.
- The expressions for γ depend, in general, on the integration constants h_s and c^i emerging from solving Eqs. (33). For BH solutions these constants are expressed in terms of k and the input parameters, so both β and γ depend on the mass, charges and input parameters.
- In the 4-E frame, one always has $\gamma = 1$. The same is true for some BH solutions in all frames with $f = N\sigma_2$, $N = \text{const}$ [18].

BH temperature. BHs are, like nothing else, strong-field gravitational objects, while the PN parameters only describe their far neighbourhood. An important observable characteristic of their strong-field behaviour is the Hawking temperature $T_{\rm H}$. One can show [18] that this quantity is *CF-independent*, at least if conformal factors that connect different frames are regular on the horizon. The conformal invariance of $T_{\rm H}$ was also discussed in another context in Ref. [54].

In particular, for the above OS BH solutions one obtains [23]

$$T_{\rm H} = \frac{1}{8\pi k k_{\rm B}} \prod_s \left(\frac{2k}{2k+P_s}\right)^{1/Y_s^2}. \tag{51}$$

where $k_{\rm B}$ is the Bolzmann constant.

The physical meaning of $T_{\rm H}$ is related to quantum evaporation, a process to be considered in the fundamental frame, while the produced particles are assumed to be observed at flat infinity, where relevant CFs do not differ. Therefore $T_{\rm H}$ should be CF-independent, and this property is obtained "by construction" [18].

All this is true for $T_{\rm H}$ in terms of the integration constant k and the charges Q_s. However, the observed mass M as a function of the same quantities is frame-dependent, see (46). Therefore $T_{\rm H}$ as a function of M and Q_s is frame-dependent as well.

Coulomb law violation is one of specific potentially observable effects of extra dimensions. Suppose in (15), (14) $d_0 = 2$ and let us try to describe the electrostatic field of a spherically symmetric source by a term $F^2\, {\rm e}^{2\overline{\lambda}\overline{\varphi}}$ in the action (21), corresponding to a true electric m-form $F_{{\rm e}I}$ with a certain set I containing 1, that is, $I = 1 \cup J$, $J \subset \{2, \ldots, n\}$.

Then the modified Coulomb law in any CF with the metric (45) can be written as follows [18]:

$$E == (|Q|/r^2)\, {\rm e}^{-2\overline{\lambda}\overline{\varphi}+\sigma(J)-\sigma(\overline{J})}, \tag{52}$$

where E is the observable electric field strength, r is the observable radius of coordinate spheres, the notations (24) are used and $\overline{J} = \{2, \ldots, n\} \setminus J$.

Deviations from the conventional Coulomb law are evidently both due to extra dimensions (and depend on the F-form structure) and due to interaction of F_s with the scalar fields. This relation (generalizing the one obtained in Ref. [12] in the framework of dilaton gravity) is valid for an arbitrary metric of the form (14) ($d_0 = 2$) and does not depend on whether or not this F-form takes part in the formation of the gravitational field.

Eq. (52) is exact and — which is remarkable — it is *CF-independent.* This is an evident manifestation of the conformal invariance of the electromagnetic field in $\mathbb{M}_{\rm ext}$ even in the present generalized framework.

5. Concluding remarks

We have seen that the properties of theoretical models look drastically different when taken in different CFs. This once again stresses the necessity of a careful reasoning for a particular choice of a CF. It even may happen (though seems unlikely) that there is an unobservable part of the Universe, separated from us by a singularity in the observational CF which is converted to a regular surface ($\mathbb{S}_{\rm trans}$) after passing to a fundamental frame. A similar thing may happen to a cosmological singularity: in some theories it can correspond to a regular bounce in a fundamental frame.

We have obtained expressions for the Eddington PN parameters β and γ for a wide range of static, spherically symmetric solutions of multidimensional gravity with the general string-inspired action (21). The experimental limits (44) on β and γ constrain certain combinations of the solution parameters. This, however, concerns only the particular system for which the measurements are carried out, in our case, the Sun's gravitational field. The main feature of the expressions for β and γ is their dependence not only on the theory (the constants entering into the action), but on the particular solution (integration constants). The PN parameters thus can be different for different self-gravitating systems, and not only, say, for stars and black holes, but even for different stars if we try to describe their external fields in models like (21).

A feature of interest is the universal prediction of $\beta > 1$ in (49) for both frames (47) and (48). The predicted deviations of γ from unity may be of any sign and depend on many integration constants. If, however, the 4-dimensional Einstein frame is adopted as the observational one, we have a universal result $\gamma = 1$ for all static, spherically symmetric solutions of the theory (21).

The BH temperature $T_{\rm H}$ also carries information about the space-time structure, encoded in Y_s^2. Being a universal parameter of a given solution to the field equations, $T_{\rm H}$ as a function of the observable BH mass and charges is still CF-dependent due to different expressions for the mass M in different frames.

One more evident consequence of extra dimensions is the Coulomb law violation, caused by a modification of the conventional Gauss theorem and also by scalar-electromagnetic interaction. A remarkable property of the modified Coulomb law is its CF independence for a given static, spherically symmetric metric.

References

1. Will, C.M. (1993) Theory and Experiment in Gravitational Physics, Cambridge University Press, Cambridge.
2. Staniukovich, K.P. and Melnikov, V.N. (1983) Hydrodynamics, Fields and Constants in the Theory of Gravitation, Energoatomizdat, Moscow (in Russian). Melnikov, V.N. (2002) Fields and Constants in the Theory of Gravitation. CBPF-MO-02/02, Rio de Janeiro, 134pp.
3. Bronnikov, K.A., Ivashchuk, V.D. and Melnikov, V.N. (1988) *Nuovo Cim.* **B102**, 209.
4. Melnikov, V.N. (1994) *Int. J. Theor. Phys.* **33**, 1569.
5. De Sabbata, V., Melnikov, V.N., and Pronin, P.I. (1992) *Prog. Theor. Phys.* **88**, 623.
6. Melnikov, V.N. (1988) in: V. de Sabbata and V.N. Melnikov (eds), *Gravitational Measurements, Fundamental Metrology and Constants*, Kluwer Academic Publishers, Dordtrecht, p. 283.
7. Damour, T. (1996) Gravitation, experiment and cosmology, gr-qc/9606079; Will, C. (2001) The confrontation between general relativity and experiment, gr-qc/0103036.
8. Webb, J.K., Murphy, M.T., Flambaum, V.V., Dzuba, V.A., Barrow, J.D., Churchill, C.W., Prochaska, J.X., and Wolfe, A.M. (2000) Further evidence for cosmological evolution of the fine structure constant, astro-ph/0012539.
9. Günther, U., Kriskiv S., and Zhuk, A. (1998) *Grav. & Cosmol.* **4**, 1; Günther, U and Zhuk, A. (1997) *Phys. Rev.* **D 56**, 6391; Günther, U and Zhuk, A. (2000) Gravitational excitons as dark matter, astro-ph/0011017.

10. Gavrilov, V.R., Ivashchuk, V.D. and Melnikov, V.N. (1995) *J. Math. Phys.* **36**, 5829.
11. Melnikov, V.N. (1995) Multidimensional Cosmology and Gravitation, CBPF-MO-002/95, Rio de Janeiro, 210 pp., also
Melnikov, V.N. (1996) Classical Solutions in Multidimensional Cosmology. In: Cosmology and Gravitation II, Ed.M.Novello, Editions Frontieres, Singapore, p.465.
12. Bronnikov, K.A. and Melnikov, V.N. (1995) *Ann. Phys. (N.Y.)* **239**, 40.
13. See recent reviews, e.g.:
Rubakov, V.A. (2001) Large and infinite extra dimensions, *Phys. Usp.* **44**, 871;
Maartens, R. (2001) Geometry and dynamics of the brane world, gr-qc/0101059;
Brax, Ph. and van de Bruck, C. (2003) Cosmology and brane worlds: a review, hep-th/0303095.
14. Magnano G. and Sokolowski, L.M. (1994) *Phys. Rev.* **D 50**, 5039.
15. Faraoni, V., Gunzig, E., and Nardone, P. (1999) Conformal transformations in classical gravitational theories and in cosmology, *Fundamentals of Cosmic Physics* **20**, 121.
16. Rainer, M. and Zhuk, A.I. (2000) *Gen. Rel. Grav.* **32**, 79.
17. Bronnikov, K.A. (1995) *Grav. & Cosmol.* **1**, 1, 67.
18. Bronnikov, K.A. and Melnikov, V.N. (2001) *Gen. Rel. Grav.* **33**, 1549.
19. Bronnikov, K.A. (2002) *J. Math. Phys.* **43**, 6096.
20. Ivashchuk, V.D. and Melnikov, V.N. (1996) *Grav. & Cosmol.* **2**, 297.
21. Ivashchuk, V.D. and Melnikov, V.N. (1996) *Phys. Lett.* **B 384**, 58.
22. Ivashchuk, V.D. and Melnikov, V.N. (1997) *Class. Quantum Grav.* **14**, 3001.
23. Bronnikov, K.A., Ivashchuk, V.D. and Melnikov, V.N. (1997) *Grav. & Cosmol.* **3**, 203.
24. Green, M.B., Schwarz, J.H., and Witten, E. (1987) Superstring Theory in 2 vols., Cambridge Univ. Press.
25. Hull C. and Townsend, P. (1995) *Nucl. Phys.* **B 438**, 109;
Horava P. and Witten, E. (1996) *Nucl. Phys.* **B 460**, 506;
Schwarz, J.M. (1997) *Nucl. Phys. Proc. Suppl.* **55B**, 1;
Stelle, K.S. (1997) Lectures on supergravity p-branes, hep-th/9701088;
Duff, M.J. (1996) *Int. J. Mod. Phys.* **A 11**, 5623.
26. Ivashchuk, V.D. and Melnikov, V.N. (1994) *Class. Quantum Grav.* **11**, 1793; (1995) *Int. J. Mod. Phys. D* **4**, 167.
27. Hull, C.M. and Khuri, R.R. (2000) *Nucl. Phys.* **B 575**, 231.
28. Khviengia, N., Khviengia, Z., Lü, H., and Pope, C.N. (1998) *Class. Quantum Grav.* **15**, 759.
29. Wagoner, R. (1970) *Phys. Rev.* **D 1**, 3209.
30. Bronnikov, K.A. (1973 *Acta Phys. Polon.* **B4**, 251.
31. Ellis, H. (1973) *J. Math. Phys.* **14**, 104.
32. Bronnikov, K.A. (2001) *Phys. Rev.* **D 64**, 064013.
33. Bekenstein, J.D. *Phys. Rev.* **D 5**, 1239; ibid., 2403.
34. Adler S. and Pearson, R.B. (1978) *Phys. Rev.* **D 18**, 2798.
35. Bronnikov, K.A. and Shikin, G.N. (1991) Self-gravitating particle models with classical fields and their stability, in *Itogi Nauki i Tekhniki ("Results of Science and Engineering)*, Subseries *Classical Field Theory and Gravitation Theory*, v. 2, p. 4, VINITI, Moscow.
36. Bronnikov, K.A., Fadeev, S.B., and Michtchenko, A.V. (2003) *Gen. Rel. Grav.* **35**, 505.
37. Torii, T., Maeda K., and Narita, M. (1999) *Phys. Rev.* **D 59**, 064027;
Lowe, D.A. (2000) *JHEP* **0004** 011;
Galloway, G.J., Surge S., and Woolgar, E. (2003) *Class. Quantum Grav.* **20**, 1635.
38. Bronnikov, K.A. and Shikin, G.N. (2002) *Grav. & Cosmol* **8**, 107.
39. Chan, K.C.K., Horne, J.H., and Mann, R.B. (1995) *Nucl. Phys.* **B 447**, 441.
40. Bronnikov, K.A. (2001) *Acta Phys. Polon.* **B32**, 357.
41. Mayo, A.E. and Bekenstein, J.D. (1996) *Phys. Rev.* **D 54**, 5059;

Saa., A. (1996) *J. Math. Phys.* **37**, 2346;
Banerjee, N. and Sen, S. (1998) *Phys. Rev.* **D 58**, 104024.

42. Fisher, I.Z. (1948) *Zh. Eksp. Teor. Fiz.* **18**, 636.
43. Fadeev, S.B., Ivashchuk, V.D., and Melnikov, V.N. (1991) *Phys. Lett.* **161A**, 98.
44. Bocharova, N.M., Bronnikov, K.A., and Melnikov, V.N. (1970) *Vestn. Mosk. Univ., Fiz. Astron.* No. 6, 706.
45. Bekenstein, J.D. (1974) *Ann. Phys. (USA)* **82**, 535.
46. Barceló, C. and Visser, M. (1999) *Phys. Lett.* **466B**, 127; (2000) *Class. Quantum Grav.* **17**, 3843–64.
47. Banks T. and O'Loughlin, M. (1993) *Phys. Rev.* **D 47**, 540.
48. Bronnikov, K.A. (1998) *Grav. & Cosmol.* **4**, 49.
49. Bronnikov, K.A. (1999) *J. Math. Phys.* **40**, 924.
50. Ivashchuk, V.D. and Melnikov, V.N. (2000) *Grav. & Cosmol.* **6**, 27; *Class. Quantum Grav.* **17**, 2073.
51. Eubanks, T.M. et al. (1999) Advances in solar system tests of gravity, preprint, available at fttp://casa.usno.navy.mil/navnet/postscript/, file prd_15.ps.
52. Nordtvedt, K. (1968) *Phys. Rev.* **169**, 1017.
53. Dickey, J.O. et al. (1994) *Science* **265**, 482.
54. Jacobson, T. and Kang, G. (1993) *Class. Quantum Grav.* **10**, L201.
55. Melnikov, V.N. (1993) Multidimensional Classical and Quantum Cosmology and Gravitation: Exact Solutions and Variations of Constants. CBPF-NF-051/93, Rio de Janeiro, 93pp., also
Melnikov, V.N. (1994) In: Cosmology and Gravitation. Ed. M.Novello, Editions Frontires,Singapore, p.147.
56. Melnikov, V.N. (2002) Exact Solutions in Multidimensional Gravity and Cosmology III. CBPF-MO-03/02, Rio de Janeiro, 297 pp.
57. Ivashchuk, V.D. and Melnikov, V.N. (2001) Exact Solutions in Multidimensional Gravity with Antisymmetric Forms. Topical Review. *Class. Quantum Grav.* **18**, R1-R66 .

Graviton Exchange and the Gravitational Constant

M J Clark*
Department of Physics and Astronomy
401 A H Nielsen Building
University of Tennessee
Knoxville
Tennessee 37996-1200 USA

Abstract. A phenomenological model for virtual graviton exchange between mass-energy establishes a link between the gravitational constant *G* and other physical constants within the bounds of Quantum Field Theory. This gives the formulation $G = \alpha_g \hbar c/u^2$, where α_g is the dimensionless coupling constant for graviton exchange between atomic units of mass energy uc^2 (u is the atomic mass unit (kg)). This formulation is identical to that obtained by dimensional analysis and is analogous to the formula for the electromagnetic coupling (fine structure) constant. Some applications are described, including a new interpretation of Planck quantities and predictions of the model are developed. The model provides an explanation of the anomalous *Pioneer* acceleration and for the occurrence of perturbations to gravitational interactions at very large (i.e., interstellar and intergalactic) distances. It provides a basis for understanding galactic rotation curves without the need for *ad hoc* modifications to Newtonian dynamics. Finally, the possibility that an enhanced graviton exchange effect could be the cause of variations in terrestrial measurements of *G* is critically examined.

Keywords: graviton exchange, gravitational constant, Planck quantities, cosmic rays, anomalous Pioneer acceleration, galactic rotation curves.

1. Introduction

This paper shows how phenomenological models for virtual graviton exchange are consistent with Newtonian gravity and give a new insight into the nature of the Newtonian gravitational constant *G*. Two models have been described previously[1] and one based on the summation of all possible virtual graviton exchange paths will be developed further here. Virtual particles, first proposed by Bethe and Fermi[2] in 1932 are a fundamental concept in quantum field theory. The scattering interaction between two charged particles arises from the continuous exchange of virtual photons (bosons) rather than from the creation of a classical

* Permanent address: National Radiological Protection Board, Didcot, Oxon, OX11 0RQ, UK (e-mail michael.clark@nrpb.org).

V. de Sabbata et al. (eds.),
The Gravitational Constant: Generalized Gravitational Theories and Experiments, 65–79.
© 2004 *Kluwer Academic Publishers. Printed in the Netherlands.*

continuous electromagnetic field. Virtual particles do not transfer energy or information, and they are not directly observable independently of their source. They are instead bound, messenger particles that mediate forces between real particles. Despite being non-observable, virtual particle exchange has real observable effects (albeit at a low, perturbative level) and the success of Quantum Electrodynamics[3] demonstrates the utility of the concept of virtual exchange processes. It will be shown that the analogous virtual graviton exchange has real effects but at an extremely low level, because the gravitational coupling constant is many orders of magnitude less than the electromagnetic (fine structure) coupling constant.

2. Exchange pathways model

The concept of virtual photon exchange mediating electromagnetic forces is an accepted and integral part of quantum electrodynamics. Any phenomenological model for virtual particle exchange must ensure that their distinct properties are incorporated fully. For example, in any virtual particle exchange process it should be impossible to distinguish between a virtual particle either being emitted or being absorbed. Virtual particles are simply exchanged[4,5] within the constraints of the Heisenberg uncertainty principle. The exchange pathways model sums the contribution of all possible virtual exchanges between mass-energy. Application of this phenomenological model to virtual photon exchange processes should be completely consistent with Coulomb's law, and this will be demonstrated below, along with an analogous application to Newton's law assuming virtual graviton exchange.

2.1 Virtual photon exchange

First, assuming that the electrostatic force F_e between two charges is mediated by virtual photon exchange, and the force is proportional to the mass-energy of two electrons involved, then $F_e \propto 2m_ec^2$, where m_e is the mass of the electron. Furthermore, considering two large negative charges Q_1 and Q_2, the electrostatic force between them should be proportional to the total number of possible virtual photon exchanges between the electrons in Q_1 and Q_2. Thus F_e is proportional to the product $(Q_1/e)(Q_2/e)$, where e is the electron charge (C). Assuming spherical symmetry and incorporating the divergence theorem gives a factor $1/4\pi X^2$, where X is the mean separation distance between Q_1 and Q_2. This implies that virtual exchange pathways follow straight lines, and therefore virtual particle exchanges between mass-energy units obey the inverse square law. Finally, because this is a model for virtual particle exchange between mass-energy, it is necessary to incorporate the length scale at which relativistic quantum field theory becomes important. This is the Compton wavelength for the electron $\lambda_C = h/m_ec$, which can be derived from the assumption that for a scattering interaction to take place, $E = h\nu = m_ec^2$. At this particular frequency ν, the corresponding wavelength is the Compton wavelength. Hence the final assumption of the model is $F_e \propto h/m_ec$.

The four basic assumptions of the model are shown schematically in Fig. 1 and in summary they are:

(a) The electrostatic force between two charges Q_1 and Q_2 is proportional to the mass-energy of electrons involved in virtual photon exchange, giving $F_e \propto 2m_ec^2$.
(b) The electrostatic force is proportional to the total number of possible virtual photon exchanges between electrons in the charges Q_1 and Q_2, giving $F_e \propto (Q_1/e)(Q_2/e)$, where e is the electron charge (C).
(c) Virtual photon obeys the inverse square law, and hence $F_e \propto 1/4\pi X^2$, where X is the mean distance between Q_1 and Q_2.
(d) Virtual photon exchange occurs at the distance scale defined by relativistic quantum field theory, hence $F_e \propto h/m_ec$.

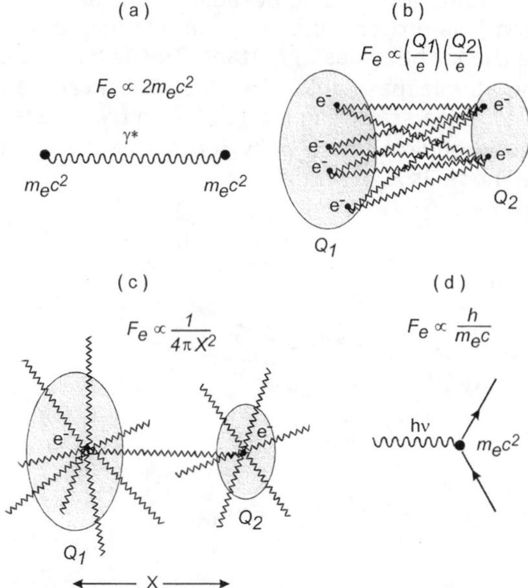

Figure 1. The basic assumptions of the exchange pathways model for virtual photons mediating the electrostatic force which, when combined, give $F_e = (\alpha\hbar c/e^2)(Q_1Q_2/X^2)$.

Combining these four basic assumptions, the composite formula for F_e becomes Coulomb's law,

$$F_e \propto 2m_ec^2\left(\frac{Q_1}{e}\right)\left(\frac{Q_2}{e}\right)\frac{1}{4\pi X^2}\frac{h}{m_ec} = \alpha\frac{\hbar c}{e^2}\frac{Q_1Q_2}{X^2} \tag{1}$$

where α is the dimensionless electromagnetic coupling (fine structure) constant and therefore the electric force constant $k = \alpha\hbar c/e^2$.

2.2 Virtual Graviton exchange

It is possible to formulate an expression for gravitational forces in an analogous manner to that described above for electromagnetic forces, by assuming that the force is mediated by the exchange of virtual gravitons between two large masses. More exactly, gravitational forces are due to a quantum field coupling via bosons exchanging between units of mass-energy within the two large mass-energies. First it is assumed that virtual graviton exchange occurs between atomic mass-energy units uc^2 in two large mass energies M_1c^2 and M_2c^2, where u is the atomic mass unit (kg). Other mass units could be chosen but it is convenient to adopt a standard mass unit, and hence the gravitational force $F_g \propto 2uc^2$. Next, the gravitational force is assumed to be proportional to the total number of virtual exchange pathways between atomic mass-energy units in the large mass-energies M_1c^2 and M_2c^2. Hence F_g is proportional to the product $(M_1/u)(M_2/u)$ which corresponds to the total number of atomic mass units (amus) in M_1 multiplied by the total number of amus in M_2. Assuming spherical symmetry, it is possible to incorporate the divergence theorem (Gauss's law) into the formulation giving $F_g \propto 1/4\pi X^2$, where X is the mean separation distance between M_1 and M_2. Again, this implies that virtual exchange pathways follow straight lines and virtual graviton exchange obeys the inverse square law. Finally, because this is a model for a relativistic quantum field process, it is necessary to incorporate the length scale at which the theory becomes important. This is the Compton wavelength for the atomic mass unit $\lambda_C = h/uc$, which can be derived from the assumption that, for a scattering interaction to take place, relativistic quantum field theory requires that $E = h\nu = uc^2$. Hence the final assumption of the model is $F_g \propto h/uc$.

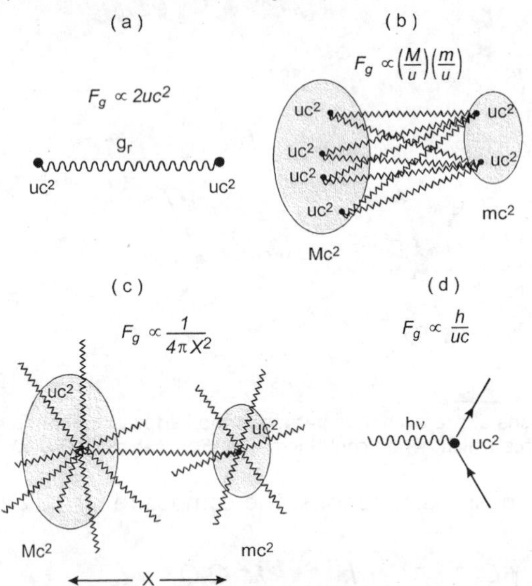

Figure 2. The basic assumptions of the exchange pathways model for virtual gravitons mediating the gravitational force which, when combined, give $F_g = (\alpha_g \hbar c/u^2)(M_1M_2/X^2)$.

The component assumptions of the graviton exchange model are shown schematically in Fig. 2 and, in summary, the four assumptions are:

(a) $F_g \propto 2uc^2$. Gravitational force is proportional to the unit mass-energy involved in virtual graviton exchange.

(b) $F_g \propto (M_1/u)(M_2/u)$. Gravitational force is proportional to the total number of possible virtual graviton exchange paths.

(c) $F_g \propto 1/4\pi X^2$. Virtual graviton exchange obeys the inverse square law, where X is the mean distance between M_1 and M_2.

(d) $F_g \propto h/uc$. Virtual graviton exchange occurs at the length scale defined by relativistic quantum field theory.

Combining these four terms, the resulting expression for F_g in Eq. (2) has a dimensionless constant of proportionality corresponding to a gravitational coupling constant α_g.

$$F_g \propto 2uc^2\left(\frac{M_1}{u}\right)\left(\frac{M_2}{u}\right)\frac{1}{4\pi X^2}\frac{h}{uc} = \alpha_g \frac{\hbar c}{u^2}\frac{M_1 M_2}{X^2} \qquad (2)$$

Equation (2) is clearly an analogue of Newton's law of gravitation, and shows that the gravitational constant can be expressed as a composite of other physical constants, $G = \alpha_g \hbar c/u^2$. Using CODATA values[6] for the constants G, $\hbar$, c and u, results in $\alpha_g = 5.82\ 10^{-39}$, which is consistent with values derived from dimensional arguments[7, 8].

Although the value of the gravitational coupling constant α_g is dependent on the choice of the mass unit u, the value of G is invariant to a change of mass unit because $G/\hbar c = \alpha_g/u^2 = \text{const.} = 2.111\ 10^{15}\ \text{kg}^{-2}$. There are an infinite number of values for α_g and mass that can satisfy this equation because there is no unique quantisation of mass-energy. This is very different to the electromagnetic coupling constant where the charge of the electron is uniquely quantised, so there is only one value for α. The two coupling constants are therefore fundamentally different.

3. Planck and Electrogravitic Quantities

Shortly after proposing his quantum theory, Planck noted[8] that the constants G, $\hbar$ and c could be combined to give quantities of length, mass and time. The formula derived here for G provides a new interpretation of Planck quantities[1]. For example, it shows that the Planck length is simply related to the Compton wavelength for the atomic mass unit,

$$l_P = \sqrt{\frac{\hbar G}{c^3}} = \sqrt{\frac{\hbar \alpha_g \hbar c}{c^3 u^2}} = \sqrt{\alpha_g}\,\frac{\hbar}{uc} \qquad (3)$$

Similarly the Planck mass is related to the atomic mass unit in a very straightforward way,

$$m_P = \sqrt{\frac{\hbar c}{G}} = \sqrt{\frac{\hbar c u^2}{\alpha_g \hbar c}} = \frac{u}{\sqrt{\alpha_g}} \qquad (4)$$

Given that α_g is a fundamental coupling constant for graviton exchange between atomic mass units, then $\sqrt{\alpha_g}$ must be the sum of probability amplitudes for this virtual particle exchange. For $\alpha_g = 5.82\ 10^{-39}$, $\sqrt{\alpha_g} = 7.63\ 10^{-20} \approx 1/1\ 10^{19}$.

In 1995 Visser[9] described electrogravitic quantities analogous to Planck quantities that include the electric force ke^2. Research has shown that these quantities are a rediscovery of quantities first proposed in 1874 by Stoney[10, 11]. It is possible to demonstrate how they are related to the Compton wavelength of the atomic mass unit, and to probability amplitudes for virtual photon and graviton exchange.

Substituting the formulae for ke^2 and G derived above in equations (1) and (2), the Stoney formula for the electrogravitic length becomes

$$L_S = \sqrt{\frac{ke^2 G}{c^4}} = \sqrt{\frac{\alpha \hbar c \alpha_g \hbar c}{c^4 u^2}} = \frac{\hbar}{uc}\sqrt{\alpha \alpha_g} \qquad (5)$$

Substituting CODATA[6] values, $L_S = 2.118\ 10^{-16}\ (7.297\ 10^{-3}\ 5.82\ 10^{-39})^{1/2} = 1.38\ 10^{-36}$ m. This shows that when $\alpha = \alpha_g = 1$, then $L_S = 1.38\ 10^{-36}$ m, a shorter distance than the Planck length. It can be argued that the Stoney length is more fundamental than the Planck length because it is the length scale at which the electromagnetic and gravitational coupling constants are equal to unity.

The electrogravitic or Stoney mass is given by

$$M_S = \sqrt{\frac{ke^2}{G}} = \sqrt{\frac{\alpha\hbar cu^2}{\alpha_g \hbar c}} = \sqrt{\frac{\alpha}{\alpha_g}}\, u \qquad (6)$$

Using CODATA[6] values, $M_S = (7.297\ 10^{-3}/\ 5.82\ 10^{-39})^{1/2}\ 1.66\ 10^{-27} = 1.859\ 10^{-9}$ kg. Again this is smaller than the Planck mass.

Finally the formula for the electrogravitic or Stoney time is,

$$T_S = \sqrt{\frac{ke^2 G}{c^6}} = \sqrt{\frac{\alpha\hbar c \alpha_g \hbar c}{c^6 u^2}} = \frac{\hbar}{uc}\frac{1}{c}\sqrt{\alpha\alpha_g} \qquad (7)$$

Substituting CODATA[6] values, $T_S = 2.118\ 10^{-16}\ (2.998\ 10^{8})^{-1}\ (7.297\ 10^{-3}\ 5.82\ 10^{-39})^{1/2} = 4.604\ 10^{-45}$ s.

The substitution of the formula for G into other units or equations can give similar new perspectives and will establish a relationship with the fundamental dimensionless coupling constant for gravitation, α_g. For example, the Schwarzschild radius $R_S = 2GM/c^2 = (1/\pi)\ \alpha_g\ (M/u)\ (h/uc)$. Hence R_S is seen to be the product of the number of mass units in M with the corresponding Compton wavelength and the gravitational coupling constant. The Schwarzschild radius can then be described as a string of Compton wavelengths. For the mass M actually to be confined within this radius, α_g=1 and u must then become the Planck mass, m_P.

4. Variations in *G*

Assuming that gravitational interactions are due to virtual graviton exchange between mass-energy, then virtual gravitons can also couple with each other and with incident radiation. This would lead to an enhanced gravitational coupling, and evidence for such an effect in astronomical and laboratory measurements is examined here.

4.1 Exchange interactions can lead to constant acceleration

It is possible to show how enhancement of virtual graviton exchange can lead to constant acceleration terms by adapting some of the models normally used to explain the inverse square law. Fig. 3 shows how, at twice the distance, the intensity of a gravitational field is spread over four times the area, thereby resulting in one fourth of the acceleration. Similarly at three times the distance, the acceleration is reduced to one ninth and so on. This assumes that graviton exchange occurs in straight lines and there is no interference with the exchange process. In other words, the process is assumed to occur in a perfect vacuum. However, if we assume a constant density of matter in the intervening space then, as the exchange distance increases, there is an increasing probability of interaction with this mass-energy. As shown in Fig. 3., at distance r there is an interaction with a unit of mass-energy leading to an enhancement of the gravitational coupling. At twice the distance another mass-energy unit is encountered and this gives four extra couplings, including all the cross interaction terms. At three times the distance another unit of mass energy interacts with the exchange pathway, and the total number of extra couplings is now nine, counting all the possible interactions. Hence, while the overall intensity of a gravitational field decreases with the inverse square law, there is a compensating quantum enhancement of the gravitational coupling with distance if there is some intervening mass-energy. This is a very small secondary effect at short distances, but at large distances it can become the dominant term because the primary acceleration decreases via the inverse square of distance. Models are developed below for this exchange enhancement effect, and estimates made of the strength of the coupling.

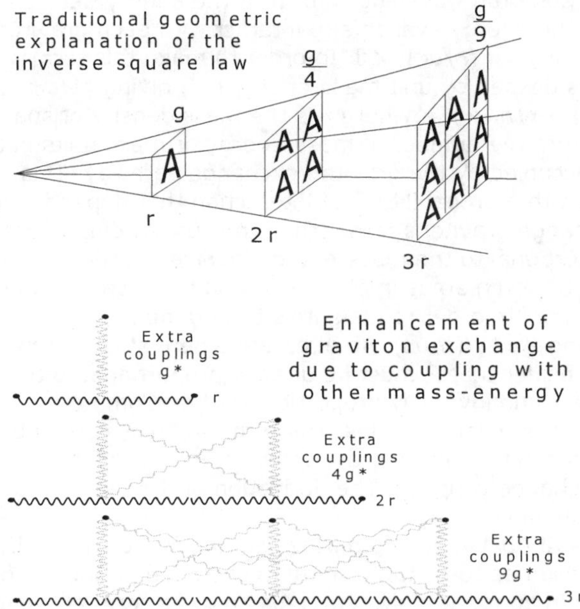

Figure 3. How a constant acceleration term can result from extra couplings with graviton exchange pathways.

4.2 The anomalous Pioneer acceleration

The development of interplanetary probes required considerable research in Newtonian celestial mechanics, and tracking the probes over many years has provided an opportunity to test this mechanics in the Solar system and beyond. An unexpected observation from extensive radiometric data from Pioneer 10/11, Galileo and Ulysses spacecraft is an anomalous constant acceleration of the spacecraft towards the Sun at distances up to $1\ 10^{13}$ m. The acceleration is approximately $8\ 10^{-10}$ m s^{-2} and was first reported in 1992 by Anderson[12]. It has been confirmed by subsequent studies[13] and various potential confounding factors due to the spacecraft itself have been eliminated as the cause. Furthermore a variety of possible perturbative forces due to celestial bodies (e.g., the Kuiper belt) cannot explain the acceleration. The possibility that it could be due to enhanced graviton exchange will now be examined.

Applying the model developed here to interplanetary distances involves virtual exchange paths of 10^{12} - 10^{13} m or more. This introduces the possibility of enhanced gravitational coupling due to interactions with mass-energy emitted by celestial bodies, the cosmic radiation, and also other virtual gravitons. This may be an insignificant effect at short distances but at large distances the likelihood of an interaction will increase. The model described in Sec. 2 is an idealised exchange of virtual bosons in isolation from other particles. Here the model will be developed to allow for extra couplings that may occur with surrounding mass-energy.

Considering two masses M_1 and M_2 separated by a large mean distance X m, the volume of space occupied by each virtual graviton exchange path between atomic mass units in M_1 and M_2 is the cylinder $\pi\,(h/uc)^2 X$ m^3, where h/uc is the Compton wavelength for u. If the energy density of space is ρ_{en} J m^{-3}, the energy available to interact with each graviton exchange pathway is $\rho_{en}\,\pi\,(h/uc)^2\,X$ J. In order to make this term dimensionless, it is divided by unit mass-energy uc^2, giving $(1/uc^2)\,\rho_{en}\,\pi\,(h/uc)^2\,X$, or $(1/u)\,\rho\,\pi\,(h/uc)^2\,X$ where ρ is the mass density of space (kg m^{-3}). This is therefore the number of mass-energy or mass units in the volume of space occupied by the graviton exchange pathway. This must now be combined with a probability P_i of interaction that depends on the nature of the exchange gravitons rather than the surrounding space. This probability is proportional to the mass-energy involved in the graviton exchange process between amus in M_1 and M_2 and the mean distance for exchange, hence uc^2X. In order to make this term dimensionless, it is divided by hc, giving $P_i \propto X/(h/uc)$. An alternative derivation of this probability is to consider that P_i must be directly proportional to the exchange distance X and inversely proportional to the Compton wavelength of the interacting particles. The term $1/(h/uc)$ corresponds to the number of nodes in the virtual graviton wave and therefore the strength of the exchange coupling. The derivation of the two terms is shown schematically in Fig. 4.

The force of gravity between two masses M_1 and M_2 separated by a large distance is assumed to be the term derived above in Sec 2.2 for graviton exchange in an isolated system, plus an extra term allowing for interactions with matter in the intervening space.

$$F_g = \alpha_g \frac{\hbar c}{u^2}\frac{M_1 M_2}{X^2} + \beta_g \frac{\hbar c}{u^2}\frac{M_1 M_2}{X^2}\frac{1}{u}\rho\pi\left(\frac{h}{uc}\right)^2 X\,P_i \tag{8}$$

Inserting the expression for P_i and re-arranging,

$$F_g = \left(\alpha_g \frac{\hbar c}{u^2} + \beta_g \frac{\hbar c}{u^2}\left(\frac{\rho}{u}\right)\pi\left(h/_{uc}\right)^2 X\left(\frac{X}{h/_{uc}}\right)\right)\frac{M_1 M_2}{X^2} \tag{9}$$

$$F_g = \alpha_g \frac{\hbar}{uc}\frac{c^2}{u}\frac{M_1 M_2}{X^2} + \beta_g\left(\frac{\hbar}{uc}\right)^2\frac{c^2}{u^2}2\pi^2\rho\,M_1 M_2 \tag{10}$$

(A) Energy available for interaction with graviton exchange pathway

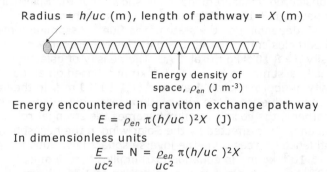

Radius = h/uc (m), length of pathway = X (m)

Energy density of space, ρ_{en} (J m^{-3})

Energy encountered in graviton exchange pathway

$$E = \rho_{en}\,\pi(h/uc)^2 X \quad (\mathrm{J})$$

In dimensionless units

$$\frac{E}{uc^2} = \mathrm{N} = \frac{\rho_{en}}{uc^2}\,\pi(h/uc)^2 X$$

(B) Enhanced exchange is dependent on distance and exchange energy

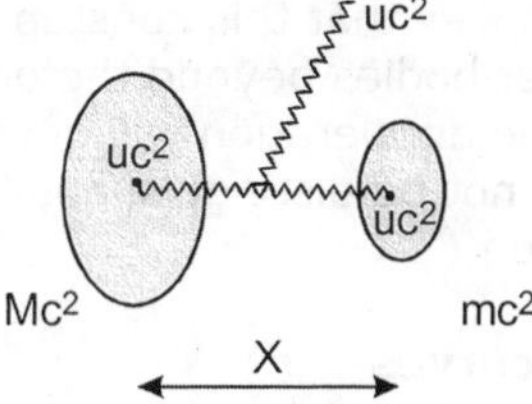

Probability of interaction $\quad P_i \propto uc^2\ X$ (kg m^3 s^{-2})

Divide by hc (kg m^3 s^{-2}) $\quad P_i \propto \dfrac{X}{(h/uc)}$

Figure 4. Assumptions required for interactions of virtual gravitons with material intervening in the exchange pathway.

The term in β_g in Eq. (10) is independent of distance when the density of space is constant. However, in the Solar system the density of space will change over the length of the exchange pathway. There are two main components to the density of this space, the Solar component and the galactic/cosmic component. The Solar component will decrease according to an inverse square law in the Solar system while the galactic/cosmic component is essentially constant. This model is therefore consistent with the Pioneer acceleration observations in the outer Solar system and is capable of explaining why the anomalous *Pioneer* acceleration is not seen in the inner Solar system. Close to the Sun, the varying Solar component of the density of space dominates and, because it is inverse square dependent, it will make the β_g term indistinguishable from the α_g term. At large distances from the Sun beyond the orbit of Jupiter, the constant galactic/cosmic density begins to dominate and the β_g term will become a constant. The extra force on a body of matter will be directed towards the Sun because it is an enhanced graviton exchange coupling with mass-energy in the Sun, but it will be independent of the distance from the Sun.

It is possible to estimate values for ρ_{en} and ρ in the Solar system. At 1 AU, the solar constant S is 1350 W m^{-2}, and therefore $S/c = \rho_{en} = 4.54$

10^{-6} J m^{-3} and $S/c^3 = \rho \approx 5\ 10^{-23}$ kg m^{-3}. These densities relate to the radiant photon energy of the Sun, so the density of the Solar wind needs also to be considered. Measurements have shown that the Solar wind has a density of between 3-10 particles cm^{-3} and its velocity varies from 200 to 1,000 km s^{-1} depending on the state of the Sun[14]. Assuming a mean density of 5 particles cm^{-3} and that the particles are primarily protons, this gives a density $\rho \approx 8\ 10^{-21}$ kg m^{-3} at 1 AU. The density of galactic/cosmic rays is normally assumed to be $\rho \approx 1\ 10^{-30}$ kg m^{-3}, based on an observed energy density of approximately 1 eV cm^{-3} or 1 10^{-13} J m^{-3}. In the inner solar system the influence of the Sun predominates, but as the exchange distance lengthens, the cosmic ray flux becomes the stronger relatively constant influence. It is affected by the Solar wind in the heliosphere but the Sun's influence decreases via the inverse square law. Substituting the value for $\rho \approx 1\ 10^{-30}$ kg m^{-3} in equation (10) implies that β_g must be of order of 1 10^{-50} to give the constant acceleration of 8.74 10^{-10} m s^{-2} observed in the *Pioneer* and other spacecraft in the outer Solar system[12].

This model therefore has the potential to explain the anomalous *Pioneer* acceleration and it implies that this constant acceleration should also affect planetary and other bodies beyond the orbit of Jupiter. At distances closer to the Sun the acceleration will governed by an inverse square law and will therefore not be distinguishable from Newtonian inverse square law acceleration.

4.3 Galactic rotation curves

At galactic scales the gravitational force acting on stars is much larger than would be expected from the distribution of matter that can be observed directly[15]. The observed gravitational force acting at these scales is significantly greater than that predicted by Newtonian gravitation or by General Relativity. This observation can be explained by postulating the existence of *dark matter* in the universe, and there is considerable effort going on to define properties and to make testable predictions for the material[16]. It represents the scientific consensus at present and is a prediction based on observation and some accepted theory. An alternative explanation is provided by Modified Newtonian Dynamics (MOND) first suggested by Milgrom[17] at large inter-stellar distances. The MOND hypothesis is extremely successful in modelling galactic rotation curves[18] but it is essentially an *ad hoc* change to Newtonian physics and is therefore subject to much fundamental criticism.

The model derived here provides an interesting alternative to dark matter and to MOND. It does not require new physics, but it does rely on a quantum field interpretation of gravity. It will be shown that the enhancement of gravitational coupling over large distances, due to adventitious encounters of exchange gravitons with intervening mass-energy, is sufficient to explain flat galactic rotation curves and the observed relationship between a galaxy's luminosity, L, and its rotation velocity, v, the Tully-Fisher relation, $L \propto v^4$. The observation that the speed of rotation of stars and clouds of hydrogen atoms in galactic disks is practically constant from the inner to outer reaches of galaxies was first made in the 1970s and the empirical Tully–Fisher law was established[19, 20] in the 1980s. The early observations have been confirmed since by a multitude of subsequent observations.

Using straightforward Newtonian dynamics, the force acting on a mass m in orbit at a distance r from the centre of a galaxy, is governed by the virial equation

$$G\frac{M_{Gal}\,m}{r^2} = m\frac{v^2}{r} \qquad (11)$$

where v is the velocity of the mass m in orbit and M_{Gal} is the mass of the galaxy. The standard solution to this Newtonian equation is a mass-rotation velocity relation, $v^2 = GM_{Gal}/r$, but a 1/r dependence is incompatible with observation. Also, because this implies $v^2 \propto M_{Gal}{}^4$, the Tully-Fisher law, $L \propto v^4$, implies the luminosity of a galaxy should be proportional to $M_{Gal}{}^2$. Application of MOND to the mass rotation velocity gives a different result, $L \propto v^4 \propto M_{Gal}$, and this is much more consistent with observation[18]. However, as concede by proponents of MOND, the relationship $M_{Gal} \propto v^4$ is built into its formalism so the result cannot strictly be called a prediction[21].

It can be shown that the application of equations (8)-(10) is consistent with flat galactic rotation curves. Considering the two terms in equation (10) at interstellar and galactic distances, it can be assumed that the inverse square term will become a small secondary term at such large values of X. The equation for the force acting on a mass m in orbit at a large distance r from the centre of the galaxy becomes simply the second term,

$$F = \beta_g\left(\frac{\hbar}{uc}\right)^2\frac{c^2}{u^2}2\pi^2\rho\,M_{Gal}\,m = m\frac{v^2}{r} \qquad (12)$$

Within framework of the model described in Sec 4.2, ρ can be assumed to be the density of a cylindrical element of the galaxy encompassing a graviton exchange pathway. For example, in a spiral galaxy the element would be within an arm of the galaxy. The major axis of this cylinder is therefore co-axial with r, and the density $\rho = m_{inter}/(\pi d^2 r)$, where m_{inter} is the amount of interstellar matter in the element and d is the radial dimension of the element. For a spiral galaxy d can be assumed to be the radius of the central bulge. Substituting this expression for the density ρ in Eq. (12) makes v independent of r and leads to $\rho_a\, M_{Gal} \propto v^2$, where ρ_a is now an area density (kg m^{-2}). The other terms in the equation are constants, as can be seen below,

$$F = \beta_g\left(\frac{\hbar}{uc}\right)^2\frac{c^2}{u^2}2\pi^2\frac{m_{inter}}{\pi d^2 r}M_{Gal}m = m\frac{v^2}{r} \qquad (13)$$

$$\text{giving}\quad \beta_g\left(\frac{\hbar}{uc}\right)^2\frac{c^2}{u^2}2\pi^2\frac{m_{inter}}{\pi d^2}M_{Gal} = v^2 \text{ and so } \quad v^2 \propto \rho_a M_{Gal} \qquad (14)$$

In terms of the Tully Fisher observation, these assumptions imply that $L \propto v^4 \propto \rho_a{}^2\, M_{Gal}{}^2$. Now M_{Gal} is the observable, luminous galactic mass whereas the mass in ρ_a is non-luminous matter. In contrast to the model developed above to explain the *Pioneer* acceleration, where interactions with cosmic rays were assumed to enhance graviton exchange, here it is assumed that all interstellar matter m_{inter} can interact with the exchange. To be consistent with observation, as the luminous galactic mass M_{Gal} and d increase, the interstellar mass m_{inter} should decrease or stay constant. Looking at the scope of possible variations, the model implies relationships in the range, $L \propto v^4 \propto M_{Gal}{}^{0\text{-}2}$. For example, assuming m_{inter} remains constant while M_{Gal} increases by a factor 2 then, if d increases by a factor 1.2, the relationship is $L \propto v^4 \propto M_{Gal}$. The assumption that ρ_a decreases as M_{Gal} increases is qualitatively reasonable at least. As the luminous mass of galaxies increases, it implies that more and more interstellar material must be involved in stellar processes. The precise relationship between the quantities d, m_{inter} and M_{Gal} for galaxies requires further investigation.

It is possible to estimate the value for β_g from measurements of the rotation velocity of our own galaxy, 220 km s^{-1}. The galactic mass is estimated at 1 10^{11} M_S, and given a radius of about 15 kiloparsecs (4.63 10^{20} m), the area density ρ_a becomes equal to 5 10^{-1} kg m^{-2} (i.e., assuming a constant interstellar volume density of 1 10^{-21} kg m^{-3}). Solving equation (14) with these values gives $\beta_g \approx 1\ 10^{-71}$. This is considerably smaller coupling constant than the values calculated in Sec 4.2 for β_g in relation to the anomalous Pioneer acceleration. However, this is to be expected because here it is assumed that interstellar material rather than cosmic rays are enhancing the gravitational coupling. If it was assumed that only cosmic rays were involved in the enhancement, then ρ_a would be equal to 5 10^{-10} kg m^{-2} (i.e., assuming a cosmic ray density of 1 10^{-30} kg m^{-3}) giving $\beta_g \approx 1\ 10^{-62}$.

A significant feature of the approach described here is that it removes the need to invoke significant quantities of *dark matter* (other than normal interstellar material) to explain flat galactic rotation curves. Also the model does not require the fundamental modifications to Newtonian dynamics that are implied by MOND.

4.4 Variations in laboratory measurements of G

The gravitational coupling between laboratory test masses is being measured in a large fluence of virtual gravitons from the relatively vast mass of the Earth. Also the Earth, Sun and stars are emitting other radiation which could interact with the exchange process and enhance the coupling. The phenomenological models developed here will be used to estimate whether such enhanced coupling could be the cause of some of the variation[22, 23] in laboratory measurements of G.

At the equator, the radius of the Earth is slightly larger than at the Poles which means that the graviton fluence at the equator would be slightly less at the equator than at the poles. If it is assumed that Earth gravitons couple with the test mass gravitons to produce a slightly enhanced gravitational coupling, measurements of G should show an increase with increasing latitude. Similarly, measurements of G should show a decrease with increasing altitude, and this should be particularly marked in satellite measurements of G, hundreds of kilometres above the Earth's surface. Equally, laboratory gravitons will couple with Solar and Lunar gravitons so there could be diurnal and seasonal variations, but these are likely to be very small effects in comparison to those from coupling with terrestrial gravitons. Furthermore, coupling with virtual photons could enhance graviton exchange, and therefore G could vary with the strength of local and terrestrial electromagnetic fields.

Mbelek and Lachièze-Rey have recently examined evidence for a latitude and longitude variation in laboratory measurements of G in relation to a hypothesis of coupling with the Earth's geomagnetic field[24].

They concluded that there is a significant variation in 17 laboratory measurements of G since 1942. For example the measurement of G at HUST, Wuhan, China[25], latitude 30.6^0, gave 6.6699(± 0.0007) 10^{-11} kg^{-1} m^3 s^{-2}, while a comparable measurement at BIPM, Sevres, France[26], latitude 48.8^0 gave 6.67559(± 0.00027) 10^{-11} kg^{-1} m^3 s^{-2}. A measurement of G near Seattle [27] (latitude 47.6^0, very close to the latitude to BIPM) gave 6.674215(± 0.000092) 10^{-11} kg^{-1} m^3 s^{-2}. Given a rigorous view of the stated measurement uncertainties, it is possible to conclude that these experiments have not measured the same quantity. The possibility that this is due to enhancement of gravitational coupling in the laboratory via terrestrial gravitons or other radiation will now be examined.

Assuming a gravitational experiment involving two masses M_1 and M_2 separated by a mean distance X m, taking place in a laboratory on the Earth's surface. The same model for interactions used in Sec. 4.1 and 4.2 will be used to evaluate the likelihood of enhanced graviton coupling due to interaction with terrestrial gravitons and other radiation. For a laboratory environment, equation (10) becomes

$$F_g = \alpha_g \frac{\hbar}{uc}\frac{c^2}{u}\frac{M_1 M_2}{X^2} + \beta_g \left(\frac{\hbar}{uc}\right)^2 \frac{c^2}{u^2} 2\pi^2 \rho_{lab} M_1 M_2 \qquad (15)$$

Here the value for the local density ρ_{lab} should be the density of material between the masses, which is of the order 1 kg m^{-3}, assuming this is air. The density should correspond to the matter that could interact with graviton exchange, so it could be a factor of 10^3 higher than this, depending on the experimental design. In order to make a comparison between the magnitude of the two terms in Eq.(15) it is helpful to consider their ratio Δ,

$$\Delta = \frac{\beta_g \left(\frac{\hbar}{uc}\right)^2 \frac{c^2}{u^2} 2\pi^2 \rho_{lab} M_1 M_2}{\alpha_g \frac{\hbar}{uc}\frac{c^2}{u}\frac{M_1 M_2}{X^2}} = \frac{\beta_g}{\alpha_g}\frac{\hbar}{uc}\frac{1}{u} 2\pi^2 \rho_{lab} X^2 = \beta_g 4.33\, 10^{50} X^2 \qquad (16)$$

In order for the enhanced acceleration term to make a significant contribution (0.1%) at a typical laboratory distance (say 20-30 cm), then β_g would have to be of the order of 1 10^{-51}. This is comparable to the coupling constant derived for cosmic rays in Sec. 4.2, and still would be if Eq. (16) incorporated a significantly higher laboratory density. It is therefore possible that the interaction mechanism involving cosmic rays which explains the anomalous *Pioneer* acceleration, could also affect terrestrial measurements of G. Also local electromagnetic fields, both natural and artificial, could affect the coupling (i.e., virtual photons rather than cosmic rays). More work is needed to assess the potential extent of such effects. The magnitude of such effects could be estimated by performing measurements of G in a mobile laboratory that could vary its position on the Earth's surface, either by changing latitude or altitude. In order to maximise the effect, the experiment should ensure that the laboratory graviton exchange is orthogonal to the terrestrial field. Perhaps the best opportunity to investigate this effect is Project SEE[28], the NASA experiment designed to make a measurement of G in orbit. Here at distances of between 150-200 km above the Earth's surface, any enhancement effect due to terrestrial and cosmic fields should become clear.

5. Conclusions

A phenomenological model based on the relativistic quantum field concept of virtual boson exchange has been developed, assuming that virtual gravitons mediate gravitational forces between mass-energy. The model establishes a relationship between G and other fundamental constants $G = \alpha_g \hbar c/u^2$, where α_g is the dimensionless coupling constant for graviton exchange between atomic units of mass energy uc^2. This relationship is identical to that obtained by dimensional analysis. The model has found application elsewhere in studying spin fluctuations at Planck energy scales[29]. Here its application to astronomical observations provides an explanation for the anomalous *Pioneer* acceleration and for flat galactic rotation curves. These effects are due to enhanced graviton coupling as a result of interactions with real and virtual particles in space. A coupling constant of $1\ 10^{-50}$ for the interaction of cosmic rays with graviton exchange can explain the *Pioneer* acceleration. In contrast, flat galactic rotation curves can be explained by a much smaller coupling constant of $1\ 10^{-71}$ for interactions of normal interstellar material with graviton exchange. Application of the model to explain variations in measurements of G is possible. At $1\ 10^{-50}$, the coupling of cosmic rays with graviton exchange is of sufficient strength to affect laboratory measurements of G. However, further work is required to show whether cosmic, terrestrial and artificial fields are of sufficient strength to effect graviton exchange in the laboratory.

Acknowledgements. I would like to thank the Directors of the International School of Cosmology and Gravitation for the invitation to present this paper at the 18th Course at the Ettore Majorana Centre for Scientific Culture, Erice, Sicily. I would like also to thank NRPB and the Department of Physics and Astronomy at the University of Tennessee for providing me with the opportunity to work on the ideas presented here. Special thanks are due to Professor G T Gillies and Professor A L Sanders for their interest, advice and practical help during the development of the graviton exchange model.

References

1. Clark, M.J. (2002). Radiation and gravitation, in P.G. Bergmann and V. de Sabbata (eds.) *Advances in the Interplay between Quantum and Gravity Physics*, Kluwer Academic Publishers, Norwell MA, pp.77-84.
2. Bethe, H.A. and Fermi, E. (1932) *Zeits. Phys.*, **77**, 296-306.
3. Wilczek, F. (2000) *Rev. Mod. Phys. Centenary*, **71**(2), S85-S95.
4. Feynman, R.P., Marinigo, F.B. and Wagner, W.G. (1995) *Feynman Lectures on Gravitation,* Ed. B. Hatfield, Addison-Wesley, Reading MA.
5. Davies, P.C.W. (1987) *The Forces of Nature*, 2nd Edition, Cambridge University Press.
6. Mohr, P.J. and Taylor, B.N. (2000) *Rev. Mod. Phys.*, **72**(2), 351-496 (See also http://www.physics.nist.gov or http://www.npl.co.uk for web versions of these data).
7. Rohlf, J.W. (1994), *Modern Physics from* α *to* Z^0, J Wiley & Sons Inc., New York.
8. Wilczek, F. (1999) *Nature* **397**, 303-306.
9. Visser, M. (1995), *Lorentzian Wormholes*, American Institute of Physics.
10. Stoney, G.J. (1881) *Philosophical Magazine* **11**, 381.
11. Barrow, J.D. (2003) *The Constants of Nature: From Alpha to Omega*, Jonathan Cape, London.
12. Nieto, M., Goldman T., Anderson, J.D., Lau, E.L. and Perez-Mercader, J. (1995) in G. Kernel, P. Krizan, and M. Mikuz (eds.), *Proc. Third Biennial Conference on Low Energy Antiproton Physics,* World Scientific, Singapore, p 606.
13. Anderson, J.D., Laing, P.A., Lau, E.L., Liu, A.S., Nieto, M.M., and Turyshev, S.G. (1998) *Phys. Rev. Lett.* **81**, 2858.
14. Harwit, M. (1998) *Astrophysical Concepts*, 3rd Edition. Springer-Verlag.
15. Bertin, G. (2000) *Dynamics of Galaxies*, Cambridge University Press.
16. (a) Fukugita, M. (2003) *Nature* **422**, 489-491 (b) Ostriker, J.P. and Steinhardt, P. (2003) *Science* **300**, 1909-1913.
17. Milgrom, M. (1983) *Astrophys. J.* **270**, 365-370.
18. Sanders, R.H. and McGaugh, S.S. (2002) *ArXiv:astro-ph*/0204521 v1.
19. Tully, R.B. and Fisher, J.R. (1977) *Astron. Astrophys.* **54**, 661-673.
20. Rubin, V.C., Ford W.K., and Thonnard N. (1980) *Astrophys. J.* **238**, 471-487.
21. Sanders, R.H. (2001) *ArXiv:astro-ph*/0106558 v1.
22. Cook, A.H. (1988) *Rep. Prog. Phys.* **51**, 707-757.
23. Gillies, G.T. (1997) *Rep. Prog. Phys.* **60**, 151-225.
24. Mbelek, J.P. and Lachièze-Rey, M. (2002) *ArXiv:gr-qc*/0204064 v1.
25. Luo, J. et alia (1998) *Phys. Rev.* **D59**, 042001.
26. Quinn, T.J. et alia (2001) *Phys. Rev. Lett.* **87**, 111101.
27. Gundlach, J.H. and Merkowitz, S.M. (2000) *Phys. Rev. Lett.* **85**, 2869.
28. Sanders, A.J., et alia (1999) *Meas. Sci. Tech.* **10**(6), 514-524.
29. De Sabbata, V., Ronchetti, L. and Clark M.J. (2002) *Nuovo Cim.* **117B**(6), 695-702.

ERICE 2003
International School of Cosmology and gravitation

SOME BASE FOR QUANTUM GRAVITY

Venzo de Sabbata and Luca Ronchetti

World Laboratory, Lausanne, Switzerland
Istituto di Fisica Nucleare, sezione di Bologna, Italy

Abstract

We will explore the possibility to quantize the Gravity. For that motivation our work will use three fundamental tools from Mathematics and from Physics. These concepts are: 1) the spin (torsion), 2) the geometric algebra, 3) a quadratic Lagrangian. We will see also that this will give some indication for supersymmetry.

1. Introduction

I would like to begin with an important work by Borzeszkowski and Treder [1,2] in which was shown that Landau-Peierls type relations

$$\Delta F(c\Delta t)^2 \geq (\hbar c)^{1/2} \qquad (1)$$

(where F denotes the magnitude of the field strenght F_{ik}, ΔF its measurement inaccuracy and Δt the time interval between two field measurements) could correspond to commutative rules of the form

$$i\,[F_{ik}\,,\, dx^i \wedge dx^k] = (\hbar c)^{1/2} \qquad (2)$$

In this case inequality (1) would have the meaning of quantum uncertainty relations, that is it would imply, via Rosenfeld's analysis [3], the uncertainty inequalities for quantum general relativity as:

V. de Sabbata et al. (eds.),
The Gravitational Constant: Generalized Gravitational Theories and Experiments, 81–111.
© 2004 *Kluwer Academic Publishers. Printed in the Netherlands.*

$$\Delta g_{ik}\Delta x^i \Delta x^k \geq \hbar G/c^3 \tag{3}$$

and

$$\Delta\Gamma_{ikl}\Delta x^i \Delta x^k \Delta x^l \geq \hbar G/c^3 \tag{4}$$

(Δg and $\Delta\Gamma$ are the corresponding inaccuracies).

However they note that in general relativity the "field strength" Γ^i_{kl} (which are the usual Christoffel symbols) are not tensorial quantities and distances cannot be defined independently of the field uantities g_{ik}.

In fact, as demonstrated in [1] (see also [4] and [5]), one should not expect to find commutation rules corresponding to Rosenfeld's inequality relations

$$\begin{aligned} g_{ik} L_o^2 &\geq \hbar G/c^3 \\ \Gamma^i_{kl} L_o^3 &\geq \hbar G/c^3 \end{aligned} \tag{5}$$

(L_o denotes the dimension of the spatial region over which the average value of g_{ik} and Γ^i_{kl} is measured) because the quantities "field strength" and "length" appearing in (5) cannot be defined independently of each other. The independent definition of those quantities for which commutation rules are to be formed is however necessary to require.

This problem may not arise if torsion is considered, since in this case the asymmetric part of the connection i.e. $\Gamma^i_{[kl]}$, i.e.the torsion tensor Q^i_{kl} is a true tensorial quantity. In fact, as we have said, with torsion one can define distances in this sense: if we consider a small closed circuit and write

$$l^i = \oint Q^i_{kl} dA^{kl} \neq 0 \tag{6}$$

(where $dA^{kl} = dx^k \wedge dx^l$ is the area element enclosed by the loop), then l^i represents the so called "closure failure", i.e. torsion has an intrinsic geometric meaning: it represents the failure of the loop to close, l^i having the dimension of length. (Q^i_{kl} has dimension of inverse length, dA^{kl} is area).

Now as is well known, the torsion tensor $Q_{ij}{}^k$ is related to the spin density tensor $J_{ij}{}^k$ (which acts as the source term for torsion, analogous to matter density being source term for curvature) through the equation [6]:

$$Q_{ij}{}^k = (8\pi G/c^3)(J_{ij}{}^k - (1/2)\delta^k_i J_{bj}{}^b - (1/2)\delta^k_j J_{ib}{}^b) \qquad (7)$$

Thus for N aligned spins per unit volume, the torsion is given as (see ref.[7]):

$$\bar{Q} = (4\pi G/c^3)\bar{J}, \qquad (8)$$

where

$$J = N\hbar/2. \qquad (9)$$

We see that we are going toward a quantization of torsion field: in fact putting at the place of J the spin of elementary particles (see eqs.(6),(8),(9)) we can write (see Appendix):

$$l^i = \oint Q^i_{kl} dA^{kl} = n\ (\hbar G/c^3)^{1/2} \qquad (10)$$

where n is an integer.

In the above relation, we see that torsion can be related to the intrinsic spin ħ, and, as the spin is quantized, we can say that the defect in space- time topology should occur in multiples of the Planck length, $(\hbar G/c^3)^{1/2}$ (notice that Planck length comes automatically in eq.(10)) [8,9]].

This is analogous to the well known $\oint pdq = n\hbar$, i.e. the

Bohr–Sommerfeld relation. As Q^i_{kl} plays the role of the field strength (analogous to F_{ik} for electromagnetism), relation (10) is analogous to that of eq.(2) (with $\sqrt{c\hbar}$ on the right hand side). So distance has been defined independently of g_{ik}. In fact eq.(10) would define a minimal fundamental length, i.e. the Planck length entering through the minimal unit of spin or action ħ. So ħ has to deal with the intrinsic defect built into torsion structure of space time through $l^i = \oint Q^i_{kl} dx^k \wedge dx^l$. Therefore we have

$$[Q^i_{kl} \, , \, dx^k \wedge dx^l] \geq (\hbar G/c^3)^{1/2} \tag{11}$$

and the corresponding uncertainty relations:

$$\Delta Q^i_{\ kl} \; (\Delta x^k \wedge \Delta x^l) \;\geq\; (\hbar G/c^3)^{1/2} \tag{12}$$

Therefore, the Einstein–Cartan theory of gravitation should, in contrast to Einstein's General Relativity Theory, provide genuine quantum–gravity effects.

But we can go a little further, trying to see if there are conjugate variables for which we can write commutation relations. Now we have in the hand two geometrical objects: the curvature R and the torsion Q; we will see that R and Q can be such conjugate variables. Consider in fact the equation of geodesic of a free particle with non–zero spin (in which coupling of spin to the space–time curvature occurs):

$$\frac{d^2x^\mu}{ds^2} + \left\{ {\mu \atop \alpha\beta} \right\} \frac{dx^\alpha}{ds}\frac{dx^\beta}{ds} = Q^\mu_{\alpha\beta}\frac{dx^\alpha}{ds}\frac{dx^\beta}{ds} \tag{13}$$

Following the above discussion, in the quantum picture the right hand side is proportional to multiples of the Planck length. As the left hand side of eq.(13) is obtained from variation of $\int g_{\alpha\beta}(dx^{\alpha}/ds)(dx^{\beta}/ds)$, we would simply have

$$\Delta\Gamma^{\mu}_{\alpha\beta}\, dx^{\alpha}\wedge dx^{\beta} = [Q^{\mu}_{\alpha\beta}\, , dx^{\alpha}\wedge dx^{\beta}] \geq (\hbar G/c^{3})^{1/2} \qquad (14)$$

In the limit of vanishing torsion, we have the usual geodesic equation. So torsion in eq.(13) can be identified with the quantum correction (in multiples of $\hbar$) to the classical equation of motion.

Now we know from the equation of geodesic deviation that curvature causes relative acceleration between neighbouring test particles we have momentum uncertainty related to curvature as:

$$ma^{\mu}ds = \Delta p^{\mu} = m\, R^{\mu}{}_{\alpha\beta\gamma}\,\frac{dx^{\alpha}}{ds}\, dx^{\beta}\eta^{\gamma} = mc\; R^{\mu}dS \qquad (15)$$

where η^{γ} is the separation vector between neighboring geodesic. So as position fluctuations are given by torsion, momentum fluctuation are due to curvature and we can interpret quantum effects (and then uncertainty principle) as consequences of space-time deformation i.e.

$$\Delta p^{\mu}\cdot\Delta x_{\mu} \geq \hbar \qquad (16)$$

where

$$\Delta x^{\mu} = Q^{\mu}\,\Delta S \qquad (17)$$

and

$$\Delta p^{\mu} = mcR^{\mu}\,\Delta S \qquad (18)$$

(i.e. uncertainty in initial and final positions and relative acceleration between them due to space-time deformations), Q (torsion) and curvature (R) thus both playing simultaneous roles as conjugate variables of the geometry (gravitational field), thus enabling us to write commutation relations between curvature and torsion (analogous to $[x,p] = i\hbar$) as:

$$\left[Q , R\right] = i(\hbar G/c^3)^{-3/2} \quad (19)$$

or

$$\Delta Q \ \Delta R \geq L_{Pl}^{-3}$$

where L_{Pl} is the Planck length.

Thus in a sense, we have written quantum commutation relations for the observables of the background geometry, rather than for the gravitational field in a fixed background as in the usual picture; thus partially removing the inconsistency between quantizing the gravitational field and not the geometry, gravity being the geometry itself!

In fact in the usual methods of quantizing the gravitational field one has the field on a fixed background, which gives rise to an inconsistency, because the gravitational field is just geometry so that how can we separate the geometry and the field? Here, on the contrary, we had not only quantized torsion on a fixed background but we have considered both torsion and background curvature. Curvature and torsion in this contest appear to be conjugate variables for which we can write commutation relations like (19).

So we have seen that if one likes to quantize the gravity, one of the basic concept is that of spin (torsion): we must pass through the concept of torsion.

Now we will go through another fundamental tool that, in our opinion, is the concept of geometric algebra.

2. Geometric algebra

We know that when we consider the early universe we have

to deal both with elementary particle physics using quantum theory and with cosmology using general relativity; but general relativity is developed in real space-time while quantum theory needs a complex manifold. How can we conciliate general relativity with quantum theory? We think that the answer lies in a reformulation of Dirac's theory in terms of space-time geometric calculus without any complex number.

At this purpose we will consider the geometric algebra that, with the multivector concept and the interpretation of imaginary units as generator of rotations, places tensors and spinors on the same foot: both are described in a real space-time.

Without going in details we can define the geometric product as [10]:

$$ab = a \cdot b + a \wedge b \tag{20}$$

We will see that Hestenes space-time algebra automatically incorporates the geometric structure of space-time. This can be done introducing first of all the *outer product* $\mathbf{a} \wedge \mathbf{b}$ which is different from the usual cross product in the sense that it has magnitude $|\mathbf{a}||\mathbf{b}|\sin\vartheta$ and shares its skew property

$$\mathbf{a} \wedge \mathbf{b} = -\mathbf{b} \wedge \mathbf{a},$$

but is not a scalar or a vector: it is a *directed area*, or *bivector*, oriented in the plane containing **a** and **b**. One can visualize the outer product as the area swept out by displacing **a** along **b** with the orientation given by traversing the so formed parallelogram first along **a** and after along **b** vector.

One can generalize this notion to products of objects with higher dimensionality or *grade* in the sense that if the bivector $\mathbf{a} \wedge \mathbf{b}$, which has grade 2, is swept along another vector **c** of grade 1, one obtain the directed volume $(\mathbf{a} \wedge \mathbf{b}) \wedge \mathbf{c}$ which is a trivector of grade 3. Thus we are led to the notion of multivector.

We are now in the position to define the geometric product: it is the sum of inner and outer product that is (dropping the convention of using bold-face type for vectors):

$$ab = a \cdot b + a \wedge b \tag{21}$$

At first sight it may seem absurd to add two directed numbers with different grades, but simply one may observe that the result of adding a scalar to a bivector is an object that has both scalar and bivector parts in exactly the same way as the addition of real and imaginary numbers leads to an object with both real and imaginary parts. The latest are called complex numbers and, in the same way, we call the former 'multivector'. We simply take separate components in the symbol for the complex number $z = x + jy$ (we denote with 'j' the imaginary unit) as in the symbol for geometric product $ab = a \cdot b + a \wedge b$. Considering for instance two dimensional space and taking two orthonormal basis vector σ_1 and σ_2 we have:

$$\sigma_1 \cdot \sigma_1 = \sigma_2 \cdot \sigma_2 = 1 \qquad \sigma_1 \cdot \sigma_2 = 0 \tag{22}$$

$$\sigma_1 \wedge \sigma_1 = \sigma_2 \wedge \sigma_2 = 0 \tag{23}$$

The outer product $\sigma_1 \wedge \sigma_2$ is a directed area. Then we have four independent basis vectors as:

$$\text{scalar } 1, \qquad \text{vectors } \sigma_1, \sigma_2 \text{ and bivector } \sigma_1 \wedge \sigma_2 \tag{24}$$

and with these four basis elements we can form a multivector:

$$A \equiv a_0 1 + a_1\sigma_1 + a_2\sigma_2 + a_3\sigma_1 \wedge \sigma_2 \tag{25}$$

In order to define the multiplication AB between two multivector A and B (where $B \equiv b_0 1 + b_1\sigma_1 + b_2\sigma_2 + b_3\sigma_1 \wedge \sigma_2$) we have to remember how to multiply the four geometric basis elements (see for instance [10]), that is:

$$\sigma_1^2 = \sigma_1\sigma_1 = \sigma_1 \cdot \sigma_1 + \sigma_1 \wedge \sigma_1 = 1 = \sigma_2^2 \qquad (26)$$

and

$$\sigma_1\sigma_2 = \sigma_1 \cdot \sigma_2 + \sigma_1 \wedge \sigma_2 = \sigma_1 \wedge \sigma_2 = - \sigma_2 \wedge \sigma_1 = -\sigma_2\sigma_1 \qquad (27)$$

from which (being the geometric product associative):

$$(\sigma_1\sigma_2)\sigma_1 = - \sigma_2\sigma_1\sigma_1 = - \sigma_2 \quad \text{and} \quad (\sigma_1\sigma_2)\sigma_2 = \sigma_1 \qquad (28)$$

also

$$\sigma_1(\sigma_1\sigma_2) = \sigma_2 \quad \text{and} \quad \sigma_2(\sigma_1\sigma_2) = - \sigma_1 \qquad (29)$$

Moreover

$$(\sigma_1 \wedge \sigma_2)^2 = \sigma_1\sigma_2\sigma_1\sigma_2 = - \sigma_1\sigma_1\sigma_2\sigma_2 = - 1 \qquad (30)$$

(being

$\sigma_1\sigma_2 = \sigma_1 \cdot \sigma_2 + \sigma_1 \wedge \sigma_2 = \sigma_1 \wedge \sigma_2 = - \sigma_2 \wedge \sigma_1 = -\sigma_2\sigma_1$)

One of the result of the above derivation is that the bivector $\sigma_1 \wedge \sigma_2$ has the geometric effect of rotating the vectors σ_1 and σ_2 by 90^0 in its own plane: this property together with that shown in the relation (30), indicates that the bivector $\sigma_1 \wedge \sigma_2$ plays the role of the unit imaginary.

So the bivector $\sigma_1 \wedge \sigma_2$ is the unit of directed area but is also the generator of rotations in the plane (see Fig. 1 and 2).

Note that we will indicate the bivector $\sigma_1 \wedge \sigma_2 = \sigma_1\sigma_2$ simply with 'i'.

We can now consider these things in 3-space, adding a third orthonormal vector σ_3 to our basis and also in four dimensional space-time.

In four dimensional space-time, from four orthonormal vectors as a basis, we can construct 16 geometrical elements, that is one scalar, four vectors, six bivectors, four trivectors and one pseudoscalar.

We can express these sixteen geometrical elements as

1	γ_μ	$\{\sigma_k, i\sigma_k\}$	$i\gamma_\mu$
1-scalar	4-vectors	6-bivectors	4-pseudovectors
i			
1-pseudoscalar			

(31)

where

$$\sigma_k \equiv \gamma_k \gamma_0 \tag{32}$$

and the unit pseudoscalar of spacetime is

$$i \equiv \gamma_0 \gamma_1 \gamma_2 \gamma_3 = \sigma_1 \sigma_2 \sigma_3 \tag{33}$$

On this ground one can proceeds to a reformulation of Dirac's theory in terms of space-time geometric calculus without any complex number. We can write Dirac's equation as [10]

$$\hbar \nabla \psi \gamma_2 \gamma_1 - (e/c)\, A\psi = mc\psi\gamma_0 \tag{34}$$

where

$$\nabla \equiv \gamma^\mu \partial_\mu , \qquad \partial_\mu \equiv \partial/\partial x^\mu , \qquad A = A_\mu \gamma^\mu = A^\mu \gamma_\mu \tag{35}$$

and ψ is connected with the Dirac column spinor Ψ by $\Psi = \psi u$, u being the unit column spinor.

As we know that in Dirac theory we have a totally antisymmetric spin density, we are in position to introduce the tri-vector

$$Q = Q^{\alpha\beta\gamma} \gamma_\alpha \wedge \gamma_\beta \wedge \gamma_\gamma \tag{36}$$

as element of space-time algebra, where $\{\gamma_\alpha\}$ are the base vectors for which we have

$$\gamma_\alpha \gamma_\beta = g_{\alpha\beta} + \gamma_\alpha \wedge \gamma_\beta . \tag{37}$$

Moreover, given the curvature bivector

$$\Omega^{\alpha\beta} = (1/2) R^{\alpha\beta\mu\nu} \gamma_\mu \wedge \gamma_\nu \tag{38}$$

one can form the tri-vectors

$$R^\alpha = \Omega^{\alpha\beta} \wedge \gamma_\beta \tag{39}$$

Now we have seen that torsion and curvature are to be considered as conjugate variables (see eq.(19)):

$$\left[Q \, , \, R \right] = i(\hbar G/c^3)^{-3/2} = i \, L_{Pl}^{-3} \qquad (19) \;)$$

and now we are in position to write eq. (19) in a real form.

First of all remember that in order to take into account both mass and spin it seems at first sight that we have to do with two different spaces: a real space-time where we describe the curvature, due to the mass, with tensors and complex space-time where we describe torsion due to the spin, with spinors. But this is not completely satisfactory: one would like to describe these two fundamental physical properties, mass and spin, in a unique manifold, the real space-time and this can be done through Hestenes algebra.

In other words we can describe at the same time bosons and fermions: curvature and torsion must be given in the same real space-time. We know that when we introduce Dirac equation in Riemann-Cartan space-time (and this is necessary if we like to describe bosons and fermions), we find that contorsion tensor is completely antisymmetric. In fact if we consider the term, in the Lagrangian for the Dirac equation in U_4, that contains interaction between spinor and torsion, we find $K_{abc}\overline{\psi}\,\gamma^{[a}\gamma^{b}\gamma^{c]}\psi$ (see [11]) (where K_{abc} is the contorsion tensor defined through torsion tensor as $= -\ Q_{abc} - Q_{cab} + Q_{bca}$) and spin density tensor is given by

$$S^{abc} \doteq (1/\sqrt{-g})\left[\partial \sqrt{-g}\,\mathcal{L}_m/\partial K_{bca}\right]$$

$$= -\ (i/4)\overline{\psi}\,\gamma^{[a}\gamma^{b}\gamma^{c]}\psi = S^{[abc]} \qquad (40)$$

that is the spin density tensor is totally antisymmetric. At this point I like to anticipate that Xin Yu [16] has found that this result can be achieved directly from equivalence principle (see ref.[16]).

Moreover we know that the transformation law for spinor field is written as

$$\psi'(x') = U(\Lambda)\psi(x) \qquad (41)$$

where $U(\Lambda)$ is the usual 4 **x** 4 constant matrix representing the Lorentz transformation (the spinor indices are not written explicity). Λ is the Lorentz matrix involved in the vector Lorentz transformation in the flat tangent space $x'^i = \Lambda^i{}_k x^k$.

Dirac equation $(i\gamma^k\partial_k\psi - m\psi = 0)$ is transformed as

$$i\gamma^i\,(\partial\psi'(x')/\partial x'^i) - m\,\psi'(x') =$$
$$= i\gamma^i\Lambda_i{}^k\partial_k U\,\psi - m\,U\,\psi = 0 \qquad (42)$$

It is well-known that multiplying from the left by U^{-1} and imposing on the Dirac equation to be invariant in form under a Lorentz transformation, we obtain the condition for the matrix U

$$U^{-1}\gamma^{i}U = \gamma^{k}\Lambda^{i}_{\ k} \qquad (43)$$

Considering an infinitesimal transformation

$$\Lambda_{ik} \simeq \eta_{ik} + \omega_{ik} \qquad (44)$$

(where $\omega_{ik} = \omega_{[ik]}$), we have

$$U = 1 + (1/2)\omega_{ik}S^{ik} \qquad (45)$$

The ω_{ik} are six constant infinitesimal parameters and S^{ik} are the generators of the infinitesimal Lorentz transformation which in order that (43) be fulfilled must satisfy

$$S^{ik} = \gamma^{[i}\gamma^{k]} = (1/2)(\gamma^{i}\gamma^{k} - \gamma^{k}\gamma^{i}) \qquad (46)$$

Here we can observe that considering Dirac equation in Riemann-Cartan space-time we have the relation (see [10])

$$g_{\mu\nu} = (1/2)(\gamma_{\mu}\gamma_{\nu} + \gamma_{\nu}\gamma_{\mu}) = \gamma_{\mu}\cdot\gamma_{\nu} \qquad (47)$$

where γ are the Dirac matrices. But we have also

$$\sigma_{\mu\nu} = (1/2)(\gamma_{\mu}\gamma_{\nu} - \gamma_{\nu}\gamma_{\mu}) \qquad (48)$$

We see that eqs. (46) and (48) are formally identical and it seems that we can pass from one to another with the help of Hestenes algebra. In fact from geometric product we have

$$\gamma_{\mu}\gamma_{\nu} = \gamma_{\mu}\cdot\gamma_{\nu} + \gamma_{\mu}\wedge\gamma_{\nu}\ .$$

So we have

$$\gamma^{\mu}\gamma^{\nu} = g^{\mu\nu} + \sigma^{\mu\nu} \qquad (49)$$

Eq.(49) seems to include automatically supersymmetry because we have commutator (fermionic field) and anticommutator (bosonic field): $g_{\mu\nu}$ is connected with bosons and $\sigma_{\mu\nu}$ is connected with fermions and they are given simultaneously. Notice that eq.(46) follows directly from Hestenes geometric product

$$\gamma_\mu \gamma_\nu = \gamma_\mu \cdot \gamma_\nu + \gamma_\mu \wedge \gamma_\nu \qquad (50)$$

where

$$\gamma_\mu \cdot \gamma_\nu = g_{\mu\nu} \qquad (51)$$

and

$$\gamma_\mu \wedge \gamma_\nu = \sigma_{\mu\nu} \qquad (52)$$

Moreover the infinitesimal variation of a spinor under the Lorentz transformation is

$$\delta\psi = \psi' - \psi = (1/2)\omega_{ik} S^{ik}\psi = (1/2)\omega_{ik}\gamma^{[i}\gamma^{k]}\psi \qquad (53)$$

The covariant differential for a spinor field is

$$D\psi = dx^k \nabla_k \psi = \psi'(x_2) - \psi(x_1) \qquad (54)$$

which can be written

$$D\psi = \psi(x_2) - \psi(x_1) - [\psi(x_2) - \psi'(x_2)] \qquad (55)$$

where we have separated the term due to translation, $\psi(x_2)-\psi(x_1)$ from the part relative to a local rotation of the tetrad $\psi'(x_2) - \psi(x_2)$. It seems that from eq.(55) we have simultaneously curvature and torsion or, in terms of Hestenes algebra, dilation and rotation (see Appendix).

We have seen that torsion and curvature are to be considered as conjugate variables (see eq.(19)) and now we are in position to write equation (19) in a real form .

In fact we have the trivectors Q and R^α. Consider now the antisymmetric part of the geometric product

$$[Q,R^\alpha] = (1/2)(QR^\alpha - R^\alpha Q) \qquad (56)$$

This type of product between two tri-vectors gives a bivector (for example it is easy to verify

$$[\gamma_0\wedge\gamma_1\wedge\gamma_2 \ , \ \gamma_0\wedge\gamma_1\wedge\gamma_3] \ = \ \gamma_2\wedge\gamma_3 \qquad (57)$$

remembering that $\gamma_0\gamma_1\gamma_2 = i\gamma_3$ where 'i' indicates the pseudoscalar unit. In the language of geometric algebra, the imaginary unit of complex numbers is substituted by a bivector; then we can have commutation relations of canonic type from eqs.(36, 39, 56):

$$[Q \ , \ R^{\alpha}] \ \equiv \ (1/2)(QR^{\alpha} - R^{\alpha}Q) =$$

$$= (1/2)\left[Q^{\mu\nu\beta}\gamma_{\mu}\wedge\gamma_{\nu}\wedge\gamma_{\beta} \ , \ R^{\alpha\tau\nu\beta}\gamma_{\nu}\wedge\gamma_{\beta}\wedge\gamma_{\tau}\right] =$$

$$= (1/4)(Q^{\mu\nu\beta}R^{\alpha\tau}{}_{\nu\beta} - R^{\alpha\mu}{}_{\nu\beta}Q^{\tau\nu\beta})\gamma_{\mu}\wedge\gamma_{\tau} = \gamma_2\wedge\gamma_1 \ L_{Pl}^{-3} \qquad (58)$$

for every α, where, being the first member a bivector, also the second member is a bivector, coherently with the fact that the imaginary unit in Dirac equation is substituted by the bivector $\gamma_2\wedge\gamma_1$

As we must be in a spin plane, we have to consider the case of $\mu,\tau = 1,2$.

In order that the commutation relation may have this form, the following six conditions should be satisfied, considering that left handed member of the commutation relation (58) is the summation of six bivector parts i.e.

$$\gamma_1\wedge\gamma_2 \ , \ \gamma_2\wedge\gamma_3 \ , \ \gamma_3\wedge\gamma_1 \ , \ \gamma_0\wedge\gamma_1 \ , \ \gamma_0\wedge\gamma_2 \ , \gamma_0\wedge\gamma_3:$$

$$(1/2) \ Q^{1\nu\beta}R^{\alpha 2}{}_{\nu\beta}\gamma_{[1}\gamma_{2]} = \gamma_1\wedge\gamma_2 L_{Pl}^{-3} \qquad (59)$$

$$Q^{\mu\nu\beta}R^{\alpha\tau}{}_{\nu\beta}\gamma_{[\mu}\gamma_{\tau]} = 0 \qquad (60)$$

Eq.(59 corresponding to the bivector $\gamma_1 \wedge \gamma_2$ gives one relation and eq.(60) corresponding to the bivectors

$$\gamma_2 \wedge \gamma_3 \,,\; \gamma_3 \wedge \gamma_1 \,,\; \gamma_0 \wedge \gamma_1 \,,\; \gamma_0 \wedge \gamma_2 \,, \gamma_0 \wedge \gamma_3 \qquad (61)$$

give five relations.

In conclusion the six conditions represent a "choice of gauge" with respect to the local Lorentz rotations, according to the fact that the choice of the spin plane is arbitrary.

So, we have seen that, in order to have tensors and spinors (bosons and fermions) in the same real space-time, we need the use of geometric algebra. Now some words on the necessity of a quadratic Lagrangian.

3. A quadratic Lagrangian

In order to give a stronger basis to the consideration of torsion and curvature as canonical conjugate variables, we try to see if it is possible to improve the theory introducing a quadratic Hamiltonian function of torsion and curvature. We use as conjugate variables the torsion and the curvature trivectors because the first Bianchi identity guarantees that it is different from zero only in the presence of torsion. Since the commutation relations between Q and R^a are defined in the geometric algebra we need to modify the usual lagrangian field theory.

We start from the results of ref.[12]: the field equation for the multivector ψ can be written in the manifestly invariant way:

$$\nabla \left(\frac{\partial \mathcal{L}}{\partial(\nabla\psi)} \right) = \frac{\partial \mathcal{L}}{\partial \psi} \qquad (62)$$

where the gradient operator $\nabla = \gamma_\mu \partial^\mu$ acts as geometric product.

Now we are able to introduce the following Lagrangian in order to have the usual expression for conjugate variables in a quantum theory:

$$\mathcal{L} = (\hbar c/2)R^a R_a - (c^4/2G)Q^a Q_a$$
$$\mathcal{L} = (\hbar c/2)\left(R^a R_a - Q^a Q_a / L_{Pl}^2\right) \tag{63}$$

The presence of ∇ in eq. (62) or, in other words, the peculiarity of the intrinsic geometric calculus, suggests that we can define the conjugate momentum field as

$$\Pi = \frac{\partial \mathcal{L}}{\partial(\nabla\psi)} \tag{64}$$

and the Hamiltonian

$$H = (\nabla\psi)\Pi - \mathcal{L} \tag{65}$$

This is different from the usual $H = \Pi\dot{\psi} - \mathcal{L}$ formula, but, following eq.(62) for the Lagrangian, we propose also a modified Hamiltonian that is manifestly invariant like eq.(62) so we cannot work with $\partial\psi/\partial x^0$ because it depends on the choice of the time-coordianate x^0. So $\Pi = \dfrac{\partial \mathcal{L}}{\partial(\nabla\psi)}$ is the natural choice; in fact eq.(62) generalizes the Lagrangian equation

$$(d/dt)(\partial\mathcal{L}/\partial\dot{q}) - (\partial\mathcal{L}/\partial q) = 0$$

simply substituting d/dt with ∇.

Notice that eq.(62) allows for vectors, tensors and spinors variables to be handled in a single equation, a considerable unification (see [12]).

So we have the Hamilton equations

$$\begin{cases} \nabla\psi = \partial H / \partial \Pi \\ \nabla\Pi = - \partial H / \partial \psi \end{cases} \tag{66}$$

For example we have the Maxwell equations in vacuum taking $\mathcal{L} = F^2/8\pi$ where the vector potential $A = A^\mu \gamma_\mu$ and the bivector $F = \nabla \wedge A$ are conjugate variables (remembering that the geometric product gives $F = \nabla A = \nabla \cdot A + \nabla \wedge A$ and Lorentz condition $\nabla \cdot A = 0$).

Moreover since we use the curved U_4 manifold in agreement with the minimal coupling principle, we substitute the covariant operator D for ∇ defining

$$D \equiv \nabla + \gamma_a [\omega^a, \quad] \tag{67}$$

where the $\{\gamma_a\}$ is an orthonormal frame of tetrads, (for which

$$\gamma_a \gamma_b = \eta_{ab} + \gamma_a \wedge \gamma_b), \tag{68}$$

(related with the coordinate system $\gamma_\mu = e_\mu^a \gamma_a$)

and

$$\omega^a = (1/2)\omega^{abc}\gamma_b \wedge \gamma_c \tag{69}$$

are the connection bivectors.

We remember that in the tetrad basis th nection is

$$\omega^{abc} = \omega^{a[bc]} = C^{cba} - C^{bac} - C^{acb} - K^{abc} \tag{70}$$

(where $C_{ab}{}^c = e^\mu{}_a e^\nu{}_b \partial_{[\mu} e_{\nu]}{}^c$ give the Riemannian part and K^{abc} is the contorsion tensor $= - Q^{abc}$ because, in this case, we are considering the totally antisymmetric torsion).

If in the eq.(62) we put D instead of ∇, we have

$$D\left(\frac{\partial \mathcal{L}}{\partial (D\psi)}\right) = \frac{\partial \mathcal{L}}{\partial \psi} \tag{71}.$$

In agreement with the Bianchi identities, the momenta conjugate to the torsion bivectors are the trivectors:

$$\Pi^a = \partial\mathcal{L}/\partial(DQ^a) = \hbar c R^a \qquad (72)$$

(in natural unit $\hbar=c=1$ we can write π^a or R^a without distinction). The Hamiltonian results:

$$H = DQ^a\Pi_a - \mathcal{L} = (\hbar c/2)\left(R^aR_a + Q^aQ_a/L^2_{Pl}\right) \qquad (73)$$

Then we have the Hamilton equations:

$$\begin{cases} DQ^a = \partial H/\partial R^a = R^a \\ DR^a = -\ \partial H/\partial Q^a = -\ Q^a/L^2_{Pl} \end{cases} \qquad (74)$$

that we can write

$$DDQ^a = -\ Q^a/L^2_{PL} \qquad (75)$$

or equivalently for R^a

$$DDR^a = -\ R^a/L^2_{Pl} \qquad (76)$$

The quadratic Lagrangian $\mathcal{L}$ is incomplete for a gravitation theory, since it does not give the Einstein equations and we know that a quantum theory of gravity must be reduced to the classical theory when one considers gravitation far away from the Planck scale.

The problem with such quadratic Lagrangian is that one has to supplement them by the Einstein-Hilbert Lagrangian in order to obtain, in the Newtonian approximation, the Laplace equation. So this Einstein-Hilbert term R seems to destroy the obtained canonical structure.

However the Lagrangian (63) completed by the

Einstein-Hilbert term (the usual linear Lagrangian - $\overset{0}{R}/2\chi$):

$$\mathcal{L}' = \mathcal{L} - \overset{0}{R}/2\chi$$

i.e.

$$\mathcal{L}' = (\hbar c/2)\left(R^aR_a - Q^aQ_a/L^2_{Pl}\right) + \overset{0}{R}/2\chi \qquad (77)$$

where $\overset{0}{R}$ is the scalar curvature related to Riemannian part of the connection and $\chi = 8\pi G/c^4$ (i.e. $L^2_{Pl} = \hbar c\chi/8\pi$), reminds of an early approach to quantum electrodynamics considered by Heisenberg, Euler and Kockel [13].

To describe quantum corrections to classical electrodynamics these authors had added the Larmor Lagrangian $\Lambda_0 = F_{ik}F^{ik}$ of the classical theory by quadratic invariants like $(F_{ik}F^{ik})^2$.

It was shown that, to some extent, these corrections represent a phenomeno-logical approximation to the exact quantum theory .At first sight, the Lagrangian (77) is a gravitational analog of the electrodynamic Heisenberg-Euler-Kockel *Ansatz*. However, in contrast to the electrodynamics, the quadratic gravitational terms are not formed from only those field quantities which occur in the R-item of the Lagrangian; besides the metric, there the connection comes into the game. This is in accordance with the above-mentioned fact that there are no genuine quantum effects in a purely metric theory. Thus, quadratic terms formed only from the metric cannot be related to quantum effects. Genuine quantum terms must contain additional fields like torsion.

The theory given by the Lagrangian (77) couples two types of gravity, metric and torsional gravity, where the first one dominates at large and the second at small distances. Its canonical structure should turn out to be of such a type that it recovers canonical quantum GRT (which is essentially classical in the above-discussed sense) for weak fields, and canonical quantum gravity (given by the Lagrangian (63) for strong fields. In terms of a cosmological scenario, in early universe one had a strong gravity era described by Riemann-Cartan geometry that later goes over into a weak gravity era described by Riemann geometry. Strong gravity can also dominate in superdense matter, e.g.in neutron stars. To found an exact theory of quantum gravity on this basis one had to construct a Hamiltonian formalism starting from the Lagrangian (63).

4. Some comments on supersymmetry

As a last comment, we like to show that with this quantum theory we have some indications of the supersymmetry.

This is done mainly because we are using the geometric algebra. In fact, for instance, we have

$$\text{a)} \qquad \gamma^{\mu}\gamma^{\nu} \quad \Rightarrow \quad g^{\mu\nu} + \sigma^{\mu\nu} \tag{49}$$

where $g^{\mu\nu}$ refers to tensors while $\sigma^{\mu\nu}$ refers to spinors (see eq.49). Eq.(49) seems to include automatically supersymmetry because we have commutator (fermionic field) and anticommutator (bosonic field): $g_{\mu\nu}$ is connected with bosons and $\sigma_{\mu\nu}$ is connected with fermions and they are given simultaneously. Notice that eq.(46) follows directly from Hestenes geometric product (see eqs.50, 51, 52)

$$\gamma_{\mu}\gamma_{\nu} = \gamma_{\mu}\cdot\gamma_{\nu} + \gamma_{\mu} \wedge \gamma_{\nu} \tag{50}$$

where

$$\gamma_{\mu}\cdot\gamma_{\nu} = g_{\mu\nu} \tag{51}$$

and

$$\gamma_{\mu} \wedge \gamma_{\nu} = \sigma_{\mu\nu} \tag{52}$$

b) $$z = \sigma_1 x \tag{78}$$

i.e. we find that we can pass from spinors to tensors: in fact, for instance, in two dimensional vector space (see ref.[10]) from the equation

$$x\sigma_1\sigma_2 = x\cdot(\sigma_1\wedge\sigma_2)$$

that is

$$x\sigma_1\sigma_2 = x\cdot\sigma_1\sigma_2 - x\cdot\sigma_2\sigma_1$$

(remember that $a\cdot(b\wedge c) = (a\cdot b)c - (a\cdot c)b$), From the product of vector σ_1 and x we obtain a spinor z in the form (see Fig.1)

$$z = \sigma_1 x \tag{79}$$

with

$$(\sigma_1 x = \sigma_1(x_1\sigma_1 + x_2\sigma_2) = x_1 + ix_2)$$

where the quantities like

$$x_1 + ix_2$$

are called complex numbers.

However it must be emphasized that besides the property $i^2 = -1$, our 'i', as already said, is a bivector, so the separation into real and imaginary parts of a complex number 'z' is equivalent to separating a spinor into scalar and pseudoscalar (bivector) parts. Conversely each spinor 'z' determines a unique vector σ_1 according to the equation (see Fig.2).

$$x = \sigma_1 z \tag{80}$$

that is we can pass from vector to spinor and viceversa.

Moreover we know that the transformation law for spinor field is written as (see eq:41)

$$\psi'(x') = U(\Lambda)\psi(x) \qquad (41)$$

where $U(\Lambda)$ is the usual 4 **x** 4 constant matrix representing the Lorentz transformation (the spinor indices are not written explicity). Λ is the Lorentz matrix involved in the vector Lorentz transformation in the flat tangent space $x'^i = \Lambda^i{}_k x^k$.

Dirac equation ($i\gamma^k \partial_k \psi - m\psi = 0$) is transformed as (see eq.42)

$$i\gamma^i\,(\partial\psi'(x')/\partial x'^i) - m\,\psi'(x') = \\ = i\gamma^i \Lambda_i{}^k \partial_k U\,\psi - m\,U\,\psi = 0 \qquad (42)$$

It is well-known that multiplying from the left by U^{-1} and imposing on the Dirac equation to be invariant in form under a Lorentz transformation, we obtain the condition for the matrix U (see eq 3)

$$U^{-1}\gamma^i U = \gamma^k\,\Lambda^i{}_k \qquad (43)$$

Considering an infinitesimal transformation (see eqs.44,45):

$$\Lambda_{ik} \simeq \eta_{ik} + \omega_{ik} \qquad (44)$$

(where $\omega_{ik} = \omega_{[ik]}$), we have

$$U = 1 + (1/2)\omega_{ik} S^{ik} \qquad (45)$$

The ω_{ik} are six constant infinitesimal parameters and S^{ik} are the generators of the infinitesimal Lorentz transformation which in order that (43) be fulfilled must satisfy (see eq.46)

$$S^{ik} = \gamma^{[i}\gamma^{k]} = (1/2)(\gamma^i\gamma^k - \gamma^k\gamma^i) \qquad (46)$$

Moreover the infinitesimal variation of a spinor under the Lorentz transformation is (see eqs.53,54,55)

$$\delta\psi = \psi' - \psi = (1/2)\omega_{ik}S^{ik}\psi = (1/2)\omega_{ik}\gamma^{[i}\gamma^{k]}\psi \quad (53)$$

The covariant differential for a spinor field is

$$D\psi = dx^k\nabla_k\psi = \psi'(x_2) - \psi(x_1) \quad (54)$$

which can be written

$$D\psi = \psi(x_2) - \psi(x_1) - [\psi(x_2) - \psi'(x_2)] \quad (55)$$

where we have separated the term due to translation, $\psi(x_2)-\psi(x_1)$ from the part relative to a local rotation of the tetrad $\psi'(x_2) - \psi(x_2)$.It seems that from eq.(63) we have simultaneously curvature and torsion or, in terms of Hestenes algebra, dilation and rotation.

c) $$Q^\alpha R_\alpha = [Q^\alpha, R_\alpha] + \{Q^\alpha, R_\alpha\} \quad (81)$$

commutat nd anticommutator

Now we must modify the commutation relations found in the section 3, because in the previous works we have considered torsion and curvature trivectors, while here we have to do with a bivector Q^α (torsion) and a trivector R^α (curvature) which are our canonical conjugate variables.

In other words we try to improve the theory and we find important developments about the possibility to have commutation relations between torsion and curvature with some consequences regarding considerations about supersymmetry.

As we have seen, the commutators in eq.(58) allow a simple interpretation in agreement with the geometric content of Dirac equation, but they are not entirely satisfactory because in the Lagrangian (77) the conjugate variables are the torsion bivector

Q^α and the curvature trivector R^α, while in eq.(58) the commutation is made between trivectors.

In fact working with the geometric algebra it is possible to have commutation relations between multivectors of different grade and not only between geometric objects of the same type. Yet eq.(58) has meaning and it is not to be rejected, because we have started with a totally antisymmetric torsion and then the torsion trivector Q is the complete and single object of geometric algebra which contains the torsion field.

Now the geometric product between a grade-r multivector A_r and a grade-s multivector B_s can be decomposed as (see ref.[14]

$$A_r B_s = (A_r B_s)_{r+s} + (A_r B_s)_{r+s-2} \quad .....+ \quad (AB)_{|r-s|} \qquad (82)$$

If r = 2 and s = 3, being in a four dimensional space-time, we are left with two terms: $(AB)_1$ and $(AB)_3$. In other words the geometric product BT, where B is a generic bivector and T a generic trivector, can be easily calculated remembering that every trivector T is dual of some vector A (see ref.[15] i.e.

$$T = iA \qquad (83)$$

where $i = \gamma_0\gamma_1\gamma_2\gamma_3$ is the unit pseudoscalar and then A is the vector dual of T (remember that the dual application, the multiplication by i transform an r-vector in (4-r) vector). Moreover

$$BA = B\cdot A + B \wedge A \qquad (84)$$

where the inner product

$$B\cdot A = (1/2)/(BA - AB) \equiv [B,A] \qquad (85)$$

has grade 1 and the outer product

$$B \wedge A = (1/2)/(BA + AB) \equiv \{B,A\} \qquad (86)$$

has grade 3. Then, using also iB = Bi, one finds

$$BT = BiA = iBA = i[B,A] + i\{B,A\} = [B,T] + \{B,T\} \tag{87}$$

where the commutator between B and T has grade 3 (trivector) and the anticommutator has grade 1 (vector).

Finally we can write

$$Q^{\alpha}R_{\alpha} = [Q^{\alpha},R_{\alpha}] + \{Q^{\alpha},R_{\alpha}\} \tag{88}$$

where

$$[Q^{\alpha},R_{\alpha}] = (1/2)Q^{\alpha\mu\nu}R_{\alpha\mu}{}^{\sigma\rho}\ \gamma_{\nu}\wedge\gamma_{\sigma}\wedge\gamma_{\rho} \tag{89}$$

and

$$\{Q^{\alpha},R_{\alpha}\} = (1/2)Q^{\alpha\mu\nu}R_{\alpha\mu\nu}{}^{\rho}\ \gamma_{\rho} \tag{90}$$

As in eq.(70), in agreement with uncertainty relation $\Delta Q\Delta R \geq L_{Pl}^{-3}$, we can put

$$[Q^{\alpha},R_{\alpha}] = L_{Pl}^{-3}\ iu \tag{91}$$

$$\{Q^{\alpha},R_{\alpha}\} = L_{Pl}^{-3}\ v \tag{92}$$

where u and v are unit vectors.

Therefore, given the conjugate variables Q and R , we have both commutator and anticommutator; we believe that this fact can be related to supersymmetry in the sense that one can treat simultaneously fermionic fields and bosonic fields if one consider, as in the second quantization procedure, the development of the fields in terms of creation and annihilation operators which present analogies with the relations (37,38) between torsion and curvature. But this is an argument for future works.

Anyway the point that we liked to put in evidence is that starting from a quantum theory of gravity based on three fundamental concepts namely the introduction of spin in General Relatvity, the use of geometric algebra and the use of a quadratic Lagrangian in torsion and curvature, we find some indication for supersymmetries.

5. Appendix

At last, to conclude this short survey on the possibility tó have a Quantum Gravity based on the three concepts aforesaid i.e. torsion, geometric algebra and quadratic Lagrangian, we like to put in evidence some question on the geometrical meaning of torsion that in general do not appear explicity.

Althoug the applications of the Cartan theory in Quantum Gravity are not so recent, we believe that the geometrical meaning of torsion is not enough clear between physicists. In particular the fact that the torsion in a manifold can be treated as defect in topology, quantizing space-time itself, remains unfortunately obscure. Therefore we need to state this point and if possible develop it.

We have often argued that the most important property of torsion is the closure failure: a closed contour in the Riemann--Cartan U_4 manifold becomes a non closed contour in the flat tangent space-time V_4, i.e. the parallel displacement of a vector along a closed path produces a translation of the vector itself.

This translation is a space-time defect (in analogy with the geometrical description of dislocations in crystals) and is expressed by the integral

$$l^\alpha = \oint Q_{\mu\nu}{}^\alpha \, dS^{\mu\nu} \neq 0 \qquad \text{(A1)}$$

where

$$Q_{\mu\nu}{}^\alpha = (1/2)(\Gamma_{\mu\nu}{}^\alpha - \Gamma_{\mu\nu}{}^\alpha) \equiv \Gamma_{[\mu\nu]}$$

is the torsion tensor, $\Gamma_{\mu\nu}{}^\alpha$ is the non-symmetric affine connection and $dS^{\mu\nu} = dx^\mu \wedge dx^\nu$ is the infinitesimal area element enclosed by the path.

This fact is evidently true, but we would expect that defects belong to the space-time manifold and not to the tangent one, if we want to quantize the space-time itself. Now we will see how in some sense also this fact can be true, when we consider an infinitesimal closed contour.

Given a vector A^β, we remember that its covariant derivative is

$$\nabla_\alpha A^\beta = \partial_\alpha A^\beta + \Gamma_{\alpha\gamma}{}^\beta A^\gamma \qquad \text{(A2)}$$

and the covariant differential evaluated at the point $x^\alpha + dx^\alpha$ is

$$DA^\beta \equiv dA^\beta - \delta A^\beta = \nabla_\alpha A^\beta \, dx^\alpha \qquad \text{(A3)}$$

where $\delta A^\beta = - \Gamma_{\alpha\gamma}{}^\beta A^\gamma dx^\alpha$ is the variation of A^β for a parallel infinitesimal displacement along dx^α.

If we have a closed path, the total variation is then

$$\Delta A^\beta = \oint \delta A^\beta \qquad \text{(A4)},$$

from eq.(A3) because $\oint dA^\beta = 0$ we can write

$$\Delta A^\beta = - \oint DA^\beta = - \oint \nabla_\nu A^\beta dx^\nu . \qquad \text{(A5)}$$

By using the Stokes theorem we have

$$\Delta A^\beta = - \oint \partial_\mu \nabla_\nu A^\beta dS^{\mu\nu} = - \oint \partial_{[\mu} \nabla_{\nu]} A^\beta dS^{\mu\nu} \qquad \text{(A6)}$$

For a sufficiently small area we can replace the integrand function by its value at some point inside the surface and the integration gives simply the area $\Delta S^{\mu\nu}$. Moreover we can replace the partial by the covariant derivatives because they differ by a second order infinitesimal quantity; finally we have

$$\Delta A^\beta = - (1/2)(\nabla_\mu \nabla_\nu - \nabla_\nu \nabla_\mu) A^\beta \Delta S^{\mu\nu} = - \nabla_{[\mu} \nabla_{\nu]} A^\beta \Delta S^{\mu\nu} \qquad \text{(A7)}$$

Now, remembering that $\nabla_\nu A^\beta$ is a 2-rank tensor and remembering the definition of the Riemann and torsion tensors, by calculus the commutators of the covariant derivatives result

$$\nabla_{[\mu}\nabla_{\nu]}A^{\beta} = (1/2)R_{\mu\nu\alpha}{}^{\beta}A^{\alpha} - Q_{\mu\nu}{}^{\alpha}\nabla_{\alpha}A^{\beta} \tag{A8}$$

that can be inserted in eq.(A7) giving

$$\Delta A^{\beta} = -(1/2)R_{\mu\nu\alpha}{}^{\beta}A^{\alpha}\Delta S^{\mu\nu} + (\nabla_{\alpha}A^{\beta})Q_{\mu\nu}{}^{\alpha}\nabla S^{\mu\nu} \tag{A9}$$

The first term represents an infinitesimal rotation: note the antisymmetry respect α and β as the generators of rotations and then $R_{\mu\nu\alpha}{}^{\beta}\Delta S^{\mu\nu}$ yields the so-called defect angle, that corresponds in crystals to disclinations or angular defects.

Instead if we consider the variation due to torsion only (as if curvature were null) and compare it to equation (A3), we have $\Delta A^{\beta} \cong DA^{\beta}$ evaluated at the point $x^{\alpha} + \Delta l^{\alpha}$ of the space-time, if $\Delta l^{\alpha} = Q_{\mu\nu}{}^{\alpha}\ \Delta S^{\mu\nu} \cong dx^{\alpha}$. Therefore it is as if the vector be displaced along $\Delta l^{\alpha} = Q_{\mu\nu}{}^{\alpha}\ \Delta S^{\mu\nu}$; we can say that, as curvature is a rotation per unit area, torsion is a translation per unit area.

References

[1] Borzeszkowski, H,-H. v. and Treder, H.J.: Ann.Phys. (Leipzig) 46, 315 (1989); see also "On Quantum Gravity", PRE-EL 88-05 Potsdam, 1988

[2] Borzeszkowski, H,-H. v and Treder, H.J. "The Meaning of Quantum Gravity"; Reidel Publishing Company Dordrecht,(1988)

[3] Rosenfeld, L., "Quanten und Gravitation" in Entstehung, Entwicklung und Perspektiven der Einsteinschen Gravitationstheorie, H.-J.Treder, ed.(Akademie-Verlag Berlin, 1966)

[4] Borzeszkowski, H.-H. v., Datta, B.K., de Sabbata,V., Ronchetti L. and Treder, H.J.: Foundations of Physics, Vol.32 No.11, 1701 (2002)

[5] Bohr, N. und Rosenfeld, L., "Zur Frage der Messbarkeit der electromagnetischen Feldgrössen", Dan.Vid.Selskab Mat-fys 12 (1933) No.8 [English translation: "On the Question of the Measurability of Electromagnetic Field Quantities", in Niels Bohr, Collected works, vol.7, J.Kalckar, ed. (North-Holland), Amsterdam et al.1996].

[6] Cartan, H.E.: C.R. Acad. Sci., 174, 437, 593 (1922); Ann. Ecole Normale 41, 1 (1924)

[7] de Sabbata, V. and Gasperini M.: Lett.Nuovo Cimento 27, 289 (1980)

[8] de Sabbata, V.: Il Nuovo Cimento 107A, 363 (1994)

[9] de Sabbata, V., "Spin and Torsion in early universe" in "Cosmology and Particle Physics" ed. by Venzo de Sabbata, and Ho Tso-Hsiu, Kluwer Academic Publishers, NATO ASI Series Dordrecht et al. Vol.427, 97-128 (1994).

[10] Datta,B.K.,de Sabbata, V.and Ronchetti L.: Il Nuovo Cimento 113B, 711 (1998)

[11] de Sabbata,V. and Gasperini,M. "Introduction to Gravitation" World Sci.Singapore 1985

[12] Lasenby, A., Doran,C.and Gull,S.,Found.Phys.23,1295 (1993)

[13] Euler H, and Kockel, B. Naturwiss. 23, 246 (1935).
Euler H., Ann.Phys. (Leipzig 26, 298 (1936).
Heisenberg, W. and Euler, H., Z.Phys. 98, 740 (1936)

[14] Hestenes D. "Clifford Algebra to Geometric Calculus" D.Reidel Publ. Co. Dordrecht Holland (1982) page 10, eq.1.36

[15] de Sabbata V., Ronchetti L. and Datta B.K. "Non Commuting Geometry and Spin Fluctuations" in "Classical and Quantum Non-Locality" ed.by Bergmann P.G., de Sabbata V. and Goldberg J.N. World Sci.Singapore 2000, Page 107 eqs,(B1)and (B2)

[16] Yu Xin (Alfred Yu), "General Relativity on Spinor-Tensor Manifold" in " Quantum Gravity" ed. by Bergmann P.G., de Sabbata V. and Treder H.-J. World Sci.Singapore 1996, Page 382-411.

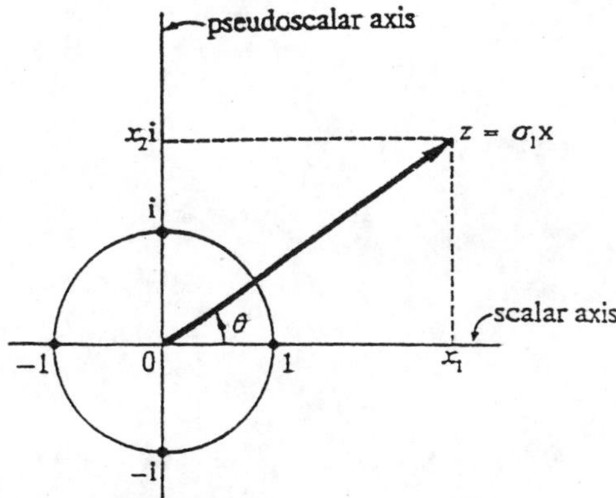

Fig. 1. – Diagram for the spinor i-plane. Each point in the spinor plane represents a rotation-dilation. Points on the unit circle represent pure rotations, while points on the positive scalar axis represent pure dilations.

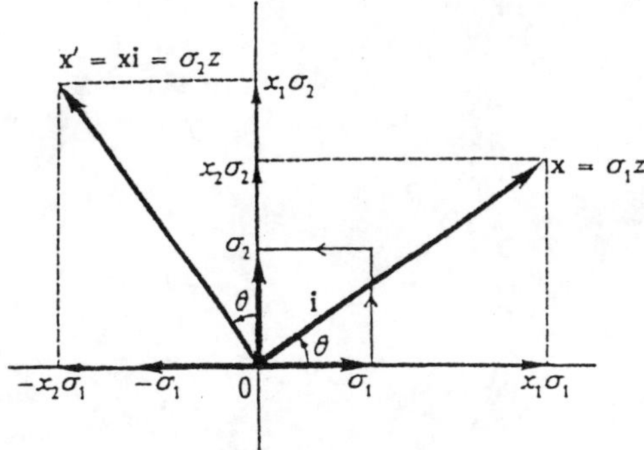

Fig. 2 – Diagram of the i-plane of vectors (real plane).

Brane-inspired models in gravitation and cosmology

Dmitri Gal'tsov*,
Department of Theoretical Physics,
Moscow State University, 119899, Moscow, Russia

October 26, 2003

Abstract

We discuss some recent development in gravitation and cosmology related to the concept of branes. These lectures include: a brief review of braneworld scenarios with an emphasis on the black hole problem, soliton and black hole solutions of the gravitating non-Abelian Born-Infeld (NBI) model, NBI homogeneous and isotropic cosmology, brane NBI cosmology, the issue of Yang-Mills chaos in the context of the NBI dynamics.

1 Introduction

During recent years various new models were proposed in gravitation and cosmology related to the concept of branes – extended objects living in multidimensional space-times. Such objects were first uncovered in eleven-dimensional supergravity, whose bosonic sector contains an antisymmetric tensor field of the third rank (three-form). Natural (and the only possible) sources for this field are an extended two-dimensional object (membrane) currently called M2-brane and its magnetic counterpart – M5-brane. In superstring theory, branes associated with Ramond-Ramond antisymmetric forms reappear as D-branes — boundary surfaces, on which the open strings can end. Such strings generate massless vector fields localized only on the brane, so one is naturally led to the class of models in which usual matter described by the standard model lives on the brane, while gravity lives in the full space (bulk).

The effective action governing the brane is the Dirac-Born-Infeld action, which describes dynamics of vector fields coupled to scalars corresponding to brane deformations.Recent development in the superstring theory [1, 2] suggests that the low-energy dynamics of a Dp-brane moving in a flat D-dimensional spacetime $z^M = z^M(x^\mu)$, $M = 0, ..., D-1$, $\mu = 0, \ldots p$ is governed by the Dirac-Born-Infeld (DBI) action

$$S_p = \int \left(1 - \sqrt{-\det(g_{\mu\nu} + F_{\mu\nu})}\right) d^{p+1}x, \tag{1}$$

*Email: galtsov@grg.phys.msu.su

V. de Sabbata et al. (eds.),
The Gravitational Constant: Generalized Gravitational Theories and Experiments, 113–137.
© 2004 *Kluwer Academic Publishers. Printed in the Netherlands.*

where

$$g_{\mu\nu} = \partial_\mu z^M \partial_\nu z^N \eta_{MN}, \tag{2}$$

is an induced metric on the brane and $F_{\mu\nu}$ is a $U(1)$ gauge field strength. Using the gauge freedom under diffeomorphisms of the world-volume, one can choose coordinates $z^M = (x^\mu, X^m)$, where X^m are transverse to the brane, and rewrite the action as

$$S_p = \int \left(1 - \sqrt{-\det(\eta_{\mu\nu} + \partial_\mu X^m \partial_\nu X^m + F_{\mu\nu})}\right) d^{p+1}x. \tag{3}$$

In the case of several branes one is led to the set of vector fields forming the non-Abelian Born-Infeld action (NBI)

Here we discuss some selected topics associated with brane-inspired models, in particular, the issue of a black hole in the braneworld setup, solitons, black holes and cosmological solutions in NBI theory, the issue of chaos in Yang-Mills homogeneous system.

2 Large extra dimensions and braneworlds

The idea of large extra dimensions suggested some time ago on a phenomenological level [3, 4], received a new impact due to the concept of branes in the superstring/M theory. In the quantum superstring theory, branes manifest themselves either as solitons, or as D-branes: boundary hypersurfaces which the open strings might attach to. In the latter case the standard model fields live only on the brane, while gravity (with its supergravity companions) exists in the whole (bulk) space. This is the key feature of the modern braneworld scenarios proposed in [5, 6, 7] (for more extensive review see [8, 9, 10, 11]).

As was argued by Arkani-Hamed, Dimopoulos and Dvali [5] (and previously discussed by Antoniadis [12]), the existing experimental data are consistent with the size of extra dimensions up to the submillimeter range. This suggests the picture of space-time consisting of a 3-brane with d flat extra dimensions of a size l. Then the Planck mass M_{Pl} will be related to the fundamental energy scale M_* on the brane as

$$M_{Pl} = M_*(M_* l)^{d/2} \tag{4}$$

(in the units $\hbar = c = 1$). The standard model is assumed to be localized on the 3-brane, while gravity lives in the full space (bulk). This construction is intended to solve the hierarchy problem of the standard model. The most radical assumption consists in taking the TeV-scale parameter M_*, this is compatible with the existing high energy physics data if $d \geq 2$. However, such a simple model suffers from the cosmological light KK particles overproduction problem, unless higher values of M_* are assumed. The distinctive features of this scenario (commonly called ADD) are the flat character of the extra dimensions and the tensionless nature of the brane, no back reaction of which on the bulk metric is taken into account.

Some modification of this construction involving the curved bulk space recently attracted particular attention: the compact hyperbolic extra dimensions [13]. An attractive feature of the compact spaces of negative curvature is that the relation between the linear size and the volume typically is exponential. This could ensure a low unification mass scale with much smaller size of extra dimensions.

In the scenario suggested by Randall and Sundrum [6], two 3-branes are present in the five-dimensional space-time: one of the positive tension σ and another of negative tension $-\sigma$ (as suggested by the Horava-Witten compactification of M-theory to the heterotic string theory [14]). The standard model matter lives on the negative tension brane, and the bulk space-time is curved due to the presence of the negative cosmological constant

$$\Lambda_5 = -\frac{4\pi}{3} G_5 \sigma^2, \tag{5}$$

(G_5 – the five-dimensional gravitational constant) being the anti-de Sitter space with curvature radius

$$l = \frac{3}{4\pi G_5 \sigma}. \tag{6}$$

The positive tension brane is located at a distance $y = R$ from the first brane, so the hierarchy problem is solved in view of the relation

$$G_4 = \frac{G_5}{l} \frac{1}{e^{2R/l} - 1} \tag{7}$$

with sufficiently large R/l. The distance R is the radion, in the cosmological setting it might be time-dependent. This model (known as RSI) has some intrinsic problems, in particular, an instability of the negative tension brane [15].

2.1 Modified Einstein's equations on the brane

An alternative model suggested by the same authors (RSII) [7] uses the positive tension brane as the physical one, in which case the negative tension brane can be moved to infinity or removed at all (an infinite extra dimension). Provided the same relation between Λ_5 and other parameters holds, this setup is compatible with both the standard model and the gravitational Newton law on the brane. Its generalization involving matter fields was given in the cosmological setting in [16] and more generally in [15]. The five-dimensional Einstein equation imply the following equations on the brane in terms of the four-dimensional quantities:

$$R_{\mu\nu} - \frac{1}{2} g_{\mu\nu} R = 8\pi G_4 T_{\mu\nu} - \Lambda_4 R_{\mu\nu} + (8\pi G_5)^2 \pi_{\mu\nu} - \mathcal{E}_{\mu\nu}, \tag{8}$$

where in the right hand side there are two additional source terms: one quadratic in the matter stress tensor $T_{\mu\nu}$,

$$\pi_{\mu\nu} = \frac{1}{12} T T_{\mu\nu} - \frac{1}{4} T_{\mu\alpha} T^{\alpha}_{\nu} + \frac{1}{24} g_{\mu\nu} \left(3 T_{\alpha\beta} T^{\alpha\beta} - T^2\right), \tag{9}$$

and another being the brane projection of the 5-dimensional Weyl tensor

$$\mathcal{E}_{\mu\nu} = C^5_{y\mu y\nu}, \tag{10}$$

currently called the dark radiation term. The four-dimensional gravitational and cosmological constants are given by

$$G_4 = \frac{4\pi\sigma G_5^2}{3}, \quad \Lambda_4 = 4\pi G_5 \left(\Lambda_5 + \frac{4}{3}\pi G_5 \sigma^2\right). \tag{11}$$

By tuning the five-dimensional cosmological constant and the brane tension one can get one of the three possibilities:

- $\Lambda_4 = 0$ — proper RSII type model (RSII_0),
- $\Lambda_4 > 0$ — RSII with the de Sitter brane (RSII_+),
- $\Lambda_4 < 0$ — RSII with the anti-de Sitter brane (RSII_-).

These three modifications of the RSII setup lead to very different predictions about the mass spectra of gravitons.

The quadratic contribution in the source term of the four-dimensional Einstein tensor tentatively may be reinterpreted (applying holographic considerations) as due to quantum corrections arising in the four-dimensional field theory dual to the bulk theory. The dark radiation term accounts for gravitational waves which may be present in the bulk space. It renders the four-dimensional picture essentially non-local: during the cosmological evolution gravitons should be emitted into the bulk feeding the dark radiation term, therefore the evolution equations at a given moment become depending on the full prehistory of the Universe. Similarly, any process on the brane, like gravitational collapse of a star, can not be described within the purely four-dimensional setting, one has to solve an entire five-dimensional problem to account properly for the dark radiation term.

2.2 KK graviton spectra

To further understand the specific features and differences of various braneworld scenarios it is useful to analyze the corresponding graviton spectra. First, recall the usual Kaluza-Klein situation, which remains intact in the ADD setup. The full graviton spectrum consists of the zero mode (coupled the brane) and an infinite number of KK gravitons with equidistant mass spectrum with the separation $1/l$, these modes interact with matter with the same Newton constant G_4. For $l \sim 1$ mm, the lowest mass is of the order of 10^{-4} eV, such particles are extensively thermally produced in the early Universe giving an unacceptable energy density at the nucleosynthesis. This is a consequence of the flatness of the bulk.

Consider now the braneworld models. In the two-brane RSI setup the bound graviton is again massless and couples to matter with the Newton constant G_4. The massive KK gravitons have the discrete spectrum as well, but now with the TeV-scale spacing. This model is therefore safe with respect to cosmological overproduction of KK gravitons. Massive modes couple to matter with the TeV-scale coupling constant.

The situation is different in the RSII type models. First, consider the case of zero four-dimensional cosmological constant (RSII_0). The graviton wave function separates in terms of the coordinates on the brane and in the bulk, the dependence on the conformal bulk coordinate z being governed by the equation

$$\left(-\frac{d^2}{dz^2} + V(z)\right) u_m(z) = m^2 u_m(z), \tag{12}$$

with the potential

$$V = \frac{15}{4}\frac{1}{(|z|+l)^2} - \frac{3}{l}\delta(z). \tag{13}$$

This potential well is infinitely deep and infinitely thin, the wave equation possesses the normalizable solutions for $m^2 \geq 0$ with the zero mode $u_0 = (|z|+l)^{-3/2}$ for $m = 0$. Since the potential tends to zero as $|z| \to \infty$, the massive spectrum is continuous, and there is no mass gap separating KK states from the bound graviton. All modes couple to matter with the four-dimensional Newton constant. The absence of the mass gap means that the KK gravitons can be produced both cosmologically and in the various local processes on the brane, like gravitational collapse of a star. Cosmological tests of this model are therefore crucial.

If the four-dimensional cosmological constant is positive (like in the RSII$_+$ scenario, or in any other situation when the matter on the brane exhibits an equation of state $p = -\rho$), the potential in the above equation is shifted up by the value $9H_0^2/4$, where H_0 is the Hubble constant associated with the brane inflation. It can be checked that the bound graviton is still massless, while the KK gravitons continuum is shifted from zero by a finite mass gap $3H_0/2$.

Finally, in the case of the negative four-dimensional cosmological constant, RSII$_-$, known also as the Karch-Randall model, the bound model is massive, while the KK spectrum is discrete. This model is interesting as a way to produce massive "ordinary" gravitons without complications associated with the Van Dam-Veltman-Zakharov discontinuity in the graviton propagator. In the RSII$_-$ setup the graviton propagator tends to the massless one when the four-dimensional cosmological constant is sent to zero, without a finite anomaly. So the anti-de Sitter braneworld may provide a sensible description of gravity once the graviton is discovered to possess a small, but finite mass.

2.3 Black holes formation and evaporation

The most intriguing prediction of the braneworld scenarios is the possibility of production of black holes in accelerators. There exists a vast literature on this subject (see for a review [17]), here we will focus on other aspects related to gravitational wave experiments. Gravitational radiation under black hole collisions is expected to be one of the most efficient astrophysical sources of gravitational waves. Already in the standard general relativity this problem is extremely complicated. In the braneworld setup the state of art is still far away from being able to provide a clear picture of brane-induced modifications to such processes. In spite of many efforts, the problem of the gravitational collapse on the brane is still not sufficiently understood even in the case of spherical symmetry. First, for the spherically symmetric (from the four-dimensional point of view) system, the Birkhoff theorem is not applicable: the metric on the brane outside the collapsing body is not necessarily the Schwarzschild metric. Moreover, the latter is incompatible with the massive KK modes in the strong field regime, and hence can not describe the final state of the collapse [18]. Secondly, the Weyl "dark radiation" term is expected to be non-zero in the realistic gravitational collapse on the brane. This term looks like a stress tensor of the Maxwell field, so the "tidal charge" Reissner-Nordström solution (with the negative charge square parameter) turns out to be appropriate for the four-dimensional spherical vacuum system [19]. This metric can not fully describe the final stage of the collapse either, but it is expected to be a good approximation in the near region where the tidal charge effect is not small. Since the charge square parameter is negative, the gravitational field is strengthened (contrary to the usual Reissner-Nordström case) with respect to the ordinary black hole picture. This should influence (expectedly enhance) gravitational radiation emitted in the strong field processes like the black hole collisions. No detailed investigation is available yet, given the absence of a clear picture of the single black hole.

Another problem associated with violation of the Birkhoff theorem is that of radiation accompanying the spherically symmetric gravitational collapse. On the general grounds, the radiation has to be present, and indeed, the four-dimensional metric of the spherical collapsing body can be matched with the radiating Vadya solution [20]. But this description takes into account only the lowest multipole in the radiation field, a complete calculation of radiation during the gravitational collapse on the brane is not available yet.

Further puzzle is associated with the Hawking evaporation of black holes in the theories with large extra dimensions. From the analysis of the graviton Green function [18] it follows that for a five-dimensional observer the black hole looks like a pancake, i.e. it is localized mainly on the brane. A more subtle question concerns the distribution of the thermal radiation from the black hole. An evaporation of small black holes, $r_h \ll l$, is dominated by the light KK modes, and large radiation into the bulk could be expected. But, as was shown by Emparan, Horowitz and Myers [21], in fact the radiation to the bulk is suppressed due to the factor $(r_h/l)^{D-4}$ appearing in the $d^{D-4}k$ phase volume. This result is relevant to black holes which might be produced on accelerators. For astrophysics this is the problem of evaporation of the "large" black holes, $r_h \gg l$, which is the most interesting. Recently it was discussed within the RSII_0 model in two papers [22, 23], using a holographic picture, with totally contradicting results. Hawking radiation in this case must be predominantly in the light KK-modes (no mass gap). The calculation in [22] was performed within the purely four-dimensional picture suggested by a holographic treatment of the bulk theory as a compactified version of the IIB theory on $AdS_5 \times S^5$. Then the dual conformal field theory is the maximally supersymmetric $\mathcal{N} = 4$, $SU(N)$ gauge theory with $N \gg 1$, with the number of massless modes of the order of N^2. In our case the number N is as large as $N \sim l/l_{Pl}$, so the very large factor $(l/l_{Pl})^2$ enters the evaporation rate reducing the lifetime of the black hole of the mass M to

$$\tau \sim 1.2 \cdot 10^2 \left(\frac{M}{M_\odot}\right)^3 \left(\frac{1\mathrm{mm}}{l}\right)^2 \ \mathrm{yr}. \tag{14}$$

Therefore the evaporation of the black hole of several solar masses could be completed within some tens of thousands years. Meanwhile, the temperature of the black hole is still as small as

$$T_H \sim 6 \cdot 10^{-8} \frac{M_\odot}{M} K^o, \tag{15}$$

so the effect seems to be purely classical. It was suggested indeed, though not demonstrated explicitly, that the evaporation might be calculable classically [24, 22].

Meanwhile, the possibility of the holographic description of the above type rather is a conjecture. Recently both the validity of the holographic picture and the rate of Hawking evaporation on the brane were tested via the computation of the trace anomaly [23]. The result of this analysis contradicts to the conclusions of [22] and suggests that the late stage of black hole evaporation on the brane is characterized by a decreasing Hawking temperature. Clearly, the problem of Hawking evaporation of large black holes in the RSII scenario requires further investigation.

Creation and evolution of primordial black holes of relatively small mass is also modified in the braneworld scenarios [25]. In particular, large fluctuations can be expected on the mesoscopic energy scale which might result in formation of black holes of lunar mass. These could serve as the cold dark matter, an expected level being about 10^{19} in the dark halo of the Milky Way.

3 Non-Abelian fields on the brane: Born-Infeld dynamics

3.1 EYM glueballs and black holes

Investigation of non-Abelian gauge theories coupled to gravity has led to many surprising discoveries. In the mid 70-ies the question was raised about the possibility of black holes in Einstein-Yang-Mills (EYM) theory

$$S_{EYM} = -\frac{1}{16\pi}\int \left(\frac{R}{G} + \frac{1}{g^2} F_{a\mu\nu}F_a^{\mu\nu}\right)\sqrt{-g}\, d^4x, \tag{16}$$

different from the Reissner-Nordström and Kerr-Newman solutions of the Einstein-Maxwell theory. At that time, the Wheeler's no-hair conjecture was believed to have a universal validity, so it could be expected that the embedded Abelian solutions might be the most general asymptotically flat black holes possessing the Coulomb charges (for more details see [26]). But after the discovery by Bartnik and McKinnon [27] (BK) of the SU(2) EYM static spherically symmetric particle-like solutions it became clear [28, 29, 30] that new black holes violating the no-hair conjecture exist which do not carry Coulomb charges but still possess a non-Abelian hair outside the event horizon. Their metric of both the BK and the corresponding EYM black hole solutions is spherically symmetric and is parameterized by two functions N, σ of the radial variable r:

$$ds^2 = N\,\sigma^2\,dt^2 - \frac{dr^2}{N} - r^2 d\Omega, \tag{17}$$

while the YM field is purely magnetic:

$$A = (1 - w)\,(T_\theta \sin\theta d\varphi - T_\varphi\, d\theta)\,, \tag{18}$$

where w is a real function of r, and T_θ, T_φ are spherical projections of the SU(2) generators. The particle-like solutions are distinguished by regularity of the metric at the origin implying that the function w takes the vacuum value $|w(0)| = 1$ (for definiteness, $w(0) = 1$). From the equations of motion one finds the series expansion

$$w = 1 - br^2 + O(r^4), \quad N = 1 - 4b^2r^2 + O(r^4), \tag{19}$$

with a free parameter b. The second condition is the finiteness of the ADM mass M, defined as the limit $m(r) \to M$, where the local mass-function is introduced via $N = 1 - 2m(r)/r$; the second metric function σ must behave asymptotically as $\sigma \to 1$. (To simplify formulas we set the Newton constant $G = 1$ everywhere if its explicit appearance is not needed for clarity.) To ensure this, the YM field asymptotically has to stay again in one of the vacua, the next term containing another free parameter p:

$$w = \pm(1 - p/r) + O(1/r^2), \tag{20}$$

The free parameters b, M, p of the asymptotic solutions can be determined via the numerical matching, one finds a discrete increasing sequence b_n on the interval $[b_1 = 0.4537, b_\infty = 0.706)$, with masses converging to unity in the units of the rescaled Planck's length m_{Pl}/g. The function w_n has n zeroes and asymptotically tends to $(-1)^n$, solutions exist for any finite n. For the lowest BK solution, $n = 1$, w is a decreasing function interpolating between $w = 1$ and $w = -1$ in exactly the same way as it does in the electroweak sphaleron. Odd-n solutions all have the sphaleron nature, while even-n ones are topologically trivial.

For black hole solutions, instead of the regularity conditions at the origin (19), the regularity of the event horizon is required. The (non-degenerate) event horizon is the largest linear zero of the metric function N : $N(r_h) = 0$, $N'(r_h) > 0$, where the YM function w starts with a finite value $w_h = w(r_h)$. This latter quantity is a parameter labelling the black hole solutions, another parameter being the horizon radius r_h. One finds the black hole solutions for any r_h and a discrete sequence of w_h within the interval $0 < w_h < 1$ [28, 29, 30]. The EYM black holes violate a naive no-hair conjecture, since the magnetic YM field outside the horizon is not associated with the conserved charges. Another expectations of the black hole physics, the uniqueness property, is not satisfied either: an infinite number of different solutions exist which possess the same ADM masses and zero Coulomb charges. They correspond to different pairs r_h, w_h.

EYM black holes have quite unusual *internal* structure [31]. There are discrete sequences of black holes which have either Schwarzschild or Reissner-Nordström (RN) type interiors, but a generic black hole exhibits a very different behavior inside the horizon. The metric function N oscillates with an exponentially growing amplitude, remaining always negative, but approaching zero closer and closer in the subsequent cycles. Each time when such an "almost" Cauchy horizon is reached, the mass function starts to "inflate", so that the true Cauchy horizon never forms. The oscillations can be well described by an approximate two-dimensional dynamical system, whose analysis reveals that oscillations can never stop. This behavior is likely to be related to the scale invariance of the YM field, in various models involving non-conformal scalar fields or a modified non-scale invariant YM lagrangian the violent oscillations are not observed [32, 33] (see Sec. 6). All solutions, except for a discrete RN-type set of a zero measure in the parameter space satisfy the strong cosmic censorship hypothesis (the absence of the internal Cauchy horizons). However the oscillation amplitude exceeds classical limits already after few cycles, so the question arises how this behavior gets modified by quantum effects. Born-Infeld type modification of the Yang-Mills action in the string theory changes this behavior substantially.

3.2 Non-Abelian Born-Infeld theory

The non-Abelian four-dimensional U(N) Dirac-Born-Infeld theory, arises as an effective theory on N D3-branes [2]. The transversal coordinates (brane deformations) play the role of the Higgs scalars. When branes are coincident, one deals with the purely massless theory, while for separated branes there is a Higgs potential with the vacuum expectation value equal to the brane separation. The simplest case is the U(1) theory on a single D3-brane with the Born-Infeld (BI) action (in the flat spacetime)

$$S = \frac{\beta}{4\pi} \int \left\{ 1 - \sqrt{-\det(\eta_{\mu\nu} + \beta^{-1} F_{\mu\nu})} \right\} d^4x. \tag{21}$$

In the non-Abelian case the field strength $F_{\mu\nu}$ is matrix-valued, and one can a priori take the trace in different ways (see, e.g., [34] for details). Two definitions are the most reliable: the symmetrized trace suggested by Tseytlin, which well covers the lower orders of the perturbative string effective action, and the "ordinary trace-square root" form, which is a direct non-Abelian generalization of the four-dimensional U(1) BI action obtained by evaluating explicitly the determinant under the square root in (50). For the static spherically symmetric SU(2) magnetic ansatz (18) these definitions lead to somewhat different one-dimensional lagrangians

$$L_{tr} = \beta^2 r^2 (1 - R), \quad \mathcal{R} = (1 + 2V^2 + 4K^2)^{1/2}, \tag{22}$$

where $V^2 = (1 - w^2)^2/(2\beta^2 r^4)$, $\quad K^2 = w'^2/(2\beta^2 r^2)$, and [34]

$$L_{str} = \beta^2 r^2 \left[1 - (1 + V^2)^{1/2} + K^2 \mathcal{A} (1 + V^2)^{-1/2} \right] \tag{23}$$

with

$$\mathcal{A} = W^{-1} \arctan W, \quad W = (1 + V^2)^{1/2} (V^2 - K^2)^{-1/2}.$$

The BI lagrangians break the scale invariance of the ordinary Yang-Mills (or Maxwell) theory, introducing a new new scale parameter — the critical field β, which might be regarded as a manifestation of non-locality of the underlying dynamical object (string). Therefore, the no-go theorem for classical glueballs in the flat space YM theory is overruled, and one has to reconsider the possibility of such solutions in the non-Abelian Born-Infeld (NBI) theory. The proof of existence and the numerical glueball solutions in the flat space NBI theory were presented in [35] for the ordinary trace (22) and in [34] for the symmetrized trace (23) theories, they have qualitatively similar properties. Solutions form a sequence of the BK type, though the values of the parameter b_n in the expansion (19) are rather large (in natural units) and rapidly growing with the node number n, for example, $b_1 = 12.7$, $b_2 = 887$ in the (22); these values are even larger for the model (23) [34]. Solutions have sphaleron features and are expected to be unstable. When gravity is taken into account [36, 37], one finds that b_n, as a function of the dimensionless parameter $G\beta$, continuously interpolates between the values of [35] at small $G\beta$, and the BK values for large $G\beta$ [37].

Black hole solutions to the Einstein-NBI equations were also constructed [36, 37], outside the event horizon they are qualitatively similar to those in the EYM case. However, an internal structure of the ENBI black holes is drastically different. In fact, one could expect that violent oscillations inside the EYM black holes should be modified in the quantum theory: already in the first several inflation cycles the mass function attains values exceeding the Planck value [38]. Although the ENBI theory is still classical, it incorporates non-perturbatively the string quantum corrections in α', so one could expect that the problem of the over-Planckian masses in the EYM black holes will be resolved. This is indeed the case [37]: the internal behavior of the mass function now is perfectly smooth, though the singularity still bears some non-analytic features. Namely, the function w attains smoothly a finite value w_0 in the singularity, and there exist a two-parameter family of solutions for which the mass function is analytic at the Schwarzschild-type singularity. This family, however, is not generic like in the EYM case. But now a generic solution (three-parametric) can be obtained in the vicinity of the singularity in the series form, which is an expansion in terms of $r^{1/2}$. Mass function again attains a finite value, and one finds that the singularity is much weaker than in the Schwarzschild case. Note that the non-analytic behavior of the mass function in the singularity is also observed in the U(1) Einstein-Born-Infeld theory coupled to dilaton [39].

Adding the (triplet) Higgs field to the NBI lagrangian, one finds the flat space monopole solutions [40, 41] exhibiting features, similar to those of the gravitating monopoles. In view of the existence of the flat-space NBI sphalerons, this is not surprising. Monopole (and dyon) solutions exist for the values of BI critical field β varying from infinity to a finite value β_{cr}, which was shown [42] to be a bifurcation point giving rise to the branch of excited monopoles with one-node w. Similarly to gravitating monopoles, there are further bifurcations where higher node sphaleron excitations pinch off. More detailed study [43] reveals that near the critical regime a large negative pressure appears in the monopole core which causes monopoles to shrink. For $\beta < \beta_{cr}$ only the pointlike magnetic monopoles exist as static solutions (these have the finite energy in the Born-Infeld theory). This behavior is reminiscent of the gravitational collapse, another aspect of similarity between the flat space NBI and the EYM theories.

3.3 NBI cosmology

It is usually assumed that large scale massless Yang–Mills (YM) fields, which could exist in the early Universe before phase transitions, do not play any significant role in cosmology. Partly this is related to scale invariance property of the YM lagrangian implying that primary YM excitations must be diluted during inflation. String theory suggests the Born–Infeld (BI) type modification of the YM action which breaks scale invariance and thus cancels the objection. Therefore, it seems reasonable to investigate the YM cosmology with the non-Abelian Born-Infeld action [44]. In the existing literature one finds several papers discussing cosmological models with the U(1) BI matter [45, 46, 47]. Such models are necessarily anisotropic (or inhomogeneous), since there is no homogeneous and isotropic configurations of the classical U(1) field. Here we discuss *non-Abelian*

Born-Infeld (NBI) homogeneous and isotropic Friedmann-Robertson-Walker (FRW) cosmology.

A notable property of the SU(2) Yang–Mills field is that it *admits* homogeneous and isotropic (G_6-invariant) configurations. Indeed, the energy can be distributed between three color components in such a way that the resulting stress-tensor is compatible with the maximal symmetry of the three-dimensional space. In the case of S^3, the corresponding ansatz was given in [48, 49, 50], two other cases (hyperbolic and flat) were treated in [51]. An immediate consequence of the breaking of scale invariance is that the FRW equation of state $p = \varepsilon/3$ arising in the usual YM theory [51] changes to some more complicated equation which now allows for a negative pressure [44]. One could wonder whether an accelerating expansion becomes possible, that is, whether the sum $\varepsilon + 3p$ can be negative. It turns out that the lower limit of acceleration which can be achieved within the present model is precisely zero: when YM field strength is much larger then the BI 'critical field', the equation of state is $p = -\varepsilon/3$. Such a state equation is typical for an averaged distribution of strings, where its origin is quite simple: each string has a *one-dimensional* equation of state $p = -\varepsilon$, averaging over all directions in the three-space gives $p = -\varepsilon/3$.

We choose the 'square root/ordinary trace' Einstein-NBI action

$$S = -\frac{1}{4\pi}\int \left\{\frac{1}{4G}R + \beta^2(\mathcal{R}-1)\right\}\sqrt{-g}\,d^4x, \tag{24}$$

where R is the scalar curvature, β is the BI critical field strength,

$$\mathcal{R} = \sqrt{1 + \frac{1}{2\beta^2}F^a_{\mu\nu}F^{\mu\nu}_a - \frac{1}{16\beta^4}(\tilde{F}^a_{\mu\nu}F^{\mu\nu}_a)^2},$$

and consider cosmological models with the homogeneous and isotropic three-space:

$$ds^2 = N^2dt^2 - a^2\,dl^2,$$

$$dl^2 = dr^2 + S^2(d\theta^2 + \sin\theta^2 d\phi^2), \tag{25}$$

where S is one of the functions $\sin r$, r or $\sinh r$ for closed, spatially flat or open cases ($k = 1, 0, -1$) respectively.

In suitable gauge the YM field tensor depends on one real function w of time

$$\begin{aligned}\mathcal{F} &= \dot{w}\,(T_r\,dt\wedge dr + T_\theta S\,dt\wedge d\theta + T_\phi S\sin\theta\,dt\wedge d\phi)\\ &\quad + S(w^2-k)\,(T_\phi\,dr\wedge d\theta - T_\theta\sin\theta\,dr\wedge d\phi + T_r S\sin\theta\,d\theta\wedge d\phi)\,.\end{aligned} \tag{26}$$

Substituting (26) into Eq. (24), we obtain the following one-dimensional action:

$$S_1 = \frac{1}{4\pi}\int dt\left[\frac{3}{2G}\frac{a(kN^2-\dot{a}^2)}{N} - Na^3\beta^2(\mathcal{R}-1)\right],$$

where now

$$\mathcal{R} = \sqrt{1 - \frac{3\dot{w}^2}{a^2\beta^2N^2} + \frac{3(w^2-k)^2}{a^4\beta^2} - \frac{9\dot{w}^2(w^2-k)^2}{a^6\beta^4N^2}}.$$

The lagrangian contains two dimensional parameters: the Newton constant G and the BI 'critical field' β. After a coordinate rescaling $t \to \beta^{-1/2}t$, $a \to \beta^{-1/2}a$, which makes all quantities dimensionless, the reduced action becomes

$$S_1 = \frac{1}{4\pi G\beta}\int dt \left\{ \frac{3}{2}\frac{a(kN^2 - \dot{a}^2)}{N} - gNa^3 \left[\sqrt{(1-K^2)(1+V^2)} - 1 \right] \right\},$$

where

$$K = \frac{\sqrt{3}\dot{w}}{aN} \quad \text{and} \quad V = -\frac{\sqrt{3}(w^2 - k)}{a^2}, \tag{27}$$

and $g = \beta G$ is the remaining dimensionless coupling constant.

Variation of the action over N gives a constraint equation; after obtaining it, we fix the gauge $N = 1$:

$$ga^2(\mathcal{P} - 1) - \frac{3}{2}(\dot{a}^2 + k) = 0, \tag{28}$$

where

$$\mathcal{P} = \sqrt{\frac{1+V^2}{1-K^2}}. \tag{29}$$

This expression can be rewritten in the standard form

$$\frac{\dot{a}^2}{a^2} + \frac{k}{a^2} = \frac{8\pi G}{3}\varepsilon, \tag{30}$$

where the energy density is given by

$$\varepsilon = \varepsilon_c(\mathcal{P} - 1), \tag{31}$$

with $\varepsilon_c = \beta/4\pi$ playing a role of the BI critical energy density.

Variation of the action with respect to a gives the acceleration equation

$$\ddot{a} = \frac{ga}{3}\mathcal{P} - \frac{\dot{a}^2 + k + 2ga^2}{2a} + \frac{2ga}{3}\mathcal{P}^{-1}. \tag{32}$$

Again, rewriting it in the standard form (using (30))

$$\frac{\ddot{a}}{a} = -\frac{4\pi G}{3}(\varepsilon + 3p), \tag{33}$$

we can read off the pressure

$$p = \frac{1}{3}\varepsilon_c\left(3 - \mathcal{P} - 2\mathcal{P}^{-1}\right). \tag{34}$$

Now comparing (31) and (34) we obtain the following equation of state

$$p = \frac{\varepsilon\,(\varepsilon_c - \varepsilon)}{3\,(\varepsilon_c + \varepsilon)}. \tag{35}$$

The critical energy density corresponds to the vanishing pressure. For larger energies the pressure becomes negative, its limiting value is $p = -\varepsilon/3$. In the opposite limit $\varepsilon \ll \varepsilon_c$

one recovers the hot matter equation of state $p = \varepsilon/3$, reflecting the scale invariance of the YM action to which the NBI action reduces at low energies.

Finally, the variation over w gives the YM (NBI) equation

$$\ddot{w} = 2\frac{w(k-w^2)}{a^2}\left(\frac{1-K^2}{1+V^2}\right) + \frac{2\dot{a}\dot{w}}{a}\left(\frac{1}{2} - \frac{1-K^2}{1+V^2}\right). \tag{36}$$

From this one can derive the following evolution equation for the energy density:

$$\dot{\varepsilon} = -2\frac{\dot{a}}{a}\frac{\varepsilon\,(\varepsilon+2\varepsilon_c)}{\varepsilon+\varepsilon_c}, \tag{37}$$

which can be easily integrated to give

$$a^4(\varepsilon+2\varepsilon_c)\varepsilon = \text{const.} \tag{38}$$

From this relation one can see that the behavior of the NBI field interpolates between two patterns: 1) for large energy densities ($\varepsilon \gg \varepsilon_c$) the energy density scales as $\varepsilon \sim a^{-2}$; 2) for small densities $\varepsilon \ll \varepsilon_c$ one has a radiation law $\varepsilon \sim\ a^{-4}$.

Using the constraint equation (28), one can express ε in terms of the scale factor a and its derivative. Substituting it into Eq. (32) we obtain a decoupled equation governing the spacetime evolution:

$$\ddot{a} = -\frac{2ga(\dot{a}^2+k)}{2ga^2+3(\dot{a}^2+k)}. \tag{39}$$

The condition of positivity of the energy density in (28) defines a boundary in the phase space:

$$\dot{a}^2 > -k. \tag{40}$$

The evolution equation (39) can be studied by means of the dynamical systems tools [44]. For this, it is convenient to introduce $b = \dot{a}$ and pass to another independent variable τ, such that $dt = h\,d\tau$, where $h = 2ga^2 + 3(b^2+k)$. This leads to the following dynamical system:

$$a' = hb, \qquad b' = -2ga(b^2+k), \tag{41}$$

where primes denote derivatives with respect to τ. Notice that the system (41) is invariant under a reflection ($a \to -a$, $b \to -b$), so further, without loss of generality, we discuss the expanding solutions. The dynamical system (41) admits a first integral:

$$3\left(b^2+k\right)^2 + 4ga^2\left(b^2+k\right) = C, \tag{42}$$

with some constant C. From this relation one can see that all solutions inside the physical boundary (40) cross the line $a = 0$, so that the singularity is unavoidable. Another observation is that, unlike the standard hot FRW cosmology, $\dot{a}$ remains finite while a tends to zero.

The global qualitative behavior of solutions does not differ substantially from that in the conformally invariant YM field model, except near the singularity, where the following power series expansion holds

$$a(t) = b_0 t - \frac{b_0 g}{9} t^3 + \frac{1}{270} \frac{b_0 g^2 (7b_0^2 + k)}{(b_0^2 + k)} t^5 + O(t^7), \quad t \to 0, \tag{43}$$

where b_0 is a free parameter. The absence of a quadratic term means that the Universe starts with zero acceleration. This is what can be expected in view of the equation of state $p \approx -\varepsilon/3$ at high densities.

For large a, the dynamics of the system (32) approaches that of the hot FRW models (for small energy densities one recovers the equation of state of radiation).

The gauge field generates the metric through the quantity $\mathcal{P}$ (29) related to the energy density which obeys the differential equation:

$$\dot{\mathcal{P}} = 2\frac{\dot{a}}{a}\left(\frac{1}{\mathcal{P}} - \mathcal{P}\right), \tag{44}$$

following from (36). This equation can be integrated

$$\mathcal{P} = \sqrt{1 + 3\left(\frac{a_0}{a}\right)^4}, \tag{45}$$

where a_0 is a constant. Recalling the definition of $\mathcal{P}$ (29), one can separate the metric and the gauge field variables as follows:

$$\frac{\dot{w}^2}{a_0^4 - (k - w^2)^2} = \frac{N^2 a^2}{a^4 + 3a_0^4}. \tag{46}$$

This equation can be solved in terms of the Jacobi elliptic functions, similarly to the case of an ordinary YM lagrangian [52]:

$$w(t) = \sqrt{k + a_0^2}\, \mathrm{cn}\left(\sqrt{2}a_0(\tau - \tau_0), \frac{\sqrt{k + a_0^2}}{\sqrt{2}a_0}\right), \tag{47}$$

where τ_0 is the integration constant, and the argument to the Jacobi function is defined as

$$\tau(t) = \int^t \frac{a(t')N(t')\, dt'}{\sqrt{a^4(t') + 3a_0^4}}. \tag{48}$$

A generic solution for $w(t)$ is of an oscillating type. In the case $a_0 < 1$, $k = 1$, the solution, oscillating near one of the values $w = \pm 1$, does not cross the line $w = 0$. In all other cases solutions oscillate around the origin. The effective frequency of oscillations is determined by the rate of growth of the argument τ. To compare the situation with the ordinary EYM cosmology, we notice that in this case the solution for $w(t)$ is still given by (47), but with a different definition of the phase variable $\tau(t)$ [52]

$$\tau_{EYM}(t) = \int^t \frac{N(t')}{a(t')}\, dt'.$$

Clearly, the main difference with the ordinary EYM cosmology relates to small values of a. One can see that near the singularity ($a \to 0$) the YM oscillations in the NBI case slow down, while in the ordinary YM cosmology the frequency remains constant in the conformal gauge $N = a$, or tends to infinity in proper time gauge $N = 1$.

When the solution starts from the metric singularity, the gauge field function has the following series expansion:

$$w = w_0 + \frac{b_0 \alpha}{6(k + b_0^2)} t^2 + \left(\frac{g^2 b_0^2 w_0 (k - w_0^2)}{9(k + b_0^2)^2} + \frac{b_0 \alpha (\lambda - 2g)}{216(k + b_0^2)} \right) t^4 + O(t^6), \quad t \to 0, \tag{49}$$

where $\alpha = \pm\sqrt{3(k + b_0^2)^2 - 4g^2(k - w_0^2)^2}$, w_0 is a free parameter (as well as the sign of α), while b_0 is a parameter from the expansion of the metric.

3.4 NBI in the RSII setup

In the RSII brane world approach [7, 15] it is assumed that the standard model fields are confined to the brane by some mechanism whose nature remains unspecified. One possibility is to invoke the D-brane picture in which vector fields are generated by open strings with Dirichlet boundary conditions on the brane. In this case one has to keep in mind that in order to incorporate the non-Abelian gauge group of the standard model one has to introduce multiple brane structure with the number of branes equal to the dimension of the fundamental representation.

Here we discuss the homogeneous and isotropic brane cosmology with the SU(2) Yang-Mills field assuming that the dynamics is governed by the non-Abelian Born-Infeld action which is an effective action for D-branes:

$$S = \lambda \widetilde{\mathrm{Tr}} \int \sqrt{-\det(g_{\mu\nu} + F_{\mu\nu}/\beta)}\, d^4x - \kappa^2 \int (R_5 + 2\Lambda_5)\sqrt{-g_5}\, d^5x, \tag{50}$$

where $g_{\mu\nu}$ is the induced metric on the brane and κ, Λ_5 and λ are the 5D gravitational and cosmological constants and the brane tension respectively. For generality, the BI critical field strength parameter β is assumed to be different from λ. The symbol $\widetilde{\mathrm{Tr}}$ denotes either the ordinary trace (with an additional prescription to be applied after evaluating the determinant with the matrix-valued F), or symmetrized trace Str.

We write the bulk metric as

$$ds_5^2 = N^2(\tau, y) - a(\tau, y)^2 \left(dr^2 + S^2(d\theta^2 + sin^2\theta d\phi^2)\right) - dy^2,$$

additionally assuming that it is invariant under reflection $y \to -y$. The induced metric on the brane is given by the first two terms with the limiting values of N and a at $y = 0$.

The computation of the reduced action is straightforward for the ordinary trace prescription [44]

$$S_1^{tr} = -\lambda \int dt N a^3 \sqrt{1 - 3K^2 + 3V^2 - 9K^2V^2}. \tag{51}$$

For the symmetrized trace one has to expand the square root into powers of the matrix-valued $F_{\mu\nu}$, to perform symmetrization in the generator products, to take the trace, and then to perform a resummation [53]:

$$S_1^{Str} = -\lambda \int dt N a^3 \frac{1 - 2K^2 + 2V^2 - 3V^2K^2}{\sqrt{1 - K^2 + V^2 - K^2V^2}}. \tag{52}$$

In these formulas

$$K^2 = \frac{\dot{w}^2}{\beta^2 a^2 N^2}, \qquad V^2 = \frac{(w^2 - k)^2}{\beta^2 a^4},$$

and functions N and a refer to the metric on the brane.

Varying the action (50) with respect to the induced metric one finds the following energy-momentum tensor of the gauge field

$$T_{\mu\nu} = \lambda \widetilde{\mathrm{Tr}}(\mathcal{R}^{-1}(g_{\mu\nu} + \beta^{-2}(\frac{1}{2}F^{\lambda\rho}F_{\lambda\rho}g_{\mu\nu} - F_{\mu\lambda}F_{\nu}{}^{\lambda})) - \lambda g_{\mu\nu}, \tag{53}$$

where

$$\mathcal{R} = \sqrt{-\det(g_{\mu\nu} + F_{\mu\nu}/\beta)},$$

which is not traceless as it is in the usual Yang-Mills theory. Within the RSII brane world setup the constraint equation replacing (28) reads

$$\left(\frac{\dot{a}}{a}\right)^2 = \frac{\kappa^2}{6}\Lambda + \frac{\kappa^4}{36}(\lambda + \varepsilon)^2 + \frac{\mathcal{E}}{a^4} - \frac{k}{a^2}, \tag{54}$$

where $\mathcal{E}$ is integration constant corresponding to the dark radiation ε is the energy density on the brane. We additionally fix the gauge by setting $N(\tau, 0) = 1$, thus identifying τ with the proper time on the brane. The constant terms in this equation, being grouped together, give the value of an effective 4D cosmological constant Λ_4 and the 4D Newton constant:

$$\Lambda_4 = \frac{1}{2}\kappa^2(\Lambda + \frac{1}{6}\kappa^2\lambda^2), \quad G_{(4)} = \frac{\kappa^4\lambda}{48\pi}. \tag{55}$$

Note that the conservation equation (37) remains valid in the brane scenario too, and consequently the energy density law

$$\varepsilon = \lambda\left(\sqrt{1 + \frac{C}{a^4}} - 1\right), \tag{56}$$

holds. Most surprisingly, the equation (54) simplifies greatly when the dependence Eq. (56) is inserted, and one obtains

$$\left(\frac{\dot{a}}{a}\right)^2 = \frac{8\pi G_{(4)}}{3}\Lambda_4 + \frac{\mathcal{C}}{a^4} - \frac{k}{a^2}, \tag{57}$$

where the constant $\mathcal{C} = \mathcal{E} + \kappa^4\lambda^2 C/36$ includes contributions from both the dark radiation and the energy density of the NBI matter. This equation fully coincides with the corresponding equation for the hot FRW cosmology in the non-brane scenario [51], and therefore we are led to the standard Tolman solutions. In particular, this means that near the singularity the scale factor grows like $t^{1/2}$.

There is, however, one essential distinction from the conventional hot cosmology: since the r.h.s. of the Eq. (57) includes the dark radiation contribution, now the total effective energy density need not be positive, i.e. the constant $\mathcal{C}$ is not a priori of a definite sign. This gives rise to solutions that could not arise in the conventional hot cosmology. One is the static flat (in the 4D sense) solution with the vanishing cosmological constant and nonzero energy density:

$$\dot{a} = 0, \qquad \text{and after simple scaling} \qquad g_{\mu\nu} = \eta_{\mu\nu}.$$

It can be obtained from the Eq. (57) with the following parameters: $\Lambda_4 = 0$, $\mathcal{C} = 0$, $k = 0$. The possibility of such a solution in the brane-world setup was investigated in [54] assuming the fluid matter content. It was found that the necessary state equation has the general form coinciding with our NBI equation (35). Therefore, the ordinary trace NBI model on the brane provides a very natural realization of the static scenario.

Another new interesting possibility is the avoidance of cosmological singularity without the cosmological constant. Such solutions can be obtained from (57) setting $\Lambda_4 = 0$, $\mathcal{C} < 0$, $k = -1$. This corresponds to the open universe. The solution reads:

$$a = \sqrt{|\mathcal{C}| + (t - t_0)^2}, \tag{58}$$

where t_0 is the integration constant. The universe contracts to the minimal radius $a = \sqrt{|\mathcal{C}|}$, bounces and reexpands.

The situation is qualitatively similar, though not precisely the same in the model with symmetrized trace (52). One has to solve the full system of equations consisting of one derived from the action (52) for the gauge field and the equation (54) for the metric. In this case one can not extract an equation of state since the pressure is no more expressible as a function of the energy only. However this can be done in the limiting cases of low and high densities, or on the particular classes of solutions. In the limiting cases one has the same equations of state as in the ordinary trace model

$$p \approx -\frac{1}{3}\varepsilon, \qquad \text{as} \quad \varepsilon \to \infty, \tag{59}$$

$$p \approx \frac{1}{3}\varepsilon, \qquad \text{as} \quad \varepsilon \to 0. \tag{60}$$

This indicates that the behavior of the solutions near the singularity (if it exists) also remains of the Tolman type. At late times the behaviour is again typical for the radiation dominated universes. So in the symmetrized trace model the main qualitative features of the hot universe pertain.

There are also some exotic solutions with negative values of dark energy contribution $\mathcal{E}$. In the case of the vanishing cosmological constant Λ_4 the static Einstein universe cannot exist. This is because the state equation (35) needed for the static universe is fulfilled in the symmetrized trace model only asymptotically. Instead, if the value of the dark radiation constant $\mathcal{E}$ is below some negative critical limit, there appears solutions starting from the initial singularity, expanding to a finite a and then collapsing back to the singularity. The analogue of the nonsingular solution (58) with $k = -1$ exists in the symmetrized trace model too.

One particular solution is worth to be mentioned. It can be checked that $w \equiv 0$ is a non-trivial solution to the field equations for $k = \pm 1$ like in the ordinary YM theory [55]. In the closed case this is the cosmological sphaleron sitting at the top of the potential barrier separating topologically distinct vacua. The scale factor starts with an expansion

$$a = (4\lambda^2 + \mathcal{E}\beta^2)^{1/4}\sqrt{\frac{2t}{\beta}} - \frac{k\beta^2}{2}(4\lambda^2 + \mathcal{E}\beta^2)^{-1/4}\left(\frac{2t}{\beta}\right)^{3/2} + O(t^{5/2}), \tag{61}$$

the global behavior of the solution being similar to that in the non-brane scenario with the ordinary YM matter [55].

4 String non-locality and chaos

In the superstring theory the Born-Infeld action describes the non-perturbative in the string slope parameter α' behavior of open strings. Therefore the above new features of the soliton solutions may be regarded as manifestation of the genuine string theory effect on the Yang-Mills dynamics. One famous property of the usual YM classical dynamics is *chaoticity*, which can be observed already on the level of homogeneous fields depending only on time [56, 57, 58, 59, 60]. The absence of classical glueballs in the ordinary YM theory is often interpreted also as related to the chaotic nature of the YM field [59] (contrary to the integrability property). In this sense the existence of glueballs in the NBI theory may serve an indication that strings render the Yang-Mills classical dynamics "less chaotic". This was checked [61] for typical chaotic YM systems previously discussed in [56]. Transition from chaos to the regular behavior was observed indeed for β sufficiently small.

The simplest non-Abelian configuration for which the ordinary YM theory predicts chaotic behavior [62] is the following

$$A = u\mathbf{T}_1 dx + v\mathbf{T}_2 dy, \tag{62}$$

where u and v are functions of time only, and $\mathbf{T}_1$,$\mathbf{T}_2$ are the gauge group generators. The corresponding field strength

$$F = \dot{u}\mathbf{T}_1 dt \wedge dx + \dot{v}\mathbf{T}_2 dt \wedge dy + uv\mathbf{T}_3 dx \wedge dy,$$

contains both electric and magnetic components, but these are mutually orthogonal, so the pseudoscalar invariant $\operatorname{tr} F\tilde{F}$, generically dominant at high density, is equal to zero. The corresponding one-dimensional lagrangian reads

$$L = \beta^2\left(1 - \sqrt{1 - \beta^{-2}\left(\dot{u}^2 + \dot{v}^2 - v^2u^2\right)}\right). \tag{63}$$

The equations of motion have the form

$$\ddot{u} = -uv^2 + \frac{2\dot{u}vu(\dot{u}v + \dot{v}u)}{\beta^2 + u^2v^2}, \tag{64}$$

$$\ddot{v} = -vu^2 + \frac{2\dot{v}vu(\dot{u}v + \dot{v}u)}{\beta^2 + u^2v^2}. \tag{65}$$

In the limit $\beta \to \infty$ of the ordinary YM action the second terms in the right hand sides of these equations vanish and one obtains the hyperbolic billiard [62], known in various problems of mathematical physics, which is a two-dimensional mechanical system with the potential u^2v^2. This potential has two valleys along the lines $u = 0$ $v = 0$, the particle motion being confined by hyperbolae $uv = \text{const}$. The existence of these valleys is crucial for emergency of chaos.

The trajectories of the system governed by the BI action depend on the energy integral E and the parameter β. Rescaling of the field variables together with time rescaling maps configurations with different E and β onto each other, so it will be enough to consider dependence of the dynamics only on β assuming for the energy any fixed value, e.g. $E = 1$. One simple (though non-rigorous) method to reveal chaoticity is the analysis of the geodesic deviation equation for a suitably defined pseudoriemannian space whose geodesics coincide with the trajectories of the system. The lagrangian (63) gives rise to the following metric tensor

$$ds^2 = \left(1 + \frac{u^2v^2}{\beta^2}\right) dt^2 - \frac{1}{\beta^2}(du^2 + dv^2), \tag{66}$$

which is regular for all finite $\beta \neq 0$. Consider the deviation equation for two close geodesics

$$\frac{D^2n^i}{ds} = R^i{}_{jkl}u^j u^k n^l, \tag{67}$$

where D^2n^i is the covariant differential of their transversal separation, $R^i{}_{jkl}$ is the Riemann-Christoffel tensor, and u^j is the three-speed in the space (66). Locally, the geodesic deviation is described by the matrix $R^i{}_{jkl}u^j u^k$ depending on points $u, v, \dot{u}, \dot{v}$ of the phase space. If at least one eigenvalue of this matrix is positive, the geodesics will exponentially diverge with time. Negative eigenvalues correspond to oscillations or to slower divergence, e.g. power-law. In our case one of the eigenvalues of the matrix $R^i{}_{jkl}u^j u^k$ vanishes (due to the static nature of the metric), two others are the roots of a quadratic equation and read

$$\lambda_{1,2} = \frac{1}{2}\left(\mathcal{B} \pm \sqrt{\mathcal{B}^2 + 4\mathcal{C}}\right), \tag{68}$$

where

$$\mathcal{B} = \frac{2uv\dot{u}\dot{v} - \beta^2(u^2 + v^2)}{\beta^2 UW} + \frac{(\dot{v}u + \dot{u}v)^2}{\beta^2 U^2 W}, \tag{69}$$

$$\mathcal{C} = \frac{v^2u^2(3\beta^2 + u^2v^2)}{\beta^2 U^2 W}, \tag{70}$$

with W and U being positive functions

$$W = 1 + \frac{u^2v^2 - \dot{u}^2 - \dot{v}^2}{\beta^2}, \qquad U = 1 + \frac{u^2v^2}{\beta^2}.$$

It is easy to see that both non-zero eigenvalues (68) are real, one positive and the other negative. Therefore, for any finite β, there exist locally divergent geodesics, and it can be expected that for any β the motion will remain chaotic. But in the limit $\beta \to 0$ these eigenvalues tend to zero and the analysis becomes inconclusive. Qualitatively, this phenomenon of decrease of the geodesic divergence with decreasing β can be attributed to the lowering of the degree of chaoticity. It is also worth noting that for $\beta \to 0$ the second terms in the equations of motion (64-65) look like singular friction terms.

Another simple tool in the analysis of chaos is the calculation of the Lyapunov exponents defined as

$$\chi = \lim_{t\to\infty} \frac{1}{t} \log \frac{|\delta x(t)|}{|\delta x(0)|}, \tag{71}$$

where $\delta x(t)$ is a solution of the linearized perturbation equation along the chosen trajectory. Here x stands for the (four-dimensional) phase space coordinates of the original system and the Cartesian metric norm is chosen. The positive value of χ signals that the close trajectories diverge exponentially with time so the motion is unstable. Stable regular motion corresponds to zero Lyapunov exponents.

The third standard tool is the construction of Poincare sections. Recall that the Poincare section is some hypersurface in the phase space. The phase space for the system (73) is four-dimensional, but the energy conservation restricts the motion to a three-dimensional manifold. Imposing one additional constraint will fix a two-dimensional surface, and we can find a set of points corresponding to intersection of the chosen trajectory with this surface. For the regular motion one typically gets some smooth curves, while in the chaotic case the intersection points fill finite regions on the surface.

Both methods reveal that with decreasing β the degree of chaoticity is diminished, though we were not able to perform calculations for very small β to decide definitely whether the chaos is stabilized or not. But in fact the ansatz (62) is not generic enough from the point of view of the BI dynamics, since the leading at high field strength term $(F\tilde{F})^2$ is zero for it. So we can hope to draw more definitive conclusions by exploring another ansatz for which the invariant $(F\tilde{F})$ is non-zero. Such an example is provided by a simple axially symmetric configuration of the $SU(2)$ gauge field which is also parameterized by two functions of time u, v:

$$A = \mathbf{T}_1 u dx + \mathbf{T}_2 u dy + \mathbf{T}_3 v dz. \tag{72}$$

The field strength contains both electric and magnetic components

$$F = \dot{u}(\mathbf{T}_1 dt \wedge dx + \mathbf{T}_2 dt \wedge dy) + \dot{v}\mathbf{T}_3 dt \wedge dz + u^2 \mathbf{T}_3 dx \wedge dy + uv(\mathbf{T}_1 dy \wedge dz + \mathbf{T}_2 dz \wedge dx),$$

and now the pseudoscalar term $(F\tilde{F})^2$ is non-zero. Substituting this ansatz into the action we obtain the following one-dimensional lagrangian:

$$L_1 = \beta^2(1-\mathcal{R}), \quad \mathcal{R} = \sqrt{1 - \frac{2\dot{u}^2 + \dot{v}^2 - u^2(u^2+2v^2)}{\beta^2} - \frac{u^2(2\dot{u}v + \dot{v}u)^2}{\beta^4}}. \tag{73}$$

In the limit of the usual Yang-Mills theory $\beta \to \infty$ we recover the system which was considered in [63]:

$$L_{YM} = \frac{1}{2}(2\dot{u}^2 + \dot{v}^2 - u^4 - 2u^2v^2). \tag{74}$$

The corresponding potential has valleys along $v = 0$. Like in the case of the hyperbolic billiard with potential v^2u^2, the dynamics governed by the usual quadratic Yang-Mills lagrangian motion exhibits chaotic character.

However, for the NBI action with finite β the situation is different. One can use again the geodesic deviation method, the corresponding metric being:

$$ds^2 = \left(1 + \frac{u^4 + 2u^2v^2}{\beta^2}\right) dt^2 - \frac{du^2 + 2dv^2}{\beta^2} - \frac{u^2(2vdu + udv)^2}{\beta^4}. \tag{75}$$

The energy integral is given by the expression

$$E = \frac{\beta^2 + u^4 + 2v^2u^2}{\mathcal{R}} - \beta^2. \tag{76}$$

One can calculate the eigenvalues of the geodesic deviation operator as before. In general one finds that the larger eigenvalue is smaller than one in the case of the previous ansatz (62). Now the regions in the configuration space appear in which both non-zero eigenvalues are negative, this corresponds to locally stable motion. With decreasing β the relative volume of the local stability regions increases, though for every β there still exist the regions of local instability as well. To obtain more definitive conclusions we performed numerical experiments using Lyapunov exponents and Poincare sections. The results look as follows. For large β the dynamics of the system (73) is qualitatively similar to that in the ordinary Yang-Mills theory (74) and exhibits typical chaotic features. The trajectories enter deep into the valleys along $u = 0$. The Poincare sections consist of the clouds of points filling the finite regions of the plane, while the Lyapunov exponents remain essentially positive. The situation changes drastically with decreasing β. For some value of this parameter, depending on a particular choice of the initial conditions, one observes after a series of bifurcations a clear transition to the regular motion. The points on the Poincare sections $u = 0$ line up along the smooth curves (sections of a torus), while the Lyapunov exponent goes to zero. The motion becomes quasiperiodic. A typical feature of such regular motion is that the trajectory no more enters the potential valleys.

Comparing with our first system (62) we can guess that the substantial role in the chaos-order transition in the NBI theory is played by the pseudoscalar term in the action. Since generically this term does not vanish, it is tempting to say that the transition observed is a generic phenomenon in the non-Abelian Born-Infeld theory. The effect is essentially non-perturbative in α', and we conjecture that it reflects the typical smothering effect of the string non-locality on the usual stiff field-theoretical behavior. A related question is whether the cosmological singularity, which was recently shown to be chaotic at the supergravity level of string models [64], will be also regularized when higher α' corrections are included. Unfortunately there is no closed form effective action for the closed strings analogous to the Born-Infeld action for the open strings. But combining gravity in the lowest order in α' with an exact in α' matter action we can probe the nature of the cosmological singularity too.

Acknowledgements

The author (DG) is grateful to the Directors of the School and especially to Prof. V. De Sabbata for an invitation, support and a very pleasant and stimulating atmosphere during this meeting.

References

[1] J. Polchinski, *String Theory*, Vol. I and II of *Cambridge Monograph on Mathematical Physics*, Cambridge University Press, 1998.

[2] A. Giveon and D. Kutasov, Rev. Mod. Phys. **71** (1999), 983, hep-th/9802067.

[3] K. Akama, *Lect.Notes.Phys.* **176**, 267 (1982).

[4] V.A. Rubakov and M.E. Shaposhnikov, *Phys. Lett.* B **75**, 4624 (1983).

[5] N. Arkani-Hamed, S. Dimopoulos and G. Dvali, *Phys. Lett.* B **429**, 263 (1998); I. Antoniadis, N. Arkani-Hamed, S. Dimopoulos and G. Dvali, *Phys. Lett.* B **436**, 257 (1998).

[6] L. Rundall and R. Sundrum, *Phys. Rev. Lett.* **83**, 3370 (1999).

[7] L. Rundall and R. Sundrum, *Phys. Rev. Lett.* **83**, 4690 (1999).

[8] V.A. Rubakov, Phys. Usp. **44**, 871-893 (2001), Usp. Fiz. Nauk **171**, 913-9 (2001).

[9] R. Maartens, Geometry and dynamics of the brane-world. Gr-qc/0101059.

[10] F. Quevedo, *Class. Quant. Grav.* **19**, 5721-5779 (2002).

[11] P. Brax and C. van de Bruck, *Class. Quant. Grav.* **20**, R201-R232 (2003).

[12] I. Antoniadis, *Phys. Lett.* B **246**, 377 (1990).

[13] N. Kaloper, J. March-Russell, J.D. Starkman and M. Trodden, *Phys. Rev. Lett.* **85**, 928 (2000).

[14] P. Horawa and E. Witten, *Nucl. Phys.* B **460**, 506 (1996), *Nucl. Phys.* B **475**, 94 (1996).

[15] T. Shiromizu, K. Maeda and M. Sasaki, *Phys. Rev.* D **62**, 024012 (2000).

[16] P. Binetruy, C. Deffayet and D. Langlois, *Nucl. Phys.* B **477**, 285 (2000).

[17] M. Cavaglia, *Int. J. Mod. Phys.* A **18**, 1841 (2003).

[18] S.B. Giddings, E. Katz and H.S. Reall, *JHEP* **03**, 023 (2000).

[19] N.K. Dadhich, R. Maartens, P. Papadopoulos and V. Rezania, *Phys. Lett.* B **487**, 1 (2000).

[20] M. Govender and N. Dadhich, *Phys. Lett.* B **538**, 233-238 (2002).

[21] R. Emparan, G.T. Horowitz and R.C. Myers, *Phys. Rev. Lett.* **85**, 499-502 (2000).

[22] R. Emparan, J. Garcia-Bellido and N. Kaloper, *JHEP* **0301**, 079 (2003).

[23] R. Casadio, Holography and trace anomaly: what is the fate of (brane-world) black holes. Hep-th/0302171.

[24] T. Tanaka, *Prog.Theor.Phys.Suppl.* **148**, 307-316 (2003).

[25] D. Clancy, R. Guedens and A. Liddle, *Phys. Rev.* D **66**, 043513, 083509 (2002); Astro-ph/0301568.

[26] M. S. Volkov and D. V. Gal'tsov, Phys. Rept. **319** (1999), 1, hep-th/9810070.

[27] R. Bartnik and J. Mckinnon, Phys. Rev. Lett. **61** (1988), 141–144.

[28] M.S. Volkov and D.V.Gal'tsov. *JETP Lett.*, **50**, 346–350, 1989; *Sov. J. Nucl. Phys.*, **51**, 747–753, 1990.

[29] H.P. Künzle and A.K.M. Masood ul Alam. *Journ.Math.Phys.*, **31**, 928–935, 1990.

[30] P. Bizon. *Phys. Rev. Lett.*, **64**, 2844–2847, 1990.

[31] E.E. Donets, D.V. Gal'tsov, and M.Yu Zotov. *Phys. Rev.*, **D 56**, 3459–3465, 1997.

[32] D.V. Gal'tsov, E.E. Donets, and M.Yu Zotov. *JETP Lett.*, **65**, 895–901, 1997.

[33] D. V. Gal'tsov and E. E. Donets, *Comptes Rend. Acad. Sci. Paris*, t.325, Serie IIB (1997) 649-657, gr-qc/9706067.

[34] V. V. Dyadichev and D. V. Gal'tsov. *Nucl. Phys.*, **B590**, 504–518, 2000, hep-th/0006242.

[35] Dmitri Gal'tsov and Richard Kerner. *Phys. Rev. Lett.*, **84**, 5955–5958, 2000, hep-th/9910171.

[36] Marion Wirschins, Abha Sood, and Jutta Kunz. *Phys. Rev.*, **D63**, 084002, 2001, hep-th/0004130.

[37] V. V. Dyadichev and D. V. Gal'tsov. *Phys. Lett.*, **B486**, 431–442, 2000, hep-th/0005099.

[38] D. V. Gal'tsov. holes. 2001, gr-qc/0101100.

[39] Gerard Clement and Dmitri Gal'tsov. *Phys. Rev.*, **D62**, 124013, 2000, hep-th/0007228.

[40] N. Grandi, R. L. Pakman, F. A. Schaposnik, and Guillermo A. Silva. *Phys. Rev.*, **D60**, 125002, 1999, hep-th/9906244.

[41] N. Grandi, E. F. Moreno, and F. A. Schaposnik. *Phys. Rev.*, **D59**, 125014, 1999, hep-th/9901073.

[42] Dmitri Galtsov and Vladimir Dyadichev. D-branes and vacuum periodicity. Proceedings of the NATO Advanced Research Workshop "Non-commutative structures in mathematics and physics", Kiev, Ukraine, September 24-28, 2000; S. Duplij and J. Wess (Eds), Kluwer publ. 2001, p. 61-78, hep-th/0012059.

[43] Vladimir Dyadichev and Dmitri Galtsov. *Phys. Rev.*, **D65**, 124026 , 2002, hep-th/0202177.

[44] Dyadichev, V. V., D. V. Gal'tsov, A. G. Zorin, and M. Y. Zotov, *Phys. Rev.* **D65**, 084007.

[45] Ricardo Garcia-Salcedo and Nora Breton. *Int. J. Mod. Phys.*, A15:4341–4354, 2000, gr-qc/0004017.

[46] B. L. Altshuler. *Class. Quant. Grav.*, 7:189–201, 1990.

[47] Dan N. Vollick. 2001, hep-th/0102187.

[48] J Cervero and L Jacobs. *Phys. Lett.*, B 78:427–429, 1978.

[49] M. Henneaux. *J. Math. Phys.*, 23:830, 1982.

[50] Y. Verbin and A. Davidson. *Phys. Lett.*, B229:364, 1989.

[51] D. V. Gal'tsov and M. S. Volkov. *Phys. Lett.*, B256:17–21, 1991.

[52] E. E. Donets and D. V. Galtsov. *Phys. Lett.*, B294:44–48, 1992, gr-qc/9209008.

[53] D.V. Gal'tsov and V.V. Dyadichev, Astrophys. and Space Sci., 273 (2003) 667-672, hep-th/0301044.

[54] Gergely, L. A. and R. Maartens: 2002, *Class. Quant. Grav.* **19**, 213–222.

[55] Gibbons, G.W. and A.R. Steif: 1995 *Phys. Lett.* **B 346**, 255–261.

[56] G. Z. Baseian, S. G. Matinian, and G. K. Savvidy. Pisma Zh. Eksp. Teor. Fiz. **29**, 641–644 (1979).

[57] B. V. Chirikov and D. L. Shepelyansky. JETP Lett. **34**, 163–166 (1981).

[58] B. V. Chirikov and D. L. Shepelyansky. Yad. Phys. **36**, 1563–1576 (1982).

[59] A.A. Birò, Matinyan S.G., and Müller B. Chaos and Gauge theory. World Scientific, Singapore, 1994.

[60] Sergei G. Matinian, Chaos in non-Abelian gauge fields, gravity, and cosmology, gr–qc/0010054.

[61] D.V. Gal'tsov and V.V. Dyadichev, JETP Letters, 2003,vol. 77, issue 4, 184-187. hep-th/0301069.

[62] S. G. Matinian, G. K. Savvidy, and N. G. Ter-Arutunian Savvidy. *JETP* **53**, 421–425, (1981).

[63] B. K. Darian and H. P. Kunzle. system. Class. Quant. Grav. **13**, 2651–2662 (1996), gr–qc/9608024.

[64] T. Damour and M. Henneaux, Phys. Rev. Lett. **85**, 920 (2000), hep–th/0003139.

EXPERIMENTAL TEST OF A TIME-TEMPERATURE FORMULATION OF THE UNCERTAINTY PRINCIPLE

G. T. GILLIES
Department of Physics
University of Virginia
Charlottesville, VA 22904-4714, U.S.A.
E-mail: gtg@virginia.edu

S. W. ALLISON
Engineering Technology Division
Oak Ridge National Laboratory
Oak Ridge, TN 37932-6472
E-mail: allisonsw@ornl.gov

ABSTRACT

A novel form of the Heisenberg uncertainty principle, as introduced by de Sabbata and Sivaram, $\Delta T \Delta t \geq \hbar/k$, was tested using laser-induced fluorescence of 30 nm particles of YAG:Ce. The temperature-dependent fluorescence decay lifetimes of this material were measured at thermal equilibrium over the range from ≈ 285 to 350 °K. The uncertainty in temperature of ≈ 4.5 °K (as derived from the relationship between temperature and lifetime) and the measured uncertainty in decay lifetime, ≈ 0.45 ns, yielded an "internal" estimate of $\Delta T \Delta t \geq 2.0 \times 10^{-9}$ °K s, which is ≈ 263 times larger than $\hbar/k = 7.6 \times 10^{-12}$ °K s. An "external" estimate of $\Delta T \Delta t \geq 4.5 \times 10^{-11}$ (which is ≈ 6 times $\hbar/k$) is derived from the measured uncertainty in the temperature of the sample and the measured uncertainty in lifetime. These results could be argued to increase by a factor of 5.6 if signal averaging is taken into account. If our approach is valid, then the findings are not inconsistent with the limitations predicted by this formulation of a time-temperature uncertainty principle and they imply the existence of a type of thermal quantum limit. The approach might thus open a path towards improved precision in the determination of the Boltzmann constant based on thermal squeezing techniques.

1. Introduction

The Heisenberg uncertainty principle is one of the fundamental tenets of quantum mechanics and many different interpretations of it have been developed, including number-phase [1], space-time [2] and information-entropy [3] versions, to list just a few. (Unnikrishnan and Gillies [4] are preparing a review of the topic which will survey the relevant literature.) In 1992, de Sabbata and Sivaram [5] developed still another formulation of it while investigating how torsion in general relativity links to defects in space-time topology. A model of the thermodynamics of the early universe served as a test bed for some of their predictions, and the general relationship between thermal energy and temperature, $E \sim kT$, led them to a phenomenological statement of the uncertainty principle of the form

V. de Sabbata et al. (eds.),
The Gravitational Constant: Generalized Gravitational Theories and Experiments, 139–147.
© 2004 *Kluwer Academic Publishers. Printed in the Netherlands.*

$$\Delta T \Delta t \geq \hbar/k \tag{1}$$

where $\hbar$ is the Planck constant and k is the Boltzmann constant. Further details of this calculation were provided later by de Sabbata [6]. Gillies *et al* [7] are in the process of examining this relationship from a quantum-mechanical as opposed to a dimensional-analysis perspective, and the work described below was motivated originally as a means of providing some experimental guidance for that theoretical effort.

Equation (1) is a rather surprising statement that predicts a fundamental limit on the precision with which temperatures may be measured. The question that arises immediately is, "the temperature of what?" In quantum mechanics, an uncertainty relation can be established between pairs of commutative variables, eg., displacement and momentum, energy and time, and so on, but is there a quantum mechanical role for temperature in that context? Kobayashi [8-10] has explored this point and has developed quantum mechanical descriptions of thermal equilibrium states and quasi-static processes in terms of eigenstates of relative phase interactions between particles or oscillators. While not placing temperature and time in the category of commutative quantum mechanical variables, his arguments do suggest that temperature can play a legitimate role in theoretical descriptions of physical phenomena at the quantum level. In a similar way, Ford and O'Connell [11] consider the effects of finite temperature on the spreading of a wave packet and show how this can lead to the calculation of decoherence rates.

Our approach to this question has been to carry out a laboratory study of a hybrid atomic/molecular process in which there is a well known coupling between time and temperature in a way that intermingles quantized atomic transitions with mesocsopic crystalline structures: the measurement of the temperature-dependent fluorescence decay lifetimes of rare-earth-doped ceramic oxides, which are often called thermographic phosphors. Fluorescence as referred to here is the emission occurring from electronic transitions and is usually in the visible region of the spectrum. It is generally true that the fluorescence spectral properties of any material will change with temperature. This is so in part because the Boltzmann distribution governs the partitioning of the populations in the various participating vibrational levels of the ground, excited and emitting states. A change in intensity distribution (including width and position of spectral lines) results since individual oscillator strengths vary in accordance with the selection rules and the Franck-Condon principle. The temperature dependence of these processes can be striking when there is competition with states which contend for nonradiative deexcitation pathways. The rate of change of the population of an emitting state, 2, to a ground state, 1, is the sum of a constant, purely radiative spontaneous emission, $A_{1,2}$, and a nonradiative component, $W_{1,2}$, which is temperature dependent. The decay rate, κ, is given by

$$\kappa = 1/\tau = A_{1,2} + W_{1,2} \tag{2}$$

where τ is the measured lifetime. One model for temperature dependence for some fluorescent materials is based on thermal promotion to a nonemitting electronic state followed by nonradiative relaxation. This is the "charge transfer state" (CTS) model and has been used by Fonger and Struck [12,13] to describe the thermal quenching of fluorescent materials. Depending on the temperature, the vibrational distribution in the excited state will be given by a Boltzmann distribution. At low temperature, states in resonance with the CTS are improbable. However, at higher temperature a significant fraction of the distribution will be at energies corresponding to the CTS and the transition to the CTS becomes likely. The rate is therefore proportional to exp(-E/kT) where E is the energy difference between the excited electronic state of the dopant atom and the charge transfer state of the host ceramic matrix. Hence, equation (2) can be extended into an expression that predicts how decay time and temperature couple, which in simplest form [12,14] is

$$\tau = [(1/\tau_0) + b\cdot e^{(-E/kT)}]^{-1} \qquad (3)$$

Here τ_0 is the unperterbed decay time and b is the rate to and from the CTS, which is on the order of the lattice vibration rate, $> 10^9\ s^{-1}$ [12,13]. A measurement of the fluorescence decay lifetime can thus be used to establish the temperature of the host lattice, with the uncertainty in the measured lifetime thus directly affecting the uncertainty in the resulting determination of lattice temperature. The measurement is intrinsically quantum mechanical in nature, in that a discrete atomic transition (or set of transitions) is being observed. One can then adopt the view that the resulting fluorescence lifetime and the associated lattice temperature are at least functionally commutative, in that a measurement of either one determines the other.

To explore what this might mean in terms of inter-related uncertainties, consider the following qualitative argument which, although non-rigorous, does illustrate a possible path toward an expression like equation (1). If the fluorescence lifetime and lattice temperature of a thermographic phosphor behave as discussed above, then the uncertainty in the energy (per dopant atom) coupled by the charge transfer state into the lattice vibrational mode, ΔE_{CTS}, would be related to the uncertainty in the measured fluorescence decay lifetime, $\Delta\tau$, via the standard uncertainty relation

$$\Delta E_{CTS}\, \Delta\tau \geq \hbar/2 \qquad (4)$$

For a crystalline structure that is otherwise at thermal equilibrium with its surroundings, this uncertainty in CTS energy will manifest as an uncertainty in the thermal energy of the atom in the corresponding lattice mode. At room temperature, the lattice vibrational energy per atom per mode, E_v, would be

$$E_v = \tfrac{1}{2}kT \qquad (5)$$

More generally, equation (5) would have a different numerical factor on the right hand side, depending on the specific solid state model that was employed. For example, the Heat Rule of Dulong and Petit [15] would require that $\tfrac{1}{2} \rightarrow 3$, and the models of Einstein [16], Debye [17], and Born and von Kármán [18] (and of course all of the more modern ones) would correct the functional form of the whole expression to take into account low temperature behavior, etc. However, our experiment was done in the range of room temperatures. Therefore, keeping to the simplest case for illustrative purposes, we then note that equation (5) leads to an uncertainty in the lattice vibrational energy of

$$\Delta E_v = \tfrac{1}{2}k\, \Delta T \qquad (6)$$

It is then straightforward to relate the uncertainty in the energy of the lattice vibrational mode to the uncertainty in the CTS energy by rewriting equation (6) as

$$\Delta E_{CTS} = \tfrac{1}{2}k\Delta T \tag{7}$$

By combining equations (4) and (7), we then arrive at

$$\Delta T \Delta \tau \geq \hbar/k \tag{8}$$

which has the form of equation (1), but with the result now being at least heuristically motivated by standard considerations of solid state physics as applied to an experimental system that has a well known coupling between quantum and thermal effects. Equation (8) suggests that the Heisenberg uncertainty principle not only establishes the familiar quantum limits encountered, eg., in gravitational wave detection, but also fixes an unavoidable *thermal quantum limit* on atomic processes that are energetically coupled to a heat bath.

In what follows, we discuss the results of our recent measurements of the laser-induced fluorescence decay lifetime of 30 nm particles of YAG:Ce [19] and use the measured uncertainties in the lifetimes and in the YAG:Ce temperature to test the predictions of equation (8). The use of nanoscale particles is of particular interest, since for small enough particles the exact ratio of dopant atoms to host matrix atoms can either be measured or estimated precisely, and this would help lead to improved versions of the model discussed above. This work has been carried out within the broader context of our ongoing studies of remote thermometry with thermographic phosphors. An extensive review of the whole field is available elsewhere [20].

2. Experimental Arrangement

The thermographic phosphor used for this work was a powder of 30 nm particles of YAG:Ce, the specific stoichiometry of which was $Y_3Al_5O_{12}$:Ce. Two different samples of it were used; in both cases, a thin layer of it was bonded onto an aluminum substrate and placed in a Petri dish of water to help keep the sample in thermal equilibrium. The temperature of this assembly was adjusted by warming it on a laboratory hot plate. A nitrogen laser operating at 337 nm was used to excite fluorescence of the phosphor, which was subsequently detected by a photomultiplier tube. A 400 MHz waveform processing oscilloscope (Tektronix TDS3052) was used to capture the decaying exponential signals transduced by the photomultiplier, and the digitized data were then stored on a laboratory computer for subsequent analysis. Fiber optic cables were used to convey the excitation and fluorescence signals to and from the instrumentation. A 700 nm optical bandpass filter was used to prevent the shorter wavelengths from saturating the photomultiplier tube. To check for effects of wavelength selection, optical bandpass filters at 550 nm and 620 nm were also used in some of the experiments.

The thermometer used to measure the phosphor temperature was a Wahl 392MX platinum resistance thermometer with a Wahl 202 immersion-tip probe. It had a resolution of 0.1 °C and its calibration was traceable to the U.S. NIST. A measurement of its stability was obtained by placing the probe into one well of a Wahl IPR-4-110 ice point reference bath. The standard deviation of 32 measurements made at 30 s intervals was < 0.05 °C. During the fluorescence experiments, the tip of the probe was kept in mechanical contact with the aluminum substrate in the immediate proximity of the phosphor layer. The thermometer data were read and recorded manually.

The fluorescence decay lifetimes were extracted from the transduced photo-optical signals following acquisition and averaging of 32 individual waveforms at each temperature tested. This was done to improve the signal-to-noise ratio and thus enable a more accurate determination of the waveform parameters. (The implications of the use of signal averaging within the context of testing the predictions of an uncertainty principle are discussed in Section 4 below.) Although the fluorescence decay signals were not pure single-exponential waveforms [19], such a fit was still good enough to permit the estimation of an equivalent single exponential decay time and the related uncertainty at each of the temperatures at which measurements were made. Further details about the preparation and use of the phosphor, and the performance characteristics of the measurement system are available elsewhere [19]. It is useful to note that the data obtained in these experiments are very similar in nature to those from many other such studies in our laboratories [20].

3. Results

The decay times measured as a function of temperature are plotted in Figure 1. Also shown there are curves of the calculated decay time estimates from equation (3), using 34.75 ns as the nominal value of the unquenched time constant and E = 1090 cm^{-1} as the electronic-state to charge-transfer-state energy difference. The standard error in the fit of the data to the model was $\Delta\tau/\tau \approx 1.3\%$, and the dashed lines show the range of models allowed by that uncertainty.

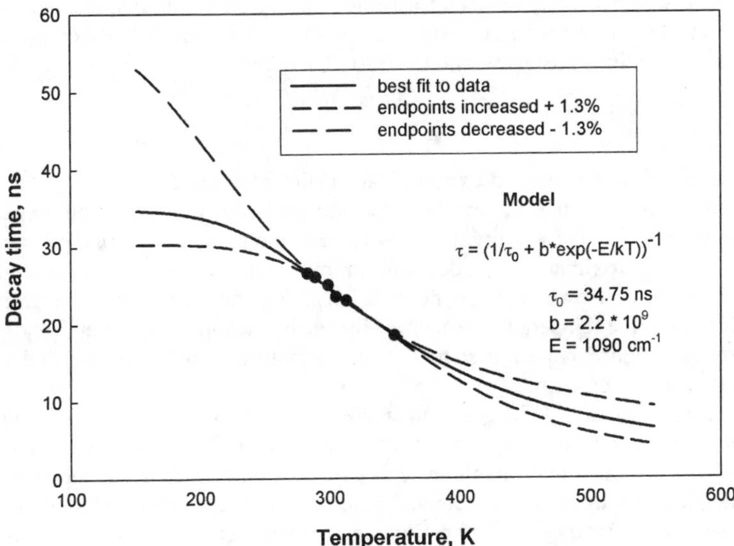

Figure 1. Decay time vs. temperature for 30 nm particles of YAG:Ce excited by a 337 nm pulsed nitrogen laser. The data are fit by equation (3) and the range of allowed models is shown by the dashed lines.

Over the range of temperatures at which the data were taken, one can approximate the roll-off in the thermal quenching with a straight line: from 285 to 350 °K, the decay time constant drops from ≈ 26 ns to ≈ 19 ns, yielding an equivalent slope having an absolute value of n = 0.1 ns $°K^{-1}$.

Systematic uncertainties were not assigned to the data points in this preliminary study, but previous work in our laboratories has shown that detailed determinations of individual overall uncertainties for this type of measurement system are often as small as $\Delta\tau/\tau < 0.03\%$ [20,21]. Instead, we used the standard error relative to the fitted curve, 1.3%, to arrive at a conservative value of $\Delta\tau = 0.013 \bullet 34.75$ ns ≈ 0.45 ns as the uncertainty of the decay time constants. The associated uncertainty in the temperature, ΔT, was found by taking $\Delta T = \Delta\tau/n \approx$ 4.5 °K. We refer to this as an "internal" estimate of the uncertainty in temperature, since it is derived directly from the uncertainty in the lifetime by way of the relationship between lifetime and temperature expressed in equation (3). The resulting product $\Delta T\Delta\tau = 2.0 \times 10^{-9}$ °K s > $\hbar/k$ = 7.6×10^{-12} °K s, thus satisfying equation (8) by a factor of 263.

An alternative approach involves what we term an "external" estimate of ΔT. In this case, the uncertainty in temperature is given by the maximum measured departure from thermal equilibrium observed during a given set of lifetime measurements. During the course of our experiments, no such fluctuations larger than the resolution of the platinum resistance thermometer were observed, hence we take ΔT = 0.1 °K to be the external estimate. The uncertainty in the decay lifetime remains the same, thus $\Delta T\Delta\tau = 4.5 \times 10^{-11}$ °K s > $\hbar/k = 7.6 \times 10^{-12}$ °K s, which also satisfies equation (8), in this case by a factor of ≈ 6. Of course, this is a much more debatable number in that it is not traceable to the quasi-commutative characteristic of the variables used to obtain the internal result. Even so, it does legitimately characterize the maximum thermal fluctuations present within the YAG:Ce sample.

4. Discussion

There is a time-honored axiom that an extraordinary claim requires extraordinary evidence to support it. We do not contend that the results presented here constitute sufficient experimental evidence to support explicitly the validity of a new form of the uncertainty principle, because of the assumptions and approximations made both in our own theoretical model and in the application of the data to it. However, the concept is intriguing and phosphor thermography does have the merit of linking the temperature of a crystalline specimen to the quantum physics of the fluorescence process. Hence we suggest that it is worth further investigation and offer the following comments and discussion.

The 30 nm crystals of YAG:Ce are an interesting material to use for this application because at a 1% dopant concentration, there is essentially one activator (Ce) atom per particle. This should simplify the modeling of the fluorescence process on the one hand, and also open the door to the application of single-atom spectroscopic methods on the other. The latter could be especially important in eliminating any competitive inter-particle effects that might act to modify the decay lifetimes. For instance, there was some evidence [19] that multi-photon processes were at work in our measurements, since the photo-optical decays were not purely single exponential. Ideally, this should be avoided and single nanocrystal spectroscopy would be a step in that direction. A number of other phosphors are also candidate materials including nanocrystalline $Y_2O_3:Eu^{3+}$, the particles of which are even smaller in size (8 to 12 nm) and which we have used in other applications [22].

Heating of the sample by the laser excitation is also a potential concern. The 337 nm nitrogen laser used here was chosen because with it, we could maximize the brightness of fluorescence and thus reduce the reliance on signal averaging for extraction of the decay time constants (for the significance of this, see below). However, the cost of this in terms of effects on the nanoscale heat balance of the samples under illumination are yet to be determined. Fortunately, any such problems can be ameliorated significantly by using low intensity sources like blue LEDs which have recently been shown to fluoresce materials of this type at very low power levels [23]. Note, however, that a switch from laser-based illumination to use of an LED would eliminate the coherence of the excitation source, with ramifications for the phasing of the non-radiative decays via the CTS.

As mentioned above, the decay lifetimes were obtained via an algorithm that used as its input an average of 32 separate fluorescence signals. The signal-to-noise ratio improvement provided by the signal averaging reduces the uncertainty in τ, but it raises an important issue that is central to this approach. In a series of concise papers on the topic, Ramsey [24-26] explored the nature of quantum mechanical limitations in precision measurement, with emphasis on the application of the uncertainty principle. He points out there that the uncertainty principle is valid for "a single measurement on a single system" [25]. The question then becomes one of defining what it means to make a single measurement of fluorescence decay lifetime. Of course, values of τ and $\Delta\tau$ can be extracted from a single exponentially decaying fluorescence pulse. However, if the detection system's shot noise on a single pulse is too large to permit extraction of a decay time constant, then signal averaging is needed. On the other hand, that does not (nor can it) mean that the intrinsic atomic decay time constant *per se* has been changed by the averaging. The central issue is really one of reducing the instrumentation system's noise floor to the point where one can observe the variations in the decay time constant arising from the presence of the charge transfer state in the crystal, which is the mechanism through which the temperature dependence of the fluorescence process arises. We would argue that it is not until that noise floor is reached that one has actually made a meaningful measurement of the decay time constant, and the uncertainty in it at that point is what is of interest. If a sufficiently noiseless instrumentation system existed, this measurement could be made with a single atomic excitation. However, ambiguity arises when considering that 32 individual atomic excitations were needed to arrive at a determination of the lifetime and its uncertainty, yet the fluctuations in that lifetime should ideally not be different from those made with a single measurement with the sufficiently noiseless system (if the system's noise is the only corrupting factor). One solution suggested by Ramsey is that the uncertainty principle's limit is still valid in such a case, but that the equivalent numerical size of it be reduced by $\sqrt{N}$ where N is the number of different systems that are measured [25]. It is not clear that this solution should be applied here, but if one does so, the right-hand side of equation (8) is reduced by a factor of $\approx$ 5.6, or equivalently, our internal and external estimates of $\Delta T \Delta\tau$ become 1472 and 33.6 times greater than $\hbar/k$, respectively. We are continuing to investigate this issue.

An interesting conjecture that derives from equation (8) is the possibility of realizing thermally squeezed states in closed systems. Aharonov and Reznik [27] have shown that conserved observables within closed systems are generally subject to an uncertainty relation. The radiative plus nonradiative components of the energy from the fluorescence decay of the dopant atom's excited state are a conserved quantity in that the sum of them must equal to the excitation energy. Therefore, in analogy with the generation of optically squeezed states of light, one should be able to generate thermally squeezed charge transfer states in the host matrix. An implication of that possibility might be the creation of phonon bundles in the lattice that have phase-coherence with the optical excitation pulses. From a metrological persepctive an interesting outcome of attaining thermal squeezing might be the creation of new techniques for determining the Boltzmann constant. The question would be this: can one manipulate the experimental conditions of an apparatus in such a way that the values of ΔT and Δt could be used to reduce the uncertainty in k since $\hbar$ is presently known to within 80 ppb? As shown in our work, a value of < 1 ns for Δt ($\equiv \Delta\tau$) is not unusual in experimental settings. In a thermally isolated system, could one then employ suitable thermal squeezing and averaging schemes to obtain an uncertainty of 10^{-3} °K or less for ΔT? If so, the right hand side of equation (8) might be re-established with a one-order-of-magnitude or more lower uncertainty, thus allowing a corresponding improvement in the knowledge of k. The ratio of the uncertainties of $\hbar$ and k as stated in the most recent re-adjustment of the fundamental constants [28] is roughly 20:1, so the uncertainty in k could conceivably be improved by a factor of 20 before the uncertainty in $\hbar$ further limits improvement in k. Any improvement in the value of the Boltzmann constant would automatically lead to an improvement in the value of the Stefan-Boltzmann constant, σ, since it is defined to be $\sigma = (\pi^2/60)\bullet(k^4/\hbar^3c^2)$, and presently known with an uncertainty of 7 ppm.

5. Conclusions

We have investigated the laser induced fluorescence of nanocrystalline YAG:Ce as a means of testing a novel time-temperature version of the Heisenberg uncertainty principle. If valid, our experimental approach to this problem leads to a value of $\Delta T\Delta t \geq 2.0 \times 10^{-9}$ °K s, which is approximately 263 times the limit called for by $\Delta T\Delta t \geq \hbar/k$ without consideration of signal averaging, and 1472 times it with $\sqrt{N}$ averaging, where N = 32 in our case. A number of fundamental questions and unsettled issues make our result open to interpretation and possible invalidation, but if ultimately found to have a workable basis, such a technique might introduce a means of achieving thermally squeezed states, with implications for improved determination of the Boltzmann constant.

Acknowledgments

We thank Olivier Pfister and Rogers C. Ritter of the University of Virginia, C. S. Unnikrishnan of the Tata Institute for Fundamental Research, and M. R. Cates of Oak Ridge National Laboratory for many useful discussions. We thank A. J. Sanders of the University of Tennessee for advice with the fitting of the data. The YAG:Ce used here was obtained from A. J. Rondinone of Oak Ridge National Laboratory. We especially acknowledge Venzo de Sabbata of the University of Bologna for his insights in developing the original statement of the time-temperature uncertainty principle studied in our work.

References

[1] Heffner H 1962 *Proc. IRE* **50** 1604
[2] Yoneya T 1989 *Mod. Phys. Lett. A* **4** 1587
[3] Abe S and Suzuki N 1990 *Phys. Rev. A* **41** 4608
[4] Unnikrishnan C S and Gillies G T 2003 *Rep. Prog. Phys.* invited for submission
[5] de Sabbata V and Sivaram C 1992 *Found. Phys. Lett.* **5** 183

[6] de Sabbata V 1994 *Nuov. Cim. A* **107** 363

[7] Gillies G T, Unnikrishnan C S, Pfister O and Ritter R C 2003 *Phys. Rev. Lett.* in preparation

[8] Kobayashi T 1995 *Phys. Lett. A* **207** 320

[9] Kobayashi T 1996 *Phys. Lett. A* **210** 241

[10] Kobayashi T 2001 *Nuov. Cim. B* **116** 493

[11] Ford G W and O'Connell R F 2002 *Am. J. Phys.* **70** 319

[12] Fonger W H and Struck C W 1970 *J. Chem. Phys.* **52** 6364

[13] Grattan K T V and Zhang Z Y 1995 *Fiber Optic Fluorescence Thermometry* (London: Chapman & Hall)

[14] Struck C W and Fonger W H 1971 *J. Appl. Phys.* **42** 4515

[15] Blackmore J S 1969 *Solid State Physics* (Philadelphia: Saunders) pp 78-81

[16] Einstein A 1907 *Ann. Phys.* **22** 180

[17] Debye P 1912 *Ann. Phys.* **39** 789

[18] Born M and von Kármán T 1912 *Z. Physik.* **13** 297

[19] Allison S W, Gillies G T, Rondinone A J and Cates M R 2002 *Nanotechnology* submitted

[20] Allison S W and Gillies G T 1997 *Rev. Sci. Instrum.* **68** 2615

[21] Dowell L J, 1989 *Investigation and Development of Phosphor Thermometry* Ph.D. Dissertation, University of Virginia

[22] Gillies G T, Allison S W and Tissue B M 2002 *Nanotechnology* **13** 484

[23] Allison S W, Cates M R and Gillies G T 2002 *Rev. Sci. Instrum.* **73** 1832

[24] Ramsey N F 1987 *IEEE Trans. Instrum. Meas.* **36** 155

[25] Ramsey N F 1992 *J. Phys. II (Paris)* **2** 573

[26] Ramsey N F 1995 *Physica Scripta* **T59** 26

[27] Aharonov Y and Reznik B 2000 *Phys. Rev. Lett.* **84** 1368

[28] Mohr P J and Taylor B N 2000 *Rev. Mod. Phys.* **72** 351

THE NEWTONIAN GRAVITATIONAL CONSTANT: PRESENT STATUS AND DIRECTIONS FOR FUTURE RESEARCH

G. T. GILLIES
Department of Physics
University of Virginia
Charlottesville, VA 22904-4714, U.S.A.
E-mail: gtg@virginia.edu

C. S. UNNIKRISHNAN
Tata Institute for Fundamental Research
Homi Bhabha Road
Mumbai 400 005, India
E-mail: unni@tifr.res.in

ABSTRACT

The recent measurements of the absolute value of the Newtonian gravitational constant, G, are discussed and some thoughts on the direction of future work in this area are presented. Although some of the recent searches for variations in G are also mentioned, special attention is paid to one branch of that work focusing on the possibility of detecting variations in G with temperature. Past experimental searches for such an effect are reviewed.

1. Introduction

Modern science finds itself in a rather anomalous situation with respect to the determination of the absolute value of the Newtonian constant of gravitation, G. While steady progress has been made in improving the accuracy of the other fundamental constants of physics, the most recent CODATA adjustment of those values actually increased the uncertainty in G to reflect ongoing disagreement among the values that were considered in that readjustment, especially the significantly different value measured at the Physikalisch-Technische Bundesanstalt (PTB) in Germany. Table 1 shows the last three accepted CODATA values of G and calls attention to this increase in the uncertainty. For comparison, we also show a value of G as derived from the original measurement by Cavendish of the mean density of the Earth, reported in his famous paper in the *Philosophical Transactions of the Royal Society of London* in 1798.

V. de Sabbata et al. (eds.),
The Gravitational Constant: Generalized Gravitational Theories and Experiments, 149–155.
© 2004 *Kluwer Academic Publishers. Printed in the Netherlands.*

Table 1. Historical listing of accepted values of G during the last 30 years.

Date	G ($m^3\ kg^{-1}\ s^{-2}$)	Reference
1798	6.75 ± 0.05	Cavendish (1798)
1973	6.672 6 ± 0.000 5	Cohen and Taylor (1973)
1986	6.672 59 ± 0.000 85	Cohen and Taylor (1987)
1998	6.673 ± 0.010	Mohr and Taylor (2000)

Many workers at laboratories around the world have taken up the challenge of improving this situation, and in the next section a short overview of the status of the recent measurements is given.

2. Overview of Recent Measurements of the Absolute Value of G

There are several reviews of the early history of the measurement of G. A comprehensive bibliographic listing of virtually all experiments on G from the time of Newton through the mid-1980s was prepared by Gillies (1987). Likewise, a detailed review of the status of the measurements of the absolute value of G covering the period from roughly 1965 through 1995, including a discussion of searches for variations in G, was also prepared by Gillies (1997). Most of the groups in the world active in the measurement of G were represented at a meeting celebrating the the Bicentennial of the Cavendish Experiment, held at the Institute of Physics in London in 1998. A large number of excellent papers on the subject were included in a proceedings of the conference, which was published by the Institute of Physics in *Measurement Science and Technology* (Speake and Quinn, editors, 1999). Gillies (1999) published a brief review of the contemporary situation in that proceedings. Another recent review has been written by Luo and Hu (2000). The anomalous result of the PTB (Michaelis *et al*, 1996), which was several standard deviations away from the 1986 CODATA value, prompted much discussion at the time. Gillies (1997, 1998) tabulated most of the experiments that were in progress at the time and which were aimed at resolving the situation by hopefully arriving at improved values of G. Since then, several of the efforts that were in progress have produced results and the relevant values of G are shown in Table 2.

Table 2. Recently measured values of G.

Experiment/Feature	G (m^3 kg^{-1} s^{-2})	Reference
Evacuated torsion balance	6.674 0 ± 0.000 7	Bagley and Luther (1997)
Adjustable mass positions	6.672 9 ± 0.000 5	Karagioz, Izmaylov, Gillies (1998)
Free-fall gravimeter	6.687 3 ± 0.009 4	Schwartz, Robertson *et al.* (1998)
High-Q torsion pendulum	6.669 9 ± 0.000 7	Luo, Hu, Fu, Fan, Tang (1999)
Rotating torsion balance	6.674 215 ± 0.000 092	Gundlach and Merkowitz (2000)
Torsion strip suspension	6.675 59 ± 0.000 27	Quinn, Speake *et al.* (2001)
Balance beam weights	6.674 07 ± 0.000 22	Schlamminger *et al.* (2002)

Preliminary results for measurements by the groups at Industrial Research Ltd. in New Zealand and at the University of Wuppertal in Germany were reported at the 2002 CPEM conference, and further reports on those experiments will be forthcoming. Also, Newman et al. (2003) are continuing their work with the cryogenic torsion balance for measuring G.

Many workers seek to use the measurements of the absolute value of the gravitational constant as a window into other aspects of physics. For instance, Lopes (2003) has explored the possibility that the experimental uncertainties in Newton's constant have implications for models of the standard evolution of the Sun. Ricci and Villante (2002) have studied related issues, and find that the solar data cannot be used to constrain the value of G to better than ≈ 1%. Unnikrishnan and Gillies (2000) have appealed to the unique characteristics of the University of Zűrich experiment to set a stringent upper limit on the Majorana gravitational absorption coefficient. They have also used the results from the same experiment to examine the equivalence of the active and passive gravitational mass of leptons (Unnikrishnan and Gillies, 2001a, 2001b). Gershteyn *et al.* (2002) argued that a directional anisotropy in the gravitational force was revealed in their measurements of G, but Unnikrishnan and Gillies (2002) pointed out that the level at which it was claimed made it incompatible with other measurements and observations.

3. Variations in G

The temporal constancy of the gravitational constant provides an important potential theoretical window on the extra-dimensional theories, as do searches for variation in G with distance at small scales of distance. Gillies (1997) has reviewed much of the relatively recent observational work on searches for a non-zero $\dot{G}/G$. Since that time, there have been a host of papers on possible theoretical implications of the time variation of G and of the fine structure constant, and the reader is referred to arXiv.org to examine the relevant literature. Ritter and Gillies (these proceedings) discuss the situation regarding the sub-millimeter measurements of G and their relationship to extra-dimensional theories.

One area of study in this field that is beginning to see growing interest is that of temperature-dependent gravity. The history of the topic goes back over 200 years, to balance-beam experiments of Count Rumford (1799) who noted that the weight of a heated body does not change by more than 10^{-6} of its unheated weight. With the advent of the theory of relativity, though, it should be acceptable that an increase in the total self-energy of a body via heating would lead to an increase in its weight. Assis and Clemente (1993) calculated the size of such an effect using the heat rule of Dulong and Petit, and found that a 100 ºC change in temperature should lead to a change in weight of an iron sphere of $\Delta W/W \approx 10^{-12}$. Interestingly, this prediction is not far outside the possibility of experimental test, though considerable improvement in sensitivity is still required. Existing limits on the size of such an effect are already quite good, see for example Poynting and Phillips (1905) who used a beam balance to arrive at a value of $\Delta G/G \approx 10^{-10}$ per ºC. In fact, the situation was somewhat complicated by Shaw (1916) in the early part of the 20th century, who claimed to find an effect at the level of $\Delta G/G \approx 1.2 \times 10^{-5}$ per ºC using a torsion balance to isolate the temperature dependent gravitational force, but it was later revealed that a systematic effect was responsible for this finding. Correction of that effect resulted in an upper limit on the size of it of $< 10^{-6}$ per ºC (Shaw and Davy, 1923). In recent times, new classes of theoretical predictions call for variations of G with thermodynamic temperature, following predictions that are generally of the form $G = G_0(1 - \alpha G_0 T^2)^{-1}$ where G_0 is the "zero temperature" (i.e., the laboratory) value of G and the constant α is approximately the ratio of the electromagnetic to gravitational coupling constants (Massa, 1989). The implications of a temperature-dependent G for interpretation of the Planck units are still under discussion.

From the relativistic relation between mass and energy one expects to find a change in the weight of a body when heated up, and this is fundamentally different from postulating that the gravitational coupling itself might depend on the thermodynamic temperature. The thermal energy has to couple to gravity exactly the same way as any other form of energy or mass if the equivalence principle has to hold. It is interesting to note that there is already direct experimental evidence that thermal energy obeys the equivalence principle, and this comes from the Lunar Laser Ranging experiment. Since a large fraction of the mass of the earth is at a temperature of several hundred degrees C, and since the moon is relatively cold, one expects an anomalous acceleration differential, $\Delta a/a$ larger than 10^{-12} between the earth and the moon towards the sun if thermal energy did not universally couple to gravity. LLR is sensitive to differential accelerations below 10^{-13} cm/s^2, one may conclude that thermal energy does couple to gravity as expected.

4. Directions for Future Work

Although many novel laboratory approaches for determining G and searching for variations in it have been conceived and tested over the years, the inability to fully isolate the gravitational force between test bodies from that due to the background field of the Earth will eventually place a limit on what can be accomplished with terrestrial experiments. The future of such measurements may then well be in space. Sanders and Gillies (1996) have reviewed the proposals for the various possible space-based measurements of G, and have discussed the Satellite-Energy Exchange (SEE) method in earlier volumes of the Proceedings of the Erice Schools. The technological possibilities for actually carrying out such a mission are good, and work towards initiation of it is underway. Of more immediate interest are the efforts aimed at measuring the value of G at inter-mass spacings below one millimeter. A number of groups are employing novel configurations of atomic force microscopes and other techniques for this purpose. As mentioned above, see the survey prepared by Ritter and Gillies (these proceedings) for a brief review of such experiments.

The scatter in the various measured values of G has always been a topic of interest, and many workers have searched for classes of systematic effects that could offer at least partial explanations for this situation, eg., variations due to anelasticity in the torsion fibers. At present, there are several independent means of measuring G that are being investigated and employed. These range from the versions of the standard torsion pendulum, to Fabry-Perot detectors of suspended masses, to precision beam balances, to free-fall gravimeters. Because it is difficult to see how the same classes of systematic effects could be at work in all of these different devices, many have instead sought for new physics as an alternative explanation of the scatter in the measurements of G. A large number of proposals for novel physical effects, including gravitational anomalies and anisotropies, have been suggested and discussed in the literature, eg., see the reviews by Gillies (1987, 1997), but nothing of that type has yet achieved any degree of substantial acceptance within the community. Even so, the weakness of the gravitational force and the extreme difficulty encountered in measuring it do leave the door open for unexpected insights into the physics of gravity that might ultimately resolve the disparities in the values of G.

Acknowledgments

The authors thank Prof. Venzo de Sabbata and the Ettore Majorana Centre for Scientific Culture for the generous hospitality offered during the XVIIIth Course of the School of Cosmology and Gravitation.

References

Assis A K T and Clemente R A 1993 *Nuov. Cim.* **108 B** 713

Bagley, C H and Luther, G G 1997 *Phys. Rev. Lett.* **78** 3047

Cavendish, H 1798 *Phil. Trans. R. Soc. London* **88** 469

Cohen E R and Taylor B N 1973 *J. Phys. Chem. Ref. Data* **2** 663

Cohen E R and Taylor B N 1987 *Rev. Mod. Phys.* **59** 1121

Gershteyn M L, Gershteyn L I, Gershteyn A and Karagioz O V 2002 physics/0202058

Gillies G T 1987 *Metrologia* **24**(Suppl.) 1

Gillies G T 1997 *Rep. Prog. Phys.* **60** 151

Gillies G T 1998 *Spin in Gravity*, P. G. Bergmann *et al.*, eds. (Singapore: World Scientific), 86

Gillies G T 1999 *Meas. Sci. Technol.* **10** 421

Gundlach J H and Merkowitz S M 2000 *Phys. Rev. Lett.* **85** 2869

Karagioz O V, Izmaylov V P and Gillies G T 1998 *Gravit. Cosmol.* **4** 239

Lopes I 2003 *Astrophys. Space Sci.* **283** 553

Luo J, Hu Z-K, Fu X-H, Fan S-H, Tang M-X 1999 *Phys. Rev. D* **59** 042001

Luo J and Hu Z-K 2000 *Class. Quantum Grav.* **17** 2351

Massa C 1989 *Helv. Phys. Act.* **62** 420

Michaelis W, Haars H and Augustin R 1996 *Metrologia* **32** 267

Mohr P J and Taylor B N 2000 *Rev. Mod. Phys.* **72** 351

Newman R 2003 gr-qc/0303027 pps. 16-17

Poynting J H and Phillips P 1905 *Proc. R. Soc. London A* **76** 445

Quinn T J, Speake C C, Richman S J, Davis R S and Picard A 2001 *Phys. Rev. Lett.* **87** 111101

Ricci B and Villante FL 2002 *Phys. Lett. B* **549** 20

Rumford (Benjamin Thompson) 1799 *Phil. Trans. Roy. Soc. London* **89** 27

Sanders A J and Gillies G T 1996 *Riv. Nuov. Cim.* **19** 1

Schlamminger S, Hozschuh E and Kűndig W 2002 *Phys. Rev. Lett.* **89** 161102

Schwarz J P, Robertson D S, Niebauer T M and Faller J E 1998 *Science* **282** 2230

Shaw P E 1916 *Phil. Trans. Roy. Soc. London* **216** 349

Shaw P E and Davy N 1923 *Phys. Rev.* **21** 680

Speake C C and Quinn, T J (eds.) 1999 *Meas. Sci. Technol.* **10**

Unnikrishnan C S and Gillies G T 2000 *Phys. Rev. D* **61** 101101

Unnikrishnan C S and Gillies G T 2001a *Phys. Lett. A* **288** 161

Unnikrishnan C S and Gillies G T 2001b *Gravit. Cosmol.* **7** 251

Unnikrishnan C S and Gillies 2002 *Phys. Lett. A* **305** 26

Toward testing the fundamental physics by SNIa data

Włodzimierz Godłowski,* Marek Szydłowski,† and Wojciech Czaja‡

Astronomical Observatory, Jagiellonian University,

30-244 Krakow, ul. Orla 171, Poland

Abstract

It is demonstrated how fundamental theory like M-theory can be tested by SNIa data. We discuss astronomical tests (redshift-magnitude relation, angular size minimum, age of the universe) for brane cosmologies with various types of matter sources on the brane. We also find the limits on exotic physics (brane models in Randall - Sundrum version) coming from CMB and BBN.

We show that in the case of dust matter on the brane, the difference between the best-fit model with a Λ-term and the best-fit brane models becomes detectable for redshifts $z > 1.2$. We show that brane models predict brighter galaxies for such redshifts that predicted by the Perlmutter model. We demonstrate that the fit to supernovae data can also be obtained if we admit super-negative dark energy on the brane instead of Λ-term. We prove that the minimum of the angular size of galaxies is very sensitive to the amount of both dark radiation and brane tension which are unique characteristics of brane models. In opposition to ordinary radiation which increases the minimum, the negative values of dark radiation and brane tension can decrease it. The minimum disappears for some large negative contribution of dark radiation. We show also that the age of the universe can increase significantly for all brane models with phantom matter on the brane. We suggest that in the near future when new data will be avaliable, the errors in estimations of the $\Omega_{i,0}$ parameters will significantly decrease, and stronger limits can be achieved.

PACS numbers: 11.15.-w,11.27.+d,98.80.Es,98.80.Cq

Keywords: brane models, dynamics, accelerating universe, testing by SNIa data

*Electronic address: godlows@oa.uj.edu.pl

†Electronic address: uoszydlo@cyf-kr.edu.pl

‡Electronic address: czja@oa.uj.edu.pl

V. de Sabbata et al. (eds.),
The Gravitational Constant: Generalized Gravitational Theories and Experiments, 157–191.
© 2004 *Kluwer Academic Publishers. Printed in the Netherlands.*

I. INTRODUCTION

Nowadays, the main points of interest in fundamental physics are superstring cosmology and brane cosmology. Superstring cosmology has already developed its successes beginning in the early nineties with its pre-big-bang scenario [1–3]. On the other hand, the idea of brane cosmology was introduced by Hořava and Witten [4] who considered strong coupling limit of heterotic $E_8 \times E_8$ superstring theory, i.e., M-theory. This limit results in 'exotic' [5, 6] Kaluza-Klein type compactification of $N = 1$, $D = 11$ supergravity on a S^1/Z_2 orbifold. Randall and Sundrum [7, 8] developed a similar scensrio to that of Hořava-Witten's. It was motivated by the hierarchy problem in particle physics [9–11]. As a result, they obtained a 5-dimensional spacetime (bulk) with Z_2 symmetry with two/one 3-brane(s) embedded in (bulk space) it.

In one-brane scenario [7] the brane appears at the $y = 0$ position, where y is an extra dimension coordinate and the 5-dimensional spacetime is an anti-de Sitter space with negative 5-dimensional cosmological constant. The extra dimension can be infinite due to the exponential 'warp' factor in the metric. In the simplest case the metric induced on a 3-brane is a Minkowski metric (energy momentum tensor of matter vanishes). However, the requirement to allow matter energy-momentum tensor on the brane, leads to breaking the conformal flatness in the bulk, which is related to the appearance of the Weyl curvature in the bulk [12, 13]. The full set of 5-dimensional and projected 4-dimensional equations has been presented in Ref. [14]. Exact analytic brane solutions have been studied by many authors [15–18], as well as the dynamical system approach is used [19–21]. Brane models admit new parameters, which are not available in standard cosmology (brane tension λ and dark radiation $\mathcal{U}$).

From the astronomical observations of supernovae Ia [22–24], one knows that the Universe is now accelerating and the best-fit flat model is for the 4-dimensional cosmological constant density parameter $\Omega_{\Lambda_{(4)},0} = 0.72$ and for the dust density parameter $\Omega_{m,0} = 0.28$ (index "0" refers to the present). In other words, only the exotic matter (negative pressure) in standard cosmology can lead to this global effect of acceleration. This is known as quintessence [25] idea. On the other hand, in brane cosmology the quadratic contribution to the energy density ϱ^2, (even for small negative pressure,) in the field equations effectively acts as the positive pressure, and makes brane models less accelerating.

In this paper we will study brane models with zero, subnegative ($p > -\rho$), and supernegative ($p < -\rho$) pressure matter (cf. Ref. [26, 27]). We will also take into account the fact that the new High z Supernovae Search Team (SNAP, GOAL) results show that the $z > 1$ supernovae are brighter than expected [28], which suggests a deviation of redshift-magnitude relation for large redshifts.

In Section 2 we introduce some notations. In Section 3 we present the formalism and write down the redshift-magnitude relations for brane models with dust $p = 0$, cosmic strings $p = -(1/3)\varrho$, domain walls $p = -(2/3)\varrho$ and phantom matter $p = -(4/3)\varrho$ on the brane. The sub-negative ($-\varrho < p < 0$) or super-negative ($p < -\varrho$) pressure is considered as an alternative to cosmological constant. In fact, super-negative pressure violates the weak energy condition, which implies violation of Lorentz invariance [26, 27]. According to standard definition, for matter not violating the weak energy condition $\rho + p \geq 0$, the following inequalities are Lorentz invariant $T^0_0 \geq 0, T^0_0 \geq |T^\mu_\nu|, R^0_0 \geq 0$ (T is an energy-momentum tensor and R is Ricci scalar). This means that if these inequalities are fulfilled in at least one reference frame, they are also fulfilled in any other reference frame. The latter equation of state is well within the current limit on the dark energy (perfect fluid which provodes to acceleration) which comes from seven cosmic microwave background experiments and reads $-1.62\varrho < p < -0.74\varrho$ [29]. An independent constraint for brane models of Ref. [30] also requires $p < -\varrho$. In Section 4 we fit the theoretical curves to supernovae data of Refs. [22–24]. In Section V we study the influence of brane parameters (dark radiation and brane tension) on other astronomical quantities, such as the angular size of galaxies and the age of the universe. This also includes the influence of super-negative pressure matter. In the Sections VI and VII other limits for brane model, going from CMB and BBN, are discussed. In Section VIII we give our conclusions.

II. DENSITY PARAMETERS Ω AND SOME EXACT SOLUTIONS

In our previous paper [21] we applied methods of dynamical systems to study the behaviour of the Randall-Sundrum models and described the formalism in which one is able to obtain the observational quantities for these models. Our formalism gives a natural base to express dynamical equations in terms of dimensionless observational density parameters $\Omega_{i,0}$ and to compare the results with data. However, before we study these quantities in detail, following the discussion of Refs. [31, 32, 34], we introduce the notation in which it is easy to tell which models are exactly integrable. The Friedmann equation for brane universes takes the form [21]

$$\frac{1}{a^2}\left(\frac{da}{dt}\right)^2 = \frac{C_\gamma}{a^{3\gamma}} + \frac{C_\lambda}{a^{6\gamma}} - \frac{k}{a^2} + \frac{\Lambda_{(4)}}{3} + \frac{C_{\mathcal{U}}}{a^4}, \tag{1}$$

where $a(t)$ is the scale factor, $\Lambda_{(4)}$ is the 4-dimensional cosmological constant, γ is the barotropic index ($p = (\gamma - 1)\varrho$), and we have defined the appropriate constants ($8\pi G = 1$)

$$C_\gamma = \frac{1}{3}a^{3\gamma}\varrho, \tag{2}$$

$$C_\lambda = \frac{1}{6\lambda}a^{6\gamma}\varrho^2, \tag{3}$$

$$C_{\mathcal{U}} = \frac{2}{\lambda}a^4\mathcal{U}. \tag{4}$$

It is easy to notice that the following cases can be exactly integrable in terms of elliptic functions [31]: $\gamma = 0$ (cosmological constant), $\gamma = 1/3$ (domain walls) and $\gamma = 2/3$ (cosmic strings). The first case is the easiest, since the first two terms on the right-hand-side of (1) play the role of cosmological constants similar to $\Lambda_{(4)}$

$$\frac{1}{a^2}\left(\frac{da}{dt}\right)^2 = \left(C_\gamma + C_\lambda + \frac{\Lambda_{(4)}}{3}\right) - \frac{k}{a^2} + \frac{C_{\mathcal{U}}}{a^4}. \tag{5}$$

The next two cases involve terms which were already integrated in the context of general relativity. For $\gamma = 1/3$ (domain walls on the brane) the term C_γ in (1) scales as domain walls in general relativity while the term with C_λ scales as cosmic strings (curvature) in general relativity, i.e.,

$$\frac{1}{a^2}\left(\frac{da}{dt}\right)^2 = \frac{C_\lambda - k}{a^2} + \frac{C_\gamma}{a} + \frac{\Lambda_{(4)}}{3} + \frac{C_{\mathcal{U}}}{a^4}. \tag{6}$$

For $\gamma = 2/3$ (cosmic strings) the term C_γ in (1) scales as cosmic strings in general relativity, while the term with C_λ scales as radiation in general relativity (compare [37, 38]), i.e.,

$$\frac{1}{a^2}\left(\frac{da}{dt}\right)^2 = \frac{C_{\mathcal{U}} + C_\lambda}{a^4} + \frac{C_\gamma - k}{a^2} + \frac{\Lambda_{(4)}}{3}, \tag{7}$$

Then, the problem of writing down the exact solutions, which are elementary, reduces to the repetition of the discussion of Refs. [31]. We will not discuss it here. For other values of $\gamma = 4/3; 1; 2$ the terms of $1/a^8$ and $1/a^{12}$ type appear, and the integration involves hyperelliptic integrals.

Now, we consider the case of $\gamma = -1/3$ (phantom matter) [26], i.e.,

$$\frac{1}{a^2}\left(\frac{da}{dt}\right)^2 = C_\gamma a + C_\lambda a^2 - \frac{k}{a^2} + \frac{C_{\mathcal{U}}}{a^4} + \frac{\Lambda_{(4)}}{3}. \tag{8}$$

Going back to observational quantities, we now define the dimensionless observational density parameters [32–34]

$$\begin{aligned} \Omega_\gamma &= \frac{1}{3H^2}\varrho, \\ \Omega_\lambda &= \frac{1}{6H^2\lambda}\varrho^2, \\ \Omega_{\mathcal{U}} &= \frac{2}{H^2\lambda}\mathcal{U}, \\ \Omega_k &= -\frac{k}{H^2a^2}, \\ \Omega_{\Lambda_{(4)}} &= \frac{\Lambda_{(4)}}{3H^2}, \end{aligned} \tag{9}$$

where the Hubble parameter H, and the deceleration parameter q read as

$$\begin{aligned} H &= \frac{\dot{a}}{a}, \\ q &= -\frac{\ddot{a}a}{\dot{a}^2}, \end{aligned} \tag{10}$$

so that the Friedmann equation (1) can be written down in the form

$$\Omega_\gamma + \Omega_\lambda + \Omega_k + \Omega_{\Lambda_{(4)}} + \Omega_{\mathcal{U}} = 1. \tag{11}$$

The exact solution for the special case of fluid with negative pressure (which in certainly available for $\gamma = 0; 1/3; 2/3$) (see Refs. [34, 35]) can be useful in writing down an explicit redshift-magnitude formula and other observational quantities for the brane models in order to study their compatibility with astronomical data, which is the subject of this paper.

It is useful to study the formalism which allows to formulate our dynamics in terms of dynamical systems, in which the coefficients are related to dimensionless density parameters Ω_i. Then we rewrite the basic equation for FRW brane models in R-S version to a new form using dimensionless quantities

$$x \equiv \frac{a}{a_0}, \quad t \to T : T \equiv |H_0|t,$$

where index "0" refers to the present moment of time and T is a new time variable, which is monotonic function of original time variable t.

The basic equations now take the form

$$\frac{\dot{x}^2}{2} = \frac{1}{2}\Omega_{k,0} + \frac{1}{2}\sum_i \Omega_{i,0} x^{2-3w_i} = -V(x), \tag{12}$$

$$\ddot{x} = \frac{1}{2}\sum_i \Omega_{i,0}(2-3w_i)x^{1-3w_i}, \tag{13}$$

where $p_i = (w_i - 1)\rho_i$ is the form of equation of state for the matter considered on the brane $\Omega_i = (\Omega_\gamma, \Omega_\lambda, \Omega_\mathcal{U}, \Omega_\Lambda)$ and $\sum_i \Omega_{i,0} = 1$.

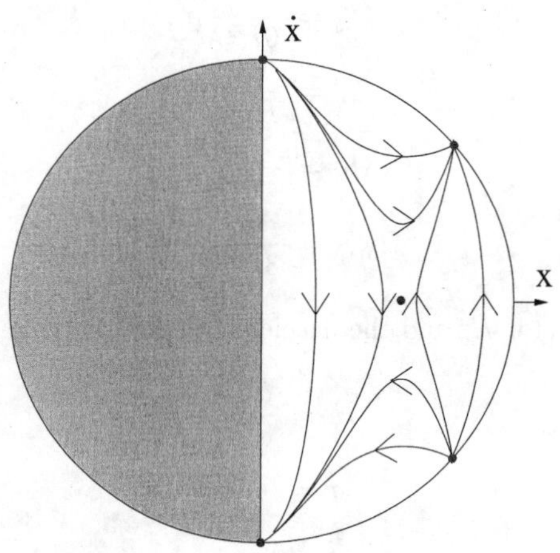

FIG. 1: The phase portrait (on compactified Poincare sphere) of the brain models with dust on the brane. We find a static critical point at finite domain and four stable and unstable nodes at infinity. Of course this system is structurally stable.

The above equations can be represented as a two-dimensional dynamical system

$$\dot{x} = y,$$
$$\dot{y} = \frac{1}{2}\sum_i \Omega_{i,0}(2-3w_i)x^{1-3w_i} = -\frac{\partial V}{\partial x}, \tag{14}$$

where $w_\mathcal{U} = 4/3$ like for radiation matter and $w_\lambda = 2\gamma$ for matter $p = (\gamma - 1)\rho$ on the brane, now dot denotes the differentiation with respect to T.

Therefore the dynamics of the brane models can be always reduced to the form of the Friedmann models of G.R. with a certain kind of noninteracting fluids (scalling with scale factor like radiation and stiff matter if $\gamma = 1$ (dust) is considered on the brane).

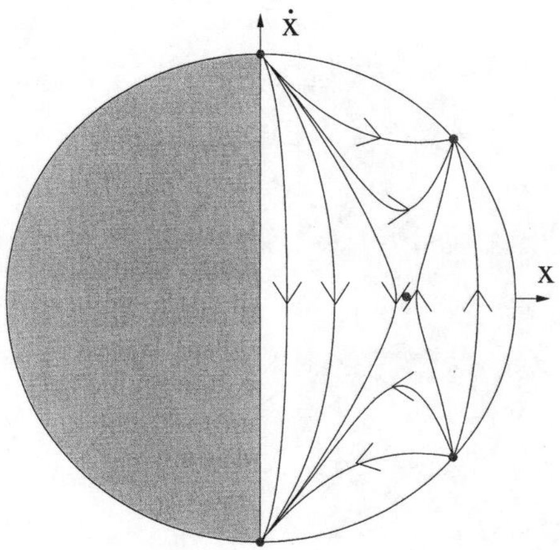

FIG. 2: The phase portrait (on compactified Poincare sphere) of the brain models with string matter on the brane. We can observe the topological equivalence, preserving the orientation of phase curves, for both portraits from Fig. 1 and Fig. 2.

Of course, the dynamical system (14) has the Hamiltonian

$$\mathcal{H} = \frac{p_x^2}{2} + V(x), \tag{15}$$

where $V(x) = -1/2(\Omega_{k,0} + \sum_i \Omega_{i,0} x^{2-3w_i})$ plays the role of potential for particle-universe $p_x = \dot{x}$.

The trajectories of the system lie on the zero energy level $\mathcal{H} \equiv 0$.

Let us consider the dynamics of brane models with dust on the brane ($w_\gamma = 1$, $w_\lambda = 2$), cosmological constant ($w_\Lambda = 0$) and dark radiation ($w_\mathcal{U} = 4/3$). The phase portrait is shown in Fig. 1.

$$\begin{aligned} \dot{x} &= y, \\ \dot{y} &= -\frac{1}{2}\Omega_{m,0}x^{-2} + \Omega_{\Lambda,0}x - 2\Omega_{\lambda,0}x^{-5} - \Omega_{\mathcal{U},0}x^{-3}, \end{aligned} \tag{16}$$

where for curvature fluid we can put $w_k = 1/3$ (like for strings) and $\sum_{\mathcal{U},0}$ can be either positive or negative.

The first integral of system (16) is:

$$\frac{\dot{x}^2}{2} + V(x) = 0.$$

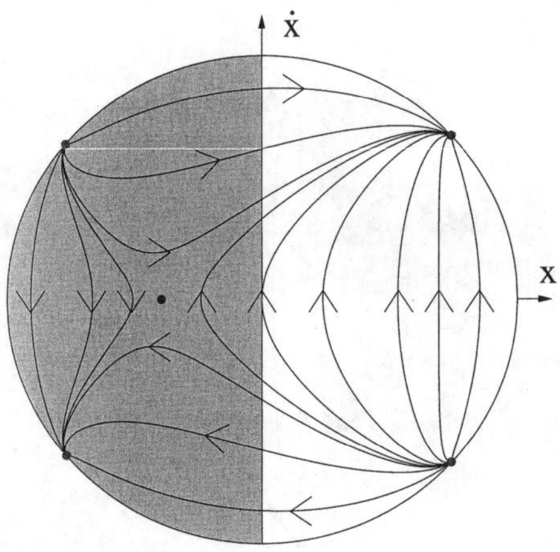

FIG. 3: This figure shows a phase portrait of the brain models with "topological defects matter" on the brane. In this case the static critical point is translated to the non-physical region (shaded region).

Thus trajectory of the flat model ($\Omega_{k,0} = 0$) divides the phase plane $(x, \dot{x})$ into the domains occupied by closed and open models respectively to $\Omega_{k,0} < 0$ and $\Omega_{k,0} > 0$.

Let us note that if we consider dust on the brane, brane effects do not drive the acceleration but if we consider matter with the negative pressure on the brane then brane effects cause the acceleration of the universe. This situation is illustrated in the Fig. 3, where instead of the dust matter, the topological effects are considered on the brane.

$$\begin{aligned} \dot{x} &= y, \\ \dot{y} &= \frac{1}{2}\Omega_{top.,0} + \Omega_{\Lambda,0}x - \Omega_{\mathcal{U},0}x^{-3}, \end{aligned} \tag{17}$$

Let us note that in this case the effects of brane are dynamically equivalent to effects of additional component of dark energy scaling like string (or curvature fluid) $p_\lambda = -1/3\rho_\lambda$.

III. REDSHIFT-MAGNITUDE RELATIONS FOR BRANE UNIVERSES

Cosmic distances measured like the luminosity distance, depends sensitively on the spatial geometry (curvature) and the dynamics. Therefore, the luminosity will depend on the present densities of different components of matter and their equations of state. For this

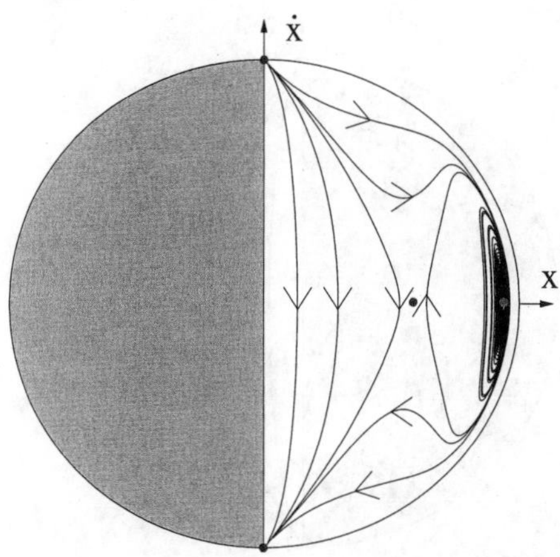

FIG. 4: The figure 4 illustrates the dynamics of brane models with phantom matter on the brane and dark radiation $\Omega_{\mathcal{U},0} = 0.2$. In this case negative brain tension is assumed. On the line $\dot{x} = 0$ we find two static critical points. They are representing by saddle point and center. Let us note that trajectories around the centre are representing oscillating models without singularity.

reason, a redshift-magnitude relation for distant galaxies is proposed as a potential test for brane cosmological models. Its point is to determine cosmological parameters as given by the formulae (9) and (10).

Let us consider an observer located at $r = 0$ at the moment $t = t_0$ who receives a light ray emitted at $t = t_1$ from the source of the absolute luminosity L located at the radial distance r_1. The redshift z of the source is related to the scale factor $a(t)$ at the two moments of evolution by $1+z = a(t_0)/a(t_1) \equiv a_0/a$. If the apparent luminosity of the source as measured by the observer is l, then the luminosity distance d_L of the source is defined by the relation

$$l = \frac{L}{4\pi d_L^2}, \tag{18}$$

where

$$d_L = (1+z)a_0 r_1. \tag{19}$$

For historical reasons, the observed and absolute luminosities are defined in terms of K-corrected observed and absolute magnitudes m and M, respectively ($l = 10^{-2m/5} \times 2.52 \times 10^{-5} erg\, cm^{-2}\, s^{-2}$, $L = 10^{-2M/5} \times 3.02 \times 10^{35} erg\, s^{-2}$) [36]. Eq.(18) expressed in terms of m

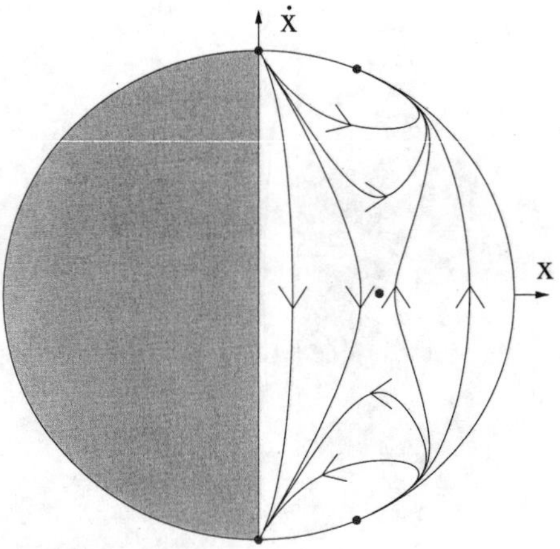

FIG. 5: The phase portrait for the brain models with phantom matter on the brane, positive dark radiation and positive cosmological constant. $(\Omega_{d,0}, \Omega_{\lambda,0}, \Omega_{\mathcal{U},0}, \Omega_{\Lambda,0}) = (0.2, 0.01, 0.09, 0.7)$.

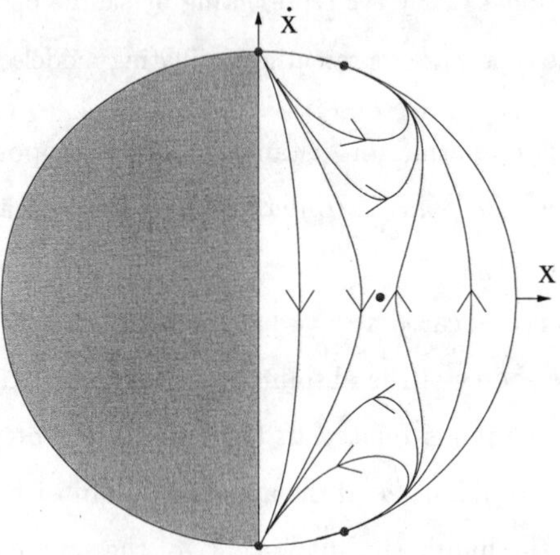

FIG. 6: The phase portrait for the brain models with phantom matter on the brane, positive dark radiation and positive cosmological constant. $(\Omega_{d,0}, \Omega_{\lambda,0}, \Omega_{\mathcal{U},0}, \Omega_{\Lambda,0}) = (0.7, 0.01, 0.09, 0.2)$.

and M yields

$$m(z) = \mathcal{M} + 5\log_{10}\{\mathcal{D}_L(z)\}, \tag{20}$$

where

$$\mathcal{M} = M - 5\log_{10} H_0 + 25, \tag{21}$$

and

$$\mathcal{D}_L(z) \equiv H_0 d_L(z), \tag{22}$$

is the dimensionless luminosity distance.

In order to compare this with the supernova data, we compute the distance modulus

$$\mu_0 = 5\log(d_L) + 25$$

where d_L is in Mps. The quality of a fit is characterised by the parameter

$$\chi^2 = \sum_i \frac{|\mu_{0,i}^0 - \mu_{0,i}^t|}{\sigma_{\mu 0,i}^2 + \sigma_{\mu z,i}^2}.$$

In above expression, $\mu_{0,i}^0$ is the measured value, $\mu_{0,i}^t$ is the value calculated in the model described above, $\sigma_{\mu 0,i}^2$ is the measurement error, $\sigma_{\mu z,i}^2$ is the dispersion in the distance modulus due to peculiar velocities of galaxies. We shall assume that SNe measurments come with uncorelated Gaussian errors, in which case the likehood functions $\mathcal{L}$ can be determined from chi-squared statistic [22, 24] $\mathcal{L} \propto \exp(-\chi^2/2)$.

By using the FRW metric we obtain coordinate distance r_1 which appears in (19) as

$$\chi(r_1) = \int\limits_{a_0/(1+z)}^{a_0} \frac{da}{a\dot{a}} = -\int\limits_{r_1}^{0} \frac{d\dot{r}}{\sqrt{1-kr^2}}, \tag{23}$$

with

$$\begin{aligned} \chi(r_1) &= \sin^{-1} r_1 \qquad for \qquad k=+1 \\ \chi(r_1) &= r_1 \qquad for \qquad k=0 \\ \chi(r_1) &= \sinh^{-1} r_1 \qquad for \qquad k=-1. \end{aligned} \tag{24}$$

From Eqs. (1) and (23) for brane models we have

$$\chi(r_1) = \frac{1}{a_0 H_0} \int\limits_0^z \left\{ \Omega_{\lambda,0} \left(1+z'\right)^{6\gamma} + \Omega_{\mathcal{U},0} \left(1+z'\right)^4 + \right.$$
$$\left. \Omega_{\gamma,0} \left(1+z'\right)^{3\gamma} + \Omega_{k,0} \left(1+z'\right)^2 + \Omega_{\Lambda_{(4)},0} \right\}^{-1/2} dz', \tag{25}$$

so, after using the constraint (11) we have

$$\mathcal{D}_L(z) = \frac{(1+z)}{\sqrt{\mathcal{K}}} \xi \left(\sqrt{\mathcal{K}} \int\limits_0^z \left\{ \Omega_{\lambda,0} \left(1+z'\right)^{6\gamma} + \Omega_{\mathcal{U},0} \left(1+z'\right)^4 + \Omega_{\gamma,0} \left(1+z'\right)^{3\gamma} + \right.\right.$$
$$\left.\left. [1 - \Omega_{\gamma,0} - \Omega_{\lambda,0} - \Omega_{\Lambda_{(4)},0} - \Omega_{\mathcal{U},0}] \left(1+z'\right)^2 + \Omega_{\Lambda_{(4)},0} \right\}^{-1/2} dz' \right), \tag{26}$$

where

$$\xi(x) = \sin x \quad with \quad \mathcal{K} = -\Omega_{k,0} \quad when \quad \Omega_{k,0} < 0$$
$$\xi(x) = x \quad with \quad \mathcal{K} = 1 \quad when \quad \Omega_{k,0} = 0$$
$$\xi(x) = \sinh x \quad with \quad \mathcal{K} = \Omega_{k,0} \quad when \quad \Omega_{k,0} > 0, \tag{27}$$

and

$$\Omega_{k,0} = -\frac{k}{\dot{a}_0^2}. \tag{28}$$

Thus, for a galaxy with a presumed absolute magnitude $\mathcal{M}$, a measured apparent magnitude m and a redshift z, we can derive the values of observational parameters $\Omega_{\gamma,0}, \Omega_{\Lambda_{(4)},0}, \Omega_{k,0}, \Omega_{\lambda,0}$ from equations (20) and (26).

In this paper we will study the redshift-magnitude relations for brane models with dust, domain walls, cosmic strings and dark energy.

For $\gamma = 1$ (dust on the brane, so that we will replace Ω_γ with Ω_m) we have

$$\mathcal{D}_L(z) = \frac{(1+z)}{\sqrt{\mathcal{K}}}\xi\left(\sqrt{\mathcal{K}}\int_0^z\left\{\Omega_{\lambda,0}\left(1+z'\right)^6 + \Omega_{\mathcal{U},0}\left(1+z'\right)^4 + \Omega_{m,0}\left(1+z'\right)^3 + \Omega_{k,0}\left(1+z'\right)^2 + \Omega_{\Lambda_{(4)},0}\right\}^{-1/2} dz'\right).$$

For $\gamma = 1/3$ (domain walls on the brane [35], so that we will replace Ω_γ with Ω_w) we have

$$\mathcal{D}_L(z) = \frac{(1+z)}{\sqrt{\mathcal{K}}}\xi\left(\sqrt{\mathcal{K}}\int_0^z\left\{\Omega_{\mathcal{U},0}\left(1+z'\right)^4 + \left[1-\Omega_{w,0}-\Omega_{\Lambda_{(4)},0}-\Omega_{\mathcal{U},0}\right]\left(1+z'\right)^2 + \Omega_{w,0}\left(1+z'\right) + \Omega_{\Lambda_{(4)},0}\right\}^{-1/2} dz'\right),$$

and we have used the constraint (11) to reduce one parameter $\Omega_{\lambda,0}$.

For $\gamma = 2/3$ (cosmic strings on the brane, so that we will write $\Omega_{s,0}$ instead of Ω_γ) we have

$$\mathcal{D}_L(z) = \frac{(1+z)}{\sqrt{\mathcal{K}}}\xi\left(\sqrt{\mathcal{K}}\int_0^z\left\{\left[\Omega_{\mathcal{U},0}+\Omega_{\lambda,0}\right]\left(1+z'\right)^4 + \Omega_{m,0}\left(1+z'\right)^3 + \left[\Omega_{k,0}+\Omega_{s,0}\right]\left(1+z'\right)^2 + \Omega_{\Lambda_{(4)},0}\right\}^{-1/2} dz'\right), \tag{29}$$

where the term $\Omega_{m,0}\left(1+z'\right)^3$ was additionally introduced in order to make reference to the recent papers about AdS-CFT motivated brane models [37, 38]. In fact, in these papers the terms with $\Omega_{\mathcal{U}}$ and Ω_λ were neglected. The presence of $\Omega_{m,0}$ term is also useful in order to study some other observational tests following Ref.[34].

For $\gamma = -1/3$ (phantom matter on the brane, we will write $\Omega_{d,0}$ instead of Ω_γ) one has

$$\mathcal{D}_L(z) = \frac{(1+z)}{\sqrt{\mathcal{K}}}\xi\left(\sqrt{\mathcal{K}}\int_0^z\left\{\Omega_{d,0}\left(1+z'\right)^{-1} + \Omega_{\lambda,0}\left(1+z'\right)^{-2}\right.\right.$$
$$\left.\left. +\Omega_{\Lambda_{(4)},0} + \Omega_{k,0}\left(1+z'\right)^2 + \Omega_{\mathcal{U},0}\left(1+z'\right)^4\right\}^{-1/2} dz'\right). \tag{30}$$

IV. BRANE MODELS TESTED BY SUPERNOVAE - OBSERVATIONAL PLOTS

In this Section we test brane models using the sample of Ref. [22]. In order to avoid any possible selection effects, we use the full sample (usually, one excludes two data points as outliers and another two points presumably reddened ones from the full sample of 60 supernovae). It means that our basic sample is the sample A of Ref. [22]. We test our model using the likelihood method [24].

Firstly, we estimated the value of $\mathcal{M}$ (cf. Eq. (21)) from the sample of 18 low redshift supernovae, also testing our result by the full sample of 60 supernovae taking $\Omega_\lambda = 0$. We obtained $\mathcal{M} = -3.39$ which is in a very good agreement with results of Refs. [39] and [42] (in Ref. [42] $\mathcal{M}_c = 24.03$ for $c = 1$, i.e., $\mathcal{M} = -3.365$). Also, we obtained for the model of Ref. [22] the same value of $\chi^2 = 96.5$.

The results of our analysis are presented on Figures 7, 8, 9, 10. On Figure 7 we study the Hubblee diagram for models with dust. In Figures 8,9 we discuss models with sub-negative pressure matter $-\varrho < p < 0$, and in Figure 9 we study the model with super-negative pressure matter $p < -\varrho$.

As for the models with dust for $\Omega_{\mathcal{U},0} = 0$ (Fig. 7, Tab I), we can say that from the formal point of view we obtain the *best fit* ($\chi^2 = 94.6$) for $\Omega_{k,0} = -0.9$, $\Omega_{m,0} = 0.59$, $\Omega_{\lambda,0} = 0.04$, $\Omega_{\Lambda,0} = 1.27$, which is completely unrealistic because $\Omega_{m,0} = 0.59$ is too large in comparison with the observational limit (also $\Omega_{k,0} = -0.9$ is not realistic from the observational point of view).

However, in fact, we obtained a 3-dimensional ellipsoid in a 3-dimensional parameter space $\Omega_{m,0}$, $\Omega_{\lambda,0}$, $\Omega_{\Lambda_{(4)},0}$. Then, we have more freedom than in the case of analysis of Ref. [22] where there was only an ellipse in 2-dimensional parameter space $\Omega_{m,0}$ and $\Omega_{\Lambda_{(4)},0}$. For

TABLE I: Results of the statistical analysis for the dust matter on the brane for Perlmutter Sample A, B, C. Two upper lines for each sample are best fit model and best fit flat model for sample. Third line is "realistic" model with $\Omega_{\rm m,0} \simeq 0.3$. We also include, for smaple A the model with $\Omega_{\lambda,0} < 0$

Sample	N	$\Omega_{k,0}$	$\Omega_{\lambda,0}$	$\Omega_{m,0}$	$\Omega_{\Lambda,0}$	χ^2
A	60	-0.9	0.04	0.59	1.27	94.7
		0.0	0.09	0.01	0.90	94.7
		0.0	0.02	0.25	0.73	95.6
		0.0	-0.01	0.35	0.66	96.3
B	56	-0.1	0.06	0.17	0.87	57.3
		0.0	0.06	0.12	0.82	57.3
		0.0	0.02	0.25	0.73	57.6
C	54	0.0	0.04	0.21	0.73	53.5
		0.0	0.04	0.21	0.73	53.5
		0.0	0.02	0.27	0.71	53.6

a flat model $\Omega_{k,0} = 0$ we obtain "corridors" of possible models (1). Formally, the best-fit flat model is $\Omega_{m,0} = 0.01$, $\Omega_{\lambda,0} = 0.09$ $\chi^2 = 95.7$ which is again unrealistic ($\chi^2 = 95.7$). It clearly shows that statistical analysis is not sufficient to discriminate between "available" models and we should investigate other extragalactic astronomy tests (especially, the estimations of $\Omega_{m,0}$ and $\Omega_{k,0}$ are useful). In the realistic case (i.e., with $\Omega_{m,0}$ which is in agreement with the observational limit for the mass of the clusters of galaxies) we obtain for a *flat* model $\Omega_{m,0} = 0.25$, $\Omega_{\lambda,0} = 0.02$, $\Omega_{\Lambda_{(4)},0} = 0.73$ ($\chi^2 = 95.6$) (see Fig. 7). One should note that all realistic brane models require also the presence of the positive 4-dimensional cosmological constant $\Omega_{\Lambda_{(4)},0} \sim 0.7$).

A question arises, if we could fit a model with negative $\Omega_{\lambda,0}$? For instance, for a flat Universe we could obtain the model with $\Omega_{m,0} = 0.35$ (which is too much in comparison with the observational limit of the mass of the cluster of galaxies) $\Omega_{\lambda,0} = -0.01$, i.e., $\Omega_{\Lambda,0} = 0.66$ ($\chi^2 = 96.3$).

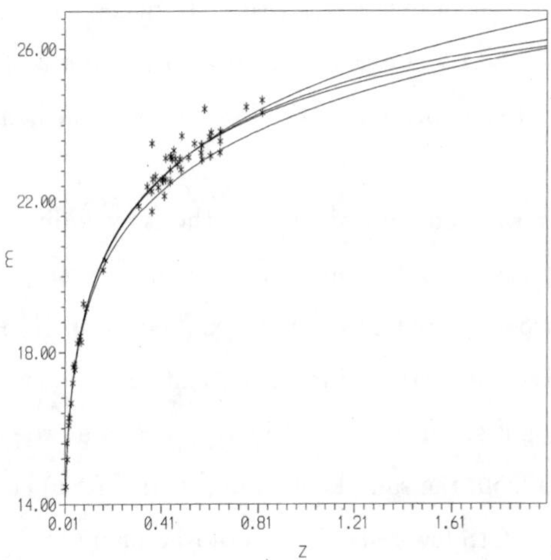

FIG. 7: Redshift-magnitude relations for $\gamma = 1$ brane universes (dust on the brane). The top line is best-fit flat model of Ref. [22] with $\Omega_{m,0} = 0.28$, $\Omega_{\Lambda_{(4)},0} = 0.72$. The bottom line is a pure flat model with the cosmological constant $\Omega_{\Lambda_{(4)},0} = 0$. Between these two lines the plots for brane models are located with $\Omega_{\lambda,0} \neq 0$: lower - the best-fit model $\Omega_{k,0} = -0.9$, $\Omega_{m,0} = 0.59$, $\Omega_{\lambda,0} = 0.04$, $\Omega_{\Lambda,0} = 1.27$; higher - the best-fit *flat* model $\Omega_{m,0} = 0.25$, $\Omega_{\lambda,0} = 0.02$, $\Omega_{\Lambda_{(4)},0} = 0.73$.

In Figure 7 we present plots of redshift-magnitude relation (29) for the supernovae data (marked with asterisks). The top line is the best-fit flat model with $\Omega_{m,0} = 0.28$, $\Omega_{\Lambda_{(4)},0} = 0.72$ (Perlmutter model [22]). The bottom line is a pure flat model without the cosmological constant ($\Omega_{\Lambda_{(4)},0} = 0$. $\Omega_{m,0} = 1$, i.e. E-deS model). The brane models with $\Omega_{\lambda,0} \neq 0$ are situated between them; the lower corresponds to the best-fit model and the upper the best-fit flat one. One can observe that the difference between brane models and a pure flat model with $\Omega_{\Lambda_{(4)},0} = 0$ is significant for z between 0.6 and 0.7, whereas it significantly decreases for the highest redshifts. It is an important difference between brane models and the Perlmutter model, where the difference between the Perlmutter model and a pure flat model (without cosmological constant) increases for high redshifts. It gives us a possibility to discriminate between models with Λ-term and $\Omega_{\lambda,0} = 0$, and the brane models ($\Omega_{\lambda,0} \neq 0$) when the data from high-redshift supernovae $z > 1$ are available. Let us note that for the present SNIa data brane models fit it only marginally better than the Perlmutter model (Ref. [22]). It should also be mentioned that brane models predict that high-redshift $z \geq 1$ supernovae should be significantly brighter than for purely relativistic Λ-term models which is in agreement with the measurement of the $z = 1.7$ supernova [28]. In other words, if the high-redshift $z > 1$ supernovae are brighter than expected by the Perlmutter model, the brane universes would be the reality.

One should note that we made our analysis without excluding any supernovae from the sample. However, from the formal point of view, when we analyze the full sample A, all analyzed models should be rejected even on the confidence level of 0.99. One of the reason could be the fact that assumed errors of measurements are too small. However, in majority of papers another solution is suggested. Usually, they exclude two supernovae as outliers and another two as reddened from the sample of 42 high-redshift supernovae and eventually two outliers from the sample of 18 low-redshift supernovae (samples B and C of Ref. [22]). We decided to use the full sample A as our basic sample because a rejection of any supernovae from the sample can be the source of lack of control for selection effects. It is the reason that we decided to make our analysis using samples B and C, too. It does not significantly change our results, however, increases the quality of the fit. Formally, the best fit for the sample B is (56 supernovae) ($\chi^2 = 57.3$): $\Omega_{k,0} = -0.1$ $\Omega_{m,0} = 0.17$, $\Omega_{\lambda,0} = 0.06$, $\Omega_{\Lambda_{(4)},0} = 0.87$. For the flat model we obtain ($\chi^2 = 57.3$): $\Omega_{m,0} = 0.12$, $\Omega_{\lambda,0} = 0.06$, $\Omega_{\Lambda_{(4)},0} = 0.82$. while for "realistic" model ($\Omega_{m,0} = 0.25$, $\Omega_{\lambda,0} = 0.02$) $\chi^2 = 57.6$. Formally, the best fit for the sample C (54 supernova) ($\chi^2 = 53.5$) gives $\Omega_{k,0} = 0$ (flat) $\Omega_{m,0} = 0.21$, $\Omega_{\lambda,0} = 0.04$, $\Omega_{\Lambda_{(4)},0} = 0.75$, while for "realistic" model ($\Omega_{m,0} = 0.27$, $\Omega_{\lambda,0} = 0.02$) $\chi^2 = 53.6$. It again confirms our conclusion that on the base of a pure statistical analysis we could only select the "corridor" of admitted models. However, if we assume that the Universe is flat $\Omega_{k,0} = 0$ (or nearly flat) we obtain estimation for $\Omega_{m,0}$ and $\Omega_{\lambda,0}$ which seem to be realistic.

One should note that we can also separately estimate the value of $\mathcal{M}$ for the sample B and C. We obtained $\mathcal{M} = -3.42$ which is again in a very good agreement with results of Ref. [39] (for a combined sample one obtains $\mathcal{M} = -3.45$). However, if we use this value in our analysis, it does not change significantly our results (χ^2 does not change more than 1 which is a marginal effect for χ^2 distribution for 53 or 55 degrees of freedom).

In Fig. 8 we present a redshift-magnitude relation (29) for brane models with $\gamma = 2/3$ (cosmic strings on the brane). We assumed $\Omega_{\mathcal{U},0} = 0$. The best-fit model is for: $\Omega_{k,0} = -0.6$, $\Omega_{s,0} = 0.12$, $\Omega_{\lambda,0} = 0.33$, $\Omega_{\Lambda_{(4)},0} = 1.15$ ($\chi^2 = 95.5$). The best-fit flat model is: $\Omega_{s,0} = 0.09, \Omega_{\lambda,0} = 0.17, \Omega_{\Lambda_{(4)},0} = 0.74$ ($\chi^2 = 95.6$). It means that we cannot avoid cosmological

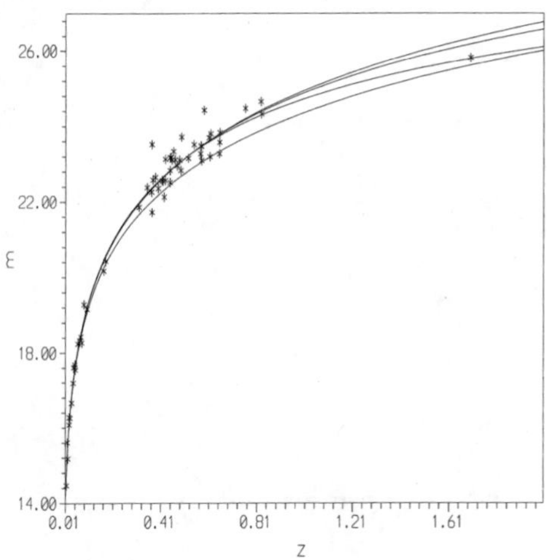

FIG. 8: A redshift-magnitude relation for $\gamma = 2/3$ brane universes (cosmic strings on the brane). The top line is the best-fit flat model of Ref. [22] with $\Omega_{m,0} = 0.28$, $\Omega_{\Lambda_{(4)},0} = 0.72$. The bottom line is a pure flat model without the cosmological constant $\Omega_{\Lambda_{(4)},0} = 0$. Between these two lines the best-fit flat model and best fit non-flat model are presented. The best-fit flat model (higher): $\Omega_{s,0} = 0.09, \Omega_{\lambda,0} = 0.17, \Omega_{\Lambda_{(4)},0} = 0.74$. The best-fit non-flat model (lower): $\Omega_{s,0} = 0.12, \Omega_{\lambda,0} = 0.33, \Omega_{\Lambda_{(4)},0} = 1.15, \Omega_{k,0} = -0.6$.

constant to fit observations, and we have checked that there is no real fit of the model with $\Omega_{\Lambda_{(4)},0} = 0$. This suggests that we need strongly negative pressure matter to avoid the presence of the cosmological term. The high-redshift supernova of Ref. [28] has also been included. Similar model has recently been studied in Ref. [38].

In Fig. 9 we present a redshift-magnitude relation (29) for brane models with $\gamma = 1/3$ (domain walls on the brane). From the formal point of view, the best fit is ($\chi^2 = 95.5$) for $\Omega_{k,0} = -1.0$, $\Omega_{\gamma,0} = 1.$, $\Omega_{\lambda,0} = 0.98$ and $\Omega_{\Lambda_{(4)},0} = 0.02$. For the best-fit flat model ($\Omega_{k,0} = 0$) we have ($\chi^2 = 96.7$): $\Omega_{\gamma,0} = 0.68$, $\Omega_{\lambda,0} = 0.3$ and $\Omega_{\Lambda_{(4)},0} = 0.7$. If it is assumed that $\Omega_{\Lambda_{(4)},0} = 0$ then the best fit would be ($\chi^2 = 95.7$) for $\Omega_{k,0} = -1.0$, $\Omega_{\gamma,0} = 1.1$, $\Omega_{\lambda,0} = 0.9$. For the best-fit flat model ($\Omega_{k,0} = 0$) we have ($\chi^2 = 97.0$): $\Omega_{\gamma,0} = 0.8$, $\Omega_{\lambda,0} = 0.2$.

In Fig. 10 we present a redshift-magnitude relation (30) for brane models with phantom matter ($\gamma = -1/3$) with $\Omega_{\mathcal{U}} > 0$ (this term is neaccesary in such type of model). To obtain an acceptable fit, $\Omega_{\mathcal{U}}$ should be as large as $\simeq 0.2$. Note that the theoretical curves are very

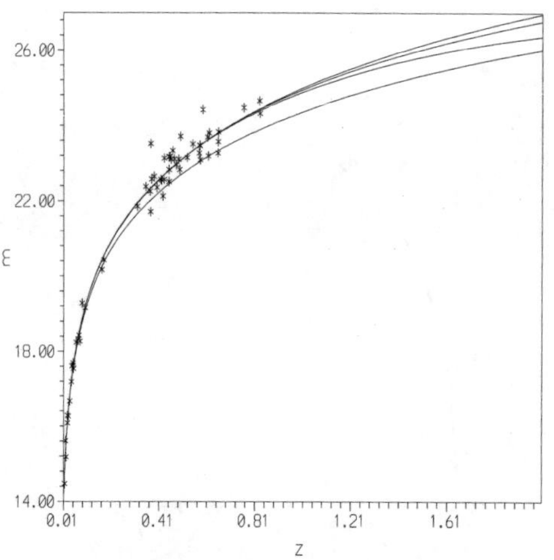

FIG. 9: A redshift-magnitude relation for $\gamma = 1/3$ brane universes (domain walls on the brane). The top line is the best-fit *flat* model $\Omega_{\gamma,0} = 0.8$, $\Omega_{\lambda,0} = 0.2$, then lower, the best-fit flat model of Ref. [22] with $\Omega_{m,0} = 0.28$, $\Omega_{\Lambda_{(4)},0} = 0.72$. The bottom line is a pure flat model with the cosmological constant $\Omega_{\Lambda_{(4)},0} = 0$, then higher, the best-fit brane model of non-zero curvature with, $\Omega_{k,0} = -1.0$, $\Omega_{\gamma,0} = 1.1$, $\Omega_{\lambda,0} = 0.9$.

close to that of [22] which means that the dark energy cancels the positive-pressure influence of the ϱ^2 term and can simulate the negative-pressure influence of the cosmological constant to cause cosmic acceleration. From the formal point of view the best fit is ($\chi^2 = 95.4$) for $\Omega_{k,0} = 0.2$, $\Omega_{d,0} = 0.7$, $\Omega_{\lambda,0} = -0.1$, $\Omega_{\mathcal{U}} = 0.2$, $\Omega_{\Lambda_{(4)},0} = 0$ which means that the cosmological constant must necessarily *vanish*. From this result we can conclude that the phantom matter $p = -(4/3)\varrho$ can *mimic* the contribution from the $\Lambda_{(4)}$-term in standard models. For the best-fit flat model ($\Omega_{k,0} = 0$) we have ($\chi^2 = 95.4$): $\Omega_{d,0} = 0.2$, $\Omega_{\lambda,0} = -0.1$, $\Omega_{\mathcal{U}} = 0.2$, $\Omega_{\Lambda_{(4)},0} = 0.7$.

However if we excluded formal possibility that $\Omega_{\lambda,0} < 0$, than for value of the parameter $\Omega_{\lambda,0} = 0.01$ we would obtain: $\Omega_{k,0} = 0.2$, $\Omega_{d,0} = 0.5$, $\Omega_{\lambda,0} = 0.01$, $\Omega_{\mathcal{U}} = 0.2$, $\Omega_{\Lambda_{(4)},0} = 0,09$ For the best-fit flat model ($\Omega_{k,0} = 0$) we have ($\chi^2 = 95.5$): $\Omega_{d,0} = 0.05$, $\Omega_{\lambda,0} = 0.01$, $\Omega_{\mathcal{U}} = 0.2$, $\Omega_{\Lambda_{(4)},0} = 0,74$ which means that the cosmological constant does not *vanish* in such type of model.

In Figures 11,12,13 we plot the values of χ^2 for brane models under studies against sample

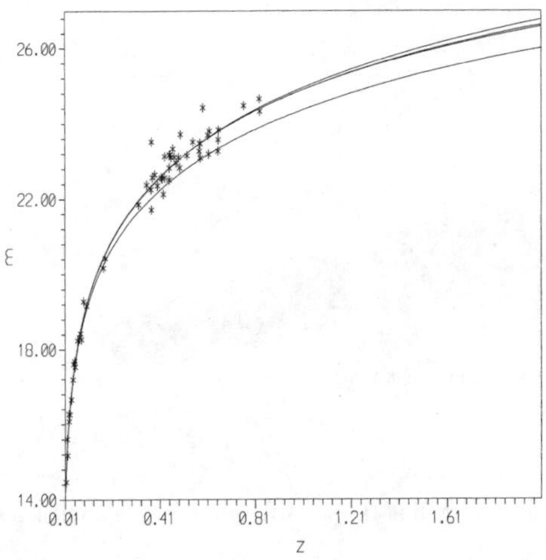

FIG. 10: A redshift-magnitude relation for $\gamma = -1/3$ brane universes (phantom matter on the brane). The top and bottom lines are the same as in Fig. 7. The brane phantom matter models plots are very close to the top line of Ref. [22]: the best-fit model for $\Omega_{k,0} = 0.2$, $\Omega_{d,0} = 0.7$, $\Omega_{\lambda,0} = -0.1$, $\Omega_{\mathcal{U}} = 0.2$, $\Omega_{\Lambda_{(4)},0} = 0$, and the best-fit flat model for $\Omega_{d,0} = 0.2$, $\Omega_{\lambda,0} = -0.1$, $\Omega_{\mathcal{U},\prime} = 0.2$, $\Omega_{\Lambda_{(4)},0} = 0.7$.

A of Ref. [22].

In the near future large number of SNIa events for one year period observations is expected [40]. It allowed us to significantly decrease erorrs in estimations. Future supernovae data could decrease significantly errors in estimation of $\Omega_{\lambda,0}$. In the near future the SNAP mission is expected to observe about 2000 SN Ia supernovae each year, over a period of three years [41].

We test how large number of new data should influence on errors in estimation of $\Omega_{\lambda,0}$. We assumed that the Universe is flat with $\Omega_{\mathrm{m},0} = 0.28$, $\Omega_{\lambda,0} = 0.01$ and $\mathcal{M} = -3.39$. For model with dust matter on the brane we generate sample of 1000 supernovae randomly distributed in the redshift range $z\varepsilon[0.01, 2]$. We assumed a Gaussian distribution of uncertaintics in measurement values m and z. The errors in redshifts $z$are of order $1\sigma = 0.002$ while uncertainty in measurement of magnitude m is assumed $1\sigma = 0.15$. The systematic uncertainty limits is $\sigma_{sys} = 0.02$ mag at $z = 1.5$ [40] that means that $\sigma_{sys}(z) = (0.02/1.5)z$. For such generated sample we should now repeat our analysis. The results of our analysis

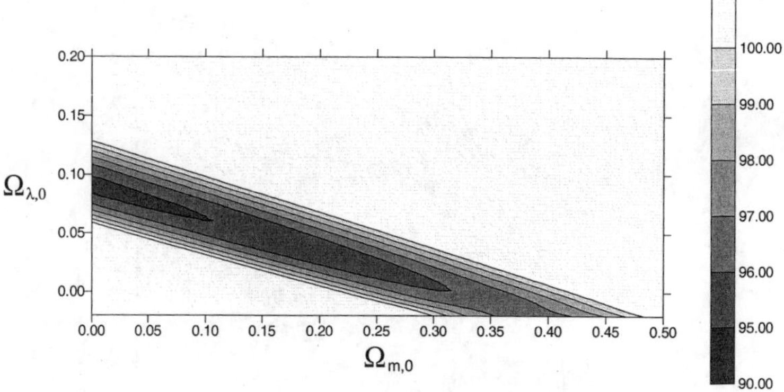

FIG. 11: The plot of χ^2 levels for flat ($\Omega_{k,0} = 0$) brane models with respect to the values of $\Omega_{m,0}$ (horizontal axis) and $\Omega_{\lambda,0}$ (vertical axis). We use the sample A of Ref. [22].

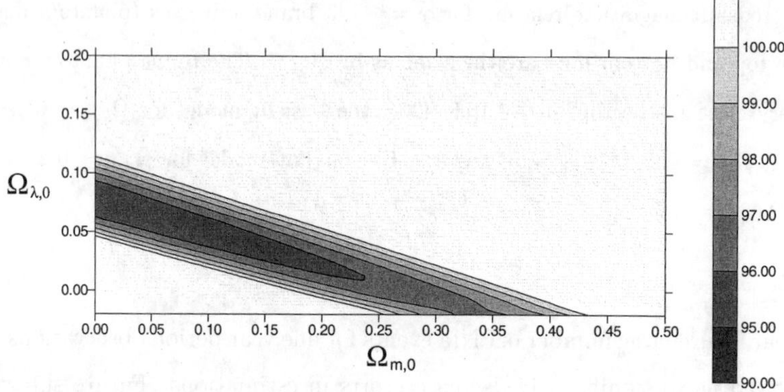

FIG. 12: The plot of χ^2 levels for $\Omega_{k,0} = 0.1$ brane models with respect to the values of $\Omega_{m,0}$ (horizontal axis) and $\Omega_{\lambda,0}$ (vertical axis).

are presented on the Fig. 14. This figure shows the confidence levels of 1-dimensional distribution of $\Omega_{\lambda,0}$ The error for $\Omega_{\lambda,0} < 0.0007$ on the confidence level of 68.3, while $\Omega_{\lambda,0} > 0.0013$ on the confidence level 95.4. It is clearly confirmed that the error in measurement of $\Omega_{\lambda,0}$ from supernovae data will decrease significantly in the new future.

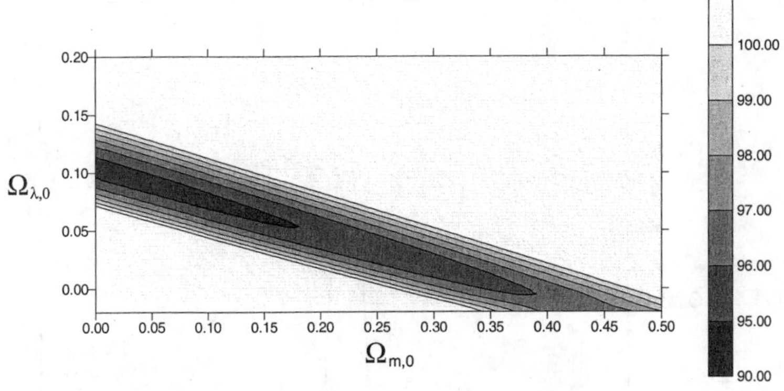

FIG. 13: The plot of χ^2 levels for $\Omega_{k,0} = -0.1$ brane models with respect to the values of $\Omega_{m,0}$ (horizontal axis) and $\Omega_{\lambda,0}$ (vertical axis).

V. OTHER OBSERVATIONAL TESTS FOR BRANE UNIVERSES

A. Angular diameter minimum value test

The angular diameter of a galaxy is defined by

$$\theta = \frac{d(z+1)^2}{d_L}, \tag{31}$$

where d is a linear size of the galaxy. In a flat, dust-filled ($\gamma = 1$) universe θ has the minimum value $z_{min} = 5/4$. In Ref. [34] it was shown that the radiation pressure lowers this value. For brane models the radiation pressure which is now present in the form of dark radiation can have both positive and negative values [14] and its influence onto the angular size minimum (31) can be different. We will study this briefly using the case $\gamma = 1$ (dust matter on the brane). It should be noticed that for flat brane models with $\Omega_{\Lambda_{(4)},0} \approx 0$, $\Omega_{\lambda,0} \approx 0$, the dark radiation can shift the minimum of $\theta(z)$ relation towards to higher z for $\Omega_{\mathcal{U},0} < 0$.

In Figs. 15-19 we present plots of the angular diameter θ of a source against redshift z for various values of parameters $\Omega_{\mathcal{U},0}, \Omega_{\lambda,0}, \Omega_{m,0}$ and $\Omega_{d,0}$. For example, it is clear that for negative dark radiation $\Omega_{\mathcal{U}} < 0$, the minimum of angular diameter gradually becomes smaller and finally disappears because d_L given by Eq. (19) becomes imaginary. One sees that the same is true for negative brane tension $\Omega_{\lambda,0} < 0$. Obviously, positive $\Omega_{\mathcal{U},0}$ and $\Omega_{\lambda,0}$ make z_{min} decrease. We can also see (Fig. 19) that the phantom matter $\Omega_{d,0}$ has very little influence onto the value of z_{min}.

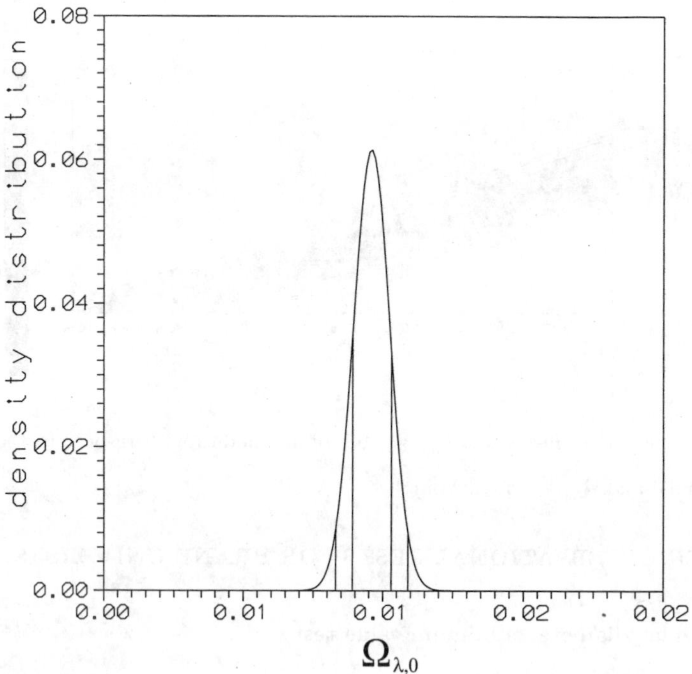

FIG. 14: The density distribution for $\Omega_{\lambda,0}$ for the flat model for the simulated data. The result of our preliminary analysis gives the following: expected error in measurement: $\Omega_{\lambda,0} = 0.7 \cdot 10^{-3}$ (at 1σ level) and $\Omega_{\lambda,0} = 1.3 \cdot 10^{-3}$ (at 2σ level).

B. Age of the universe

Now let us briefly discuss the effect of brane parameters and dark energy onto the age of the universe which according to (1) is given by

$$H_0 t_0 = \int_0^1 \left\{ \Omega_{\gamma,0} x^{-3\gamma+4} + \Omega_{\lambda,0} x^{-6\gamma+4} + \Omega_{\mathcal{U},0} + \Omega_{k,0} x^2 + \Omega_{\Lambda_{(4)},0} x^4 \right\}^{-\frac{1}{2}} x dx, \tag{32}$$

where $x = (1+z)^{-1} = a/a_0$. We made a plot for the dust $\gamma = 1$ on the brane in Figs.20 and 21 which show that the effect of quadratic term in energy density term represented by $\Omega_{\lambda,0}$ is to *lower* significantly the age of the universe. The problem can be avoided, if we accept the phantom matter $\gamma = -1/3$ [26] on the brane, since the dark energy has a very strong influence on increasing the age. In Fig. 22 we made a plot for this case which shows how the phantom matter enlarges the age. Finally, in Figs. 23, 24, 25 we study some other

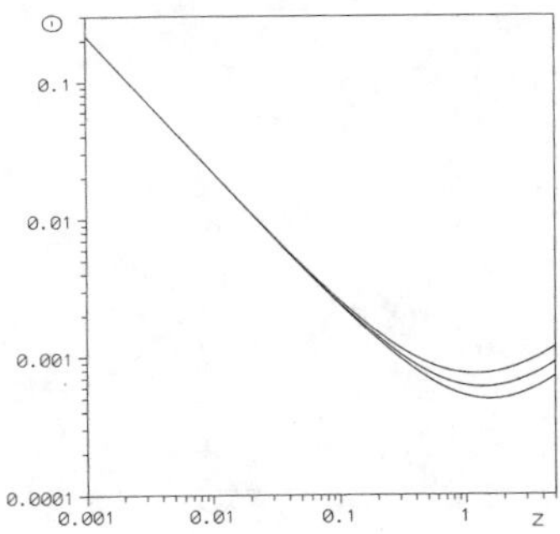

FIG. 15: The angular diameter $\theta(z)$ relation for $\Omega_{k,0} = 0$, $\Omega_{\mathcal{U},0} = 0.1, \Omega_{\lambda,0} = 0$, and $\Omega_{m,0} = 0.9, \Omega_{\Lambda_{(4)},0} = 0$ (top); $\Omega_{m,0} = 0.3, \Omega_{\Lambda_{(4)},0} = 0.6$ (middle); $\Omega_{m,0} = 0, \Omega_{\Lambda_{(4)},0} = 0.9$ (bottom). The minima are at $z_{min} = 1.20, 1.36, 1.54$ respectively. It is important to notice that in the case $\Omega_{m,0} = 0, \Omega_{\Lambda_{(4)},0} \approx 1$ for $\Omega_{\mathcal{U},0} = -0.02$ $z_{min} = 1.66$ - for smaller and possibly negative contribution from dark radiation the minimum disappears because d_L becomes imaginary.

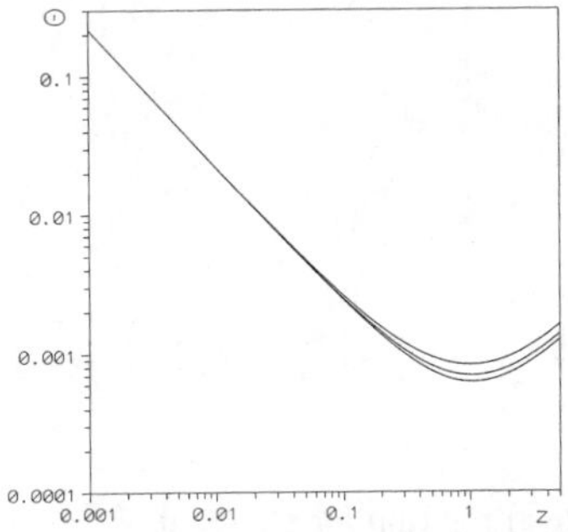

FIG. 16: The angular diameter $\theta(z)$ relation for $\Omega_{k,0} = 0$, $\Omega_{\mathcal{U},0} = 0, \Omega_{\lambda,0} = 0.1$, and $\Omega_{m,0} = 0.9, \Omega_{\Lambda_{(4)},0} = 0$ (top); $\Omega_{m,0} = 0.3, \Omega_{\Lambda_{(4)},0} = 0.6$ (middle); $\Omega_m = 0, \Omega_{\Lambda_{(4)}} = 0.9$. The minima are at $z_{min} = 1.012, 1.035, 1.045$ respectively. It is important to notice that in the case $\Omega_{m,0} = 0, \Omega_{\Lambda_{(4)},0} \approx 1$ for $\Omega_{\lambda,0} = -0.01$ the minimum disappears because d_L becomes imaginary.

possible plots for the age. In every case of this subsection we restricted ourselves to flat models which means we used the constraint (11) for $\Omega_k = 0$.

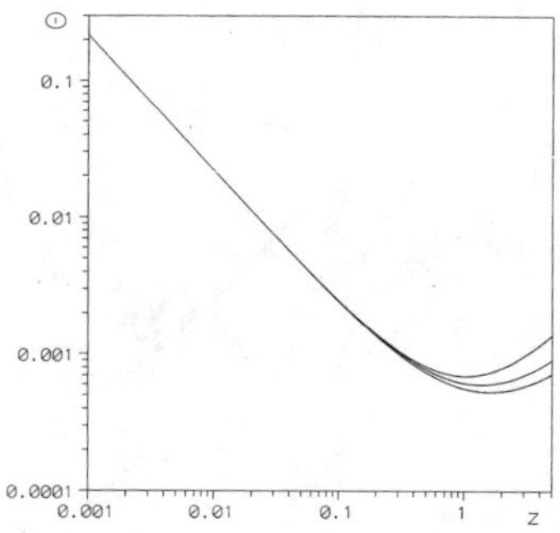

FIG. 17: The angular diameter $\theta(z)$ relation for $\Omega_{k,0} = 0$, $\Omega_{m,0} = 0.3, \Omega_{\Lambda_{(4)},0} = 0.6$, and for a) $\Omega_{\lambda,0} = 0.1, \Omega_{\mathcal{U},0} = 0$ (top); b) $\Omega_{\lambda,0} = 0, \Omega_{\mathcal{U},0} = 0.1$ (middle) in comparison with the model of Ref. [22] with $\Omega_{m,0} = 0.28, \Omega_{\Lambda_{(4)},0} = 0.72$ (bottom).

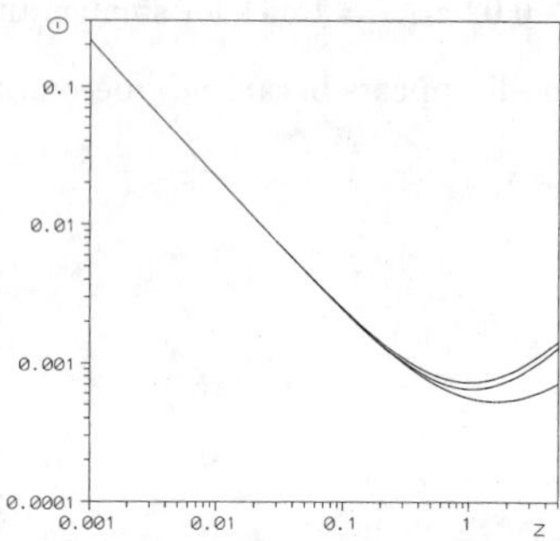

FIG. 18: The angular diameter $\theta(z)$ relation for $\Omega_{k,0} = 0$, $\Omega_{\lambda,0} = 0.1, \Omega_{m,0} = 0.3$, ($\Omega_{\Lambda_{(4)},0} \neq 0$) and the two values of $\Omega_{\mathcal{U},0} = 0.1, -0.1$ (top, middle) in comparison with the model of Ref. [22] with $\Omega_{m,0} = 0.28, \Omega_{\Lambda_{(4)},0} = 0.72$ (bottom).

VI. CMB PEAKS IN THE BRANE MODEL

The CMB peaks arise from acoustic oscillations of the primeval plasma. Physically these oscillations represent hot and cold spots. Thus, the wave which has a density maximum at time of the last scattering corresponds to a peak in the power spectrum. In the Legendre multipole space this corresponds to the angle subtended by the sound horizon at the last scattering. Higher harmonics of the principal oscillations which have oscillated more than

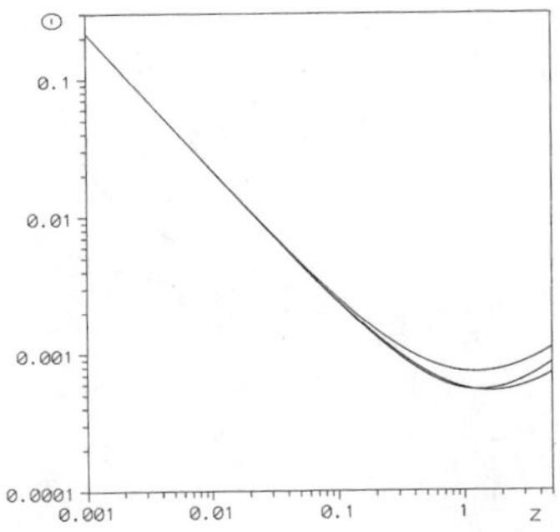

FIG. 19: The angular diameter $\theta(z)$ relation for the model with $\Omega_{m,0} = 1.0$ and $\Omega_{k,0} = \Omega_{\lambda,0} = \Omega_{\Lambda_{(4)},0} = \Omega_{\mathcal{U},0} = \Omega_{d,0} = 0$ (top), the model of Ref. [22] (bottom) in comparison with the phantom model with $\Omega_{d,0} = 1.0$ ($\Omega_{m,0} = 0$) (middle) which shows that the phantom matter has little influence onto the value of z_{min}.

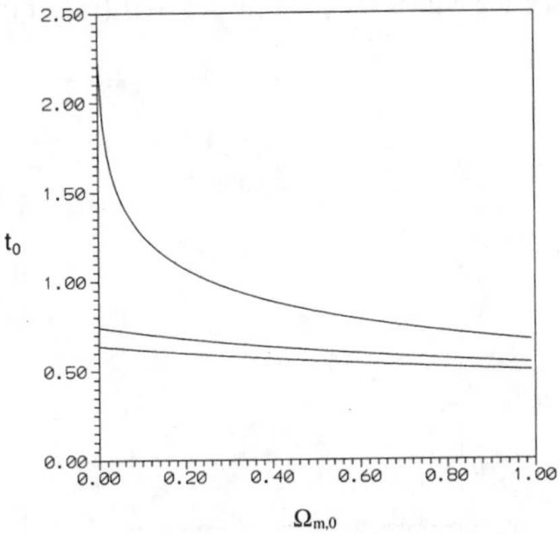

FIG. 20: The age of the universe t_0 in units of H_0^{-1} for the brane models with dust ($0 \leq \Omega_{m,0} \leq 1$ on the horizontal axis). Here $\Omega_{\mathcal{U},0} = \Omega_{k,0} = 0$, $\Omega_{\lambda,0} = 0, 0.05, 0.1$ (top, middle, bottom). The age decreases with increasing contribution from the brane tension Ω_{λ}.

once correspond to secondary peaks.

Finally it is very important that locations of these peaks are very sensitive to the variations in the model parameters. Therefore, it can be served as a sensitive probe to constrain the cosmological parameters and discriminate among various models.

The locations of the peaks are set by the acoustic scale l_A which can be defined as the

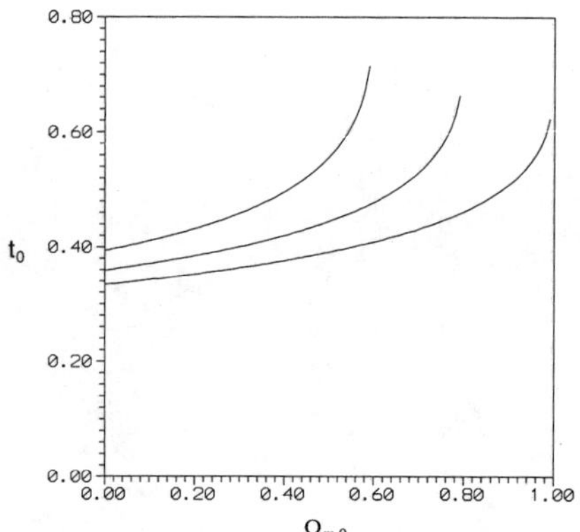

FIG. 21: The age of the universe t_0 in units of H_0^{-1} for the brane models with dust ($0 \leq \Omega_{m,0} \leq 1$ on the horizontal axis). Here $\Omega_{\mathcal{U},0} = \Omega_{k,0} = 0$, $\Omega_{\lambda,0} = 1 - \Omega_{m,0} - \Omega_{\Lambda_{(4)},0}$ and $\Omega_{\Lambda_{(4)},0} = 0.4, 0.2, 0$ (top, middle, bottom). The brane effects lower the age of the universe. This can easily be seen if one takes $\Omega_{m,0} = 0$ in the plot for which $\Omega_{\lambda,0} = 0,6; 0,8; 1,0$ (top, middle, bottom).

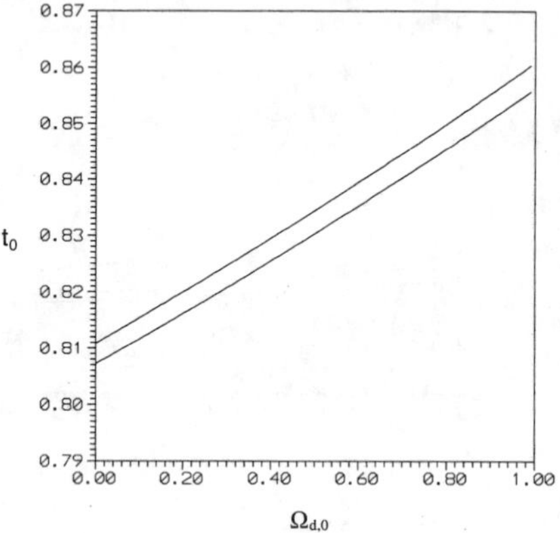

FIG. 22: The age of the universe t_0 in units of H_0^{-1} for the brane models with phantom matter on the brane ($0 \leq \Omega_{d,0} \leq 1$ on the horizontal axis). Here $\Omega_{k,0} = 0$, $\Omega_{\mathcal{U},0} = 0.2$, $\Omega_{\lambda,0} = 0.05, 0$ (top, bottom). It is obvious that the phantom matter increases the age.

angle θ_A subtended by the sound horizon at the last scattering surface. The acoustic scale $l_A = \pi/\theta_A$ in the flat model is given by

$$l_A = \pi \frac{\int_0^{z_{\rm dec}} \frac{dz'}{H(z')}}{\int_{z_{\rm dec}}^{\infty} c_s \frac{dz'}{H(z')}}$$

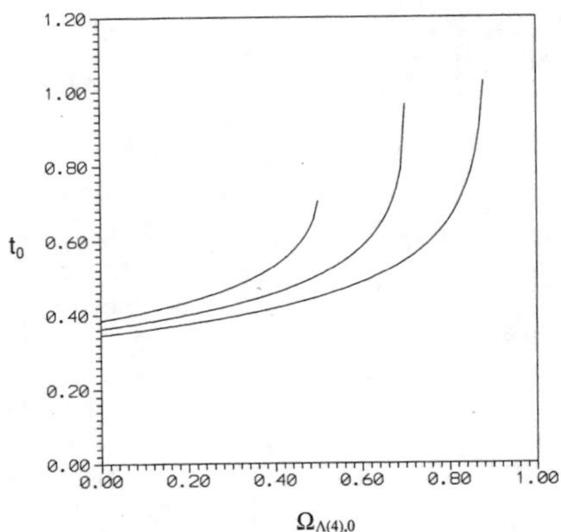

FIG. 23: The age of the universe t_0 in units of H_0^{-1} for the brane models with dust ($\Omega_{\Lambda_{(4)},0}$ on the horizontal axis). Here $\Omega_{k,0} = 0$, $\Omega_{m,0} = 0.3$ so that $\Omega_{\lambda,0} = 0.7 - \Omega_{\mathcal{U},0} - \Omega_{\Lambda_{(4)},0}$ and $\Omega_{\mathcal{U},0} = 0.2, 0, -0.2$ (top, middle, bottom). As one can see that in present case positive dark radiation contribution increases the age and negative dark radiation contribution lowers the age. In other words, the smaller $\Omega_{\mathcal{U},0}$ require the larger $\Omega_{\lambda,0}$ and the age decreases.

where in the case of dust matter on the brane

$$H(z) = H_0\sqrt{\Omega_{\mathrm{r},0}(1+z)^4 + \Omega_{\lambda,0}(1+z)^6 + \Omega_{\mathrm{m},0}(1+z)^3}$$

and c_s is the speed of sound in the plasma (we assume additionally the presence of radiation in the model). The sound velocity can be calculated from the formula

$$c_{\mathrm{eff}}^2 \equiv \frac{dp_{\mathrm{eff}}}{d\rho_{\mathrm{eff}}} = c_s^2 + \frac{\Omega_{b,0}\,(1+z)^3 + \frac{4}{3}\Omega_{r,0}\,(1+z)^4}{\Omega_{\lambda,0} + \Omega_{b,0}\,(1+z)^3 + \Omega_{r,0}\,(1+z)^4}, \tag{33}$$

As a reference system we choose the model of primeval plasma, where there is simple relation

$$l_m \approx l_A(m - \phi_m)$$

between the location of m-th peak and the acoustic scale [45, 46]. The prior assumptions in our calculations are as follows $\Omega_{\mathrm{r},0} = 9.89 \cdot 10^{-5}$, $\Omega_{\mathrm{b},0} = 0.05$, and the spectral index for initial density perturbations $n = 1$, and $h = 0.65$.

The phase shift is caused by the prerecombination physics (plasma driving effect) and, hence, has no significant contribution from the term containing brane in that epoch. Because

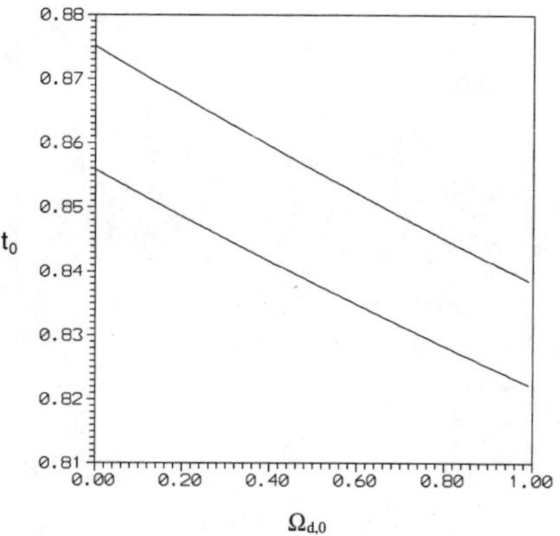

FIG. 24: The age of the universe t_0 in units of H_0^{-1} for the brane models with phantom matter on the brane ($\Omega_{d,0}$ on the horizontal axis). Here $\Omega_{k,0} = 0$, $\Omega_{\mathcal{U},0} = 0.2$ so that $\Omega_{\lambda,0} = 0.8 - \Omega_{d,0} - \Omega_{\Lambda_{(4)},0}$ and $\Omega_{\Lambda_{(4)},0} = 0, 0.2$ (top, bottom).

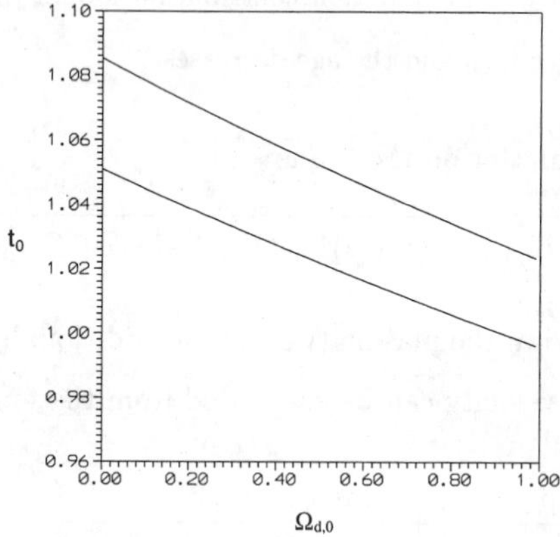

FIG. 25: The age of the universe t_0 in units of H_0^{-1} for the brane models with phantom matter on the brane ($\Omega_{d,0}$ on the horizontal axis). Here $\Omega_{k,0} = 0$, $\Omega_{\mathcal{U},0} = 0.1$ so that $\Omega_{\lambda,0} = 0.9 - \Omega_{d,0} - \Omega_{\Lambda_{(4)},0}$ and $\Omega_{\Lambda_{(4)},0} = 0, 0.2$ (top, bottom).

of above assumptions the phase shift ϕ_m can be taken from standard cosmology [46]

$$\phi_m \approx 0.267 \left[\frac{r(z_{\rm dec})}{0.3}\right]^{0.1}$$

where $\Omega_{\rm b,0}h^2 = 0.02$, $r(z_{\rm dec}) \equiv \rho_{\rm r}(z_{\rm dec})/\rho_{\rm m}(z_{\rm dec}) = \Omega_{\rm r,0}(1 + z_{\rm dec})/\Omega_{\rm m,0}$, $\Omega_{\rm r,0} = \Omega_{\gamma,0} + \Omega_{\nu,0}$, $\Omega_{\gamma,0} = 2.48h^{-2} \cdot 10^{-5}$, $\Omega_{\nu,0} = 1.7h^{-2} \cdot 10^{-5}$, $r(z_{\rm dec})$ is the ratio of radiation to matter at the last scattering.

The influence of brane on the location of the peaks is to shift them towards higher values of l. For example, for $\Omega_{\mathrm{m},0} = 0.3$, $\Omega_{\mathrm{b},0} = 0.05$, $h = 0.65$, the different choices of $\Omega_{\lambda,0}$ yield the following

$$\Omega_{\lambda,0} = 1.4 \cdot 10^{-10}: \qquad l_{\mathrm{peak},1} = 227, \qquad l_{\mathrm{peak},2} = 541, \qquad l_{\mathrm{peak},3} = 854$$

$$\Omega_{\lambda,0} = 1.5 \cdot 10^{-10}: \qquad l_{\mathrm{peak},1} = 232, \qquad l_{\mathrm{peak},2} = 552, \qquad l_{\mathrm{peak},3} = 873.$$

We could also analyze the influence of possible dark radiation term in the brane world cosmology. The corresponding term in brane cosmology scales just like radiation with a constant $\rho_{\mathrm{r},0}$ and both positive and negative $\rho_{\mathrm{r},0}$ are mathematically possible. Dark radiation should strongly affect both the Big-Bang nucleosynthesis (BBN) and the cosmic microwave background (CMB). Ichiki *et al.* [47] used such observations to constrain both the magnitude and the sign of dark radiation in the case when term ρ^2 coming from the brane is negligible (it rapidly decays as a^{-8} in the early radiation dominated universe). Therefore, the presence of the term is insignificant during the later nucleosynthesis. Let us note that negative contribution coming from the dark radiation presence can reconcile the tension between the observed ^{4}He and D abundance [47].

$$\Omega_{\mathcal{U},0} = 0.11\Omega_{\gamma,0} \qquad \Omega_{\lambda,0} = 1.5 \cdot 10^{-10}: \qquad l_{\mathrm{peak},1} = 233, \qquad l_{\mathrm{peak},2} = 555, \qquad l_{\mathrm{peak},3} = 879$$

$$\Omega_{\mathcal{U},0} = -\Omega_{\gamma,0} \qquad \Omega_{\lambda,0} = 1.5 \cdot 10^{-10}: \qquad l_{\mathrm{peak},1} = 219, \qquad l_{\mathrm{peak},2} = 519, \qquad l_{\mathrm{peak},3} = 826.$$

We conclude that the influence of dark radiation is not very significant in our case.

On the other hand from the Boomerang observations [48] we obtain $l_{\mathrm{peak},1} = 200 - 223$, $l_{\mathrm{peak},2} = 509 - 561$. Therefore, uncertainties in values l_{peak} can be used in constraining cosmology with brane effect, namely

$$\Omega_{\lambda,0} \leq 1.4 \cdot 10^{-10}$$

from the location of the first peak. However, the phase shift ϕ is taken from the standard cosmology, i.e. we assumed that contribution from the brane term is not significant at the pre-recombination epoch.

We also compare results from the above procedure with recent bounds on the location of the first two peaks obtained by WMAP collaboration [49, 50] namely $l_{\text{peak},1} = 220.1 \pm 0.8$, $l_{\text{peak},2} = 546 \pm 10$, together with the bound of the location of the third peak obtained by Boomerang collaborations $l_{\text{peak},3} = 825^{+10}_{-13}$ leads to quite strong constraints on the model parameters. These constraints can be summarized as follows. If we assumed no dark radiation, the brane model would be in agreement with observations when

$$\Omega_{\lambda,0} \leq 1.3 \cdot 10^{-10}$$

With such a value of $\Omega_{\lambda,0}$ we obtain

$$\Omega_{\mathcal{U},0} = 0 \qquad \Omega_{\lambda,0} = 1.3 \cdot 10^{-10}: \qquad l_{\text{peak},1} = 222, \qquad l_{\text{peak},2} = 529, \qquad l_{\text{peak},3} = 835.$$

However, phase shift ϕ is taken from the standard cosmology, i.e., we assume that the contribution from the brane term is insignificant at the pre-recombination epoch. If this assumption is not valid then the limit from CMB will change.

From the other hand, the best fit for flat brane model with cosmological model obtained from SN IA data gives $\Omega_{\lambda,0} \simeq 0.01$. For value of $\Omega_{\lambda,0} \simeq 0.01$ the position of peaks is unrealistic: $l_{\text{peak},1} = 724942$, $l_{\text{peak},2} = 1720942$, $l_{\text{peak},3} = 2716941$ what is cleary ruled out. Therefore CMB observations ruled out the brane model with such a value of $\Omega_{\lambda,0}$. From the supernovae data we obtain the best fit for flat model with brane $\Omega_{\lambda,0} = 0.004 \pm 0.016$, what means that the error is significant in comparision with obtained value of $\Omega_{\lambda,0}$. Therefore the value $\Omega_{\lambda,0} = 0$ is still admissible from SN Ia data. In the near future, when now supernovae data at high redshift will be available, the errors in estimation of $\Omega_{\lambda,0}$ will significantly decrease. Then more precise limit on $\Omega_{\lambda,0}$ can be achieved, and hypothesis $\Omega_{\lambda,0} > 0$ can be verified.

However, in our approxomation the phase shift ϕ is taken from the standard cosmology, i.e., we assume that the contribution from the brane term is insignificant at the pre-recombination epoch. If this assumption is not valid then the limit from CMB will change.

VII. BBN IN THE MODEL WITH BRANE FLUID

The model considers the non-relativistic matter on the brane. In this case the brane effects scale like $(1+z)^6$. It is clear that such a term can lead to accelerated expansion and the detailed analysis of supernovae Ia data requires it to be very subdominant today.

However, going back in time, the term $\Omega_{\lambda,0} \simeq 0.01$ would be dominant at redshift $z < 2$. In such a model, radiation domination would never occur and all the BBN predictions would be lost.

If we assumed that brane model does not change the prerecombination physics, BBN would allow only small deviation from the standard expansion law, because BBN is very well tested area of cosmology and does not allow for significant deviation apart from very early times before the onset of the BBN. The consistency with the BBN seems to be crucial issue in brane fluid cosmology [51, 52]. For this reason, not to suffer from the contradiction with the BBN, the contribution of brane fluid $\Omega_{\lambda,0}$ cannot dominate over the standard radiation and dark radiation term before the onset of BBN

$$\Omega_{\lambda,0}(1+z)^6 < (\Omega_{r,0} + \Omega_{\mathcal{U},0})(1+z)^4.$$

where $\Omega_{\mathcal{U},0} < 0.11 \cdot \Omega_{r,0}$ Therefore, the term $\Omega_{\lambda,0}(1+z)^6$, describing the brane effects, the brane effects, is constrained by the Big-Bang nucleosynthesis because it requires the change of expansion rate due to this term to be sufficiently small, so that an acceptable helium-4 abundance is produced. We obtain the following limit in this case $\Omega_{\lambda,0} \leq 10^{-20}$ if $z \simeq 10^8$.

In practice, in such approach the limit value obtained from SN Ia data gives the term ρ_λ^2 which is far too large to be compatible with the Big-Bang Nucleosynthesis. However one should note that the status of prediction from BBN (and CMB) is different from that obtained for low redshift (SNIa). Then limits obtained from BBN and CMB based on assumption that brane model does not change the prerecombination physics i.e. these limits are strongly model dependent.

VIII. CONCLUSIONS

We discussed astronomical tests (redshift-magnitude relation, angular size minimum, age of the universe) for brane cosmologies with various types of matter sources on the brane. Apart from dust $p = 0$, we imposed "sub-negative" pressure matter $p = -1/3\varrho$ (cosmic strings), and $p = -2/3\varrho$ (domain walls), and "super-negative" pressure $p = -4/3\varrho$ (phantom matter) on the brane.

We showed that in the case of dust matter on the brane, the difference between the best-fit model with Λ-term and the best-fit brane models becomes detectable for redshifts

$z > 0.6$ [44]. It is interesting that brane models predict brighter galaxies for such redshifts which is in agreement with the measurement of the $z = 1.7$ supernova [28]. The admission of sub-negative pressure matter on the brane does not exclude the necessity for the positive cosmological constant $\Lambda_{(4)}$. All realistic models require presence of the positive cosmological constant $\Omega_{\Lambda_{(4)},0} \sim 0.7$.

We also demonstrated that the fit to supernovae data, as given in Ref. [22], can be obtained provided that we admit the "super-negative" phantom matter $p = -(4/3)\varrho$ on the brane, where the phantom matter *mimics* the influence of the cosmological constant and is well within the observational limit coming from cosmic microwave background observations [29]. This means the redshift-magnitude plots of brane models with dark energy are very close to the redshift-magnitude plots for Perlmutter model.

We also studied the minimum of the angular size of galaxies test and showed that it is very sensitive onto the amount of dark radiation $\Omega_{\mathcal{U},0}$ and brane tension $\Omega_{\lambda,0}$ which are unique characteristics of brane models. In spite of ordinary radiation which increases the minimum, dark radiation and brane tension can decrease it. The minimum disappears for some large negative dark radiation and brane tension contribution where the values of luminosity distance d_L becomes imaginary. It is interesting to notice that the super-negative pressure matter (phantom matter) has rather small influence onto the value of the minimum.

Finally, the zero or sub-negative pressure matter imposed on the brane, because of its quadratic in the energy density term ϱ^2, lowers the age of the universe. However, the age can be increased provided that we accept the super-negative pressure matter on the brane.

Let us summarize our main results.

1. Our general conclusion is that some exotic physics coming from superstrinng or M-theory can be tested by astronomical tests.

2. If a brane term is added to FRW models with Λ-term then brane models fit better SN Ia data than the Perlmutter model with Λ.

3. We demonstrate that at high redshifts the expected luminosity of SN Ia should be brighter in the brane model then in the Perlmutter model

4. For the flat model with $\Omega_{m,0} = 0.3$ we obtain $\Omega_{\lambda,0} = 0.004 \pm 0.016$ as a best fit value.

5. We also discuss other limits coming from masurement of CMB anisotropies and from BBN. In this case we obtain most stronger limits $\Omega_{\lambda,0} < 1.5 \cdot 10^{-10}$ and $\Omega_{\lambda,0} < 10^{-20}$ respectively.

6. In the near future, if the new supernovae Ia observations at high redshift will available, then errors in estimation of $\Omega_{\lambda,0}$ parameter should be significantly smaller. In this way the hypothesis that brane exist ($\Omega_{\lambda,0} > 0$) can be verified. At present day brane theory cannot be confirmed nor ruled out. The value of SN Ia observations should be stressed out because they there are model independent whereas in the limits for CMB and BBN some theoretical assumptions are used at the very begining. The value $\Omega_{\lambda,0} \sim 0.1$ sugest by SNIa is the observational limit which is not based on theoretical assumptions. It is important, becouse on purely theoretical grounds limits are most stronger.

IX. ACKNOWLEDGEMENTS

The authors are very grateful to Prof M.Dąbrowski for his comments and interesting remarks.

---

[1] G. Veneziano, Phys. Lett. B**265** (1991), 287.

[2] M. Gasperini and G. Veneziano, Astropart. Phys. **1** (1993), 317.

[3] J.E. Lidsey, D.W. Wands, and E.J. Copeland, Phys. Rep. **337** (2000), 343.

[4] P. Hořava and E. Witten, Nucl. Phys. B**460** (1996), 506; *ibid* B**475**, 94.

[5] M. Visser, Phys. Lett. B**159** (1985), 22.

[6] C. Barcelo and M. Visser, Phys. Lett. B**482** (2000) 183.

[7] L. Randall and R. Sundrum, Phys. Rev. Lett., **83** (1999), 3370.

[8] L. Randall and R. Sundrum, Phys. Rev. Lett., **83** (1999), 4690.

[9] N. Arkani-Hamed, S. Dimopoulos, and G. Dvali, Phys. Lett. B**516** (1998) 70.

[10] I. Antoniadis, N. Arkani-Hamed, S. Dimopoulos, G. Dvali, Phys. Lett. B**436** (1998) 257.

[11] N. Arkani-Hamed, S. Dimopoulos, and G. Dvali, Phys. Rev. D**59** (1999) 086004.

[12] P. Binétruy, C. Deffayet and D. Langlois, Nucl. Phys. B**565** (2000), 269.

[13] P. Binétruy, C. Deffayet and D. Langlois, Phys. Lett. B**477** (2000), 285.

[14] T. Shiromizu, K. Maeda, and M. Sasaki, Phys. Rev. D**62**, 024012 (2000), M. Sasaki, T. Shiromizu and K. Maeda, *ibid* D**62** (2000), 024008S, Mukhoyama, T. Shiromizu and K. Maeda, *ibid* D**62** (2000), 024028.

[15] C.M. Chen, T. Harko and M.K. Mak, Phys. Rev. D**64** (2001), 044013.

[16] C.M. Chen, T. Harko and M.K. Mak, Phys. Rev. D**64** (2001), 124017.

[17] L. Anchordoqui, C. Nuñez and K. Olsen, JHEP 0010:050 (2000).

[18] L. Anchordoqui *et al.*, Phys. Rev. D**64** (2001) 084027.

[19] A. Campos and C.F. Sopuerta, Phys. Rev. D**63** (2001), 104012.

[20] A. Campos and C. F. Sopuerta, Phys. Rev. D**64** (2001), 104011.

[21] M. Szydłowski, M.P. Dąbrowski, and A. Krawiec, Phys. Rev. D**66** (2002), 0640XX (hep-th/0201066).

[22] S. Perlmutter *et al.*, Ap. J. **517**, (1999) 565.

[23] P. M. Garnavich *et al.* Ap. J. Lett. **493** (1998) L53.

[24] A. G. Riess *et al.* Astron. J. **116** (1998) 1009.

[25] R.R. Caldwell, R. Dave and P.J. Steinhardt, Phys. Rev. Lett. **80** (1998), 1582; I. Zlatev, L. Wang and P.J. Steinhardt, Phys. Rev. Lett. **82** (1999), 896; P.J. Steinhardt, L. Wang and I. Zlatev, Phys. Rev. D**59** (1999), 123504.

[26] R.R. Caldwell, astro-ph/9908168, S. Hannestad and E. Mörstell, astro-ph/0205096, P.H. Frampton, astro-ph/0209037, P.H. Frampton and T. Takahashi, astro-ph/0211544.

[27] B. McInnes, e-print:astro-ph/0210321.

[28] A.G. Riess et al., Ap. J. **560** (2001), 49.

[29] A. Melchiorri *et al.*, astro-ph/0211522.

[30] U. Alam and V. Sahni, astro-ph/0209443.

[31] M.P. Dąbrowski and J. Stelmach, Ann. Phys (N. Y.) 166 (1986) 422. , M.P. Dąbrowski, Ann. Phys (N. Y.) **248** (1996) 199.

[32] M.P. Dąbrowski and J. Stelmach, Astron. Journ. **92** (1986), 1272.

[33] M.P. Dąbrowski and J. Stelmach, Astron. Journ. **93** (1987), 1373.

[34] M.P. Dąbrowski and J. Stelmach, Astron. Journ. **97** (1989), 978.

[35] M. P. Dąbrowski and J. Stelmach, in *Large Scale Structure in the Universe* IAU Symposium No 130 eds. Jean Adouze, Marie-Christine Palletan and Alex Szalay, (Kluwer Academic Publishers, 1988), 566.

[36] S. Weinberg, *Gravitation and Cosmology* (Wiley, New York, 1972).

[37] P. Singh, R.G. Vishwakarma and N. Dadhich, hep-th/0206193.

[38] R.G. Vishwakarma and P. Singh, Can brane cosmology with a vanishing Λ explain the obser-

vations?, astro-ph/0211285.

[39] G. Efstathiou, Mon. Not. R. Astr. Soc. **303** (1999), 147.

[40] U.Alam 2003 astro-ph/0303009,

[41] http://www-supernova.lbl.gov, http:/cfa-www.harvard.edu/cfa/our/Research, http://snfactory.lbl.gov.

[42] Vishwakarma, Gen. Rel. Grav. **33** (2001), 1973.

[43] G.F.R. Ellis and M.A.H. MacCallum, Comm. Math. Phys. **19** (1970), 31.

[44] M.P. Dąbrowski, W. Godłowski and M. Szydłowski, astro-ph/0210156.

[45] M. Doran, M. Lilley, J. Schwindt, and C. Wetterich, *Astrophys. J.* **559**, 501 (2001).

[46] W. Hu, M. Fukugita, M. Zaldarriaga, and M. Tegmark, *Astrophys. J.* **549**, 669 (2001).

[47] K. Ichiki, M. Yahiro, T. Kajino, M. Orito, and G. J. Mathews, *Phys. Rev. D* **66**, 043521 (2002).

[48] P. de Bernardis et al., *Astrophys. J.* **564**, 559 (2002).

[49] D. N. Spergel et al. (2003), astro-ph/0302209.

[50] L. Page et al. (2003), astro-ph/0302220.

[51] N. Arkani-Hamed, S. Dimopoulos, and G. Dvali, *Phys. Rev. D* **59**, 086004 (1999).

[52] P. Binetruy, C. Deffayet, and D. Langlois, *Nucl. Phys. B* **565**, 269 (2000).

INVESTIGATION OF SCHMUTZER'S EXACT EXTERNAL SPHERICALLY SYMMETRIC STATIC SOLUTION FOR A CENTRAL BODY WITHIN THE FRAMEWORK OF THE 5-DIMENSIONAL PROJECTIVE UNIFIED FIELD THEORY

A. GORBATSIEVICH
Belorussian State University, Minsk, Belarus

Abstract. The present paper is devoted to the investigation of the spherically symmetric static vacuum solution within the framework of Schmutzer's 5-dimensional Projective Unified Field Theory (PUFT). By an appropriate coordinate transformation we could present the Schmutzer solution (implicit form) in an explicit form. Further it was shown that this solution can be obtained with the help of a special conformal transformation from the well-know Heckmann-Jordan-Fricke solution. We mention that the Schmutzer solution describes the exterior field of a spherically symmetric static cosmic body as well as a naked singularity without any horizons. We could show that within the framework of PUFT in the exterior of the spherically symmetric body time-dependent vacuum solutions can exist.

1. Introduction

As it is well known, the 5-dimensional idea of a unified field theory goes back to the works of Kaluza and Klein [1,2]. The pioneers of the 5-dimensional projective approach were Veblen and van Dantzig [3,4]. Later this approach to the so-called projective relativity theories was developed further by many other authors.

An essential progress in this field of projective theories was done by Jordan [5] who first took into consideration the occurring scalar field which inevitably appears in this type of theories. However, the field equations used by him were unacceptable.

A basically different approach to a particular projective field theory was proposed by E.Schmutzer [6] who had no longer considered the scalar field mentioned above to be an auxiliary one. On the contrary, he hypothetically associated this field with a new phenomenon of nature existing on the same level as grav-

V. de Sabbata et al. (eds.),
The Gravitational Constant: Generalized Gravitational Theories and Experiments, 193–204.
© 2004 *Kluwer Academic Publishers. Printed in the Netherlands.*

itation and electromagnetism. In 1980 he introduced the new term **"scalarism"** [7] for this phenomenon. For this scalar field considered as of fundamental importance Schmutzer introduced the term **"scalaric field"** in order to distinguish it from the various other scalar fields in physics.

Since 1995 a series of papers by E.Schmutzer on a closed homogeneous isotropic cosmological model of the universe and on the influence of the expansion of such a model on cosmogony and astrophysics appeared (see [8] where further literature is quoted). Many very interesting results were published by him.

As it is well known, for astrophysical applications it is very important to know solutions which describe external fields of compact cosmic objects. Usually it is accepted that for simplicity such objects may be electrically neutral. Therefore main interest refers to solutions of bodies exhibiting only gravitational and scalaric fields. Here we should remember that in PUFT the scalaric mass (one source of the scalaric field) is inseparably linked to the concept of mass. Therefore we denote the field equations for the exterior of matter, producing only gravitational and scalaric fields as vacuum field equations.

2. Field equations

Now the 4-dimensional vacuum field equations of PUFT [9] take the form

$$R_{mn} - \frac{1}{2} g_{mn} R = \varkappa_0 S_{mn} \,, \tag{1a}$$

$$\sigma^{,k}{}_{;k} = 0 \,. \tag{1b}$$

Here $\varkappa_0 = \dfrac{8\pi\gamma_N}{c^4}$ is the Einsteinian gravitational constant (γ_N Newtonian gravitational constant) and σ is the scalaric field. Further

$$S_{mn} = \frac{2}{\varkappa_0} \left(\sigma_{,m} \sigma_{,n} - \frac{1}{2} g_{mn} \sigma_{,k} \sigma^{,k} \right) \tag{2}$$

is the energy-momentum tensor of the scalaric field. Latin indices run from 1 to 4, the signature of the space-time metric is $(+++-)$, comma means the partial derivative and semicolon the covariant derivative. Let us mention that our definition of the Ricci tensor, $R_{mn} = R^k{}_{mnk}$, differs in sign from the definition, later mostly used.

For further investigation let us cast the field equations (1) into the following form

$$R_{mn} - 2\sigma_{,m}\sigma_{,n} = 0 \,, \tag{3a}$$

$$\sigma^{,k}{}_{;k} = 0 \,. \tag{3b}$$

As it is well known, for the spherically symmetric space-time the metric can be written in Schwarzschild coordinates as follows

$$\mathrm{d}s^2 = e^{\lambda(r,t)} \mathrm{d}r^2 + r^2 (\mathrm{d}\theta^2 + \sin^2\theta \mathrm{d}\varphi^2) - e^{\nu(r,t)} \mathrm{d}(x^4)^2 \,, \tag{4}$$

where $x^4 = ct$ and further $\nu(r,t)$ and $\lambda(r,t)$ are the two unknown functions wanted. Using the metric (4), from the gravitational field equations (3a) we obtain

$$e^{\lambda}\Big[-(\dot\lambda)^2+\dot\lambda\dot\nu-2\ddot\lambda\Big]+e^{\nu}\Big[-\lambda'\Big(\frac{4}{r}+\nu'\Big) \\ +(\nu')^2-8(\sigma')^2+2\nu''\Big]=0\,, \tag{5a}$$

$$\dot\lambda+2r\dot\sigma\sigma'=0\,, \tag{5b}$$

$$1-e^{\lambda}+\frac{r}{2}\,(\nu'-\lambda')=0\,, \tag{5c}$$

$$e^{\lambda}\Big[-(\dot\lambda)^2+\dot\lambda\dot\nu-2\ddot\lambda+8(\dot\sigma)^2\Big]+e^{\nu}\Big[\frac{4\nu'}{r}-\lambda'\nu'+(\nu')^2+2\nu''\Big]=0 \tag{5d}$$

(prime and dot mean derivative with respect to r and t, respectively). From the equation (5b) one could conclude that because of the presence of the scalaric field in PUFT there do not exist static spherically symmetric vacuum solutions. If so, the well–known Birkhoff theorem of general relativity would be violated. Though this question of the existence of non–static spherically symmetric vacuum solutions in PUFT from the physical point of view is of course very important, we restrict our consideration here to the static spherically symmetric solutions.

3. Spherically symmetric static vacuum solution of Schmutzer

3.1. SCHWARZSCHILD COORDINATES

For the static case the equation (4) reduces to

$$ds^2=e^{\alpha(r)}dr^2+r^2(d\theta^2+\sin^2\theta d\varphi^2)-e^{\beta(r)}d(x^4)^2\,. \tag{6}$$

This line element exhibits the two unknown function $\alpha(r)$ and $\beta(r)$. In this case the field equations (5) (because of $\dot\lambda=\dot\nu=0$) become much simpler and can be integrated by elementary functions. Some years ago E. Schmutzer [9] succeeded in finding the exact solution in implicit form. This solution which contains three constants of integration (α_1, τ_1 and $\hat r$) can be presented in the following form:

$$e^{\alpha}=\frac{1}{\tau_1}\Big[\Lambda^2+\tau_1-1\Big]\,, \tag{7a}$$

$$e^{\beta}=\frac{\alpha_1^2\tau_1}{4r^2}\,\frac{\Lambda^2+\tau_1-1}{(\Lambda-1)^2}\,, \tag{7b}$$

$$\sigma'=\frac{d\sigma}{dr}=-\frac{1}{r\sqrt{\tau_1}}\Big[\Lambda-1\Big]\,, \tag{7c}$$

where

$$\Lambda=\sqrt{1+\tau_1(e^{\alpha}-1)}\qquad(\Lambda>1). \tag{8}$$

The explicit form of the function $\Lambda(r)$ depends on the value of the constant τ_1 and can implicitly be written as follows:

1) $\underline{0 < \tau_1 < 1:}$

$$r = \hat{r}\,\frac{\sqrt{\Lambda^2 + \tau_1 - 1}}{(\Lambda - 1)}\left(\frac{\Lambda - \sqrt{1-\tau_1}}{\Lambda + \sqrt{1-\tau_1}}\right)^{\dfrac{1}{2\sqrt{1-\tau_1}}}, \tag{9a}$$

2) $\underline{\tau_1 = 1:}$

$$r = \hat{r}\,\frac{\Lambda}{\Lambda - 1}\,\exp\left\{-\frac{1}{\Lambda}\right\}, \tag{9b}$$

3) $\underline{\tau_1 > 1:}$

$$r = \hat{r}\,\exp\left\{\frac{1}{\sqrt{\tau_1 - 1}}\arctan\left(\frac{\Lambda}{\sqrt{\tau_1 - 1}}\right)\right\}\frac{\sqrt{\Lambda^2 + \tau_1 - 1}}{\Lambda - 1}\,. \tag{9c}$$

Let us notice that the constant α_1 can be determined from the boundary conditions. In particular if we claim that $\lim\limits_{r\to\infty} e^{\beta(r)} = 1$, then we obtain e.g. in the case $0 < \tau_1 < 1$ the relation

$$\alpha_1 = \frac{2\hat{r}}{\sqrt{\tau_1}}\left(\frac{1 - \sqrt{1-\tau_1}}{1 + \sqrt{1-\tau_1}}\right)^{\dfrac{1}{2\sqrt{1-\tau_1}}}. \tag{10}$$

3.2. EXPLICIT FORM OF THE SCHMUTZER SOLUTION

As it will be shown below, with the help of a transformed new radial coordinate $\varrho = \varrho(r)$ the Schmutzer solution (7) can be rewritten into an explicit form. It is easy to verify that the coordinate transformation

$$r(\varrho) = \varrho\,\sqrt{\omega(\varrho) f(\varrho)} \tag{11}$$

with

$$f(\varrho) = \begin{cases} 1 + \dfrac{r_g}{\varrho}\sqrt{1-\tau_1}, & \text{if } 0 < \tau_1 < 1, \\ 1, & \text{if } \tau_1 = 1, \\ 1 + \dfrac{r_g^2(\tau_1 - 1)}{4\varrho^2}, & \text{if } \tau_1 > 1, \end{cases} \tag{12}$$

and

$$\omega(\varrho)=\begin{cases}\left(1+\frac{r_g}{\varrho}\sqrt{1-\tau_1}\right)^{\frac{1}{\sqrt{1-\tau_1}}}, & \text{if } 0<\tau_1<1,\\ \exp\!\left(\frac{r_g}{\varrho}\right), & \text{if } \tau_1=1,\\ \exp\left\{\frac{1}{\tau_1-1}\left[\pi-2\arctan\left(\frac{2\varrho}{\sqrt{\tau_1-1}r_g}\right)\right]\right\}, & \text{if } \tau_1>1,\end{cases} \tag{13}$$

presents the Schmutzer solution (6) with (7) in the following explicit form:

$$ds^2=\omega(\varrho)\left[d\varrho^2+\varrho^2 f(\varrho)\left(d\vartheta^2+\sin^2\vartheta d\varphi^2\right)\right]-\frac{1}{\omega(\varrho)}\left(dx^4\right)^2, \tag{14a}$$

$$\sigma(\varrho)=\sigma_0+\frac{\sqrt{\tau_1}}{2}\ln\omega(\varrho). \tag{14b}$$

Here σ_0 is a constant of integration. It is easy to check that $\lim\limits_{\varrho\to\infty}\sigma(\varrho)=\sigma_0$. The new constant r_g can be expressed in terms of $\hat{r}$ and τ_1. For example in the physically important case $0<\tau_1<1$

$$r_g=\frac{2\hat{r}\left(\frac{1-\sqrt{1-\tau_1}}{1+\sqrt{1-\tau_1}}\right)^{\frac{1}{2\sqrt{1-\tau_1}}}\sqrt{1-\tau_1}}{\sqrt{\tau_1}}. \tag{15}$$

Let us mention that the Schwarzschild solution (but not in the Schwarzschild coordinates, see (19)) is the special case

$$\tau_1\longrightarrow 0\,,\qquad (\lim_{\tau_1\to 0} r_g=\hat{r}) \tag{16}$$

of the Schmutzer solution presented by the formulas (14), (12) and (13). Let us denominate the constant r_g occurring in the last formula as the “gravitational radius of PUFT”. Recently Schmutzer could show that this constant differs from the Einstein gravitational radius of the Einstein theory. Furthermore, the investigations of the non-relativistic limit of PUFT (for details see [10]) led to the following identification of the constants:

$$\text{a)}\quad r_g=\frac{2\gamma_N\mathcal{M}_c\sigma_c}{c^2}\,,\qquad \text{b)}\quad \sqrt{\tau_1}=\frac{1}{\sigma_c}\,,\qquad \text{c)}\quad \sigma_c=\sigma_0\,. \tag{17}$$

Here $\mathcal{M}_c$ is the scalmass of the center and $M_c=\mathcal{M}_c\sigma_c$ is the inertial mass of the center. Further σ_c is the present value of the scalaric world function. The cosmological investigations show [10] that because of the cosmological uncertainties (Hubble constant, age of the world) σ_0 lies in the interval (20 to 75), i.e. in any case $\tau_1 \ll 1$ is valid. Hence for physical application follows that the case 1 ($0 < \tau_1 < 1$) is important only. Therefore we restrict our further investigation to this case.

As it is well-known, the post-Newtonian approximation can be easiest obtained in harmonic coordinates. Therefore we transform the solution to these coordinates.

First of all we make the following substitution:

$$\varrho = \bar{\varrho} + \varrho_0 \,, \quad \text{where} \quad \varrho_0 = r_g \sqrt{1-\tau_1} \,. \tag{18}$$

After this transformation the solution (14) leads to

$$\begin{aligned} \mathrm{d}s^2 = &\left(\frac{1}{1-\varrho_0/\bar{\varrho}}\right)^{\frac{1}{\sqrt{1-\tau_1}}} \left[\mathrm{d}\bar{\varrho}^2 + \bar{\varrho}^2\left(1-\frac{\varrho_0}{\bar{\varrho}}\right)(\mathrm{d}\theta^2 + \sin^2\theta \mathrm{d}\varphi^2)\right] \\ &- \left(1-\frac{\varrho_0}{\bar{\varrho}}\right)^{\frac{1}{\sqrt{1-\tau_1}}} c^2 \mathrm{d}t^2 \,; \end{aligned} \tag{19a}$$

$$\sigma'(\bar{\varrho}) = -\frac{\varrho_0 \sqrt{\frac{\tau_1}{1-\tau_1}}}{2\bar{\varrho}^2\left(1-\frac{\varrho_0}{\bar{\varrho}}\right)} \,, \qquad \sigma = \sigma_0 + \frac{1}{2}\sqrt{\frac{\tau_1}{1-\tau_1}} \ln\left|\frac{\bar{\varrho}}{\bar{\varrho}-\varrho_0}\right| . \tag{19b}$$

It is easy to show that the new coordinates $\{X, Y, Z, T\}$ defined by

$$\begin{cases} \varrho = \dfrac{\varrho_0}{2} + R \,; \\ \theta = \arccos\left(\dfrac{Z}{R}\right); \\ \varphi = \arctan\left(\dfrac{Y}{X}\right); \\ t = T \,. \end{cases} \quad \text{with} \quad R = \sqrt{X^2 + Y^2 + Z^2} \tag{20}$$

are harmonic, e.g. these coordinates are in accordance with

$$X_{,}{}^{i}{}_{;i} = 0 \,, \quad Y_{,}{}^{i}{}_{;i} = 0 \,, \quad Z_{,}{}^{i}{}_{;i} = 0 \,, \quad T_{,}{}^{i}{}_{;i} = 0 \,. \tag{21}$$

After the substitution (20) the solution (19) takes the form

$$\mathrm{d}s^2 = \frac{\left(1-\dfrac{\varrho_0^2}{4R^2}\right)}{\Omega(R)}\,\mathrm{d}\boldsymbol{R}^2 + \frac{\varrho_0^2}{4R^4\Omega(R)}\left(\mathrm{d}\boldsymbol{R}\cdot\boldsymbol{R}\right)^2 - \Omega(R)c^2\mathrm{d}t^2\,, \tag{22a}$$

$$\sigma'(R) = -\frac{\varrho_0\sqrt{\dfrac{\tau_1}{1-\tau_1}}}{2\big(1-\varrho_0/(4R^2)\big)}\,, \tag{22b}$$

$$\sigma = \sigma_0 + \frac{1}{2}\ln\left|\frac{1+\varrho_0/(2R)}{1-\varrho/(2R)}\right|\sqrt{\frac{\tau_1}{1-\tau_1}}\,, \tag{22c}$$

where the following abbreviations are used:

$$\boldsymbol{R} = X\,\boldsymbol{e}_X + Y\,\boldsymbol{e}_Y + Z\,\boldsymbol{e}_Z\,; \quad \Omega(R) = \left(1-\frac{\varrho_0}{R+\varrho_0/2}\right)^{\dfrac{1}{\sqrt{1-\tau_1}}}. \tag{23}$$

4. Relationship of the Schmutzer solution to the Heckmann–Jordan–Fricke solution

4.1. THE HECKMANN–JORDAN–FRICKE SOLUTION

Let us mention that for special applications, particularly for the vacuum Jordan's projective relativity theory [5] is allied to PUFT: for example in both theories the scalar field appears inevitably. Therefore it is very interesting to compare both the spherically symmetric static vacuum solutions — the Heckmann–Jordan–Fricke solution [11] and the Schmutzer solution presented above.

The vacuum field equations of the Jordan theory read (we follow the abbreviations which are introduced in the book [12])

$$\varkappa_{,}{}^{i}{}_{;i} = 0\,, \tag{24a}$$

$$G_{kl} + \frac{\varkappa_{,k;l}}{\varkappa} - \zeta\frac{\varkappa_{,k}\varkappa_{,l}}{\varkappa^2} = 0\,, \tag{24b}$$

where $G_{kl} = R_{kl}$ is the Ricci tensor and $\varkappa$ is the scalar field which corresponds to Schmutzer's scalaric field σ.

The field equations (24) for the spherically symmetric static metric (6) were solved by Heckmann, Jordan and Fricke. This solution reads

$$r = \frac{r_0}{\sqrt{\tau}\left(\tau^{-h} + \tau^{h}\right)}, \tag{25a}$$

$$e^{\lambda(r)} = \frac{4h^2}{\left[\left(\frac{1}{2}+h\right)\tau^{h} - \left(\frac{1}{2}-h\right)\tau^{-h}\right]^2}, \tag{25b}$$

$$e^{\nu(r)} = \tau^{\frac{1}{B}}, \tag{25c}$$

$$\varkappa = \varkappa_0 \tau^{\frac{\beta_0}{B}}, \tag{25d}$$

where the constants of integration r_0, τ, h, A, B, β_0 satisfy the following conditions:

$$h^2 = \frac{1}{4} - \frac{A}{B^2}, \qquad A = \frac{\beta_0}{2}\,(1+\beta_0\zeta)\,, \qquad B = 1 + 2\beta_0\,. \tag{25e}$$

4.2. OBTAINING THE SCHMUTZER SOLUTION FROM THE HECKMANN–JORDAN–FRICKE SOLUTION BY MEANS OF A CONFORMAL TRANSFORMATION

The vacuum field equations (3) obviously differ from the field equation (24). However, we can show that the equations (3) can be obtained from the equations (24) by means of a special conformal transformation.

If we write down the equations (24) in the space with the metric $\hat{g}_{ab}$,

$$\text{a)}\quad \hat{g}_{ab} = e^{2u}\, g_{ab}\,, \qquad \text{b)}\quad \hat{g}^{ab} = e^{-2u}\, g^{ab}\,, \tag{26}$$

we find

$$\hat{\nabla}_i \hat{\varkappa}^{,i} = 0\,, \tag{27a}$$

$$\hat{G}_{kl} + \frac{1}{\hat{\varkappa}}\,\hat{\nabla}_l \hat{\varkappa}_{,k} - \zeta \frac{\hat{\varkappa}_{,k}\hat{\varkappa}_{,l}}{\hat{\varkappa}^2} = 0\,, \tag{27b}$$

where $\hat{\nabla}_i$ means the covariant derivative with respect to the metric $\hat{g}_{mn}$. Detailed calculation shows that the field equations (27) (i.e. the field equations (24) in the space with the metric $\hat{g}_{mn}$) are equivalent to Schmutzer's vacuum field equations (1) if the following relations are valid:

$$\hat{\varkappa}^{,i} = \hat{g}^{ik}\,\hat{\varkappa}_{,k}\,, \qquad \hat{\varkappa}_{,k} = \varkappa_{,k}\,, \tag{28a}$$

$$\hat{\varkappa} = \varkappa = \varkappa_0 e^{\alpha\sigma}\,, \qquad u = \varepsilon\,\sigma\,, \tag{28b}$$

$$\varepsilon = \pm\frac{1}{\sqrt{2\zeta - 3}}\,, \qquad \alpha = -2\varepsilon\,. \tag{28c}$$

With respect to the case $0 < \tau_1 < 1$ (19) of the Schmutzer solution let us consider the following special case of the Heckmann-Jordan-Fricke solution (25):

$$h = \frac{1}{2}\,, \quad A = 0\,, \quad B = 1 - \frac{2}{\zeta}\,, \quad \beta_0 = -\frac{1}{\zeta}\,; \tag{29a}$$

$$\tau(r) = 1 - \frac{r_0}{r}\,, \qquad e^{\lambda(r)} = \frac{1}{1 - \frac{r_0}{r}}\,, \tag{29b}$$

$$e^{\nu(r)} = \left(1 - \frac{r_0}{r}\right)^{\frac{\zeta}{\zeta-2}}\,, \qquad \varkappa = \varkappa_0\left(1 - \frac{r_0}{r}\right)^{\frac{1}{2-\zeta}}\,. \tag{29c}$$

With the help of the conformal transformation described above (see (26) and (28)), we immediately obtain from (29) the following solution of the vacuum field equation within the framework of PUFT:

$$\begin{aligned} \mathrm{d}s^2 = {} & \left(1 - \frac{r_0}{r}\right)^{-\frac{\zeta-1}{\zeta-2}} \left[\mathrm{d}r^2 + r^2\left(1 - \frac{r_0}{r}\right)\left(\mathrm{d}\theta^2 + \sin^2\theta\mathrm{d}\varphi^2\right)\right] \\ & - \left(1 - \frac{r_0}{r}\right)^{\frac{\zeta-1}{\zeta-2}} (\mathrm{d}x^4)^2\,, \end{aligned} \tag{30a}$$

$$\sigma = \pm\frac{\sqrt{2\zeta - 3}}{2(2 - \zeta)} \ln\left(1 - \frac{r_0}{r}\right). \tag{30b}$$

Obviously both solutions (19) and (30) coincide if

$$r \leftrightarrow \bar{\varrho} \quad \text{and} \quad r_0 \leftrightarrow \varrho_0\,; \quad \zeta = \frac{2 - \sqrt{1 - \tau_1}}{1 - \sqrt{1 - \tau_1}} \quad \text{and} \quad \tau_1 = \frac{2\zeta - 3}{(\zeta - 1)^2}\,. \tag{31}$$

Let us mention that the following correspondence between both constants τ_1 and ζ exists:

$$\tau_1 = 0 \leftrightarrow \zeta = \infty\,; \quad \text{and} \quad \tau_1 = 1 \leftrightarrow \zeta = 2\,. \tag{32}$$

5. Investigation of the spherically symmetric vacuum solution of Schmutzer with respect to a singularity

The space-time geometry described by the solution (19) appears to behave badly near $\bar{\varrho} = \varrho_0$, since g_{11}, g_{22} and g_{33} become infinite, and g_{44} becomes zero. However, without careful study one cannot be sure, whether this pathology in the metric is due to a pathology in the space-time geometry itself, or merely to a pathology of the coordinate system near $\bar{\varrho} = \varrho_0$, as it takes place in the Schwarzschild geometry near $r = r_g$ using Schwarzschild coordinates.

Since here is not space enough for treating the problem mentioned in detail, we limit our presentation to a summary of the basic results.

5.1. CURVATURE INVARIANT

First from equation (19) we calculate the curvature invariant:

$$I \equiv R^{ijkl} R_{ijkl} = \frac{-{\varrho_0}^2}{4\,\bar{\varrho}^8\,(1-\tau_1)^2} \left(\frac{1}{1-\dfrac{\varrho_0}{\bar{\varrho}}} \right)^{\lambda} f(\bar{\varrho})\,, \tag{33a}$$

where

$$\lambda = 4 - \frac{2}{\sqrt{1-\tau_1}}\,, \tag{33b}$$

and

$$\begin{aligned} f(\bar{\varrho}) = &-48\,\bar{\varrho}^2\,(1-\tau_1) - 16\,\bar{\varrho}\,\varrho_0 \left[-3\,\left(1+\sqrt{1-\tau_1}\right)\right. \\ &+ \left.\left(3+\sqrt{1-\tau_1}\right)\,\tau_1\right] + {\varrho_0}^2 \left[\left(-24\,\left(1+\sqrt{1-\tau_1}\right)\right.\right. \\ &+ \left.4\,\left(5+2\sqrt{1-\tau_1}\right)\,\tau_1 - 3\,{\tau_1}^2\right]. \end{aligned} \tag{33c}$$

From the last equations (33) we obtain the following values of I at $\bar{\varrho} = \varrho_0$:

$$\lim_{\bar{\varrho}\to\infty} I = \begin{cases} \infty & \text{if} \quad 0 < \tau_1 < \dfrac{3}{4}\,, \\ \dfrac{27}{4{\varrho_0}^4} & \text{if} \quad \tau_1 = \dfrac{3}{4}\,, \\ 0 & \text{if} \quad \dfrac{3}{4} < \tau_1 < 1\,, \\ \dfrac{12}{{\varrho_0}^4} & \text{if} \quad \tau_1 = 0\,. \end{cases} \tag{34}$$

Let us mention that the metric determinant diverges for all values of τ_1 from the considered interval ($0 < \tau_1 < 1$):

$$\det(g_{ij}) = -\bar{\varrho}^{\,4} \left(1 - \frac{\varrho_0}{\bar{\varrho}}\right)^{1-\dfrac{1}{\sqrt{1-\tau}}} \sin^2\theta\,, \tag{35a}$$

$$\lim_{\bar{\varrho}\to\varrho_0} \det(g_{ij}) = \pm\infty\,. \tag{35b}$$

5.2. TIDAL FORCES AND GEODESICS

5.2.1. *Time-like radial geodesics*

Let us consider an explorer radially moving in the gravitational and scalaric field described by (19). As it is well know from the equation of geodesic deviation, the tidal forces felt by the explorer, when he passes a given radius $\bar{\varrho}$, are measured by the component of the Riemann curvature tensor with respect to his orthonormal frame. The calculation of tetradic components of the curvature tensor in the explorer's frame shows that the tidal forces go to infinity if the explorer passes the point $\bar{\varrho} = \varrho_0$ for both types of his world lines: for radial geodesics and for free fall (in PUFT the equation of motion of a point-like test body contains the term with the gradient of the scalaric field [9]).

5.2.2. *Isotropic circular and radial geodesics*

It can be shown that in the space-time with the metric (19) there are circular light-like geodesics with the radius R_0, where

$$R_0 = \frac{1}{2}\,\varrho_0\,\frac{2+\sqrt{1-\tau_1}}{\sqrt{1-\tau_1}}\,, \tag{36a}$$

$$\lim_{\tau_1\to 0} R_0 = \frac{3}{2}\,\varrho_0 \quad (\varrho_0 \Leftrightarrow r_g \text{ if } \tau_1 \to 0). \tag{36b}$$

We can interpret this value of R_0 as the radius of a "photonic sphere" within the framework of PUFT.

The detailed investigation of the behavior of geodesics as well as Kruskal-like diagrams show that at $\bar{\varrho} = \varrho_0$ (at $\varrho = 0$ in (14)) exists a singularity. Therefore the region $\bar{\varrho} = \varrho_0$ is a physical singularity of infinite tidal gravitational forces (the scalaric σ-field becomes an infinite value at $\bar{\varrho} = \varrho_0$, too) and infinite Riemann curvature. The Schmutzer solution (19) (or (14)) consequently describes an exterior field of the spherically symmetric static cosmic body as well as a naked singularity without any horizons. However, taking into account the quantum effects because of possible birth of particles nearby the singularity can obviously destroy this conclusion. Let us finally mention that for $\tau_1 = 0$ this solution goes over into the Schwarzschild solution and describes therefore the Schwarzschild black hole.

I would like to express my thanks to the Friedrich-Schiller-University of Jena for hospitality as well as to professor E. Schmutzer (University of Jena) for scientific discussions and help.

6. References

1. Kaluza, Th. (1921) *Sitzungsber. d. preuss. Akad. d. Wiss., Phys.-math. Kl.* 541
2. Klein, O. (1926) *Z. Phys.* **37**, 895
3. Dantzig, D. van (1932) *Math. Ann.* **106**, 400
4. Veblen, O. (1933) *Projektive Relativitaetstheorie*, Springer-Verlag, Berlin
5. Jordan, P. (1955) *Schwerkraft und Weltall*, Vieweg Braunschweig

6. Schmutzer, E. (1968) *Relativistische Physik*, Teubner Verlagsgesellschaft, Leipzig
7. Schmutzer, E. (1983) *Proceedings of the 9th International Conference on General Relativity and Gravitation*, Deutscher Verlag der Wissenschaften, Berlin
8. Schmutzer, E. (1999) *Astronomische Nachrichten.* **320**, 1 (in AN further papers appeared)
9. Schmutzer, E. (1995) *Fortschr. Phys.* **43**, 613; (1995) *Ann. Physik* **4**, 251
10. Schmutzer, E. (2002) *Proceedings of the 17th Erice Course of the International School of Cosmology and Gravitation (Eds. P.G. Bergmann and V. de Sabbata)*, Kluwer Academic Publishers, Dordrecht
11. Heckmann, D., Jordan, P. and Fricke, W. (1951) *Z. f. Astrophysik* **28** 113
12. Ludwig, G. (1951) *Fortschritte der projektiven Relativitätstheorie*, F. Vieweg & Sohn, Braunschweig

Alexander Gorbatsievich e-mail: gorbatsievich@bsu.by
Scoriny Av., 4
220080 Minsk
Belarus

On exact solutions in multidimensional gravity with antisymmetric forms

V. D. Ivashchuk[1]

Center for Gravitation and Fundamental Metrology, VNIIMS, 3-1 M. Ulyanovoy Str., Moscow, 119313, Russia and
Institute of Gravitation and Cosmology, Peoples' Friendship University of Russia, 6 Miklukho-Maklaya St., Moscow 117198, Russia

Abstract

This short review deals with a multidimensional gravitational model containing dilatonic scalar fields and antisymmetric forms. The manifold is chosen in the form $M = M_0 \times M_1 \times \ldots \times M_n$, where M_i are Einstein spaces ($i \geq 1$). The sigma-model approach and exact solutions in the model are reviewed and the solutions with p-branes (e.g. solutions with harmonic functions, "cosmological", spherically symmetric and black-brane ones) are considered.

1 Introduction

In these lectures we consider several classes of the exact solutions for the multidimensional gravitational model governed by the action

$$\begin{aligned} S_{act} = & \frac{1}{2\kappa^2} \int_M d^D z \sqrt{|g|} \{ R[g] - 2\Lambda - h_{\alpha\beta} g^{MN} \partial_M \varphi^\alpha \partial_N \varphi^\beta \\ & - \sum_{a \in \Delta} \frac{\theta_a}{n_a!} \exp[2\lambda_a(\varphi)](F^a)^2_g \} + S_{GH}, \end{aligned} \tag{1.1}$$

where $g = g_{MN} dz^M \otimes dz^N$ is the metric on the manifold M, $\dim M = D$, $\varphi = (\varphi^\alpha) \in \mathbf{R}^l$ is a vector from dilatonic scalar fields, $(h_{\alpha\beta})$ is a non-degenerate symmetric $l \times l$ matrix ($l \in \mathbf{N}$), $\theta_a \neq 0$,

$$F^a = dA^a = \frac{1}{n_a!} F^a_{M_1 \ldots M_{n_a}} dz^{M_1} \wedge \ldots \wedge dz^{M_{n_a}}$$

[1]e-mail: ivas@rgs.phys.msu.su

V. de Sabbata et al. (eds.),
The Gravitational Constant: Generalized Gravitational Theories and Experiments, 205–231.
© 2004 *Kluwer Academic Publishers. Printed in the Netherlands.*

is a n_a-form ($n_a \geq 2$) on a D-dimensional manifold M, Λ is a cosmological constant and λ_a is a 1-form on $\mathbf{R}^l$: $\lambda_a(\varphi) = \lambda_{a\alpha}\varphi^\alpha$, $a \in \Delta$, $\alpha = 1, \ldots, l$. In (1.1) we denote $|g| = |\det(g_{MN})|$, $(F^a)^2_g = F^a_{M_1 \ldots M_{n_a}} F^a_{N_1 \ldots N_{n_a}} g^{M_1 N_1} \ldots g^{M_{n_a} N_{n_a}}$, $a \in \Delta$, where Δ is some finite set, and $S_{\rm GH}$ is the standard (Gibbons-Hawking) boundary term. In the models with one time all $\theta_a = 1$ when the signature of the metric is $(-1, +1, \ldots, +1)$.

For certain field contents with distinguished values of total dimension D, ranks n_a, dilatonic couplings λ_a and $\Lambda = 0$ such Lagrangians appear as "truncated" bosonic sectors (i.e. without Chern-Simons terms) of certain supergravitational theories or low-energy limit of superstring models (see [1, 2] and references therein).

2 The model

2.1 Ansatz for composite p-branes

Let us consider the manifold

$$M = M_0 \times M_1 \times \ldots \times M_n, \tag{2.1}$$

with the metric

$$g = e^{2\gamma(x)} \hat{g}^0 + \sum_{i=1}^{n} e^{2\phi^i(x)} \hat{g}^i, \tag{2.2}$$

where $g^0 = g^0_{\mu\nu}(x) dx^\mu \otimes dx^\nu$ is an arbitrary metric with any signature on the manifold M_0 and $g^i = g^i_{m_i n_i}(y_i) dy_i^{m_i} \otimes dy_i^{n_i}$ is a metric on M_i satisfying the equation

$$R_{m_i n_i}[g^i] = \xi_i g^i_{m_i n_i}, \tag{2.3}$$

$m_i, n_i = 1, \ldots, d_i$; $\xi_i = \text{const}$, $i = 1, \ldots, n$. Here $\hat{g}^i = p_i^* g^i$ is the pullback of the metric g^i to the manifold M by the canonical projection: $p_i : M \to M_i$, $i = 0, \ldots, n$. Thus, (M_i, g^i) are Einstein spaces, $i = 1, \ldots, n$. The functions $\gamma, \phi^i : M_0 \to \mathbf{R}$ are smooth. We denote $d_\nu = \dim M_\nu$; $\nu = 0, \ldots, n$; $D = \sum_{\nu=0}^{n} d_\nu$. We put any manifold M_ν, $\nu = 0, \ldots, n$, to be oriented and connected. Then the volume d_i-form

$$\tau_i \equiv \sqrt{|g^i(y_i)|}\, dy_i^1 \wedge \ldots \wedge dy_i^{d_i}, \tag{2.4}$$

and signature parameter

$$\varepsilon(i) \equiv \text{sign}(\det(g^i_{m_i n_i})) = \pm 1 \tag{2.5}$$

are correctly defined for all $i = 1, \dots, n$.

Let $\Omega = \Omega(n)$ be a set of all non-empty subsets of $\{1, \dots, n\}$. The number of elements in Ω is $|\Omega| = 2^n - 1$. For any $I = \{i_1, \dots, i_k\} \in \Omega$, $i_1 < \dots < i_k$, we denote

$$\tau(I) \equiv \hat{\tau}_{i_1} \wedge \dots \wedge \hat{\tau}_{i_k}, \tag{2.6}$$

$$\varepsilon(I) \equiv \varepsilon(i_1) \dots \varepsilon(i_k), \tag{2.7}$$

$$d(I) \equiv \sum_{i \in I} d_i. \tag{2.8}$$

Here $\hat{\tau}_i = p_i^* \hat{\tau}_i$ is the pullback of the form τ_i to the manifold M by the canonical projection: $p_i : M \to M_i$, $i = 1, \dots, n$. We also put $\tau(\emptyset) = \varepsilon(\emptyset) = 1$ and $d(\emptyset) = 0$.

For fields of forms we consider the following composite electromagnetic ansatz

$$F^a = \sum_{I \in \Omega_{a,e}} \mathcal{F}^{(a,e,I)} + \sum_{J \in \Omega_{a,m}} \mathcal{F}^{(a,m,J)} \tag{2.9}$$

where

$$\mathcal{F}^{(a,e,I)} = d\Phi^{(a,e,I)} \wedge \tau(I), \tag{2.10}$$

$$\mathcal{F}^{(a,m,J)} = e^{-2\lambda_a(\varphi)} * (d\Phi^{(a,m,J)} \wedge \tau(J)) \tag{2.11}$$

are elementary forms of electric and magnetic types respectively, $a \in \Delta$, $I \in \Omega_{a,e}$, $J \in \Omega_{a,m}$ and $\Omega_{a,v} \subset \Omega$, $v = e, m$. In (2.11) $* = *[g]$ is the Hodge operator on (M, g). For scalar functions we put

$$\varphi^\alpha = \varphi^\alpha(x), \quad \Phi^s = \Phi^s(x), \tag{2.12}$$

$s \in S$. Thus φ^α and Φ^s are functions on M_0.

Here and below

$$S = S_e \sqcup S_m, \quad S_v = \sqcup_{a \in \Delta} \{a\} \times \{v\} \times \Omega_{a,v}, \tag{2.13}$$

$v = e, m$. Here and in what follows $\sqcup$ means the union of non-intersecting sets. The set S consists of elements $s = (a_s, v_s, I_s)$, where $a_s \in \Delta$ is colour index, $v_s = e, m$ is electro-magnetic index and set $I_s \in \Omega_{a_s, v_s}$ describes the location of brane.

Due to (2.10) and (2.11)

$$d(I) = n_a - 1, \quad d(J) = D - n_a - 1, \tag{2.14}$$

for $I \in \Omega_{a,e}$ and $J \in \Omega_{a,m}$ (i.e. in electric and magnetic case, respectively). The sum of worldvolume dimensions for electric and magnetic branes corresponding to the same form is equal to $D-2$, it does not depend upon the rank of the form.

2.2 The sigma model

Let $d_0 \neq 2$ and

$$\gamma = \gamma_0(\phi) \equiv \frac{1}{2-d_0} \sum_{j=1}^{n} d_j \phi^j, \tag{2.15}$$

i.e. the generalized harmonic gauge (frame) is used. As we shall see below the equations of motions have a rather simple form in this gauge.

2.2.1 Restrictions on p-brane configurations.

Here we present two restrictions on the sets of p-branes that guarantee the block-diagonal form of the energy-momentum tensor and the existence of the sigma-model representation (without additional constraints) [7] (see also [6]).

Restriction 1 reads

$$\textbf{(R1)} \quad d(I \cap J) \leq d(I) - 2, \tag{2.16}$$

for any $I, J \in \Omega_{a,v}$, $a \in \Delta$, $v = e, m$ (here $d(I) = d(J)$).

Restriction 2 has the following form

$$\textbf{(R2)} \quad d(I \cap J) \neq 0 \ for \ d_0 = 1, \qquad d(I \cap J) \neq 1 \quad for \ d_0 = 3 \tag{2.17}$$

(see (2.14)).

2.2.2 Sigma-model action for harmonic gauge

It was proved in [7] that equations of motion for the model (1.1) and the Bianchi identities: $d\mathcal{F}^s = 0$, $s \in S_m$, for fields from (2.2), (2.9)–(2.12), when Restrictions 1 and 2 are imposed, are equivalent to equations of motion for the σ-model governed by the action

$$\begin{aligned} S_{\sigma 0} = \ & \frac{1}{2\kappa_0^2} \int d^{d_0}x \sqrt{|g^0|} \Big\{ R[g^0] - \hat{G}_{AB} g^{0\mu\nu} \partial_\mu \sigma^A \partial_\nu \sigma^B \\ & - \sum_{s \in S} \varepsilon_s \exp(-2U_A^s \sigma^A) g^{0\mu\nu} \partial_\mu \Phi^s \partial_\nu \Phi^s - 2V \Big\}, \end{aligned} \tag{2.18}$$

where $(\sigma^A) = (\phi^i, \varphi^\alpha)$, $k_0 \neq 0$, the index set S is defined in (2.13),

$$V = V(\phi) = \Lambda e^{2\gamma_0(\phi)} - \frac{1}{2}\sum_{i=1}^{n} \xi_i d_i e^{-2\phi^i + 2\gamma_0(\phi)} \tag{2.19}$$

is the potential,

$$(\hat{G}_{AB}) = \begin{pmatrix} G_{ij} & 0 \\ 0 & h_{\alpha\beta} \end{pmatrix}, \tag{2.20}$$

is the target space metric with

$$G_{ij} = d_i \delta_{ij} + \frac{d_i d_j}{d_0 - 2}, \tag{2.21}$$

and co-vectors

$$U_A^s = U_A^s \sigma^A = \sum_{i \in I_s} d_i \phi^i - \chi_s \lambda_{a_s}(\varphi), \quad (U_A^s) = (d_i \delta_{iI_s}, -\chi_s \lambda_{a_s \alpha}), \tag{2.22}$$

$s = (a_s, v_s, I_s)$. Here $\chi_e = +1$ and $\chi_m = -1$;

$$\delta_{iI} = \sum_{j \in I} \delta_{ij} \tag{2.23}$$

is an indicator of i belonging to I: $\delta_{iI} = 1$ for $i \in I$ and $\delta_{iI} = 0$ otherwise; and

$$\varepsilon_s = (-\varepsilon[g])^{(1-\chi_s)/2} \varepsilon(I_s) \theta_{a_s}, \tag{2.24}$$

$s \in S$, $\varepsilon[g] \equiv \mathrm{sign} \det(g_{MN})$. More explicitly (2.24) reads

$$\varepsilon_s = \varepsilon(I_s)\theta_{a_s} \text{ for } v_s = e; \qquad \varepsilon_s = -\varepsilon[g]\varepsilon(I_s)\theta_{a_s}, \text{ for } v_s = m. \tag{2.25}$$

For finite internal space volumes V_i (e.g. compact M_i) and electric p-branes the action (2.18) coincides with the action (1.1) when $\kappa^2 = \kappa_0^2 \prod_{i=1}^n V_i$.

Sigma-model with constraints. In [7] a general proposition concerning the sigma-model representation when the **Restrictions 1** and **2** are removed is presented. In this case the stress-energy tensor is not identically block-diagonal and several additional constraints on the field configurations appear [7].

We note that the symmetries of target space metric were studied in [17]. Sigma-model approach for models with non-diagonal metrics was suggested in [15].

3 Solutions governed by harmonic functions

3.1 Solutions with block-orthogonal U^s

Here we consider a special class of solutions to equations of motion governed by several harmonic functions when all factor spaces are Ricci-flat and cosmological constant is zero, i.e. $\xi_i = \Lambda = 0$, $(i = 1, \dots, n)$ [13, 16]. In certain situations these solutions describe extremal p-brane black holes charged by fields of forms.

The solutions crucially depend upon scalar products of U^s-vectors $(U^s, U^{s'})$; $s, s' \in S$, where

$$(U, U') = \hat{G}^{AB} U_A U'_B, \tag{3.1}$$

for $U = (U_A), U' = (U'_A) \in \mathbf{R}^N$, $N = n + l$ and

$$(\hat{G}^{AB}) = \begin{pmatrix} G^{ij} & 0 \\ 0 & h^{\alpha\beta} \end{pmatrix} \tag{3.2}$$

is matrix inverse to the matrix (2.20). Here (as in [29])

$$G^{ij} = \frac{\delta^{ij}}{d_i} + \frac{1}{2 - D}, \tag{3.3}$$

$i, j = 1, \dots, n$.

The scalar products (3.1) for vectors U^s were calculated in [7]

$$(U^s, U^{s'}) = d(I_s \cap I_{s'}) + \frac{d(I_s) d(I_{s'})}{2 - D} + \chi_s \chi_{s'} \lambda_{a_s \alpha} \lambda_{a_{s'} \beta} h^{\alpha\beta}, \tag{3.4}$$

where $(h^{\alpha\beta}) = (h_{\alpha\beta})^{-1}$; and $s = (a_s, v_s, I_s)$, $s' = (a_{s'}, v_{s'}, I_{s'})$ belong to S. This relation is a very important one since it encodes p-brane data (e.g. intersections) in scalar products of U-vectors.

Let

$$S = S_1 \sqcup \ldots \sqcup S_k, \tag{3.5}$$

$S_i \neq \emptyset$, $i = 1, \dots, k$, and

$$(U^s, U^{s'}) = 0 \tag{3.6}$$

for all $s \in S_i$, $s' \in S_j$, $i \neq j$; $i, j = 1, \dots, k$. Relation (3.5) means that the set S is a union of k non-intersecting (non-empty) subsets $S_1, \dots, S_k$.

Here we consider exact solutions in the model (1.1), when vectors $(U^s, s \in S)$ obey the block-orthogonal decomposition (3.5), (3.6) with scalar products defined in (3.4) [13]. These solutions may be obtained from the following proposition.

Proposition [13]. *Let (M_0, g^0) be Ricci-flat: $R_{\mu\nu}[g^0] = 0$. Then the field configuration*

$$g^0, \qquad \sigma^A = \sum_{s \in S} \varepsilon_s U^{sA} \nu_s^2 \ln H_s, \qquad \Phi^s = \frac{\nu_s}{H_s}, \tag{3.7}$$

$s \in S$, satisfies to field equations corresponding to action (2.18) with $V = 0$ if (real) numbers ν_s obey the relations

$$\sum_{s' \in S} (U^{:}, \qquad) \varepsilon_{s'} \nu_{s'}^2 = -1 \tag{3.8}$$

$s \in S$, functions $H_s > 0$ are harmonic, i.e. $\Delta[g^0] H_s = 0$, $s \in S$, and H_s are coinciding inside blocks: $H_s = H_{s'}$ for $s, s' \in S_i$, $i = 1, \ldots, k$.

Using the sigma-model solution from Proposition and relations for contravariant components [7]:

$$U^{si} = \delta_{iI_s} - \frac{d(I_s)}{D-2}, \qquad U^{s\alpha} = -\chi_s \lambda_{a_s}^{\alpha}, \tag{3.9}$$

$s = (a_s, v_s, I_s)$, we get [13]:

$$g = \left(\prod_{s \in S} H_s^{2d(I_s)\varepsilon_s \nu_s^2} \right)^{1/(2-D)} \left\{ \hat{g}^0 + \sum_{i=1}^{n} \left(\prod_{s \in S} H_s^{2\varepsilon_s \nu_s^2 \delta_{iI_s}} \right) \hat{g}^i \right\}, \tag{3.10}$$

$$\varphi^\alpha = - \sum_{s \in S} \lambda_{a_s}^{\alpha} \chi_s \varepsilon_s \nu_s^2 \ln H_s, \tag{3.11}$$

$$F^a = \sum_{s \in S} \mathcal{F}^s \delta_{a_s}^a, \tag{3.12}$$

where $i = 1, \ldots, n$, $\alpha = 1, \ldots, l$, $a \in \Delta$ and

$$\mathcal{F}^s = \nu_s dH_s^{-1} \wedge \tau(I_s) \text{ for } v_s = e, \tag{3.13}$$

$$\mathcal{F}^s = \nu_s (*_0 dH_s) \wedge \tau(\bar{I}_s) \text{ for } v_s = m, \tag{3.14}$$

H_s are harmonic functions on (M_0, g^0) coinciding inside blocks (i.e. $H_s = H_{s'}$ for $s, s' \in S_i$, $i = 1, \ldots, k$) and relations (3.8) on parameters ν_s are imposed.

Here the matrix $((U^s, U^{s'}))$ and parameters ε_s, $s \in S$, are defined in (3.4) and (2.24), respectively; $\lambda_a^\alpha = h^{\alpha\beta}\lambda_{\beta a}$, $*_0 = *[g^0]$ is the Hodge operator on (M_0, g^0) and $\bar{I}$ is defined as follows

$$\bar{I} \equiv I_0 \setminus I, \qquad I_0 = \{1, \ldots, n\}. \tag{3.15}$$

In (3.14) we redefined the sign of ν_s-parameter compared to (2.11).

3.1.1 Solutions with orthogonal U^s

Let us consider the orthogonal case [7]

$$(U^s, U^{s'}) = 0, \qquad s \neq s', \tag{3.16}$$

$s, s' \in S$. Then relation (3.8) reads as follows

$$(U^s, U^s)\varepsilon_s \nu_s^2 = -1, \tag{3.17}$$

$s \in S$. This implies $(U^s, U^s) \neq 0$ and

$$\varepsilon_s (U^s, U^s) < 0, \tag{3.18}$$

for all $s \in S$. For $d(I_s) < D - 2$ and $\lambda_{a_s\alpha}\lambda_{a_s\beta}h^{\alpha\beta} \geq 0$ we get from (3.4) $(U^s, U^s) > 0$, and, hence, $\varepsilon_s < 0$, $s \in S$. If $\theta_a > 0$ for all $a \in \Delta$, then

$$\varepsilon(I_s) = -1 \text{ for } v_s = e; \qquad \varepsilon(I_s) = \varepsilon[g] \text{ for } v_s = m. \tag{3.19}$$

For pseudo-Euclidean metric g all $\varepsilon(I_s) = -1$ and, hence, all p-branes should contain time manifold. For the metric g with the Euclidean signature only magnetic p-branes can exist in this case.

From scalar products (3.4) and the orthogonality condition (3.16) we get the "orthogonal" intersection rules [6, 7]

$$d(I_s \cap I_{s'}) = \frac{d(I_s) d(I_{s'})}{D - 2} - \chi_s \chi_{s'} \lambda_{a_s\alpha}\lambda_{a_{s'}\beta}h^{\alpha\beta} \equiv \Delta(s, s'), \tag{3.20}$$

for $s = (a_s, v_s, I_s) \neq s' = (a_{s'}, v_{s'}, I_{s'})$. (For pure electric case see also [4, 5].)

Certain supersymmetric solutions in $D = 11$ supergravity defined on product of Ricci-flat spaces were considered in [18].

3.2 General Toda-type solutions with one harmonic function

It is well known that geodesics of the target space equipped with some harmonic function on a three-dimensional space generate a solution to the σ-model equations [27, 28]. Here we apply this null-geodesic method to our sigma-model and obtain a new class of solutions in multidimensional gravity with p-branes governed by one harmonic function H.

3.2.1 Toda-like Lagrangian

Action (2.18) may be also written in the form

$$S_{\sigma 0} = \frac{1}{2\kappa_0^2} \int d^{d_0}x \sqrt{|g^0|} \{ R[g^0] - \mathcal{G}_{\hat{A}\hat{B}}(X) g^{0\mu\nu} \partial_\mu X^{\hat{A}} \partial_\nu X^{\hat{B}} - 2V \} \quad (3.21)$$

where $X = (X^{\hat{A}}) = (\phi^i, \varphi^\alpha, \Phi^s) \in \mathbf{R}^N$, and minisupermetric $\mathcal{G} = \mathcal{G}_{\hat{A}\hat{B}}(X) dX^{\hat{A}} \otimes dX^{\hat{B}}$ on minisuperspace $\mathcal{M} = \mathbf{R}^N$, $N = n + l + |S|$ ($|S|$ is the number of elements in S) is defined by the relation

$$(\mathcal{G}_{\hat{A}\hat{B}}(X)) = \begin{pmatrix} G_{ij} & 0 & 0 \\ 0 & h_{\alpha\beta} & 0 \\ 0 & 0 & \varepsilon_s \exp(-2U^s(\sigma)) \delta_{ss'} \end{pmatrix}. \quad (3.22)$$

Here we consider exact solutions to field equations corresponding to the action (3.21)

$$R_{\mu\nu}[g^0] = \mathcal{G}_{\hat{A}\hat{B}}(X) \partial_\mu X^{\hat{A}} \partial_\nu X^{\hat{B}} + \frac{2V}{d_0 - 2} g^0_{\mu\nu}, \quad (3.23)$$

$$\frac{1}{\sqrt{|g^0|}} \partial_\mu [\sqrt{|g^0|} \mathcal{G}_{\hat{C}\hat{B}}(X) g^{0\mu\nu} \partial_\nu X^{\hat{B}}] - \frac{1}{2} \mathcal{G}_{\hat{A}\hat{B},\hat{C}}(X) g^{0,\mu\nu} \partial_\mu X^{\hat{A}} \partial_\nu X^{\hat{B}} = V_{,\hat{C}} \quad (3.24)$$

$s \in S$. Here $V_{,\hat{C}} = \partial V / \partial X^{\hat{C}}$.

We put

$$X^{\hat{A}}(x) = F^{\hat{A}}(H(x)), \quad (3.25)$$

where $F : (u_-, u_+) \to \mathbf{R}^N$ is a smooth function, $H : M_0 \to \mathbf{R}$ is a harmonic function on M_0 (i.e. $\Delta[g^0]H = 0$), satisfying $u_- < H(x) < u_+$ for all

$x \in M_0$. Let all factor spaces are Ricci-flat and cosmological constant is zero, i.e. relation $\xi_i = \Lambda = 0$ is satisfied. In this case the potential is zero : $V = 0$. It may be verified that the field equations (3.23) and (3.24) are satisfied identically if $F = F(u)$ obey the Lagrange equations for the Lagrangian

$$L = \frac{1}{2}\mathcal{G}_{\hat{A}\hat{B}}(F)\dot{F}^{\hat{A}}\dot{F}^{\hat{B}} \tag{3.26}$$

with the zero-energy constraint

$$E = \frac{1}{2}\mathcal{G}_{\hat{A}\hat{B}}(F)\dot{F}^{\hat{A}}\dot{F}^{\hat{B}} = 0. \tag{3.27}$$

This means that $F : (u_-, u_+) \to \mathbf{R}^N$ is a null-geodesic map for the minisupermetric $\mathcal{G}$. Thus, we are led to the Lagrange system (3.26) with the minisupermetric $\mathcal{G}$ defined in (3.22).

The problem of integrability will be simplified if we integrate the Lagrange equations corresponding to Φ^s:

$$\frac{d}{du}\left(\exp(-2U^s(\sigma))\dot{\Phi}^s\right) = 0 \Longleftrightarrow \dot{\Phi}^s = Q_s \exp(2U^s(\sigma)), \tag{3.28}$$

where Q_s are constants, $s \in S$. Here $(F^{\hat{A}}) = (\sigma^A, \Phi^s)$. We put $Q_s \neq 0$ for all $s \in S$.

For fixed $Q = (Q_s, s \in S)$ the Lagrange equations for the Lagrangian (3.26) corresponding to $(\sigma^A) = (\phi^i, \varphi^\alpha)$, when equations (3.28) are substituted, are equivalent to the Lagrange equations for the Lagrangian

$$L_Q = \frac{1}{2}\hat{G}_{AB}\dot{\sigma}^A\dot{\sigma}^B - V_Q, \tag{3.29}$$

where

$$V_Q = \frac{1}{2}\sum_{s \in S}\varepsilon_s Q_s^2 \exp[2U^s(\sigma)], \tag{3.30}$$

the matrix $(\hat{G}_{AB})$ is defined in (2.20). The zero-energy constraint (3.27) reads

$$E_Q = \frac{1}{2}\hat{G}_{AB}\dot{\sigma}^A\dot{\sigma}^B + V_Q = 0. \tag{3.31}$$

3.2.2 Toda-type solutions

Here we are interested in exact solutions for a case when $K_s = (U^s, U^s) \neq 0$, for all $s \in S$, and the quasi-Cartan matrix

$$A_{ss'} \equiv \frac{2(U^s, U^{s'})}{(U^{s'}, U^{s'})}, \tag{3.32}$$

is a non-degenerate one. Here some ordering in S is assumed.

It follows from the non-degeneracy of the matrix (3.32) that the vectors $U^s, s \in S$, are linearly independent and, hence, $|S| \leq n + l$.

The exact solutions were obtained in [22]:

$$g = \Bigl(\prod_{s \in S} f_s^{2d(I_s)h_s/(D-2)} \Bigr) \Bigl\{ \exp(2c^0 H + 2\bar{c}^0) \hat{g}^0 \tag{3.33}$$

$$+ \sum_{i=1}^{n} \Bigl(\prod_{s \in S} f_s^{-2h_s \delta_{iI_s}} \Bigr) \exp(2c^i H + 2\bar{c}^i) \hat{g}^i \Bigr\},$$

$$\exp(\varphi^\alpha) = \left(\prod_{s \in S} f_s^{h_s \chi_s \lambda^\alpha_{a_s}} \right) \exp(c^\alpha H + \bar{c}^\alpha), \tag{3.34}$$

$\alpha = 1, \dots, l$ and $F^a = \sum_{s \in S} \mathcal{F}^s \delta^a_{a_s}$ with

$$\mathcal{F}^s = Q_s \left(\prod_{s' \in S} f_{s'}^{-A_{ss'}} \right) dH \wedge \tau(I_s), \qquad s \in S_e, \tag{3.35}$$

$$\mathcal{F}^s = Q_s (*_0 dH) \wedge \tau(\bar{I}_s), \qquad s \in S_m, \tag{3.36}$$

where $*_0 = *[g^0]$ is the Hodge operator on (M_0, g^0). Here

$$f_s = f_s(H) = \exp(-q^s(H)), \tag{3.37}$$

where $q^s(u)$ is a solution to Toda-like equations

$$\ddot{q}^s = -B_s \exp(\sum_{s' \in S} A_{ss'} q^{s'}), \tag{3.38}$$

with $B_s = K_s \varepsilon_s Q_s^2$, $s \in S$, and $H = H(x)$ $(x \in M_0)$ is a harmonic function on (M_0, g^0). Vectors $c = (c^A)$ and $\bar{c} = (\bar{c}^A)$ satisfy the linear constraints

$$U^s(c) = \sum_{i \in I_s} d_i c^i - \chi_s \lambda_{a_s \alpha} c^\alpha = 0, \quad U^s(\bar{c}) = 0, \tag{3.39}$$

$s \in S$, and

$$c^0 = \frac{1}{2-d_0} \sum_{j=1}^{n} d_j c^j, \quad \bar{c}^0 = \frac{1}{2-d_0} \sum_{j=1}^{n} d_j \bar{c}^j. \tag{3.40}$$

The zero-energy constraint reads

$$2E_{TL} + h_{\alpha\beta} c^\alpha c^\beta + \sum_{i=1}^{n} d_i (c^i)^2 + \frac{1}{d_0 - 2} \left(\sum_{i=1}^{n} d_i c^i \right)^2 = 0, \tag{3.41}$$

where

$$E_{TL} = \frac{1}{4} \sum_{s,s' \in S} h_s A_{ss'} \dot{q}^s \dot{q}^{s'} + \sum_{s \in S} A_s \exp(\sum_{s' \in S} A_{ss'} q^{s'}), \tag{3.42}$$

is an integration constant (energy) for the solutions from (3.38) and $A_s = \frac{1}{2}\varepsilon_s Q_s^2$.

We note that the equations (3.38) correspond to the Lagrangian

$$L_{TL} = \frac{1}{4} \sum_{s,s' \in S} h_s A_{ss'} \dot{q}^s \dot{q}^{s'} - \sum_{s \in S} A_s \exp(\sum_{s' \in S} A_{ss'} q^{s'}), \tag{3.43}$$

where $h_s = K_s^{-1}$.

Thus, the solution is presented by relations (3.33)-(3.37) with the functions q^s defined in (3.38) and the relations on the parameters of solutions c^A, $\bar{c}^A$ $(A = i, \alpha, 0)$, imposed in (3.39), (3.40) and (3.41).

4 Classical and quantum cosmological-type solutions

Here we consider the case $d_0 = 1$, $M_0 = \mathbf{R}$, i.e. we are interesting in applications to the sector with one-variable dependence. We consider the manifold

$$M = (u_-, u_+) \times M_1 \times \ldots \times M_n \tag{4.1}$$

with the metric

$$g = w e^{2\gamma(u)} du \otimes du + \sum_{i=1}^{n} e^{2\phi^i(u)} \hat{g}^i, \tag{4.2}$$

where $w = \pm 1$, u is a distinguished coordinate which, by convention, will be called "time"; (M_i, g^i) are oriented and connected Einstein spaces (see (2.3)), $i = 1, \ldots, n$. The functions γ, ϕ^i: $(u_-, u_+) \to \mathbf{R}$ are smooth.

Here we adopt the p-brane ansatz from Sect. 2. putting $g^0 = w du \otimes du$.

4.1 Lagrange dynamics

It follows from sect. 2.2 that the equations of motion and the Bianchi identities for the field configuration under consideration (with the restrictions from subsect. 2.2.1 imposed) are equivalent to equations of motion for 1-dimensional σ-model with the action

$$S_\sigma = \frac{\mu}{2}\int du \mathcal{N}\Big\{G_{ij}\dot\phi^i\dot\phi^j + h_{\alpha\beta}\dot\varphi^\alpha\dot\varphi^\beta + \sum_{s\in S}\varepsilon_s \exp[-2U^s(\phi,\varphi)](\dot\Phi^s)^2 - 2\mathcal{N}^{-2}V_w(\phi)\Big\}, \tag{4.3}$$

where $\dot x \equiv dx/du$,

$$V_w = -wV = -w\Lambda e^{2\gamma_0(\phi)} + \frac{w}{2}\sum_{i=1}^{n}\xi_i d_i e^{-2\phi^i + 2\gamma_0(\phi)} \tag{4.4}$$

is the potential with $\gamma_0(\phi) \equiv \sum_{i=1}^n d_i\phi^i$, and $\mathcal{N} = \exp(\gamma_0 - \gamma) > 0$ is the lapse function, $U^s = U^s(\phi,\varphi)$ are defined in (2.22), ε_s are defined in (2.24) for $s = (a_s, v_s, I_s) \in S$, and $G_{ij} = d_i\delta_{ij} - d_i d_j$ are components of the "pure cosmological" minisupermetric, $i,j = 1,\dots,n$ [29].

In the electric case ($\mathcal{F}^{(a,m,I)} = 0$) for finite internal space volumes V_i the action (4.3) coincides with the action (1.1) if $\mu = -w/\kappa_0^2$, $\kappa^2 = \kappa_0^2 V_1 \dots V_n$.

Action (4.3) may be also written in the form

$$S_\sigma = \frac{\mu}{2}\int du \mathcal{N}\left\{\mathcal{G}_{\hat A\hat B}(X)\dot X^{\hat A}\dot X^{\hat B} - 2\mathcal{N}^{-2}V_w\right\}, \tag{4.5}$$

where $X = (X^{\hat A}) = (\phi^i, \varphi^\alpha, \Phi^s) \in \mathbf{R}^N$, $N = n + l + |S|$, and minisupermetric $\mathcal{G}$ is defined in (3.22).

Scalar products. The minisuperspace metric (3.22) may be also written in the form $\mathcal{G} = \hat G + \sum_{s\in S}\varepsilon_s e^{-2U^s(\sigma)} d\Phi^s \otimes d\Phi^s$, where $\sigma = (\sigma^A) = (\phi^i, \varphi^\alpha)$,

$$\hat G = \hat G_{AB} d\sigma^A \otimes d\sigma^B = G_{ij} d\phi^i \otimes d\phi^j + h_{\alpha\beta} d\varphi^\alpha \otimes d\varphi^\beta, \tag{4.6}$$

is truncated minisupermetric and $U^s(\sigma) = U^s_A\sigma^A$ is defined in (2.22). The potential (4.4) reads

$$V_w = (-w\Lambda)e^{2U^\Lambda(\sigma)} + \sum_{j=1}^{n}\frac{w}{2}\xi_j d_j e^{2U^j(\sigma)}, \tag{4.7}$$

where

$$U^j(\sigma) = U^j_A\sigma^A = -\phi^j + \gamma_0(\phi), \qquad (U^j_A) = (-\delta^j_i + d_i, 0), \tag{4.8}$$

$$U^\Lambda(\sigma) = U^\Lambda_A\sigma^A = \gamma_0(\phi), \qquad (U^\Lambda_A) = (d_i, 0). \tag{4.9}$$

The integrability of the Lagrange system (4.5) crucially depends upon the scalar products of co-vectors U^Λ, U^j, U^s (see (3.1)). These products are defined by (3.4) and the following relations [7]

$$(U^i, U^j) = \frac{\delta_{ij}}{d_j} - 1, \tag{4.10}$$

$$(U^i, U^\Lambda) = -1, \qquad (U^\Lambda, U^\Lambda) = -\frac{D-1}{D-2}, \tag{4.11}$$

$$(U^s, U^i) = -\delta_{iI_s}, \qquad (U^s, U^\Lambda) = \frac{d(I_s)}{2-D}, \tag{4.12}$$

where $s = (a_s, v_s, I_s) \in S$; $i, j = 1, \dots, n$.

Toda-like representation. We put $\gamma = \gamma_0(\phi)$, i.e. the harmonic time gauge is considered. Integrating the Lagrange equations corresponding to Φ^s (see (3.28)) we are led to the Lagrangian from (3.29) and the zero-energy constraint (3.31) with the modified potential

$$V_Q = V_w + \frac{1}{2}\sum_{s \in S} \varepsilon_s Q_s^2 \exp[2U^s(\sigma)], \tag{4.13}$$

where V_w is defined in (4.4).

4.2 Classical solutions with $\Lambda = 0$

Here we consider classical solutions with $\Lambda = 0$.

4.2.1 Solutions with Ricci-flat spaces

Let all spaces be Ricci-flat, i.e. $\xi_1 = \ldots = \xi_n = 0$.

Since $H(u) = u$ is a harmonic function on (M_0, g^0) with $g^0 = w du \otimes du$ we get for the metric and scalar fields from (3.33), (3.34) [22]

$$g = \Big(\prod_{s \in S} f_s^{2d(I_s)h_s/(D-2)}\Big)\Big\{\exp(2c^0 u + 2\bar{c}^0) w du \otimes du \tag{4.14}$$

$$+ \sum_{i=1}^{n} \Big(\prod_{s \in S} f_s^{-2h_s \delta_{iI_s}}\Big) \exp(2c^i u + 2\bar{c}^i) \hat{g}^i \Big\},$$

$$\exp(\varphi^\alpha) = \Big(\prod_{s \in S} f_s^{h_s \chi_s \lambda_{a_s}^\alpha}\Big) \exp(c^\alpha u + \bar{c}^\alpha), \tag{4.15}$$

$\alpha = 1, \ldots, l$, where $f_s = f_s(u) = \exp(-q^s(u))$ and $q^s(u)$ obey Toda-like equations (3.38).

Relations (3.40) and (3.41) take the form

$$c^0 = \sum_{j=1}^{n} d_j c^j, \qquad \bar{c}^0 = \sum_{j=1}^{n} d_j \bar{c}^j, \tag{4.16}$$

$$2E_{TL} + h_{\alpha\beta} c^\alpha c^\beta + \sum_{i=1}^{n} d_i (c^i)^2 - \left(\sum_{i=1}^{n} d_i c^i \right)^2 = 0, \tag{4.17}$$

with E_{TL} from (3.42) and all other relations (e.g. constraints (3.39) and relations for forms (3.35) and (3.36) with $H = u$) are unchanged. In a special $\mathbf{A_m}$ Toda chain case this solution was considered previously in [19].

4.2.2 Solutions with one curved space

The cosmological solution with Ricci-flat spaces may be also modified to the following case: $\xi_1 \neq 0, \quad \xi_2 = \ldots = \xi_n = 0$, i.e. one space is curved and others are Ricci-flat and $1 \notin I_s$, $s \in S$, i.e. all "brane" submanifolds do not contain M_1.

The potential (3.30) is modified for $\xi_1 \neq 0$ as follows (see (4.13))

$$V_Q = \frac{1}{2} \sum_{s \in S} \varepsilon_s Q_s^2 \exp[2U^s(\sigma)] + \frac{1}{2} w \xi_1 d_1 \exp[2U^1(\sigma)], \tag{4.18}$$

where $U^1(\sigma)$ is defined in (4.8) ($d_1 > 1$).

For the scalar products we get from (4.10) and (4.12)

$$(U^1, U^1) = \frac{1}{d_1} - 1 < 0, \qquad (U^1, U^s) = 0 \tag{4.19}$$

for all $s \in S$. The solution in the case under consideration may be obtained by a little modification of the solution from the previous section (using (4.19) and relations $U^{1i} = -\delta_1^i / d_1$, $U^{1\alpha} = 0$). We get [22]

$$g = \left(\prod_{s \in S} [f_s(u)]^{2d(I_s)h_s/(D-2)} \right) \Big\{ [f_1(u)]^{2d_1/(1-d_1)} \exp(2c^1 u + 2\bar{c}^1) \tag{4.20}$$
$$\times [w du \otimes du + f_1^2(u) \hat{g}^1] + \sum_{i=2}^{n} \Big(\prod_{s \in S} [f_s(u)]^{-2h_s \delta_{iI_s}} \Big) \exp(2c^i u + 2\bar{c}^i) \hat{g}^i \Big\}.$$

$$\exp(\varphi^\alpha) = \left(\prod_{s \in S} f_s^{h_s \chi_s \lambda_{as}^\alpha} \right) \exp(c^\alpha u + \bar{c}^\alpha), \tag{4.21}$$

and $F^a = \sum_{s \in S} \delta^a_{a_s} \mathcal{F}^s$ with forms

$$\mathcal{F}^s = Q_s \left(\prod_{s' \in S} f_{s'}^{-A_{ss'}} \right) du \wedge \tau(I_s), \qquad s \in S_e, \tag{4.22}$$

$$\mathcal{F}^s = Q_s \tau(\bar{I}_s), \qquad s \in S_m \tag{4.23}$$

$Q_s \neq 0$, $s \in S$. Here $f_s = f_s(u) = \exp(-q^s(u))$ where $q^s(u)$ obeys Toda-like equations (3.38) and

$$f_1(u) = R\,\mathrm{sh}(\sqrt{C_1}(u - u_1)),\ C_1 > 0,\ \xi_1 w > 0; \tag{4.24}$$

$$R \sin(\sqrt{|C_1|}(u - u_1)),\ C_1 < 0,\ \xi_1 w > 0; \tag{4.25}$$

$$R\,\mathrm{ch}(\sqrt{C_1}(u - u_1)),\ C_1 > 0,\ \xi_1 w < 0; \tag{4.26}$$

$$|\xi_1(d_1 - 1)|^{1/2},\ C_1 = 0,\ \xi_1 w > 0, \tag{4.27}$$

u_1, C_1 are constants and $R = |\xi_1(d_1 - 1)/C_1|^{1/2}$.

Vectors $c = (c^A)$ and $\bar{c} = (\bar{c}^A)$ satisfy the linear constraints

$$U^r(c) = U^r(\bar{c}) = 0, \qquad r = s, 1, \tag{4.28}$$

(for $r = s$ see (3.39)) and the zero-energy constraint

$$C_1 \frac{d_1}{d_1 - 1} = 2E_{TL} + h_{\alpha\beta} c^\alpha c^\beta + \sum_{i=2}^{n} d_i (c^i)^2 + \frac{1}{d_1 - 1} \left(\sum_{i=2}^{n} d_i c^i \right)^2. \tag{4.29}$$

Restriction 1 (see subsect. 2.2.1) forbids certain intersections of two p-branes with the same color index for $n_1 \geq 2$. **Restriction 2** is satisfied identically in this case.

This solution in a special case of $\mathbf{A_m}$ Toda chain was obtained earlier in [19] (see also [12]). For special sets of parameters one can get so-called S-brane and flux-brane solutions [31, 32] (for pioneering flux-brane solution see also [15].)

4.2.3 Block-orthogonal solutions

Let us consider block-orthogonal case: (3.5), (3.6). In this case we get $f_s = \bar{f}_s^{b_s}$ where $b_s = 2 \sum_{s' \in S} A^{ss'}$, $(A^{ss'}) = (A_{ss'})^{-1}$ and

$$\bar{f}_s(u) = R_s\,\mathrm{sh}(\sqrt{C_s}(u - u_s)),\ C_s > 0,\ \eta_s \varepsilon_s < 0; \tag{4.30}$$

$$R_s \sin(\sqrt{|C_s|}(u - u_s)),\ C_s < 0,\ \eta_s \varepsilon_s < 0; \tag{4.31}$$

$$R_s \,\mathrm{ch}(\sqrt{C_s}(u - u_s)),\ C_s > 0,\ \eta_s \varepsilon_s > 0; \tag{4.32}$$

$$\frac{|Q^s|}{|\nu_s|}(u - u_s),\ C_s = 0,\ \eta_s \varepsilon_s < 0, \tag{4.33}$$

where $R_s = |Q_s|/(|\nu_s||C_s|^{1/2})$, $\eta_s \nu_s^2 = b_s h_s$, $\eta_s = \pm 1$, C_s, u_s are constants, $s \in S$. The constants C_s, u_s are coinciding inside blocks: $u_s = u_{s'}$, $C_s = C_{s'}$, $s, s' \in S_i$, $i = 1, \ldots, k$. The ratios $\varepsilon_s Q_s^2/(b_s h_s)$ are also coinciding inside blocks, or, equivalently,

$$\frac{\varepsilon_s Q_s^2}{b_s h_s} = \frac{\varepsilon_{s'} Q_{s'}^2}{b_{s'} h_{s'}}, \tag{4.34}$$

$s, s' \in S_i$, $i = 1, \ldots, k$.

Here

$$E_{TL} = \sum_{s \in S} C_s \eta_s \nu_s^2. \tag{4.35}$$

The solution (4.20)-(4.23) with block-orthogonal set of vectors was obtained in [21] (for non-composite case see also [14]). In the special orthogonal case when: $|S_1| = \ldots = |S_k| = 1$, the solution was obtained in [11].

4.3 Quantum solutions.

4.3.1 Wheeler–De Witt equation.

Here we fix the gauge as follows

$$\gamma_0 - \gamma = f(X), \quad \mathcal{N} = e^f, \tag{4.36}$$

where f: $\mathcal{M} \to \mathbf{R}$ is a smooth function. Then we obtain the Lagrange system with the Lagrangian

$$L_f = \frac{\mu}{2} e^f \mathcal{G}_{\hat{A}\hat{B}}(X) \dot{X}^{\hat{A}} \dot{X}^{\hat{B}} - \mu e^{-f} V_w \tag{4.37}$$

and the energy constraint

$$E_f = \frac{\mu}{2} e^f \mathcal{G}_{\hat{A}\hat{B}}(X) \dot{X}^{\hat{A}} \dot{X}^{\hat{B}} + \mu e^{-f} V_w = 0. \tag{4.38}$$

Using the standard prescriptions of (covariant and conformally covariant) quantization (see, for example, [29]) we are led to the Wheeler-DeWitt (WDW) equation

$$\hat{H}^f \Psi^f \equiv \left(-\frac{1}{2\mu} \Delta \left[e^f \mathcal{G} \right] + \frac{a}{\mu} R \left[e^f \mathcal{G} \right] + e^{-f} \mu V_w \right) \Psi^f = 0, \tag{4.39}$$

where $a = a_N = (N-2)/8(N-1)$ and $N = n + l + |S|$.

Here $\Psi^f = \Psi^f(X)$ is the so-called "wave function of the universe" corresponding to the f-gauge (4.36) and satisfying the relation

$$\Psi^f = e^{bf} \Psi^{f=0}, \quad b = (2-N)/2, \tag{4.40}$$

($\Delta[\mathcal{G}_1]$ and $R[\mathcal{G}_1]$ denote the Laplace-Beltrami operator and the scalar curvature corresponding to $\mathcal{G}_1$, respectively).

Harmonic-time gauge The WDW equation (4.39) for $f = 0$

$$\hat{H} \Psi \equiv \left(-\frac{1}{2\mu} \Delta[\mathcal{G}] + \frac{a}{\mu} R[\mathcal{G}] + \mu V_w \right) \Psi = 0, \tag{4.41}$$

where

$$R[\mathcal{G}] = -\sum_{s \in S} (U^s, U^s) - \sum_{s,s' \in S} (U^s, U^{s'}). \tag{4.42}$$

and

$$\Delta[\mathcal{G}] = e^{U(\sigma)} \frac{\partial}{\partial \sigma^A} \left(\hat{G}^{AB} e^{-U(\sigma)} \frac{\partial}{\partial \sigma^B} \right) + \sum_{s \in S} \varepsilon_s e^{2U^s(\sigma)} \left(\frac{\partial}{\partial \Phi^s} \right)^2, \tag{4.43}$$

with $U(\sigma) = \sum_{s \in S} U^s(\sigma)$.

4.3.2 Solutions with one curved factor space and orthogonal U^s

Here as in subsect. 4.2.2 we put $\Lambda = 0$, $\xi_1 \neq 0$, $\xi_2 = \ldots = \xi_n = 0$, and $1 \notin I_s$, $s \in S$, i.e. the space M_1 is curved and others are Ricci-flat and all "brane" submanifolds do not contain M_1. We also put orthogonality restriction on the vectors U^s: $(U^s, U^{s'}) = 0$ for $s \neq s'$ and $K_s = (U^s, U^s) \neq 0$ for all $s \in S$. In this case the potential (4.7) reads $V_w = \frac{1}{2} w \xi_1 d_1 e^{2U^1(\sigma)}$. The truncated minisuperspace metric (4.6) may be diagonalized by the linear transformation $z^A = S^A{}_B \sigma^B$, $(z^A) = (z^1, z^a, z^s)$ as follows

$$\hat{G} = -dz^1 \otimes dz^1 + \sum_{s \in S} \eta_s dz^s \otimes dz^s + dz^a \otimes dz^b \eta_{ab}, \tag{4.44}$$

where $a, b = 2, \ldots, n + l - |S|$; $\eta_{ab} = \eta_{aa}\delta_{ab}; \eta_{aa} = \pm 1$, $\eta_s = \mathrm{sign}(U^s, U^s)$ and $q_1 z^1 = U^1(\sigma)$, $q_1 \equiv \sqrt{|(U^1, U^1)|} = \sqrt{1 - d_1^{-1}}$, $q_s z^s = U^s(\sigma)$, $q_s = \nu_s^{-1} \equiv \sqrt{|(U^s, U^s)|}$ $s = (a_s, v_s, I_s) \in S$.

We are seeking the solution to WDW equation (4.41) by the method of the separation of variables, i.e. we put

$$\Psi_*(z) = \Psi_1(z^1) \left(\prod_{s \in S} \Psi_s(z^s) \right) \mathrm{e}^{\mathrm{i} P_s \Phi^s} \mathrm{e}^{\mathrm{i} p_a z^a}. \tag{4.45}$$

It follows from the relation for the Laplace operator (4.43) that $\Psi_*(z)$ satisfies WDW equation (4.41) if

$$\left\{ \left(\frac{\partial}{\partial z^1} \right)^2 + \mu^2 w \xi_1 d_1 \mathrm{e}^{2 q_1 z^1} \right\} \Psi_1 = 2\mathcal{E}_1 \Psi_1; \tag{4.46}$$

$$\left\{ -\eta_s \mathrm{e}^{q_s z^s} \frac{\partial}{\partial z^s} \left(\mathrm{e}^{-q_s z^s} \frac{\partial}{\partial z^s} \right) + \varepsilon_s P_s^2 \mathrm{e}^{2 q_s z^s} \right\} \Psi_s = 2\mathcal{E}_s \Psi_s, \tag{4.47}$$

$s \in S$, and

$$2\mathcal{E}_1 + \eta^{ab} p_a p_b + 2 \sum_{s \in S} \mathcal{E}_s + 2aR[\mathcal{G}] = 0, \tag{4.48}$$

where $a = a_N = (N-2)/8(N-1)$ and $R[\mathcal{G}] = -2\sum_{s \in S}(U^s, U^s) = -2\sum_{s \in S} \eta_s q_s^2$.

We obtain the following linearly independent solutions to (4.46) and (4.47), respectively

$$\Psi_1(z^1) = B^1_{\omega_1} \left(\sqrt{-w\mu^2 \xi_1 d_1} \frac{\mathrm{e}^{q_1 z^1}}{q_1} \right), \tag{4.49}$$

$$\Psi_s(z^s) = \mathrm{e}^{q_s z^s/2} B^s_{\omega_s} \left(\sqrt{\eta_s \varepsilon_s P_s^2} \frac{\mathrm{e}^{q_s z^s}}{q_s} \right), \tag{4.50}$$

where $\omega_1 = \sqrt{2\mathcal{E}_1}/q_1$; $\omega_s = \sqrt{\frac{1}{4} - 2\eta_s \mathcal{E}_s \nu_s^2}$, $s \in S$, and $B^1_\omega, B^s_\omega = I_\omega, K_\omega$ are the modified Bessel function.

The general solution to the WDW equation (4.41) is a superposition of the "separated" solutions (4.45):

$$\Psi(z) = \sum_B \int dp dP d\mathcal{E} C(p, P, \mathcal{E}, B) \Psi_*(z|p, P, \mathcal{E}, B), \tag{4.51}$$

where $p = (p_a)$, $P = (P_s)$, $\mathcal{E} = (\mathcal{E}_s, \mathcal{E}_1)$, $B = (B^1, B^s)$, $B^1, B^s = I, K$; and $\Psi_* = \Psi_*(z|p, P, \mathcal{E}, B)$ is given by relation (4.45), (4.49)–(4.50) with $\mathcal{E}_1$ from (4.48). Here $C(p, P, \mathcal{E}, B)$ are smooth enough functions.

Solution with n Ricci-flat spaces. In the case $\xi_1 = 0$ the solution to WDW equation is given by the following modification of the relation (4.45) and (4.48), respectively

$$\Psi_*(z) = \left(\prod_{s\in S} \Psi_s(z^s)\right) e^{iP_s\Phi^s} e^{ip_a z^a}, \tag{4.52}$$

$$\eta^{ab} p_a p_b + 2\sum_{s\in S} \mathcal{E}_s + 2aR[\mathcal{G}] = 0, \tag{4.53}$$

where here $a, b = 1, \ldots, n+l-|S|$. In this case (4.44) should be also modified

$$\hat{G} = \sum_{s\in S} \eta_s dz^s \otimes dz^s + dz^a \otimes dz^b \eta_{ab}, \tag{4.54}$$

(with $a, b = 1, \ldots, n+l-|S|$). Here restriction $1 \notin I_s$, $s \in S$, should be removed and z^1 is not obviously related to U^1.

4.4 Spherically-symmetric solutions with a horizon

Here we consider the spherically-symmetric case of the metric (4.20), i.e. we put $w = 1$, $M_1 = S^{d_1}$, $g^1 = d\Omega^2_{d_1}$, where $d\Omega^2_{d_1}$ is the canonical metric on a unit sphere S^{d_1}, $d_1 \geq 2$. In this case $\xi^1 = d_1 - 1$. We put $M_2 = \mathbf{R}$, $g^2 = -dt \otimes dt$, i.e. M_2 is a time manifold. We also assume that $(U^s, U^s) \neq 0$, $s \in S$, and

$$\det((U^s, U^{s'})) \neq 0. \tag{4.55}$$

We put $C_1 \geq 0$. When the matrix $(h_{\alpha\beta})$ is positive definite and

$$2 \in I_s, \quad \forall s \in S, \tag{4.56}$$

i. e. all p-branes have a common time direction t, the horizon condition (of infinite time of light propagation) singles out the unique solution with $C_1 > 0$ and linear asymptotics at infinity $q^s = -\beta^s u + \bar{\beta}^s + o(1)$, $u \to +\infty$, where $\beta^s, \bar{\beta}^s$ are constants, $s \in S$, [25, 26].

The solutions with horizon have the following form [24, 25, 26]

$$g = \Bigl(\prod_{s\in S} H_s^{2h_s d(I_s)/(D-2)}\Bigr)\Bigl\{\Bigl(1 - \frac{2\mu}{R^{\bar{d}}}\Bigr)^{-1} dR \otimes dR + R^2 d\Omega^2_{d_1} \tag{4.57}$$

$$-\Big(\prod_{s\in S} H_s^{-2h_s}\Big)\Big(1-\frac{2\mu}{R^{\bar d}}\Big)\, dt\otimes dt + \sum_{i=3}^{n}\Big(\prod_{s\in S} H_s^{-2h_s\delta_{iI_s}}\Big)\hat g^i\Big\},$$

$$\exp(\varphi^\alpha) = \prod_{s\in S} H_s^{h_s\chi_s\lambda^\alpha_{a_s}}, \tag{4.58}$$

where $F^a = \sum_{s\in S}\delta^a_{a_s}\mathcal{F}^s$, and

$$\mathcal{F}^s = -\frac{Q_s}{R^{d_1}}\left(\prod_{s'\in S} H_{s'}^{-A_{ss'}}\right) dR\wedge\tau(I_s), \tag{4.59}$$

$s\in S_e$,

$$\mathcal{F}^s = Q_s\tau(\bar I_s), \tag{4.60}$$

$s\in S_m$. Here $Q_s\neq 0$, $h_s = K_s^{-1}$, $s\in S$, and the quasi-Cartan matrix $(A_{ss'})$ is non-degenerate.

Functions H_s obey the following differential equations

$$\frac{d}{dz}\,\backslash\;\; \frac{-\,2\mu z)}{H_s}\frac{d}{dz}H_s\Big) = \bar B_s\prod_{s'\in S} H_{s'}^{-A_{ss'}}, \tag{4.61}$$

where $H_s(z)>0$, $\mu>0$, $z=R^{-\bar d}\in(0,(2\mu)^{-1})$ and $\bar B_s = \varepsilon_s K_s Q_s^2/\bar d^2\neq 0$. equipped with the boundary conditions

$$H_s((2\mu)^{-1}-0) = H_{s0}\in(0,+\infty), \tag{4.62}$$

$$H_s(+0) = 1, \tag{4.63}$$

$s\in S$.

Equations (4.61) are equivalent to Toda-like equations. The first boundary condition leads to a regular horizon at $R^{\bar d} = 2\mu$. The second condition (4.63) guarantees the asymptotical flatness (for $R\to+\infty$) of the $(2+d_1)$-dimensional section of the metric.

There exist solutions to eqs. (4.61)-(4.63) of polynomial type. The simplest example occurs in orthogonal case [8, 9, 11, 10]: $(U^s, U^{s'}) = 0$, for $s\neq s'$, $s,s'\in S$. In this case $(A_{ss'}) = \mathrm{diag}(2,\ldots,2)$ is a Cartan matrix for semisimple Lie algebra $\mathbf{A_1}\oplus\ldots\oplus\mathbf{A_1}$ and

$$H_s(z) = 1 + P_s z, \tag{4.64}$$

with $P_s\neq 0$, satisfying

$$P_s(P_s+2\mu) = -\bar B_s, \tag{4.65}$$

$s \in S$.

In [14, 20] this solution was generalized to a block-orthogonal case (3.5), (3.6). In this case (4.64) is modified as follows

$$H_s(z) = (1 + P_s z)^{b_s}, \tag{4.66}$$

where $b_s = 2\sum_{s' \in S} A^{ss'}$ and parameters P_s and are coinciding inside blocks, i.e. $P_s = P_{s'}$ for $s, s' \in S_i$, $i = 1, \ldots, k$. Parameters $P_s \neq 0$ satisfy the relations

$$P_s(P_s + 2\mu) = -\bar{B}_s/b_s, \tag{4.67}$$

$b_s \neq 0$, and parameters $\bar{B}_s/b_s$ are also coinciding inside blocks.

Conjecture. *Let $(A_{ss'})$ be a Cartan matrix for a semisimple finite-dimensional Lie algebra $\mathcal{G}$. Then the solution to eqs. (4.61)-(4.63) (if exists) is a polynomial*

$$H_s(z) = 1 + \sum_{k=1}^{n_s} P_s^{(k)} z^k, \tag{4.68}$$

where $P_s^{(k)}$ are constants, $k = 1, \ldots, n_s$; $n_s = b_s = 2\sum_{s' \in S} A^{ss'} \in \mathbf{N}$ and $P_s^{(n_s)} \neq 0$, $s \in S$.

In this case all powers n_s are natural numbers coinciding with the components of twice the dual Weyl vector in the basis of simple coroots [23]. In extremal case ($\mu = +0$) an a analogue of this conjecture was suggested previously in [12]. Conjecture 1 was verified for $\mathbf{A_m}$ and $\mathbf{C_{m+1}}$ series of Lie algebras in [25, 26]. Explicit relations for $\mathbf{C_2}$ and $\mathbf{A_3}$ algebras were obtained in [33] and [34], respectively.

5 Billiard representation near the singularity

It was found in that [36] the cosmological models with p-branes may have a "never ending" oscillating behaviour near the cosmological singularity as it takes place in Bianchi-IX model [35]. Remarkably, this oscillating behaviour may be described using the so-called billiard representation near the singularity. In [36] the billiard representation for a cosmological model with a set of electro-magnetic composite p-branes in a theory with the action (1.1) was obtained (see also [37, 38] and references therein).

In terms of the Kasner parameters $\alpha = (\alpha^A) = (\alpha^i, \alpha^\gamma)$, satisfying relations

$$\sum_{i=1}^{n} d_i \alpha^i = \sum_{i=1}^{n} d_i (\alpha^i)^2 + \alpha^\beta \alpha^\gamma h_{\beta\gamma} = 1, \tag{5.1}$$

the existence of never ending oscillating behaviour near the singularity takes place if for any α there exists $s \in S$ such that $(U^s, U^s) > 0$ and

$$U^s(\alpha) = U^s_A \alpha^A = \sum_{i \in I_s} d_i \alpha^i - \chi_s \lambda_{a_s \gamma} \alpha^\gamma \leq 0 \tag{5.2}$$

[36]. Thus, U-vectors play a key role in determination of possible oscillating behaviour near the singularity. In [36] the relations (5.2) were also interpreted in terms of illumination of a (Kasner) sphere by point-like sources.

General "collision law". The set of Kasner parameters (α'^A) after the collision with the s-th wall $(s \in S)$ is defined by the Kasner set before the collision (α^A) according to the following formula

$$\alpha'^A = \frac{\alpha^A - 2U^s(\alpha) U^{sA} (U^s, U^s)^{-1}}{1 - 2U^s(\alpha)(U^s, U^\Lambda)(U^s, U^s)^{-1}}, \tag{5.3}$$

where $U^{sA} = \bar{G}^{AB} U^s_B$, $U^s(\alpha) = U^s_A \alpha^A$ and co-vector U^Λ is defined in (4.9).

The formula (5.3) follows just from the reflection "law"

$$v'^A = v^A - 2U^s(v) U^{sA} (U^s, U^s)^{-1}, \tag{5.4}$$

for Kasner free "motion": $\sigma^A = v^A t + \sigma^A_0$, ($t$ is harmonic time) and the definition of Kasner parameters: $\alpha^A = v^A / U^\Lambda(v)$. In the special case of one scalar field and 1-dimensional factor-spaces (i.e. $l = d_i = 1$) this formula was suggested earlier in [37].

6 Conclusions

Here we considered several rather general families of exact solutions in multidimensional gravity with a set of scalar fields and fields of forms. These solutions describe composite (non-localized) electromagnetic p-branes defined on products of *Ricci-flat* (or sometimes Einstein) spaces of *arbitrary signatures.* The metrics are block-diagonal and all scale factors, scalar fields and fields of forms depend on points of some (mainly Ricci-flat) manifold M_0. The solutions include those depending upon harmonic functions, cosmological and spherically-symmetric solutions (e.g. black-brane ones). Our scheme is based on the sigma-model representation obtained in [7].

Here we considered also the Wheeler-DeWitt equation for p-brane cosmology in d'Alembertian (covariant) and conformally covariant form and integrated it for orthogonal U-vectors (and $n - 1$

Ricci-flat internal spaces). We also overviewed general classes of "cosmological" and spherically symmetric solutions governed by Toda-type equations, e.g. black brane ones. An interesting point here is the appearance of polynomials for black brane solutions when brane intersections are governed by Cartan matrices of finite-dimensional simple Lie algebras [24]. It should be noted that post-Newtonian parameters corresponding to certain 4-dimensional section of metrics were calculated in [20, 25].

Another topic of interest appearing here is the possible oscillating behaviour in the models with p-branes, that has a description by billiards in hyperbolic (Lobachevsky) spaces [36].

Acknowledgments

This work was supported in part by the DFG grant 436 RUS 113/236/O(R), by the Russian Ministry of Science and Technology and Russian Foundation for Basic Research grant. The author thanks Prof. V. De Sabbata and his colleagues for kind hospitality during the School in Erice (in May 2003).

References

[1] K.S. Stelle, Lectures on supergravity p-branes, hep-th/9701088.

[2] V.D. Ivashchuk and V.N. Melnikov, Exact solutions in multidimensional gravity with antisymmetric forms, topical review, *Class. Quantum Grav.* **18**, R82-R157 (2001); hep-th/0110274.

[3] H. Lü and C.N. Pope, $SL(N+1, R)$ Toda solitons in supergravities, hep-th/9607027; *Int. J. Mod. Phys.* **A 12**, 2061 -2074 (1997).

[4] V.D. Ivashchuk and V.N. Melnikov, Intersecting p-Brane Solutions in Multidimensional Gravity and M-Theory, hep-th/9612089; *Grav. and Cosmol.* **2**, No 4, 297-305 (1996).

[5] V.D. Ivashchuk and V.N. Melnikov, *Phys. Lett. B* **403**, 23-30 (1997).

[6] I.Ya. Aref'eva and O.A. Rytchkov, Incidence Matrix Description of Intersecting p-brane Solutions, hep-th/9612236.

[7] V.D. Ivashchuk and V.N. Melnikov, Sigma-model for the Generalized Composite p-branes, hep-th/9705036; *Class. Quantum Grav.* **14**, 3001-3029 (1997); Corrigenda *ibid.* **15** (12), 3941 (1998).

[8] I.Ya. Aref'eva, M.G. Ivanov and I.V. Volovich, Non-extremal intersecting p-branes in various dimensions, hep-th/9702079; *Phys. Lett.* **B 406**, 44-48 (1997).

[9] N. Ohta, Intersection rules for non-extreme p-branes, hep-th/9702164; *Phys. Lett.* **B 403**, 218-224 (1997).

[10] K.A. Bronnikov, V.D. Ivashchuk and V.N. Melnikov, The Reissner-Nordström Problem for Intersecting Electric and Magnetic p-Branes, gr-qc/9710054; *Grav. and Cosmol.* **3**, No 3 (11), 203-212 (1997).

[11] V.D. Ivashchuk and V.N. Melnikov, Multidimensional classical and quantum cosmology with intersecting p-branes, hep-th/9708157; *J. Math. Phys.*, **39**, 2866-2889 (1998).

[12] H. Lü, J. Maharana, S. Mukherji and C.N. Pope, Cosmological Solutions, p-branes and the Wheeler De Witt Equation, hep-th/9707182; *Phys. Rev.* **D 57** 2219-2229 (1997).

[13] V.D. Ivashchuk and V.N. Melnikov, Madjumdar-Papapetrou Type Solutions in Sigma-model and Intersecting p-branes, *Class. Quantum Grav.* **16**, 849 (1999); hep-th/9802121.

[14] K.A. Bronnikov, Block-orthogonal Brane systems, Black Holes and Wormholes, hep-th/9710207; *Grav. and Cosmol.* **4**, No 1 (13), 49 (1998).

[15] D.V. Gal'tsov and O.A. Rytchkov, Generating Branes via Sigma models, hep-th/9801180; *Phys. Rev.* **D 58**, 122001 (1998).

[16] M.A. Grebeniuk and V.D. Ivashchuk, Sigma-model solutions and intersecting p-branes related to Lie algebras, hep-th/9805113; *Phys. Lett.* **B 442**, 125-135 (1998).

[17] V.D. Ivashchuk, On symmetries of Target Space for σ-model of p-brane Origin, hep-th/9804102; *Grav. and Cosmol.*, **4**, 3(15), 217-220 (1998).

[18] V.D. Ivashchuk, On supersymmetric solutions in $D = 11$ supergravity on product of Ricci-flat spaces, hep-th/0012263; *Grav. Cosmol.* **6**, No. 4(24), 344-350 (2000); hep-th/0012263.

[19] V.R. Gavrilov and V.N. Melnikov, Toda Chains with Type A_m Lie Algebra for Multidimensional Classical Cosmology with Intersecting p-branes; hep-th/9807004.

[20] V.D. Ivashchuk and V.N.Melnikov. Multidimensional cosmological and spherically symmetric solutions with intersecting p-branes; gr-qc/9901001.

[21] V.D. Ivashchuk and V.N.Melnikov, Cosmological and Spherically Symmetric Solutions with Intersecting p-branes. *J. Math. Phys.*, 1999, **40** (12), 6558-6576.

[22] V.D.Ivashchuk and S.-W. Kim. Solutions with intersecting p-branes related to Toda chains, *J. Math. Phys.*, **41** (1) 444-460 (2000); hep-th/9907019.

[23] J. Fuchs and C. Schweigert, Symmetries, Lie algebras and Representations. A graduate course for physicists (Cambridge University Press, Cambridge, 1997).

[24] V.D.Ivashchuk and V.N.Melnikov. P-brane black Holes for General Intersections. *Grav. and Cosmol.* **5**, No 4 (20), 313-318 (1999); gr-qc/0002085.

[25] V.D.Ivashchuk and V.N.Melnikov, Black hole p-brane solutions for general intersection rules. *Grav. and Cosmol.* **6**, No 1 (21), 27-40 (2000); hep-th/9910041.

[26] V.D.Ivashchuk and V.N.Melnikov, Toda p-brane black holes and polynomials related to Lie algebras. *Class. and Quantum Gravity* **17** 2073-2092 (2000); math-ph/0002048.

[27] G. Neugebauer and D. Kramer, *Ann. der Physik (Leipzig)* **24**, 62 (1969).

[28] D. Kramer, H. Stephani, M. MacCallum, and E. Herlt, Ed. Schmutzer, Exact solutions of the Einstein field equations, Deutscher Verlag der Wissenshaften, Berlin, 1980.

[29] V.D. Ivashchuk, V.N. Melnikov and A.I. Zhuk, *Nuovo Cimento* **B 104**, 575 (1989).

[30] K.A. Bronnikov and V.N. Melnikov, p-Brane Black Holes as Stability Islands, *Nucl. Phys.* **B 584**, 436-458 (2000).

[31] V.D. Ivashchuk, Composite S-brane solutions related to Toda-type systems, *Class. Quantum Grav.* **20**, 261-276 (2003); hep-th/0208101.

[32] V.D. Ivashchuk, Composite fluxbranes with general intersections, *Class. Quantum Grav.*, **19**, 3033-3048 (2002); hep-th/0202022.

[33] M.A. Grebeniuk, V.D. Ivashchuk and S.-W. Kim, Black-brane solutions for C_2 algebra, *J. Math. Phys.* **43**, 6016-6023 (2002); hep-th/0111219.

[34] M.A. Grebeniuk, V.D. Ivashchuk and V.N. Melnikov Black-brane solution for A_3 algebra, *Phys. Lett.* , **B 543**, 98-106 (2002); hep-th/0208083.

[35] V.A. Belinskii, E.M. Lifshitz and I.M. Khalatnikov, *Usp. Fiz. Nauk* **102**, 463 (1970) [in Russian]; *Adv. Phys.* **31**, 639 (1982).

[36] V.D. Ivashchuk and V.N. Melnikov, Billiard representation for multidimensional cosmology with intersecting p-branes near the singularity. *J. Math. Phys.*, **41**, No 8, 6341-6363 (2000); hep-th/9904077.

[37] T. Damour and M. Henneaux, Chaos in Superstring Cosmology, *Phys. Rev. Lett.* **85**, 920-923 (2000); hep-th/0003139

[38] T. Damour, M. Henneaux and H. Nicolai, Cosmological billiards, topical review, *Class. Quantum Grav.* **20**, R145-R200 (2003); hep-th/0212256.

5D gravity and the discrepant G measurements

J. P. Mbelek
Service d'Astrophysique, C.E. Saclay
F-91191 Gif-sur-Yvette Cedex, France

Abstract

It is shown that 5D Kaluza-Klein theory stabilized by an external bulk scalar field may solve the discrepant laboratory G measurements. This is achieved by an effective coupling between gravitation and the geomagnetic field. Experimental considerations are also addressed.

1 Introduction

Although the methods and techniques have been greatly improved since the late nineteenth century, the precision on the measurement of the gravitational constant, G, is still the less accurate in comparison with the other fundamental constants of nature [1]. Moreover, given the relative uncertainties of most of the individual experiments (reaching about 10^{-4} for the most precise measurements), they show an incompatibility which leads to an overall precision of only about 1 part in 10^3. Thus the current status of the G terrestrial measurements (see [2]) implies either an unknown source of errors (not taken into account in the published uncertainties), or some new physics [3]. In the latter spirit, many theories which include extradimensions have been proposed as candidates for the unification of physics. As such, they involve a coupling between gravitation and electromagnetism (GE coupling), as well as with other gauge fields present.

V. de Sabbata et al. (eds.),
The Gravitational Constant: Generalized Gravitational Theories and Experiments, 233–245.
© 2004 *Kluwer Academic Publishers. Printed in the Netherlands.*

Here we show that the discrepancy between the present results of the G-measurements may be understood as a consequence of the GE coupling. Also, this theory predicts a variation of the effective fine structure "constant" α with the gravitational field, and thus with the cosmological time (see [4]).

2 Theoretical background

An argument initially from Landau and Lifshitz [5] may be applied to the pure Kaluza-Klein (KK) action (see [6]): the negative sign of the kinetic term of the five dimensional (5D) KK internal scalar field, Φ, leads to inescapable instability. The question to know which of the two conformally related frames (Einstein-Pauli frame or Jordan-Fierz frame) is the best remains debated in the literature, each frame having its own advantage. The negative kinetic energy density for the Φ-field, and thus the instability, occur in both frames. In the following, we perform calculations in the Jordan-Fierz frame, where, as we show, the discrepant laboratory measurements of G find a natural explanation. Also, it is true that the 5D KK theory may yield a zero kinetic term (and thus a zero kinetic energy density, which would also be unusual in 4D), but this occurs only when the electromagnetic (EM) field is identically zero everywhere. This may be relevant for some cosmological solutions, but not for our discussion. On the contrary, we argue here that the Φ-field is tightly related to the EM field via the link between the compactified space of the fifth dimension and the U(1) gauge group.

Since stabilization may be obtained if an external field is present, we assume here a version, KKψ, which includes an external bulk scalar field minimally coupled to gravity (like the radion in the brane world scenario). After dimensional reduction ($\alpha = 0, 1, 2, 3$), this bulk field reduces to a four dimensional scalar field $\psi = \psi(x^\alpha)$ and, in the Jordan-Fierz frame, the low energy effective action takes the form (up to a total divergence)

$$S = -\int \sqrt{-g}\,[\frac{c^4}{16\pi}\frac{\Phi}{G}R + \frac{1}{4}\,\varepsilon_0\,\Phi^3\,F_{\alpha\beta}\,F^{\alpha\beta} + \frac{c^4}{4\pi G}\frac{\partial_\alpha\Phi\,\partial^\alpha\Phi}{\Phi}]\,d^4x$$

$$+ \int \sqrt{-g}\,\Phi\,[\frac{1}{2}\partial_\alpha\psi\,\partial^\alpha\psi - U - J\psi]\,d^4x, \qquad (1)$$

where A^α is the potential 4-vector of the EM field, $F_{\alpha\beta} = \partial_\alpha A_\beta - \partial_\beta A_\alpha$ the EM field strength tensor, U the self-interaction potential of ψ of the symmetry breaking type and J its source term.

The source term of the ψ-field, J, includes the contributions of the ordinary matter, of the EM field and of the internal scalar field Φ. For each, the coupling is defined by a function (temperature dependent, as for the potential U) $f_X = f_X(\psi\,,\,\Phi)$, where the subscript X stands for "matter", "EM" and "Φ". In addition, the necessity to recover the Einstein-Maxwell equations in the weak fields limit, implies the following conditions: $U(v) = 0$ and $f_{EM}(v\,,\,1) = f_{matter}(v\,,\,1) = 0$, where v denotes the vacuum expectation value (VEV) of the ψ-field.

The contributions of matter and Φ are proportional to the traces of their respective energy-momentum tensors. A contribution of the form $\varepsilon_0\,f_{EM}\,F_{\alpha\beta}\,F^{\alpha\beta}$ accounts for the coupling with the EM field. Besides, though the fit to the data involves $\frac{\partial f_{EM}}{\partial\Phi}\,(v\,,\,1)\,v \gg 4\pi G/c^4$, we may infer that $\frac{\partial f_{EM}}{\partial\Phi}\,(v\,,\,1)\,v$ is negligibly small at very high temperature (e.g., like in the core of the Sun or at the big bang nucleosynthesis) and even vanishes beyond the critical temperature (say $T_c = 6000$ K) of the potential $U = U(\psi, T)$ as one gets $v = 0$ in that case.

Following Lichnerowicz [7], let us interpret the quantity

$$G_{eff} = \frac{G}{\Phi} \qquad (2)$$

of the Einstein-Hilbert term, and the factor $\varepsilon_{0eff} = \varepsilon_0\,[\,\Phi^3 + 4f_{EM}(\psi\,,\,\Phi)]$ of the Maxwell term respectively as the effective gravitational "constant" and the effective vacuum dielectric permittivity. The effective vacuum magnetic permeability reads

$\mu_{0eff} = \mu_0 [\, \Phi^3 \, + \, 4 f_{EM}(\psi \, , \, \Phi)]^{-1}$, so that the velocity of light in vacuum remains a true universal constant. Both terms depend on the local (for local physics) or global (at cosmological scale) value of the KK scalar field Φ, assumed to be positive defined.

The least action principle applied to the action (1) yields the generalized Einstein-Maxwell equations and the scalar fields equations

$$\nabla_\nu \nabla^\nu \psi = - \, J \, - \, \frac{\partial J}{\partial \psi} \psi \, - \, \frac{\partial U}{\partial \psi} \tag{3}$$

and

$$\nabla_\nu \, \nabla^\nu \, \Phi = - \, \frac{4\pi G}{c^4} \, \varepsilon_0 \, F_{\alpha\beta} \, F^{\alpha\beta} \, \Phi^3 \, + \, U \, \Phi \, + \, J \psi \, \Phi \, + \, \frac{\partial J}{\partial \Phi} \, \Phi^2 \, \psi \, - \, \frac{1}{2} (\, \partial_\alpha \psi \, \partial^\alpha \psi \,) \, \Phi. \tag{4}$$

3 Solutions in presence of a dipolar magnetic field

Let us study the *spatial* variation of Φ in the weak fields conditions out of the fields' source, but in presence of a static dipolar magnetic field, $\vec{B} \, = \, \vec{\nabla} \, V(r, \varphi, \theta)$. We denote r, φ and θ respectively the radius from the centre, the azimuth angle and the colatitude. Thus, writing $\Phi \, = \, \Phi(r, \varphi, \theta)$, and taking into account that $\frac{\partial U}{\partial \psi}(v) \, = \, 0$ (definition of the VEV), equation (4) simplifies after linearization as

$$\Delta \Phi = - \, \frac{2}{\mu_0} [\, \frac{\partial f_{EM}}{\partial \Phi} \, (v \, , \, 1) \, v \, + \, \frac{4\pi G}{c^4} \,] \, (\, \vec{\nabla} V \,)^2. \tag{5}$$

Since $\Delta V \, = \, div \, \vec{B} \, = 0$ and $(\vec{\nabla} V)^2 \, = \, \frac{1}{2} \, \Delta(V^2) \, - \, V \, \Delta V$ identically, the solution of equation (5) above reads merely

$$\Phi = 1 \, - \, \frac{1}{\mu_0} [\, \frac{\partial f_{EM}}{\partial \Phi} \, (v \, , \, 1) \, v \, + \, \frac{4\pi G}{c^4} \,] \, V^2. \tag{6}$$

For our purpose, it is sufficient to limit the expansion of the scalar potential, V, to the terms of the Legendre function of degree one ($n \, = \, 1$) and order one ($m \, = \, 1$). Hence $V \, = \, (a^3 / r^2) \, [\, g_1^0 \, \cos \theta \, + \, g_1^1 \, \sin \theta \, \cos \varphi \, + \, h_1^1 \, \sin \theta \, \sin \varphi \,]$, where g_1^0, g_1^1 and h_1^1 are the relevant Gauss coefficients, a is the Earth's radius and $M = \frac{4\pi}{\mu_0} a^3 \sqrt{(g_1^0)^2 + (g_1^1)^2 + (h_1^1)^2}$ denotes its magnetic moment. Setting $\cos\varphi_1 = -g_1^1/\sqrt{(g_1^1)^2 + (h_1^1)^2}$, $\sin\varphi_1 = h_1^1/\sqrt{(g_1^1)^2 + (h_1^1)^2}$ and $\tan\lambda = g_1^0/\sqrt{(g_1^1)^2 + (h_1^1)^2}$, the solution of equation (5) then reads (and similarly for ψ by making the substitution $\frac{\partial f_{EM}}{\partial \Phi} \rightarrow -\frac{\partial f_{EM}}{\partial \psi}$)

$$\Phi = 1 - \frac{1}{\mu_0} \frac{\partial f_{EM}}{\partial \Phi} (v\,,1)\, v\, (\frac{\mu_0\, M}{4\pi\, r^2})^2\, x(\theta,\varphi), \tag{7}$$

where we have set

$$x(\theta,\varphi) = \cos^2\theta + \cot^2\lambda \sin^2\theta \cos^2(\varphi + \varphi_1) - \cot\lambda \sin 2\theta \cos(\varphi + \varphi_1). \tag{8}$$

Thence, one derives the expression of $G_{eff}(r,\theta,\varphi)$ by inserting the solution (7) above in relation (2).

It is worth noticing that the magnetic potential, V, scales as $B\,r$, where B is the magnitude of the geomagnetic field at radius r. Indeed, because of this scaling effect, the small spatial variations of the geomagnetic field will influence significantly the laboratory measurements of G whereas the large local magnetic fields present in the laboratory (*e. g.*, the magnetic suspension used to support the balance beam, the magnetic damper, etc...) will not. A rough estimate shows that, even using a 30 Tesla superconducting magnet, one still needs to gain at least one order of magnitude with the most precise G measuring apparatus avalaible yet. Hence, our prediction is consistent with the earlier conclusion of Lloyd [11].

4 Comparison with laboratory measurements

There are presently almost 45 results of G measurements published since 1942 [12]. Because of the too numerous uncontrolled systematic biases, the mine measurements are excluded from the present study (including them will not change our fit because of their lack of precision, typically less than 1%). Also, the more discordant laboratory measurement (high PTB value [8]) is excluded, since it may suffer from a systematic effect. The " official " values are presently $G = 6.67259 \pm 0.00085\ 10^{-11}$ (CODATA 86, [9]) and $G = 6.670 \pm 0.010\ 10^{-11}$ (CODATA 2000, [10]) in MKS unit. In the following, all the measurements are weighted equally in the fit. Fitting the 44 data with these values gives respectively $\chi^2_\nu = 11.128$ and $\chi^2_\nu = 62.498$ ($\chi^2_\nu = \chi^2$ per degrees of freedom). If we forget the official values and try a best fit, assuming an arbitrary constant value of G, we obtain $G = 6.6741\ 10^{-11}\ m^3\ kg^{-1}\ s^{-2}$ with $\chi^2_\nu = 2.255$. The fit to the same sample of 44 measurements (figure 1), on account of the GE coupling, yields (in MKS units) with $\chi^2_\nu = 1.669$:

$$\frac{1}{10^{11}\ G_{eff}} = (\,0.149929 \pm 0.000017\,) \;-\; (\,0.0001509 \pm 0.0000252\,)\, x(L,l). \qquad (9)$$

From the above fit, one derives both estimates of the true gravitational constant

$$G = (\,6.6696 \pm 0.0008\,)\ 10^{-11}\ m^3\ kg^{-1}\ s^{-2}, \qquad (10)$$

and the coupling parameter

$$\frac{\partial f_{EM}}{\partial \Phi}(v\,,\,1)\,v = (\,5.44 \pm 0.66\,)\ 10^{-6}\ fm\ TeV^{-1}. \qquad (11)$$

The latter quantity, expressed in the canonical form $\frac{\partial f_{EM}}{\partial \Phi}(v\,,\,1)\,v = \hbar c\, M_5^{-2}$, yields a 5D Planck scale $M_5 \simeq 5.9$ TeV of the order of the value that is invoked in the literature to solve the hierarchy problem.

Location [reference]	*Latitude (°)*	*Longitude (°)*	G_{lab} $(10^{-11}\ m^3\ kg^{-1}\ s^{-2})$
Lower Hutt (MSL) [16, 15]	-41.2	174.9	6.6742 ± 0.0007 6.6746 ± 0.0010
Wuhan (HUST) [17]	30.6	106.88	6.6699 ± 0.0007
Los Alamos [18]	35.88	-106.38	6.6740 ± 0.0007
Gaithersburg (NBS) [19, 20]	38.9	-77.02	6.6726 ± 0.0005 6.6720 ± 0.0041
Boulder (JILA) [21]	40	-105.27	6.6873 ± 0.0094
Gigerwald lake [22, 23]	46.917	9.4	6.669 ± 0.005 (at 112 m) 6.678 ± 0.007 (at 88 m) 6.6700 ± 0.0054
Zurich [24, 25]	47.4	8.53	6.6754 ± 0.0005 ± 0.0015 6.6749 ± 0.0014
Budapest [26]	47.5	19.07	6.670 ± 0.008
Seattle [14]	47.63	- 122.33	6.674215 ± 0.000092
Sevres (BIPM) [27, 28]	48.8	2.13	6.67559 ± 0.00027 6.683 ± 0.011
Fribourg [29]	46.8	7.15	6.6704 ± 0.0048 (Oct. 84) 6.6735 ± 0.0068 (Nov. 84) 6.6740 ± 0.0053 (Dec. 84) 6.6722 ± 0.0051 (Feb. 85)
Magny-les-Hameaux [30]	49	2	6.673 ± 0.003
Wuppertal [31]	51.27	7.15	6.6735 ± 0.0011 ± 0.0026
Braunschweig (PTB) [8, 32]	52.28	10.53	6.71540 ± 0.00056 6.667 ± 0.005
Moscow [33, 34]	55.1	38.85	6.6729 ± 0.0005 6.6745 ± 0.0008
Dye 3, Greenland [35]	65.19	-43.82	6.6726 ± 0.0027
Lake Brasimone [36]	43.75	11.58	6.688 ± 0.011

Table 1 : Results of the most precise laboratory measurements of G published during the last sixty years and location of the laboratories.

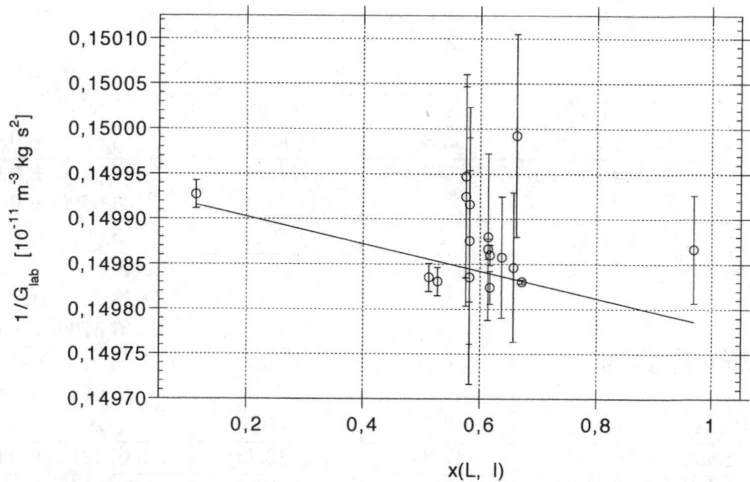

Figure 1: Laboratory measurements with relative uncertainty $\frac{\delta G_{lab}}{G_{lab}} < 10^{-3}$ and measuring time $\Delta t < 200\ s$ (sample S1, 17 points [16], [17] - [20], [23], [14], [29] - [31], [32] - [35]). The line indicates the best fit G_{lab} versus the mixed variable x ($\chi^2_\nu = 1.327$). Assuming a constant G would yield a bad fit to the data ($\chi^2_\nu = 3.607$), mostly because of the HUST value.

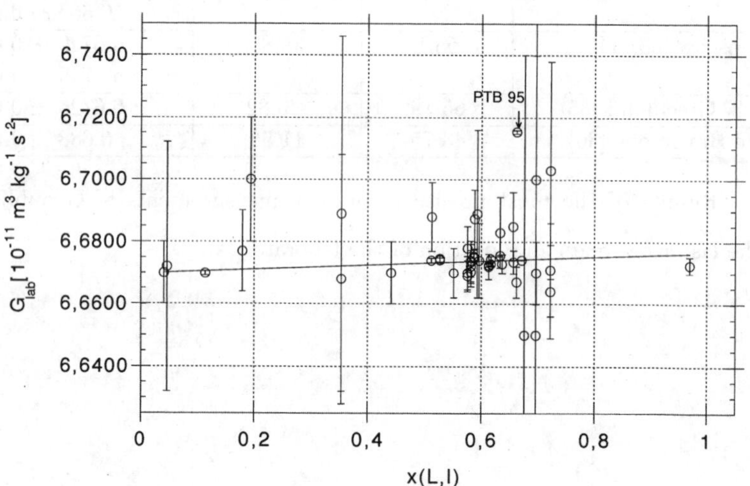

Figure 2: G_{lab} versus x (whole sample plus the PTB 95 value, 45 points [15] - [37]).

Sample	H0	H1
S1 17 points (Fig.1)	$\chi^2_\nu = 3.607$ (best fit) $\chi^2_\nu = 21.523$ (mean of CODATA 86) $\chi^2_\nu = 141.46$ (mean of CODATA 2000)	$\chi^2_\nu = 1.327$
Whole [16] - [37] 44 points (Fig.2)	$\chi^2_\nu = 2.255$ (best fit) $\chi^2_\nu = 11.128$ (mean of CODATA 86) $\chi^2_\nu = 62.498$ (mean of CODATA 2000)	$\chi^2_\nu = 1.669$

Table 2 : Reduced χ^2 for the two different hypothesis H0 (Hypothesis of a constant G) and H1 (Hypothesis of an effective G), and different samples S1 and whole (except the high PTB value [8], see text).

Considering the whole sample, we check the relevance of our result under H1 compared to the best value of G under H0, by applying the F test (Fisher law). This yields $F_\chi = \frac{\Delta\chi^2}{\chi^2_\nu} = 16.09$, which indicates that, independently of the number of parameters (two instead of one), our fit is better with a significance level greater than 99.9% [13]. Let us emphasize that the most precise value of G today [14] contributes to χ^2_ν to less than 0.0006 in the above fit. Likewise, the last published value of G [38] would contribute to less than 0.02. This suggests that the agreement may be better than purely indicated by the χ^2 values. Moreover, if one substitutes $G = (6.6731 \pm 0.0002)\ 10^{-11}\ m^3\ kg^{-1}\ s^{-2}$ [39] announced by the MSL team in June 2002 (CPEM, Ottawa, Canada) for the prior $G = (6.6742 \pm 0.0007)\ 10^{-11}\ m^3\ kg^{-1}\ s^{-2}$ published in 1999 [16], one would obtain with the sample S1 :

H0 : $\chi^2_\nu = 5.168$

H1 : $\chi^2_\nu = 1.162$

and with the whole sample of 44 measurements :

H0 : $\chi^2_\nu = 2.836$

H1 : $\chi^2_\nu = 1.488$

5 Discussion and Conclusion

It is worth noticing that the scalar fields under considerations identify neither to the dilaton nor to the inflaton of higher dimensional theories, without further assumptions. In particular, the computation of both scalar fields, as given by equations (3) and (4), involves only (see the right hand sides) quantities related to fields sources, and not to the test bodies. Hence, the effective G_{eff} given by relation (2) does not depend on the composition of the test bodies. Besides, let us emphasize that the equation of motion of a neutral point-like particle in the genuine KK theory reduces to the 4D geodesic equation after dimensional reduction [40], although the KK scalar field is coupled to the Maxwell invariant $F_{\alpha\beta}F^{\alpha\beta}$. In a forthcoming paper [41], we address the effect of the varying effective coupling constants on the masses of composit particles. On account of the Higgs mechanism of quarks and leptons masses generation, by promoting the Yukawa coupling constants to effective parameters (on an equal footing with G or α) that depend on both scalar fields ψ and Φ, we prove that the KKψ model is actually consistent with the current experimental bounds on the violation of the equivalence principle. Hence, we conclude that present laboratory experiments may not measure a true constant of gravitation. Instead, in addition to all other possible biases (*e. g.*, anelasticity in the wire of torsion pendulum as pointed out by Kuroda [42], and which since has been generically taken into account in the experiments), they may be pointing out an effective one depending on the geomagnetic field at the laboratory position.

References

[1] J. Luo and Z. K. Hu, (2000), Class. Quantum Grav. **17**, 2351.

[2] G. T. Gillies, Rep. Prog. Phys. **60**, (1997), 151, and references therein.

[3] V. N. Melnikov, gr-qc/9903110.

[4] J.-P. Mbelek and M. Lachièze-Rey, (2003), A & A **397**, 803.

[5] L. D. Landau and E. M. Lifshitz, *The Classical Theory of Fields* (Addison-Wesley Publishing Company Inc., Reading, Massachusetts, 1959), pp. 289-293.

[6] J.-P. Mbelek and M. Lachièze-Rey, gr-qc/0012086, and references therein.

[7] A. Lichnerowicz, *Théories relativistes de la gravitation et de l'électromagnétisme*, (Masson et Cie, Paris, France, 1955), pp. 201-206.

[8] W. Michaelis, H. Haars and R. Augustin, (1995/96), metrologia **32**, 267.

[9] E. R. Cohen and B. N. Taylor, (1987), Rev. Mod. Phys. **59**, 1121.

[10] P. J. Mohr and B. N. Taylor, (2000), Rev. Mod. Phys. **72**, 351.

[11] M. G. Lloyd, (1909), Terrestrial Magnetism and Atmospheric Electricity **14**, 67.

[12] J.-P. Mbelek and M. Lachièze-Rey, (2002), Grav. Cosmol. **8**, 331.

[13] P. R. Bevington, *Data reduction and error analysis for the physical sciences*, (McGraw-Hill Book Company, New York, 1969), pp. 200-203 and pp. 317-323.

[14] J. H. Gundlach and S. M. Merkowitz, (2000), Phys. Rev. Lett. **85**, 2869.

[15] M. P. Fitzgerald and T. R. Armstrong, 1995, IIIE Trans. Instrum. Meas. **44**, 494.

[16] M. P. Fitzgerald and T. R. Armstrong, (1999), Meas. Sci. Technol. **10**, 439.

[17] J. Luo *et al.*, (1998), Phys. Rev. **D59**, 042001.

[18] C. H. Bagley and G. G. Luther, (1997), Phys. Rev. Lett. **78**, 3047.

[19] G. G. Luther and W. Towler, (1982), Phys. Rev. Lett. **48**, 121.

[20] P. R. Heyl and P. Chrzanowski, (1942), J. Res. Nat. Bur. Standards **29**, 1.

[21] J. P. Schwarz *et al.*, (1998), Science **282**, 2230.

[22] B. Hubler, A. Cornaz and W. Kündig, (1995), Phys. Rev. **D51**, 4005.

[23] A. Cornaz, B. Hubler and W. Kündig, (1994), Phys. Rev. Lett. **72**, 1152.

[24] J. Schurr, F. Nolting and W. Kündig, (1998), Phys. Rev. Lett. **80**, 1142.

[25] F. Nolting, J. Schurr, S. Schlamminger and W. Kündig, (1999), Meas. Sci. Technol. **10**, 487.

[26] J. Renner, in *Determination of Gravity Constant and Measurement of Certain Fine Gravity Effects*, Y. D. Boulanger and M. U. Sagitov (Eds.), (National Aeronautics and Space Administration, Washington, 1974), pp. 26-31.

[27] T. J. Quinn, C. C. Speake, S. J. Richman, R. S. Davis and A. Picard, (2001), Phys. Rev. Lett. **87**, 111101.

[28] S. J. Richman, T. J. Quinn, C. C. Speake and R. S. Davis, (1999), Meas. Sci. Technol. **10**, 460.

[29] J. -Cl. Dousse and Ch. Rhême, (1987), Am. J. Phys. **55**, 706.

[30] L. Facy and C. Pontikis, (1971), Comptes Rendus des Scéances de l'Académie des Sciences de Paris **272**, Série B, 1397.

[31] U. Kleinevoss, H. Meyer, A. Schumacher and S. Hartmann, (1999), Meas. Sci. Technol. **10**, 492.

[32] H. de Boer, H. Haars and W. Michaelis, (1987), metrologia **24**, 171.

[33] O. V. Karagioz, V. P. Izmaylov and G. T. Gillies, (1998), Grav. Cosmol. **4**, 239.

[34] M. U. Sagitov *et al.*, (1979), Dok. Acad. Nauk SSSR **245**, 567.

[35] M. A. Zumberge *et al.*, (1990), J. Geophys. Res. **95**, 15483.

[36] P. Baldi *et al.*, (2001), Phys. Rev. **D64**, 082001.

[37] G. Müller *et al.*, (1990), Geophys. J. Int. **101**, 329 ; M. A. Zumberge *et al.*, (1991), Phys. Rev. Lett. **67**, 3051 ; M. Oldham, F. J. Lowes and R. J. Edge, (1993), Geophys. J. Int. **113**, 83 ; J. K. Hoskins *et al.*, (1985), Phys. Rev. **D32**, 3084 ; Y. T. Chen *et al.*, (1984), Proc. R. Soc. Lond. **A394**, 47 ; R. D. Rose *et al.*, (1969), Phys. Rev. Lett. **23**, 655 ; G. I. Moore *et al.*, (1988), Phys. Rev. **D38**, 1023 ; F. D. Stacey *et al.*, (1987), Rev. Mod. Phys. **59**, 157 ; X. Yang *et al.*, (1991), Chinese Phys. Lett., **8**, 329 ; M. S. Saulnier and D. Frisch, (1989), Am. J. Phys., **57**, 417 ; G. J. Tuck *et al.*, in *The Fith Marcel Grossmann Meeting*, D. G. Blair and M. J. Buckingham (Eds.), (World Scientific, Singapore, 1989), pp. 1605-1612 ; K. Kuroda and H. Hirakawa, (1985), Phys. Rev. **D32**, 342.

[38] St. Schlamminger, E. Holzschuh and W. Kündig, (2002), Phys. Rev. Lett. **89**, 161102.

[39] R. Newman, *in* J. Pullin (Ed.), gr-qc/0303027.

[40] P. S. Wesson and J. Ponce de Leon, (1995), A & A **294**, 1 ; J. Ponce de Leon, (2002), Grav. Cosmol. **8**, 272.

[41] J.-P. Mbelek and M. Lachièze-Rey, to appear.

[42] K. Kuroda, (1995), Phys. Rev. Lett. **75**, 2796.

2-COMPONENT COSMOLOGICAL MODELS WITH PERFECT FLUID AND SCALAR FIELD: EXACT SOLUTIONS

V.N. MELNIKOV
Center for Gravitation and Fundamental Metrology, VNIIMS, 3-1 M. Ulyanovoy Str., Moscow 117313, Russia
Institute of Gravitation and Cosmology, PFUR, Michlukho-Maklaya Str. 6, Moscow 117198, Russia

AND

V.R. GAVRILOV
Russian Gravitational Society, 3-1 M. Ulyanovoy Str., Moscow 117313, Russia
Moscow State Technical University, 2-nd Bauman Str. 5, Moscow 107005, Russia

Abstract. We study integrability by quadrature of a spatially flat Friedmann model containing both a perfect fluid with barotropic equation of state $p = (1-h)\rho$ and a minimally coupled scalar field φ with either a single exponential potential $V(\varphi) \sim \exp[-\sqrt{6}\sigma\kappa\varphi]$, $\kappa = \sqrt{8\pi G_N}$, of arbitrary sign or a simplest multiple exponential potential $V(\varphi) = W_0 - V_0 \sinh\left(\sqrt{6}\sigma\kappa\varphi\right)$, where the parameters W_0 and V_0 are arbitrary. From the mathematical view point the model is pseudo-Euclidean Toda-like system with 2 degrees of freedom. We apply the methods developed in our previous papers, based on the Minkowsky-like geometry for 2 characteristic vectors depending on the parameters σ and h. For the single exponential potential we present 4 classes of general solutions with the parameters obeying the following relations: **A**. σ is arbitrary, $h = 0$; **B**. $\sigma = 1 - h/2$, $0 < h < 2$; **C1**. $\sigma = 1 - h/4$, $0 < h \leq 2$; **C2**. $\sigma = |1-h|$, $0 < h \leq 2$, $h \neq 1, 4/3$. The properties of the exact solutions near the initial singularity and at the final stage of evolution are analyzed. For the multiple exponential potential the model is integrated with $h = 1$ and $\sigma = 1/2$ and all exact solutions describe the recollapsing universe. We single out the exact solution describing the evolution within the time approximately equal to $2H_0^{-1}$ with the present-day values of the acceleration parameter $q_0 = 0.5$ and the density parameter $\Omega_{\rho 0} = 0.3$.

V. de Sabbata et al. (eds.),
The Gravitational Constant: Generalized Gravitational Theories and Experiments, 247–268.
© 2004 *Kluwer Academic Publishers. Printed in the Netherlands.*

1. Introduction

Scalar fields play an essential role in modern cosmology. They are attributed to inflation models of the early universe and the models describing the present stage of the accelerated expansion as well. There is no a unique candidate for the potential of minimally coupled scalar field. Typically a potential is a sum of exponentials. Such potentials appear quite generically in a large class of theories (multidimensional [1], Kaluza-Klein models , supergravity and string/M - theories, see, for instance, [2], [3] and references therein).

A number of authors [2]-[9] (see also references therein) have studied a spationally homogeneous and isotropic Friedmann model containing both a perfect fluid subject to the linear equation of state

$$p = (1 - h)\rho, \tag{1}$$

where the constant h satisfied $0 \leq h \leq 2$, and a weakly coupled scalar field φ with a potential of the form

$$V(\varphi) = \frac{V_0}{\kappa^2}\, e^{-\sqrt{6}\sigma\kappa\varphi}, \tag{2}$$

where $\kappa = \sqrt{8\pi G_N}$, σ is a dimensionless positive constant, characterizing the steepness of the potential, and the constant V_0 may be positive and negative. The attention was mainly focussed on the qualitative behaviour of solutions, stability of the exceptional solutions to curvature and shear perturbations and their possible applications within the known cosmological scenaria such as inflation and scaling ("tracking") . In particular, it was found by a phase plane analysis [5, 6, 7] that for "flat" positive potentials ($V_0 > 0,\ 0 < \sigma^2 < 1 - h/2$) there exists an unique late-time attractor in the form of the scalar dominated solution. It is stable within homogeneous and isotropic models with non-zero spatial curvature with respect to spatial curvature perturbations for $\sigma^2 < 1/3$ and provides the power-law inflation. For "intermediate" positive potentials ($V_0 > 0,\ 1-h/2 < \sigma^2 < 1$) an unique late-time attractor is the scaling solution, where the scalar field "mimics" the perfect fluid, adopting its equation of state. The energy-density of the scalar field scales with that of the perfect fluid. For $h > 4/3$ this solution is stable within generic Bianchi models to curvature and shear perturbations and provides the power-low inflation. The scaling solution is unstable to curvature perturbations, when $0 < h < 4/3$, although it is stable to shear perturbations. Regions on (σ^2, h) parametrical plane corresponding to various qualitative evolution are presented on Figure 1. Integrability of the model with this potential is not well studied yet. Only for the special case with $h = 1$ (dust) and $\sigma = 1/2$ the procedure of getting a solution to the model was given in [9].

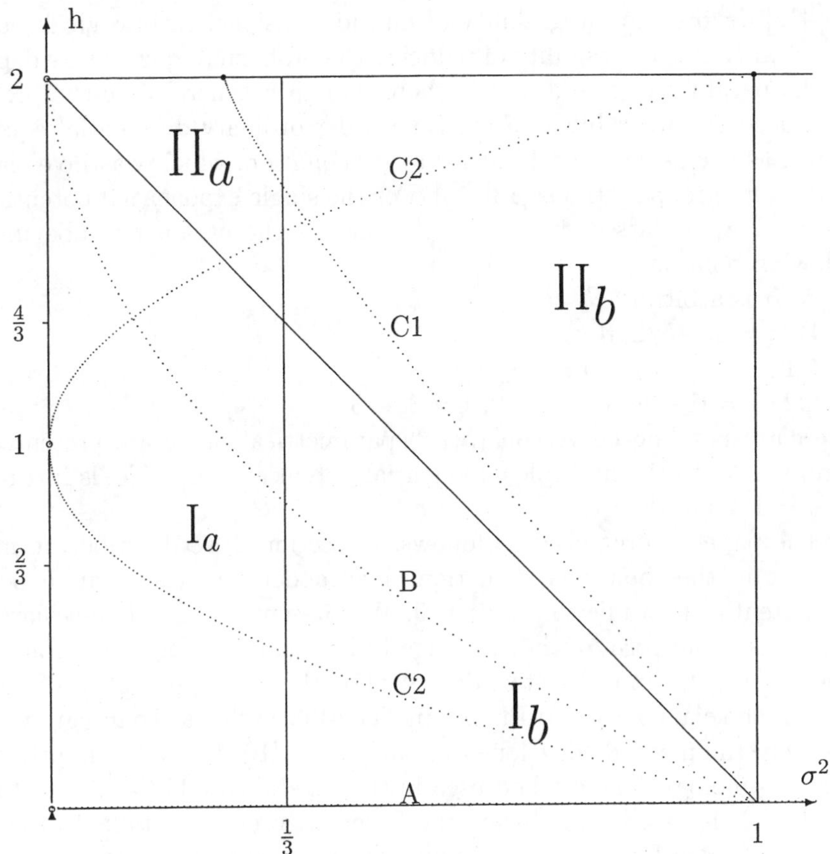

Figure 1. The domains **Ia, Ib, IIa and IIb** are bounded by the lines $h = 0$, $h = 2$, $\sigma^2 = 0$, $\sigma^2 = 1/3$, $\sigma^2 = 1 - h/2$. The solutions corresponding to **Ia and IIa** are inflationary on late times with an attractor stable to curvature perturbations. The late-time attractor for **Ia and Ib** is the scalar dominated solution and for **IIa and IIb** it is the scaling solution. The dotted curves **A,B,C1** and **C2** are identified in the text and present the integrable by quadrature classes of the model.

Here we study the problem of integrability by quadrature of a spatially flat Friedmann model containing both a perfect fluid with the equation of state (1) and a minimally coupled scalar field with either the single exponential potential (2) or a simplest multiple exponential potential

$$V(\varphi) = W_0 - V_0 \sinh\left(\sqrt{6}\sigma\kappa\varphi\right), \tag{3}$$

where the parameters W_0 and V_0 are arbitrary, for several classes of sets of (σ, h)-parameters. We apply the methods developed in our previous papers

[10]-[12] devoted to integrability of multidimensional cosmological models. It is clear that the possibility of reducing the problem to quadrature depends on the parameters σ and h. We show that in a general case the problem is reduced to integrability of a second order ordinary differential equation known as the generalized Emden-Fowler equation, which was investigated by discrete-group methods [13],[14]. For the single exponential potential we present here 4 classes of general solutions for the parameters obeying the following relations.

A. σ is arbitrary, $h = 0$.

B. $\sigma = 1 - h/2$, $0 < h < 2$.

C1. $\sigma = 1 - h/4$, $0 < h \leq 2$.

C2. $\sigma = |1 - h|$, $0 < h \leq 2$, $h \neq 1, 4/3$.

The corresponding curves on (h, σ^2) parametrical plane are presented on Figure 1. As to the multiple exponential potential, the model is integrated for $\sigma = 1/2$ and $h = 1$.

The paper is organized as follows. In section 2 we describe the model and obtain the equations of motion. The model with the single exponential potential is studied in section 3. We present the Einstein-scalar field equations in the form of the Lagrange-Euler equations following from some Lagrangian. Dynamical system described by the Lagrangian of this form belongs to the class of pseudo-Euclidean Toda-like systems. To integrate them we apply the methods developed in our papers [10, 11, 12] on multidimensional cosmology. The method used in the general case **C** ($\sigma^2 \neq (1-h/2)^2$, $0 < h \leq 2$) is based on reducing the Euler-Lagrange equations to the generalized Emden-Fowler (second-order ordinary differential) equation. We discuss the physical properties of the obtained exact solutions. Asymptotic behavior of these solutions at early and late times is analyzed. Example of the explicit solution in the cosmic time is presented.

The model with the multiple exponential potential is studied in section 4. The equations are integrated and the general properties of these exact solutions are studied. All of them describe the recollapsing of the universe within the finite time. The time as well as the intermediate behavior of the model crucially depend on the parameters W_0 and V_0. Then we fit the model to the observational parameters.

2. General model

We start with Einstein equations in a spatially flat Friedmann metric

$$\mathrm{d}s^2 = -\,\mathrm{e}^{2\alpha(t)}\mathrm{d}t^2 + \mathrm{e}^{2x(t)}\left[\mathrm{d}r^2 + r^2\left(\mathrm{d}\theta^2 + \sin^2\theta\mathrm{d}\phi^2\right)\right], \tag{4}$$

where $\exp[x(t)] \equiv a(t)$ is the scale factor and the function $\alpha(t)$ determines a time gauge ($\alpha(t) \equiv 0$ corresponds to the cosmic time t_c gauge). We assume that the universe contains both a self-interacting scalar field φ with a potential $V(\varphi)$ and a separately conserved perfect fluid with the barotropic equation of state (1). The governing set of equations, which follows from coupled Einstein and scalar field equations, reads

$$\dot{x}^2 = \frac{\kappa^2}{3}\left[\frac{1}{2}\dot{\varphi}^2 + (V(\varphi) + \rho)\,\mathrm{e}^{2\alpha}\right], \tag{5}$$

$$\ddot{x} = \frac{1}{2}(\dot{\alpha} - 3\dot{x})\dot{x} - \frac{\kappa^2}{2}\left[\frac{1}{2}\dot{\varphi}^2 - (V(\varphi) - p)\,\mathrm{e}^{2\alpha}\right], \tag{6}$$

$$\ddot{\varphi} = (\dot{\alpha} - 3\dot{x})\dot{\varphi} - V'(\varphi)\,\mathrm{e}^{2\alpha}, \tag{7}$$

$$\dot{\rho} = -3\dot{x}(p + \rho). \tag{8}$$

Using the barotropic equation of state we get immediately from equation (8)

$$\rho = \frac{\rho_0}{\kappa^2}\,\mathrm{e}^{-3(2-h)x}, \tag{9}$$

where ρ_0 is an arbitrary positive constant.

3. Single exponential potential

It can be easily verified, that the set of equations (5)-(7), where the presence of a perfect fluid density ρ and its pressure p is cancelled by equations (1)and (9), is equivalent to the set of Euler-Lagrange equations obtained from the Lagrangian

$$L = \frac{1}{2}\,\mathrm{e}^{3x-\alpha}\left(-\dot{x}^2 + \dot{y}^2\right) - \frac{\kappa^2}{6}\,\mathrm{e}^{\alpha-3x}\left[\frac{\rho_0}{\kappa^2}\,\mathrm{e}^{3hx} + \mathrm{e}^{6x}V\left(\frac{\sqrt{6}}{\kappa}y\right)\right], \tag{10}$$

where we introduced the following dimensionless variable

$$y = \frac{\kappa}{\sqrt{6}}\varphi.$$

The equation $\partial L/\partial\alpha = \mathrm{d}(\partial L/\partial\dot{\alpha})/\mathrm{d}t = 0$ leads to the constraint equation (5). Fixing the gauge $\alpha \equiv F(x, y)$, we can consider equations (6),(7) as the Euler-Lagrange equations obtained from the Lagrangian (10) under the zero-energy constraint (5).

In what follows we consider the potential of the form (2). As the system is symmetric under the transformation $\sigma \to -\sigma$, $\varphi \to -\varphi$, without a loss of generality we will consider only the case $\sigma > 0$. For the exponential potential (2) the Lagrangian (10) looks as follows

$$L = \frac{1}{2}\,\mathrm{e}^{3x-\alpha}\left(-\dot{x}^2 + \dot{y}^2\right) - \frac{1}{6}\,\mathrm{e}^{\alpha-3x}\left[\rho_0\,\mathrm{e}^{3hx} + V_0\,\mathrm{e}^{6(x-\sigma y)}\right]. \tag{11}$$

Dynamical system described by the Lagrangian of this form belongs to the class of pseudo-Euclidean Toda-like systems investigated in our previous papers [10, 11, 12]. Methods for integrating of pseudo-Euclidean Toda-like systems are based on the Minkowsky-like geometry for characteristic vectors $(\alpha_0 + 3(h-1), 0) \in \mathbb{R}^2$ and $(\alpha_0 + 3, -6\sigma) \in \mathbb{R}^2$ appearing when one puts the gauge $\alpha = \alpha_0 x$ with α_0 =const. Here we do not describe the methods and refer to the above mentioned papers.

3.1. GENERAL SOLUTIONS

A. σ is arbitrary, $h = 0$.
Here we suppose that the perfect fluid pressure p is equal to its density ρ (Zeldovich-type, or stiff matter). In this case the system with Lagrangian (11) is integrable for arbitrary parameter σ in the so-called harmonic time gauge defined by

$$\alpha(x) = 3x. \tag{12}$$

Let us consider two different cases: $\sigma \neq 1$ and $\sigma = 1$.

If $\sigma \neq 1$ we introduce the following variables

$$u = -\sigma x + y, \quad v = x - \sigma y.$$

In the terms of u and v the equations of motion and the zero-energy constraint look as follows

$$\begin{aligned} \ddot{u} &= 0, \\ \ddot{v} &= V_0(1-\sigma^2)\,e^{6v}, \\ -\dot{u}^2 + \dot{v}^2 &= \frac{1-\sigma^2}{3}\left[\rho_0 + V_0\,e^{6v}\right]. \end{aligned}$$

We notice that for v we get the Liouville equation. Integrating this set of equations and inverting the linear transformation, we get the following general solution:

the scale factor

$$a \equiv e^x = a_0\left[f(t-t_0)\,e^{\sigma A(t-t_0)}\right]^{1/[3(\sigma^2-1)]}, \tag{13}$$

the scalar field

$$\varphi = \frac{\sqrt{6}}{\kappa}\left\{\ln\left[f^{\sigma}(t-t_0)\,e^{A(t-t_0)}\right]^{1/[3(\sigma^2-1)]} + y_0\right\}, \tag{14}$$

where we introduced the function

$$f(t-t_0) \quad = \sinh(\sqrt{B}|t-t_0|)/\sqrt{B},\ V_0(1-\sigma^2) > 0,\ B > 0,$$

$$= \cosh(\sqrt{B}|t - t_0|)/\sqrt{B},\ V_0(1-\sigma^2) < 0,\ B > 0,$$
$$= \sin(\sqrt{B}|t - t_0|)/\sqrt{B},\ V_0(1-\sigma^2) > 0,\ B < 0,$$
$$= |t - t_0|,\ V_0(1-\sigma^2) > 0,\ B = 0.$$

The constant B is defined by

$$B = A^2 + 3(1-\sigma^2)\rho_0.$$

The general solution contains 2 arbitrary constants t_0, A and 2 constants a_0,y_0, obeying the constraint:

$$e^{6\sigma y_0} = 3a_0^6|V_0(1-\sigma^2)|.$$

For the second case $\sigma = 1$ the equations of motion and the zero-energy constraint with respect to the harmonic time gauge read in the old variables

$$\ddot{x} = V_0\, e^{6(x-y)},$$
$$\ddot{y} = V_0\, e^{6(x-y)},$$
$$-\dot{x}^2 + \dot{y}^2 = -\frac{1}{3}\left[\rho_0 + V_0\, e^{6(x-y)}\right].$$

We immediately find the integral of motion $\dot{x} - \dot{y} = \text{const} \equiv 2A$.

If $V_0 > 0$ the constant A is nonzero due to the zero-energy constraint. The general solution in this case for arbitrary V_0 looks as follows

$$a \equiv e^x = \exp\left\{\left(A + \frac{\rho_0}{12A}\right)(t - t_0) + V_0\left[e^{12A(t-t_0)} - 1\right] + x_0\right\}, \quad (15)$$

$$\varphi = \frac{\sqrt{6}}{\kappa}\left\{\left(-A + \frac{\rho_0}{12A}\right)(t - t_0) + V_0\left[e^{12A(t-t_0)} - 1\right] + y_0\right\}, \quad (16)$$

where the constants A, t_0 are arbitrary and the constants x_0, y_0 obey the relation $y_0 - x_0 = 2At_0$.

For $V_0 < 0$ the additional solution appears (corresponding to $A = 0$, $\dot{x} = \dot{y}$)

$$a \equiv e^x = \exp\left\{B(t - t_0) - \frac{\rho_0}{2}(t - t_0)^2 + x_0\right\},$$
$$\varphi = \frac{\kappa}{\sqrt{6}}\left[\ln a + \frac{1}{6}\ln(-\rho_0/V_0)\right],$$

where B and x_0 are arbitrary constants. We remind that t is the harmonic time. It is connected with the cosmic time t_c by the differential equation $dt_c = a^3 dt$.

B. $\sigma = 1 - h/2$, $0 < h < 2$.
We notice that the model for $\sigma = 1/2$ and $h = 1$ has been integrated in [9]. Here we study a more general case. We fix the time gauge choosing the following function $\alpha(x)$:

$$\alpha(x) = 3(1-h)x.$$

Then, with respect to the new variables u and v defined by

$$u = \exp\left[\frac{3h}{2}(x-y)\right], \quad v = \exp\left[\frac{3h}{2}(x+y)\right]$$

the equations of motion and the zero-energy constraint may be written in a rather simple form

$$\begin{aligned} \ddot{u} &= 0, \\ \ddot{v} &= \frac{3}{2}h(2-h)V_0 u^{(4-3h)/h}, \\ \dot{u}\dot{v} &= \frac{3}{4}h^2\left[\rho_0 + V_0 u^{2(2-h)/h}\right]. \end{aligned}$$

The set of equations is easily integrable. We obtain

$$\begin{aligned} u &= A(t-t_0) > 0, \\ v &= \frac{3h^2}{4A^2}\left\{\rho_0 A(t-t_0) + \frac{h}{4-h}V_0[A(t-t_0)]^{(4-h)/h} + B\right\} > 0, \end{aligned}$$

where A is an arbitrary nonzero constant and B is an arbitrary nonnegative constant, V_0 has arbitrary sign. If $V_0 < 0$, then, the following additional special solution arises

$$\begin{aligned} u &= (-\rho_0/V_0)^{h/[2(2-h)]}, \\ v &= \frac{3}{4}h(2-h)\rho_0^{h/(2-h)]}\left[T^2 - (t-t_0)^2\right], \end{aligned}$$

where the integration constant $T \neq 0$. Then, one easily gets the scale factor

$$a \equiv \mathrm{e}^x = (uv)^{1/(3h)} \tag{17}$$

and the scalar field

$$\varphi = \frac{\sqrt{2/3}}{\kappa h}\ln\frac{v}{u}. \tag{18}$$

C. $\sigma^2 \neq (1-h/2)^2$, $0 < h \leq 2$.
We introduce the following variables for the factorization of the potential in the Lagrangian

$$u = 3[\sigma x - (1-h/2)y],$$
$$v = 3[(h/2-1)x + \sigma y] - \ln\sqrt{|V_0|/\rho_0}.$$

Then, the equations of motion and the zero-energy constraint for u and v in the harmonic time gauge defined by equation (12) look as follows

$$\ddot{u} = \frac{3}{2}h\sigma A_0\left(\mathrm{e}^{2v} + \varepsilon\right)\exp\left\{\frac{h\sigma u + (2-h-2\sigma^2)v}{\sigma^2 - (1-h/2)^2}\right\}, \tag{19}$$

$$\begin{aligned}\ddot{v} &= -\frac{3}{2}A_0\left(h(1-\frac{h}{2})\,\mathrm{e}^{2v} + (2-h-2\sigma^2)\varepsilon\right)\\ &\times \exp\left\{\frac{h\sigma u + (2-h-2\sigma^2)v}{\sigma^2 - (1-h/2)^2}\right\},\end{aligned} \tag{20}$$

$$\begin{aligned}&-\dot{u}^2 + \dot{v}^2\\ &= 3A_0[\sigma^2 - (1-h/2)^2]\left(\mathrm{e}^{2v} + \varepsilon\right)\exp\left\{\frac{h\sigma u + (2-h-2\sigma^2)v}{\sigma^2 - (1-h/2)^2}\right\},\end{aligned} \tag{21}$$

where we denoted

$$A_0 = \rho_0^{-\frac{2-h-2\sigma^2}{2[\sigma^2 - (1-h/2)^2]}}|V_0|^{\frac{h(1-h/2)}{2[\sigma^2 - (1-h/2)^2]}},$$
$$\varepsilon = \mathrm{sgn}(V_0).$$

Let us express $\dot{v}^2$ from the zero-energy condition (21) as follows

$$\begin{aligned}\dot{v}^2 &= \left[\left(\frac{\mathrm{d}u}{\mathrm{d}v}\right)^2 - 1\right]^{-1} 3A_0[\sigma^2 - (1-h/2)^2]\left(\mathrm{e}^{2v} + \varepsilon\right)\\ &\times \exp\left\{\frac{h\sigma u + (2-h-2\sigma^2)v}{\sigma^2 - (1-h/2)^2}\right\}.\end{aligned} \tag{22}$$

Substituting $\ddot{u}$, $\ddot{v}$ and $\dot{v}^2$ into the relation

$$\frac{\mathrm{d}^2u}{\mathrm{d}^2v} = \frac{\ddot{u} - \ddot{v}\mathrm{d}u/\mathrm{d}v}{\dot{v}^2}, \tag{23}$$

we obtain the following second order ordinary differential equation

$$\frac{d^2u}{dv^2} = \left\{\frac{1}{2}\left(-\frac{\sigma^2 - (1-h^2/4)}{\sigma^2 - (1-h/2)^2} + \frac{e^{2v}-\varepsilon}{e^{2v}+\varepsilon}\right)\frac{du}{dv} + \frac{h\sigma/2}{\sigma^2 - (1-h/2)^2}\right\} \times \left[\left(\frac{du}{dv}\right)^2 - 1\right]. \tag{24}$$

The procedure is valid if $\dot{v} \not\equiv 0$.

The exceptional solution with $\dot{v} \equiv 0$ appears only for the positive potential when $\sigma^2 > 1 - h/2$ and $0 < h < 2$. In the terms of the cosmic time t_c it reads

$$a = \left\{\sqrt{\frac{3\rho_0}{4(\sigma^2 - (1-h/2)}}\sigma(2-h)|t_c - t_c^0|\right\}^{2/[3(2-h)]}, \tag{25}$$

$$\varphi = \frac{\sqrt{2/3}}{\kappa\sigma}\ln\left(\sqrt{\frac{3}{h}(2-h)V_0\sigma|t_c - t_c^0|}\right). \tag{26}$$

It should be mentioned, that the set of the equations (19)-(21) does not admit static solutions $\dot{u} = \dot{v} \equiv 0$ as well as the solutions with $\dot{u} = \pm\dot{v}$. So, using the relations (22) and (23) we do not lose any solutions except, possibly, the exceptional solution (25),(26).

Let us suppose that one is able to obtain the general solution to the equation (24) in the parametrical form $v = v(\tau)$, $u = u(\tau)$, where τ is a parameter. Then, we obtain the scale factor $a \equiv \exp[x]$ and the scalar field $\varphi = (\sqrt{6}/\kappa)y$ as functions of the parameter τ. The relation between the parameter τ and the harmonic time t may be always derived by integration of the zero-energy constraint written in the form of the following separable equation

$$dt^2 = \frac{\left(\frac{du}{d\tau}\right)^2 - \left(\frac{dv}{d\tau}\right)^2}{3A_0[\sigma^2 - (1-h/2)^2]\,(e^{2v}+\varepsilon)} \times \exp\left\{-\frac{h\sigma u + (2-h-2\sigma^2)v}{\sigma^2 - (1-h/2)^2}\right\} d\tau^2. \tag{27}$$

Thus, the problem of the integrability by quadrature of the model is reduced to the integrability of the equation (24). For du/dv it represents the first-order nonlinear ordinary differential equation. Its right hand side is the third-order polynomial (with coefficients depending on v) with respect to the du/dv. An equation of such type is called Abel's equation (see, for instance, [13],[14]).

First of all let us notice that the equation (24) has the partial solution $u = \pm v$+const that make the relation (22) singular. However, as was already mentioned, the set of equations (19)-(21) does not admit the solutions with $\dot{u} = \pm\dot{v}$. Existence of this partial solution to the Abel equation (24) allows one to find the following nontrivial transformation

$$e^{2v} = -\varepsilon \frac{X}{Y}\frac{dY}{dX}, \tag{28}$$

$$u = \delta\left[v + \ln\left|\frac{Y}{X}\right| + \ln C\right], \quad \delta = \pm 1, \quad C > 0, \tag{29}$$

that reduces it to the so-called generalized Emden-Fowler equation

$$\frac{d^2Y}{dX^2} = \mathrm{sgn}[\sigma^2 - (1 - h/2)^2]\left(-\varepsilon\frac{dY}{dX}\right)^l Y^m X^n, \tag{30}$$

where the constant parameters l, m and n read

$$l = \frac{2(\delta\sigma - 1 + h/4)}{\delta\sigma - 1 + h/2}, \quad m = -\frac{\delta\sigma + 1 - h}{\delta\sigma + 1 - h/2}, \quad n = -m - 3. \tag{31}$$

In the special case $l = 0$ equation (30) is known as the Emden-Fowler equation.

There are no methods for integrating of the generalized Emden-Fowler equation with arbitrary independent parameters l, m and n. However, the discrete-group methods developed in [14] allow to integrate by quadrature 3 two-parametrical classes, 11 one-parametrical classes and about 90 separated points in the parametrical space (l, m, n) of the generalized Emden-Fowler equation. Further we consider the following integrable classes.

C1. $\sigma = 1 - h/4$, $0 < h \leq 2$, $\delta = 1$.
The parameters l and m given by equation (31) are the following

$$l = 0, \quad m = -1 + \frac{2h}{8 - 3h} \in (-1, 1]. \tag{32}$$

The general solution to the generalized Emden-Fowler equation (30) with these parameters reads

$$Y = \frac{\tau}{F_m(\tau)} > 0, \quad X = \frac{1}{F_m(\tau)} > 0,$$

where we introduced the following function

$$F_m(\tau) = \pm\int\left[\frac{2}{m+1}\tau^{m+1} + C_1\right]^{-1/2} d\tau + C_2. \tag{33}$$

The variable τ changes on the interval which follows from $G_m(\tau) > 0$, where we used the function

$$G_m(\tau) = \varepsilon \left[\frac{F_m(\tau)}{\tau F'_m(\tau)} - 1 \right] \tag{34}$$

equal to the right hand side of equations (28) with substitutions of X, Y and (33). Finally, using equations (28),(29) we find the scale factor

$$a = a_0 \tau^{\frac{4(4-h)}{3h(8-3h)}} G_m^{\frac{2}{3h}}(\tau) \tag{35}$$

and the scalar field

$$\varphi = \frac{\sqrt{6}}{\kappa} \left\{ \ln \left[\tau^{\frac{8(2-h)}{3h(8-3h)}} G_m^{\frac{2}{3h}}(\tau) \right] + y_0 \right\}, \tag{36}$$

where

$$a_0 = C^{\frac{4(4-h)}{3h(8-3h)}} (|V_0|/\rho_0)^{\frac{4(2-h)}{3h(8-3h)}}, \; y_0 = \ln \left[C^{\frac{8(2-h)}{3h(8-3h)}} (|V_0|/\rho_0)^{\frac{2(4-h)}{3h(8-3h)}} \right].$$

The relation between the variable τ and the cosmic time t_c is the following

$$\mathrm{d}t_c^2 = I_0 \tau^{\frac{4(2-h)(4-3h)}{h(8-3h)}} G_m^{\frac{2(2-h)}{h}}(\tau) \left[F'_m(\tau) \right]^2 \mathrm{d}\tau^2, \tag{37}$$

where

$$I_0 = 16 \rho_0^{-\frac{(4-h)^2}{h(8-3h)}} |V_0|^{\frac{4(2-h)^2}{h(8-3h)}} C^{\frac{4(2-h)(4-h)}{h(8-3h)}} / [3h(8-3h].$$

The general solution (35),(36) (with l and m given by (32)) contains 3 arbitrary constants C, C_1 and C_2 as required.

C2. $\sigma = |1-h|$, $0 < h \leq 2$, $h \neq 1, 4/3, \delta = \mathrm{sgn}(1-h)$.

Now we have the following parameters in the generalized Emden-Fowler equation

$$l = 3, \quad m = -1 + \frac{h}{4-3h}. \tag{38}$$

Its general solution looks as follows

$$X = \frac{\tau}{R_m(\tau)} > 0, \quad Y = \frac{1}{R_m(\tau)} > 0,$$

where the function $R_m(\tau)$ is defined by

$$R_m(\tau) = \pm \int \left[\frac{-2\varepsilon}{|m+2|} \tau^{-m-2} + C_1 \right]^{-1/2} \mathrm{d}\tau + C_2, \quad m \neq -2, \tag{39}$$

$$= \pm \int \left[\varepsilon \ln \tau^2 + C_1 \right]^{-1/2} \mathrm{d}\tau + C_2, \quad m = -2. \tag{40}$$

The variable τ changes on an interval, where

$$S_m(\tau) = \varepsilon \left[\frac{R_m(\tau)}{\tau R'_m(\tau)} - 1\right] > 0. \tag{41}$$

Finally, we get

$$a = a_0 \tau^{\frac{4(1-h)}{3h(4-3h)}} S_m^{\frac{1}{3h}}(\tau), \tag{42}$$

$$\varphi = \frac{\sqrt{6}}{\kappa}\mathrm{sgn}(1-h)\left\{\ln\left[\tau^{\frac{2(2-h)}{3h(4-3h)}} S_m^{\frac{1}{3h}}(\tau)\right] + y_0\right\}, \tag{43}$$

where

$$a_0 = C^{\frac{4(1-h)}{3h(3h-4)}}(|V_0|/\rho_0)^{\frac{2(2-h)}{3h(3h-4)}},\ y_0 = \ln\left[C^{\frac{2(2-h)}{3h(3h-4)}}(|V_0|/\rho_0)^{\frac{4(1-h)}{3h(3h-4)}}\right].$$

The variable τ and the cosmic time t_c are related by

$$\mathrm{d}t_c^2 = U_0 \tau^{\frac{4(2-3h)}{3h}} S_m^{\frac{2(1-h)}{h}}(\tau)\left[R'_m(\tau)\right]^2 \mathrm{d}\tau^2, \tag{44}$$

where

$$U_0 = 4\rho_0^{-\frac{4(1-h)^2}{h(3h-4)}}|V_0|^{\frac{(2-h)^2}{h(3h-4)}} C^{\frac{4(2-h)(1-h)}{h(3h-4)}}/(3h|3h-4|).$$

So, this general solution is given in the parametrical form by (42),(43) with l and m from (38). The transition to the cosmic time may be done by solving (44).

3.2. PROPERTIES OF SOLUTIONS

Here we study properties of the obtained exact solutions for positive potentials, though our solutions are valid for any sign of the potential. As the system is symmetrical under the time reflection $t \to -t$, without loss of generality we only consider expanding near the singularity cosmologies with the Hubble parameter $H = \dot{x}\exp(-\alpha) > 0$.

We introduce the following notation for the relative energy densities

$$\Omega_\rho = \frac{\kappa^2\rho}{3H^2},\ \Omega_{\varphi K} = \frac{\kappa^2(\mathrm{d}\varphi/\mathrm{d}t_c)^2}{6H^2},\ \Omega_{\varphi P} = \frac{\kappa^2 V(\varphi)}{3H^2}.$$

Due to the constraint (5) the relative energy densities obey the relation

$$\Omega_\rho + \Omega_{\varphi K} + \Omega_{\varphi P} = 1.$$

Also we introduce the scalar field barotropic parameter

$$w_\varphi = \frac{p_\varphi}{\rho_\varphi} = \frac{\frac{1}{2}\left(\frac{d\varphi}{dt_c}\right)^2 - V(\varphi)}{\frac{1}{2}\left(\frac{d\varphi}{dt_c}\right)^2 + V(\varphi)}.$$

A1. $0 < \sigma < 1$, $h = 0$.
The general solution is given by equations (13),(14) for $V_0(1 - \sigma^2) > 0$, $B \geq 0$. Near the initial singularity ($t_c \to +0$) the scale factor is in the main order $a \propto t_c^{1/3}$. There exists the special solution for $A = -\sigma\sqrt{B}$ with $\Omega_\rho \to 1$ as $t_c \to +0$. It corresponds to the fluid-dominated solution mentioned in [5, 7, 8]. It gives $w_\varphi \to 1$ (stiff matter) as $t_c \to +0$.

For all remaining solutions we obtain $\Omega_\rho \to 1 - \Omega^0_{\varphi K}$, $\Omega_{\varphi K} \to \Omega^0_{\varphi K}$, $\Omega_{\varphi P} \to 0$ as $t_c \to +0$, where $\Omega^0_{\varphi K} = (\sigma\sqrt{B} - A)^2/(\sqrt{B} - \sigma A)^2$. Then the barotropic parameter w_φ tends to -1, i.e the scalar field is vacuum-like (de Sitter) near the initial singularity.

There is an unique late-time attractor in the form of the scalar field dominated solution. It corresponds to $B = 0$, and $\rho_0 = 0$ in formulas (13),(14) presenting the general solution. The attractor may be written down for the cosmic time t_c

$$a = \tilde{a}_0 t_c^{1/(3\sigma^2)},$$

$$\varphi = \frac{\sqrt{2/3}}{\kappa\sigma}\left[\ln t_c + \ln\sqrt{\frac{3V_0\sigma^4}{1-\sigma^2}}\right].$$

It is easy to see that for this solution

$$\Omega_\rho = 0, \quad \Omega_{\varphi K} = \sigma^2, \quad \Omega_{\varphi P} = 1 - \sigma^2, \quad \frac{p_\varphi}{\rho_\varphi} = 2\sigma^2 - 1.$$

This attracting solution according to [8] may be called kinetic-potential scaling. For $\sigma^2 < 1/3$ all solutions provide the power law inflation at late times.

A2. $1 < \sigma$, $h = 0$.
The general solution is given by (13),(14) for $V_0(1 - \sigma^2) < 0$, $B > 0$. Behavior near the initial singularity is the same for both general and special solutions with the only difference that the constant $\Omega^0_{\varphi K}$ is the following $\Omega^0_{\varphi K} = (\sigma\sqrt{B} + A)^2/(\sqrt{B} + \sigma A)^2$.

At the final stage of evolution as $t_c \to +\infty$ we obtain $\Omega_\rho \to 1 - \Omega^f_{\varphi K}$, $\Omega_{\varphi K} \to \Omega^f_{\varphi K}$, $\Omega_{\varphi P} \to 0$, $w_\varphi \to 1$, $a \propto t_c^{1/3}$, where

$\Omega^f_{\varphi K} = (\sigma\sqrt{B}-A)^2/(\sqrt{B}-\sigma A)^2$. Such behaviour of the scalar field, when it adopts the usual perfect fluid equation of state and its energy-density scales with that of the perfect fluid, is called scaling (or sometimes "tracking").

B and **C2** in regions **Ia** and **Ib**.

Behaviour near the initial singularity was described in [5, 7, 8] using qualitative methods. There exists the fluid dominated solution with $a \propto t_c^{2/[3(2-h)]}$ and $\Omega_\rho \to 1$ as $t_c \to +0$. All remaining solutions describe the domination of the kinetic contribution of the scalar field: $\Omega_\rho \to 0, \quad \Omega_{\varphi K} \to 1, \quad \Omega_{\varphi P} \to 0, \quad w_\varphi \to 1, \quad a \propto t_c^{1/3}$.

The behaviour at late times is the same as one for **A1**.

C1 and **C2** in regions **IIa** and **IIb**.

Behaviour near the initial singularity is the same as in the previous case. The late-time attractor for $h \in (0, h)$ is the special solution described by equations (25) and (26). For these solution we have

$$\Omega_\rho = 1 - \frac{2-h}{2\sigma^2}, \; \Omega_{\varphi K} = \frac{(2-h)^2}{4\sigma^2}, \; \Omega_{\varphi P} = \frac{h(2-h)}{4\sigma^2}, \; \frac{p_\varphi}{\rho_\varphi} = 1-h.$$

This is a typical scaling behaviour.

One of the examples of this behaviour may be given explicitly: the exact solution of the class **C1** with $h = 2$ and $\sigma = 1/2$ for the positive potential reads with respect to the cosmic time

$$\frac{a}{a_0} = \mathrm{e}^{\sqrt{\Lambda/3}t_c} \left\{ \frac{1 - A\,\mathrm{e}^{-\sqrt{3\Lambda}(t_c-t_c^0)}}{1 + A\,\mathrm{e}^{-\sqrt{3\Lambda}(t_c-t_c^0)}} \right. \times \left. \left[\frac{\sqrt{3\Lambda}}{2}(t_c - t_c^0) - \frac{2A\left(1 - A\,\mathrm{e}^{-\sqrt{3\Lambda}(t_c-t_c^0)}\right)}{(1+A)\left(1 + A\,\mathrm{e}^{-\sqrt{3\Lambda}(t_c-t_c^0)}\right)} \right] \right\}^{1/3},$$

$$\varphi = \frac{\sqrt{6}}{\kappa} \ln \left\{ \frac{V_0}{\Lambda} \left[\frac{1 + A\,\mathrm{e}^{-\sqrt{3\Lambda}(t_c-t_c^0)}}{1 - A\,\mathrm{e}^{-\sqrt{3\Lambda}(t_c-t_c^0)}} \left(\frac{\sqrt{3\Lambda}}{2}(t_c - t_c^0) - \frac{1-A}{1+A} \right) - 1 \right] \right\}^{1/3},$$

where $\Lambda \equiv \rho_0$ is the cosmological constant. The solution contains 3 integration constants: arbitrary t_c^0, positive a_0 and A obeying $|A| < 1$. The late-time attractor corresponds to $A = 0$. For this attracting solution we get: $\Omega_\rho \to 1, \quad \Omega_{\varphi K} \to 0, \quad \Omega_{\varphi P} \to 0, \quad w_\varphi \to -1, \quad a \propto t_c^{1/3} \exp\sqrt{\Lambda/3}t_c$, $H \to \sqrt{\Lambda/3}$ as $t_c \to +\infty$.

4. Multiple exponential potential

In what follows we consider the scalar field potential of the form (3) with $\sigma = 1/2$ and the perfect fluid with the barotropic parameter $h = 1$ (dust).

As the system is symmetrical under the transformation $\varphi \to -\varphi$, $V_0 \to -V_0$, without loss of generality we consider only the case $V_0 > 0$.

Now we introduce new variables x and y by the following transformation

$$a^3 = xy, \; \kappa\varphi = \sqrt{2/3}\log(y/x), \; x > 0, \; y > 0. \tag{45}$$

Then the equations of motion with respect to the cosmic time gauge $(\alpha(t) \equiv 0)$ results in

$$\ddot{x} = (\omega_1^2 - \omega_2^2)x - 2\omega_1\omega_2 y, \tag{46}$$
$$\ddot{y} = 2\omega_1\omega_2 x + (\omega_1^2 - \omega_2^2)y, \tag{47}$$

where we introduced the positive parameters ω_1 and ω_2 by

$$\omega_1^2 - \omega_2^2 = 3/4\kappa^2 W_0, \; 2\omega_1\omega_2 = 3/4\kappa^2 V_0.$$

Then zero-energy constraint, where the presence of ρ is cancelled by

$$\rho = \rho_0(a_0/a)^3 \tag{48}$$

takes the following form

$$\dot{x}\dot{y} - (\omega_1^2 - \omega_2^2)xy - \omega_1\omega_2(x^2 - y^2) = 3/4\kappa^2\rho_0 a_0^3. \tag{49}$$

The set of equations (46),(47) may be presented in the following complex form

$$\ddot{z} = \omega^2 z, \tag{50}$$

where we introduced the complex variable

$$z = x + \imath y$$

and the complex parameter

$$\omega = \omega_1 + \imath\omega_2.$$

It is easy to see that equation (50) implies the following complex integral of motion

$$\dot{z}^2 - \omega^2 z^2 = \text{const},$$

The presence of this complex integral is equivalent to the existence of 2 real integrals of motion. One of them $\Im(\dot{z}^2 - \omega^2 z^2) = -3/2\kappa^2\rho_0 a_0^3$ represents the constraint given by equation (49) and the other integral $\Re(\dot{z}^2 - \omega^2 z^2)$ has an arbitrary value.

4.1. EXACT SOLUTIONS AND BEHAVIOR OF THE MODEL

The equations of motion in form equation (50) are easily integrable. The result is

$$z = C\,e^{\imath\alpha}\left(e^{\omega_1(t-t_0)+\imath\omega_2(t-t_0-\delta)} + e^{-\omega_1(t-t_0)-\imath\omega_2(t-t_0-\delta)}\right), \tag{51}$$

where $C > 0$. The constants t_0 and δ are arbitrary, C and α obey the relation

$$\Im\left(\omega^2 C^2\,e^{\imath 2\alpha}\right) = \frac{3}{8}\kappa^2\rho_0 a_0^3$$

following from equation (49). Then substituting $x = \Re z$ and $y = \Im z$ into the relations given by equation (45) one easily gets the explicit expressions for the scale factor a and the scalar field φ.

Let us now consider the general properties of the exact solutions by analyzing the corresponding to equation (51) trajectories (orbits) of the moving point on the Cartesian xy plane. It follows from the definition of the variables x and y given by equation (45) that the physical segments of a trajectory belong to the angular domain with $x > 0$ and $y > 0$. Hereafter we show that each segments is of a finite length and its end-points are attached either to one coordinate axis or the both axes. From the physical viewpoint it means that all solutions describe the universe evolution within a finite time interval. Moreover, as the equations of motion are invariant with respect to the time reflection $t \to -t$ each segment of the trajectory may be passed in both directions. We notice that the constant factor $\exp(\imath\alpha)$ leads to the rotation of the trajectory about the origin $(0,0)$ through the angle α. Further we consider only the case when $\alpha = 0$ taking into account that all remaining trajectories may be obtained by rotation.

If the time variable t is positive and large enough the last term in equation (51) is negligible ($\omega_1 > 0$). Then the motion is confined to the repelling logarithmic spiral with $x = C\exp[\omega_1(t-t_0)]\cos[\omega_2(t-T_0)]$ and $y = C\exp[\omega_1(t-t_0)]\sin[\omega_2(t-T_0)]$, where $T_0 \equiv t_0 + \delta$. As the time t grows the point clockwise rotates around the origin $(0,0)$ and its distance from the origin increases as $\exp(\omega_1 t)$. It means that the evolution starts at some moment $t = T_0$ when the spiral intersects the axis x and finishes when it further intersects the axis y. The finite time of evolution is about the value $T = \pi/(2\omega_2)$. In this case the zeros of the functions $x(t)$ and $y(t)$ are different. Then near the initial point the expansion of the universe can be approximately described by the power-law equation $a(t) \sim (t-T_0)^{1/3}$. The similar equation $a(t) \sim [T-(t-T_0)]^{1/3}$ approximately describes the collapsing near the final point of evolution.

If $t < 0$ and $|t|$ is large enough the first term in equation (51) vanishes and the last term dominates. In this case the behavior of the model is similar

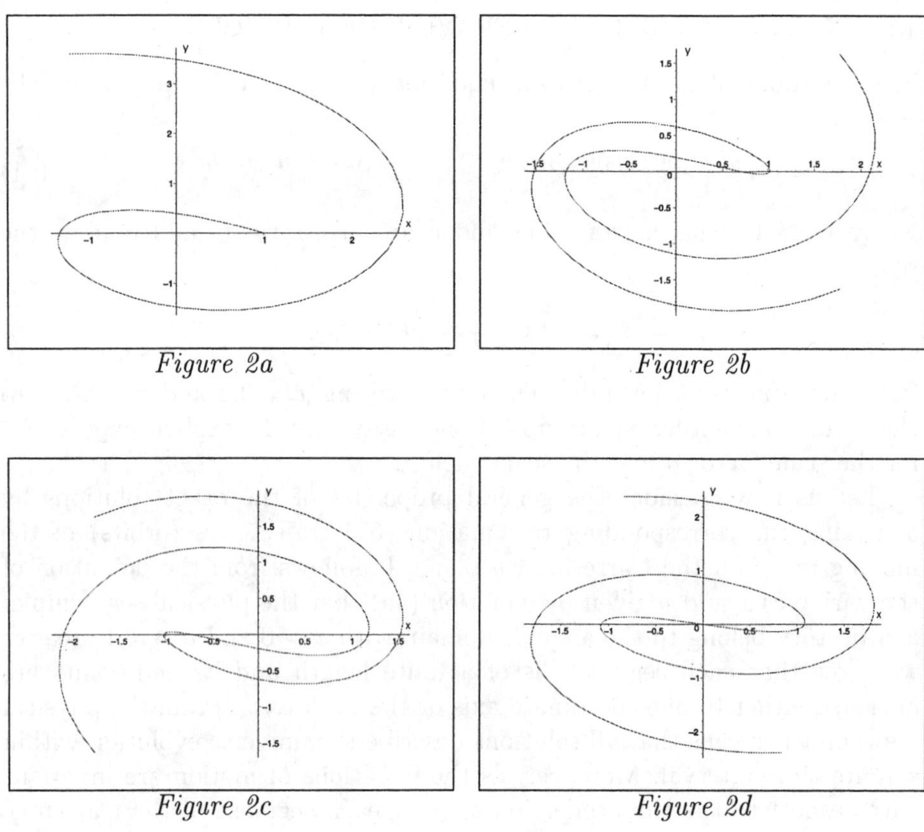

Figure 2a *Figure 2b*

Figure 2c *Figure 2d*

Figure 2. The corresponding to the exact solutions trajectories on the xy plane. All remaining trajectories may be obtain by rotation about the origin $(0,0)$. Each solution is defined on some finite time interval which corresponds to a lying in the angular domain with $x > 0$ and $y > 0$ segment of the trajectory. End-points of such segment correspond to big bang and big crunch.

with the only difference: the corresponding spiral is attracting. It describes the counterclockwise rotation with approaching to the origin of coordinates x and y.

The form of trajectories for the intermediate values of the time variable t is more complicated. To analyze the behavior one needs to superpose both attracting and repelling spirals. The result depends on the constant δ. Evidently it's enough to consider $\omega_2\delta \in [0, \pi)$. If $\delta = 0$ we get $x = C\cosh[\omega_1(t - t_0)]\cos[\omega_2(t - t_0)]$ and $y = C\sinh[\omega_1(t - t_0)]\sin[\omega_2(t - t_0)]$. The corresponding trajectory with $C = 1$ is presented on Figure 2a. The moving point passes the curve twice in both directions. It's not difficult to prove that for $\delta \neq 0$ any trajectory has no selfintersection points. On Figures 2b,2c the constant $\omega_2\delta$ ranges from 0 to $\pi/2$ and from $\pi/2$ to π, respectively. For $\omega_2\delta = \pi/2$ the trajectory is the central curve (see Figure 2d).

This analysis shows that all exact solutions describes the evolution from big bang to big crunch within some finite time. Typically this time is about $T = \pi/(2\omega_2)$, however it may be arbitrarily shorter (by choosing trajectories obtained after rotation). Each solution is defined on some finite time interval which corresponds to a lying in the angular domain with $x > 0$ and $y > 0$ segment of the trajectory. End-points of such segment correspond to big bang and big crunch. If the end-points of the segment are attached to the different coordinate axes then the scalar field goes from $-\infty$ to $+\infty$ or vice versa during the evolution. If both end-points are attached to the coordinate axis y the scalar field diverges to $+\infty$ as the time tends to the initial or final value. For almost all solutions the scale factor is proportional to the time in power 1/3 near the initial and the final points of evolution. But if the end-point of the segment coincides with the origin $(0,0)$ the roots of the functions $x(t)$ and $y(t)$ coincide and the scale factor is proportional to the time in power 2/3 near the corresponding point.

4.2. FITTING MODEL TO THE OBSERVATIONAL PARAMETERS

Putting the present-day value of time t equal to zero we present the solution to the equations of motion equation (50) in the form

$$z = (x_0 + \imath y_0)\cosh\omega t + \frac{\dot{x}_0 + \dot{y}_0}{\omega}\sinh\omega t \tag{52}$$

with the values $x_0 \equiv x(0)$, $y_0 \equiv y(0)$, $\dot{x}_0 \equiv \dot{x}(0)$ and $\dot{y}_0 \equiv \dot{y}(0)$ expressed in the terms of the observational parameters: the Hubble constant H_0, the present-day values q_0 and $\Omega_{\rho 0}$ of the acceleration parameter $q = a\ddot{a}/\dot{a}^2$ and the density parameter $\Omega_\rho = \kappa^2\rho/(3H^2)$. One easily gets these expressions using the definition of the variables x and y given by equation (45)

$$x_0 = \left(\sqrt{1+\varepsilon_0^2} + \varepsilon_0\right)^{1/2} a_0^{3/2}, \tag{53}$$

$$y_0 = \left(\sqrt{1+\varepsilon_0^2} - \varepsilon_0\right)^{1/2} a_0^{3/2}, \tag{54}$$

$$\dot{x}_0 = \frac{3}{2}x_0 H_0 \left(1 \mp \sqrt{\frac{1}{3}(1-q_0) - \frac{1}{2}\Omega_{\rho 0}}\right), \tag{55}$$

$$\dot{y}_0 = \frac{3}{2}y_0 H_0 \left(1 \pm \sqrt{\frac{1}{3}(1-q_0) - \frac{1}{2}\Omega_{\rho 0}}\right), \tag{56}$$

where we denoted

$$\varepsilon_0 = \frac{9H_0^2}{8\omega_1\omega_2}\left[\frac{1}{3}(2+q_0) - \frac{1}{2}\Omega_{\rho 0}\right] - \frac{\omega_1^2 - \omega_2^2}{2\omega_1\omega_2}. \qquad (57)$$

The upper sign in equations (55),(56) corresponds to the positive value $\dot{\varphi}_0 \equiv \dot{\varphi}(0)$, the lower sign appears when $\dot{\varphi}_0 < 0$. We notice that the model implies the value $(1-q_0)/3-\Omega_{\rho 0}/2$ to be nonnegative. It can be expressed via the present-day value $w_{\varphi 0}$ of the scalar field effective barotropic parameter. One easily obtain $(1-q_0)/3-\Omega_{\rho 0}/2 = (1+w_{\varphi 0})(1-\Omega_{\rho 0})/2$. Therefore the model is consistent with the observational data if the present-day value of the scalar field effective barotropic parameter is not less than -1.

The solution presented by equations (52)-(56) with the scale factor a and the scalar field φ obtained from equation (45) exists in some finite time interval (t_{01}, t_{02}) with negative t_{01} and positive t_{02} where $x >$ and $y > 0$. Obviously, for given observational parameters H_0, q_0 and $\Omega_{\rho 0}$ the values t_{01} and t_{02} and, consequently, the full time $(t_{02}-t_{01})$ of the universe evolution depends on parameters ω_1 and ω_1 determining the potential of the model. As we already mentioned, the typical time of evolution may be approximately estimated by the value $\pi/(2\omega_2)$. Then, instead of ω_1 and ω_2 we use further the following dimensionless model parameters: the typical time $\pi H_0/(2\omega_2)$ of the evolution in units of $H_0^{-1} \approx 14$ billions years and the present-day value $\kappa\varphi(0) \equiv \kappa\varphi_0 = \sqrt{2/3}\log(\sqrt{1+\varepsilon_0^2}-\varepsilon_0)$ of the scalar field φ in units of $M_p = \kappa^{-1}$. Evidently, the model parameter ω_1 may be found for given ω_2 and φ_0 from equation (57). On Figure 3 we present the exact solution for the following observational parameters: $q_0 = 0.5$, $\Omega_{\rho 0} = 0.3$ and the following model parameters: $\pi H_0/(2\omega_2) = 2$, $\varphi_0 = 0$, $\dot{\varphi}_0 > 0$. The interval of definition of the solution is turned to be with the following end-points $t_{01} \approx -0.68 H_0^{-1}$ and $t_{02} \approx 1.22 H_0^{-1}$. The scale factor presented on Figure 3a may be described in the main order by $a \sim (t-t_{01})^{1/3}$ and $a \sim (t-t_{02})^{1/3}$ near the initial and final points. The scalar field presented on Figure 3b diverges to $+\infty$ as $t \to t_{01}$ or $t \to t_{02}$. The scalar field potential given diverges to $-\infty$ as $t \to t_{01}$ or $t \to t_{02}$ (see Figure 3c). Figure 3d shows the domination of the scalar field near the initial and final singularities. Comparing Figures 3c,3d one easily concludes that the kinetic term $\dot{\varphi}^2/2$ dominates the potential $V(\varphi)$ near these points. Then $w_\varphi \to 1$ as $t \to t_{01}$ or $t \to t_{02}$, i.e. the scalar field is like stiff matter.

4.3. DISCUSSION

We studied a spatially flat Friedmann model containing a pressureless perfect fluid (dust) and a minimally coupled scalar field with an unbounded from below potential of the form $V(\varphi) = W_0 - V_0 \sinh\left(\sqrt{3/2}\kappa\varphi\right)$, where

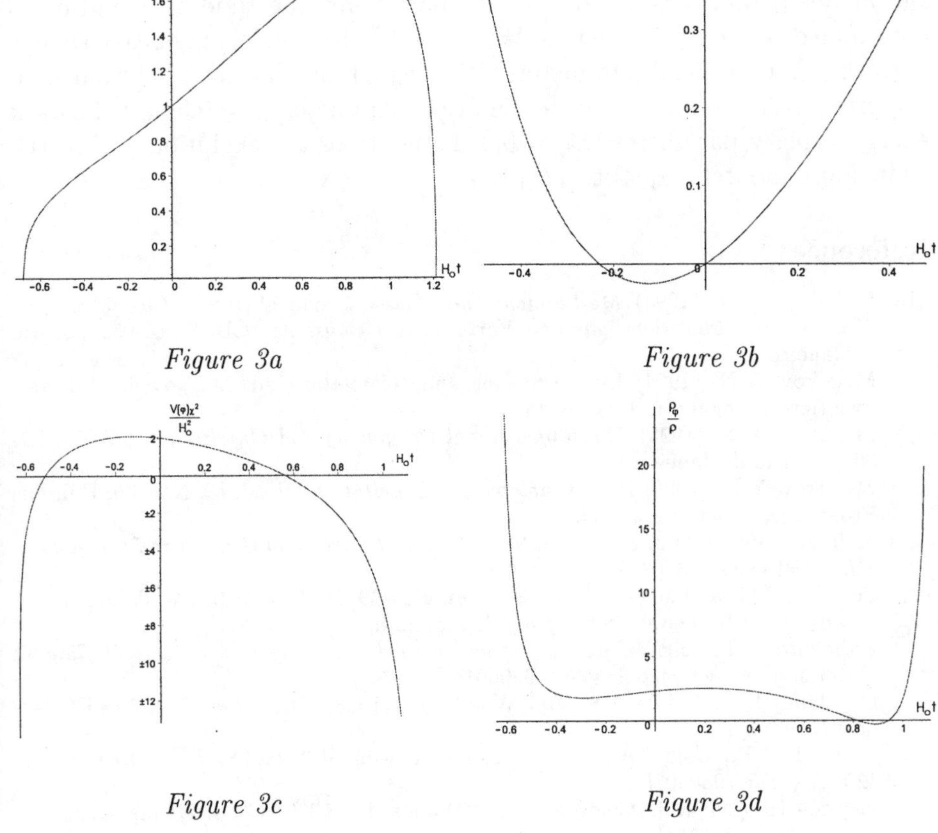

Figure 3a

Figure 3b

Figure 3c

Figure 3d

Figure 3. We present the exact solution for the following observational parameters: $q_0 = 0.5$, $\Omega_{\rho 0} = 0.3$ and the following model parameters: $\pi H_0/(2\omega_2) = 2$, $\varphi_0 = 0$, $\dot{\varphi}_0 > 0$.

the parameters W_0 and V_0 are arbitrary. All exact solutions describe the recollapsing universe. The behavior of the model near both initial and final points of evolution is analyzed. Near the singularity the scale factor is proportional to the time in power 1/3 for almost all solutions. In this case the zero-energy constraint is dominated by the kinetic energy of the scalar field. Such solutions are called kinetic dominated. There exists a special solution with $a \sim t^{2/3}$ near the singularity. It appears when the zero-energy constraint is dominated by the density of dust. It's so called fluid dominated solution. The scalar field typically diverges to $\pm\infty$ as the time tends to the initial or final value. However, there is a special solution where the scalar field tends to some constant.

The evolution between big bang and big crunch crucially depends on the model parameters ω_1 and ω_2. In particular, the time of evolution is determined by these parameters. We show that the model may be consistent with the observational parameters. We singled out the exact solution with the present-day values of the acceleration parameter $q_0 = 0.5$ and the dark energy density parameter $\Omega_{\rho 0} = 0.3$. It describes the evolution within the time approximately equal to $2H_0^{-1}$.

References

1. Melnikov, V.N. (1993) *Multidimensional Classical and Quantum Cosmology and Gravitation. Exact Solutions and Variations of Constants*, CBPF-NF-051/93, Rio de Janeiro;
   Melnikov, V.N. (1994) In: *Cosmology and Gravitation*, ed. M. Novello, Editions Frontieres, Singapore, 1994, p. 147;
   Melnikov, V.N. (1995) *Multidimensional Cosmology and Gravitation*, CBPF-MO-002/95, Rio de Janeiro;
   Melnikov, V.N. (1996) In: *Cosmology and Gravitation. II*, ed. M. Novello, Editions Frontieres, Singapore, p. 465;
   Melnikov, V.N. (2002) *Exact Solutions in Multidimensional Gravity and Cosmology. III*, CBPF-MO-003/02, Rio de Janeiro.
2. Ferreira, P.G. and Joice, M. (1998) *Phys. Rev.* **D 58**, 023503 (astro-ph/9711102).
3. Townsend, P.K. *Quintessence from M-theory*, hep-th/0110072.
4. Chimento, L.P. and Jakubi, A.S. *Scalar field cosmologies with perfect fluid in Robertson-Walker metric*, gr-qc/9506015.
5. Copeland, E.J., Liddle A.R. and Wands, D. (1998) *Phys. Rev.* **D 57**, 4686 (gr-qc/9711068).
6. Billyard, A.P., Coley, A.A. and van den Hoogen, R.J. (1998) *Phys. Rev.* **D 57**, 123501 (gr-qc/9805085).
7. van den Hoogen, R.J., Coley A.A. and Wands, D. (1999) *Class. Quant. Grav.* **16**, 1843 (gr-qc/9901014).
8. Heard, I.P. C and Wands, D. *Cosmology with positive and negative exponential potentials*, gr-qc/0206085.
9. Rubano, C. and Scudellaro, P. *On some exponential potentials for a cosmological scalar field as quintessence*, gr-qc/01033335.
10. Ivashchuk, V.D. and Melnikov, V.N. (1994) *Int. J. Mod. Phys.* D **3**, 795 (gr-qc/9403064).
11. Gavrilov, V.R., Ivashchuk, V.D. and Melnikov, V.N. (1995) *J. Math. Phys.* **36**, 5829 (gr-qc/9407019).
12. Gavrilov, V.R. and Melnikov, V.N. (1998) *Theor. Math. Phys.* **114**, 454 (gr-qc/9801042).
13. Polyanin, A.D. and Zaitsev, V.F. (1995) *Handbook on Exact Solutions for Ordinary Differential Equations*, CRC Press, Boca Raton.
14. Zaitsev, V.F. and Polyanin, A.D. (1994) *Discrete-Groups Methods for Integrating Equations of Nonlinear Mechanics*, CRC Press, Boca Raton.
15. Kallosh, R. and Linde, A. *Dark Energy and the fate of the universe*, astro-ph/0301087.

CONSTRAINTS ON NON-NEWTONIAN GRAVITY FROM RECENT CASIMIR FORCE MEASUREMENTS

V.M. MOSTEPANENKO
Departamento de Física, Universidade Federal da Paraíba, C.P. 5008, CEP 58059-970, João Pessoa, Pb-Brazil.
On leave from A. Friedmann Laboratory for Theoretical Physics, St.Petersburg, Russia

Abstract. Corrections to Newton's gravitational law inspired by extra dimensional physics and by the exchange of light and massless elementary particles between the atoms of two macrobodies are considered. These corrections can be described by the potentials of Yukawa-type and by the power-type potentials with different powers. The strongest up to date constraints on the corrections to Newton's gravitational law are reviewed following from the Eötvos- and Cavendish-type experiments and from the measurements of the Casimir and van der Waals force. We show that the recent measurements of the Casimir force gave the possibility to strengthen the previously known constraints on the constants of hypothetical interactions up to several thousand times in a wide interaction range. Further strengthening is expected in near future that makes Casimir force measurements a prospective test for the predictions of fundamental physical theories.

1. Introduction

It is common knowledge that the gravitational interaction is described on a different basis than all the other physical interactions. Up to the present there is no unified description of gravitation and gauge interactions of the Standard Model which would be satisfactory both physically and mathematically. Gravitational interaction persistently avoids unification with the other interactions. In addition, there is an evident lack of experimental data in gravitational physics. Newton's law of gravitation, which is also valid with high precision in the framework of the Einstein General Relativity Theory, is not verified with a sufficient precision at the separations less than 1 mm. Surprisingly, at the separations less than 1 μm corrections to the Newton's gravitational law are not excluded experimentally that are many orders of magnitude greater than the Newtonian force itself. What this means is the general belief, that the Newton's law of gravitation is obeyed up to Planckean separation distances, is nothing more than a large scale extrapolation. It

V. de Sabbata et al. (eds.),
The Gravitational Constant: Generalized Gravitational Theories and Experiments, 269–288.
© 2004 *Kluwer Academic Publishers. Printed in the Netherlands.*

is meaningful also that the Newton's gravitational constant G is determined with much less accuracy than the other fundamental physical constants. In spite of all attempts the results of recent experiments on the precision measurement of G are contradictory [1].

Prediction of non-Newtonian corrections to the law of gravitation comes from the extra dimensional unification schemes of High Energy Physics. According to this schemes, which go back to Kaluza [2] and Klein [3], the true dimensionality of physical space is larger than 3 with the extra dimensions being spontaneously compactified at the Planckean length-scale. At the separation distances several times larger than a compactification scale, the Yukawa-type corrections to the Newtonian gravitational potential do arise. This prediction would be of only academic interest if to take account of the extreme smallness of the Planckean length $l_{Pl} = \sqrt{G} \sim 10^{-33}$ cm (we use units with $\hbar = c = 1$) and the excessively high value of the Planckean energy $M_{Pl} = 1/\sqrt{G} = 10^{19}$ GeV. Recently, however, the low energy (high compactification length) unification schemes were proposed [4,5]. In the framework of these schemes the "true", multidimensional, Planckean energy takes a moderate value $M_* \sim 10^3$ GeV=1 TeV and the value of a compactification scale belongs to a submillimeter range. It is amply clear that in the same range the Yukawa-type corrections to the Newtonian gravitation are expected [6,7] and this prediction can be verified experimentally.

Much public attention given to non-Newtonian gravitation is generated not only by the extra dimensional physics. The new long-range forces which can be considered as corrections to the Newton's law of gravitation are produced also by the exchange of light and massless hypothetical elementary particles between the atoms of closely spaced macrobodies. Such particles (like axion, scalar axion, dilaton, graviphoton, moduli, arion etc.) are predicted by many extensions to the Standard Model and practically inavoidable in the modern theory of elementary particles and their interactions [8]. The long-range forces produced due to the exchange of hypothetical particles can be considered as some corrections to the Newton's gravitational law leading to the same phenomenological consequences as in the case of extra spatial dimensions.

The constraints on the constants characterizing the magnitude and interaction range of hypothetical forces are usually obtained from the gravitational experiments of Cavendish- and Eötvos-type. These experiments lead to the most strong constraints in the interaction range $10^{-5}\,\mathrm{m} < \lambda < 10^{6}\,\mathrm{km}$ (see Ref. [9] and also some recent results in Refs. [10–13]). In nanometer and micrometer interaction range the best constraints on the constants of hypothetical interactions follow from the van der Waals and Casimir force measurements which provide the dominant background force at so small separations. The first results in this direction were obtained in Refs. [14,15] (see also Refs. [16,17] for details).

During the last years, the new experiments on measuring the Casimir force with an increased precision were performed [18–25]. They gave the possibility to considerably increase the strength of constraints on hypothetical interactions within a submillimeter interaction range [26–32]. Thus, from the measurement of the Casimir force by the use of an atomic force microscope [19–21] the strengthening of the previously known constraints up to 4500 times was obtained (the dynamical Casimir force measurements [23,33] lead to weaker constraints than those mentioned above). The increased experimental precision calls for a more accurate theory taking into account corrections to the Casimir force due to surface roughness, finite conductivity of the boundary metal and nonzero temperature. New constraints were obtained from a comparision between the more precise experimental data and improved theory (for the recent review of both experimental and theoretical developments in the Casimir effect see Ref. [34]).

In the present paper we report the most strong constraints on the hypothetical long-range interactions obtained from the Casimir effect. In Sec. 2 the hypothetical long-range forces are discussed originating from both extra dimensional physics and exchange of light elementary particles predicted by the unified gauge theories of fundamental interactions. In Sec. 3 the constraints from gravitational experiments are briefly summarized. Sec. 4 contains constraints following from Lamoreaux experiment [18] on measuring the Casimir force by means of a torsion pendulum. In Sec. 5 Mohideen et al experiments [19,20] on measurement of the Casimir force between an aluminum disk and a sphere by means of an atomic force microscope are considered. Sec. 6 is devoted to the results of Ederth experiment [22]. In Sec. 7 the most conclusive and reliable results are presented following from Mohideen et al experiment on measuring the Casimir force between gold surfaces [21]. In Sec. 8 reader will find the most recent results obtained from the lateral Casimir force measurement and from the new experiment using a microelectromechanical torsional oscillator. Sec. 9 contains conclusions and discussion.

Throughout the paper units are used in which $\hbar = c = 1$.

2. Origination of the hypothetical long-range interactions

The usual Newton's law of gravitation is only valid in a 4-dimensional spacetime. If the extra dimensions exist, it will be modified by some corrections. In models with large but compact extra dimensions (like those proposed in Ref. [4]) the gravitational potential between two point particles with masses m_1 and m_2 separated by a distance $r \gg R_*$, where R_* is a compactification scale, is given by [6,7]

$$V(r) = -\frac{Gm_1m_2}{r}\left(1+\alpha_G e^{-r/\lambda}\right). \tag{1}$$

The first term in the right-hand side of Eq. (1) is the Newtonian contribution, whereas the second term represents the Yukawa-type correction. Here G is the

Newton's gravitational constant, α_G is a dimensionless constant depending on the nature of extra dimensions and λ is the interaction range of a correction.

The dimensionless constant α_G in (1) depends on the nature of the extra dimensions. By way of example, for a toroidal compactification with all extra dimensions having equal size, $\alpha_G = 2n$ [6,7]. If extra dimensions have the topology of n-sphere $\alpha_G = n+1$ [6,7].

In fact the search of corrections to Newtonian gravity, like in Eq. (1), is the simplest way to check the predictions of the models with low compactification scale. The thing is that, according to these models, all interactions and particles of the Standard Model are considered as living on a $(3+1)$-dimensional wall. They remain almost unchanged as this wall has a thickness only of order $M_*^{-1} \sim 10^{-17}$ cm in the extra dimensions. Only gravitational interaction penetrates freely into extra dimensions and can serve as a test for their existence.

At small separation distances $r \ll R_*$ the usual Newton's law of gravitation should be generalized to

$$V(r) = -\frac{G_{4+n}m_1m_2}{r^{n+1}} \tag{2}$$

in order to preserve the continuity of the force lines in a $(4+n)$-dimensional space-time. Here G_{4+n} is the underlying multidimensional gravitational constant connected with the usual one by the relation $G_{4+n} \sim GR_*^n$.

In fact the characteristic energy scale in multidimensional space-time is given by the multidimensional Planckean mass $M_* = 1/G_{4+n}^{1/(2+n)}$, and the compactification scale is given by [4]

$$R_* = \frac{1}{M_*}\left(\frac{M_{Pl}}{M_*}\right)^{2/n} \sim 10^{\frac{32}{n}-17}\,\mathrm{cm}, \tag{3}$$

where $M_{Pl} = 1/\sqrt{G}$ is the usual Planckean mass, $n \geq 1$, and $M_* \sim 10^3$ GeV as was told in Introduction. Then, for $n = 1$ (one extra dimension) one finds from Eq. (3) $R_* \sim 10^{15}$ cm. If to take into account that, as was shown in Refs. [6,7], $\alpha_G \sim 10$ and $\lambda \sim R_*$, this possibility must be rejected on the basis of the solar system tests of Newton's gravitational law [9]. If, however, $n = 2$ one obtains from Eq. (3) $R_* \sim 1\,\mathrm{mm}$, and for $n = 3$ $R_* \sim 5\,\mathrm{nm}$. For these scales the corrections of form (1) to Newton's gravitational law are not excluded experimentally.

The other type of multidimensional models considers noncompact but warped extra dimensions. In these models the leading contribution to the gravitational potential is given by [5,35]

$$U(r) = -\frac{Gm_1m_2}{r}\left(1+\frac{2}{3k^2r^2}\right), \tag{4}$$

where $r \gg 1/k$ and $1/k$ is the so-called warping scale. Here the correction to the Newton's gravitational law depends on the separation distance inverse proportionally to the third power of separation.

As was mentioned in Introduction, many extensions to the Standard Model predict the hypothetical long-range forces, distinct from gravitation and electromagnetism, caused by the exchange of light and massless elementary particles between the atoms of macrobodies. Under appropriate parametrization of the interaction constant these forces also can be considered as some corrections to the Newton's gravitational law. The velocity independent part of the effective potential due to the exchange of hypothetical particles between two atoms can be calculated by means of Feynman rules. For the case of massive particles with mass $\mu = 1/\lambda$ (λ is their Compton wavelength) the effective potential takes the Yukawa form

$$V_{Yu}(r) = -\alpha N_1 N_2 \frac{1}{r} e^{-r/\lambda}, \tag{5}$$

where $N_{1,2}$ are the numbers of nucleons in the atomic nuclei, α is a dimensionless interaction constant. If to introduce a new constant $\alpha_G = \alpha/(Gm_p^2) \approx 1.7 \times 10^{38}\alpha$ (m_p being a nucleon mass) and consider the sum of potential (5) and Newton's gravitational potential one returns back to the potential (1).

For the case of exchange of one massless particle the effective potential is just the usual Coulomb potential which is inverse proportional to separation. The effective potentials inverse proportional to higher powers of a separation distance appear if the exchange of even number of pseudoscalar particles is considered. The power-type potentials with higher powers of a separation are obtained also in the exchange of two neutrinos, two goldstinos or other massless fermions [16]. The resulting interaction potential acting between two atoms can be represented in the form [36]

$$U(r) = -\Lambda_l N_1 N_2 \frac{1}{r} \left(\frac{r_0}{r}\right)^{l-1}, \tag{6}$$

where $r_0 = 1\,\text{F}=10^{-15}$ m is introduced for the proper dimensionality of potentials with different l, and Λ_l with $l = 1, 2, 3, \ldots$ are the dimensionless constants.

If to introduce a new set of constants $\Lambda_l^G = \Lambda_l/(Gm_p^2)$ and consider the sum of (6) and Newton's gravitational potential one obtains

$$U_l(r) = -\frac{Gm_1 m_2}{r}\left[1 + \Lambda_l^G \left(\frac{r_0}{r}\right)^{l-1}\right]. \tag{7}$$

This equation represents the power-type hypothetical interaction as a correction to the Newton's gravitational law. The potential (4) following from the extra dimensional physics is obtained from Eq. (7) with $l = 3$. Note that the case $l = 3$ corresponds also to two arions exchange between electrons [16].

3. Constraints from gravitational experiments

Constraints on the corrections to Newton's gravitational law can be obtained from the experiments of Eötvos- and Cavendish-type. In the Eötvos-type experiments the difference between inertial and gravitational masses of a body is measured, i.e. the equivalence principle is verified. The existence of an additional hypothetical force which is not proportional to the masses of interacting bodies can lead to the appearance of the effective difference between inertial and gravitational masses. Therefore some constraints on hypothetical interactions emerge from the experiments of Eötvos type.

The typical result of the Eötvos-type experiments is that the relative difference between the accelerations imparted by the Earth, Sun or some laboratory attractor to various substances of the same mass is less than some small number. Many Eötvos-type experiments were performed (see, e.g., Refs. [37–40]). By way of example, in Ref. [39] the above relative difference of accelerations was to be less than 10^{-11}.

The results of the two precise Eötvos-type experiments can be found in Refs. [10,41]. They permit to obtain the best constraints on the constants of hypothetical long-range interactions inspired by extra dimensions or by the exchange of light and massless elementary particles (see Fig. 1).

The constraints under consideration can be obtained also from the Cavendish-type experiments. In these experiments the deviations of the gravitational force F from Newton's law are measured (see, e.g., Refs. [42–47]). The characteristic value of deviations in the case of two point-like bodies a distance r apart can be described by the parameter

$$\varepsilon = \frac{1}{rF}\frac{d}{dr}\left(r^2 F\right), \tag{8}$$

which is equal exactly to zero in the case of pure Newton's gravitational force. For example, in Refs. [44,45] $|\varepsilon| \leq 10^{-4}$ at the separation distances $r \sim 10^{-2} - 1\,\mathrm{m}$. This can be used to constrain the size of corrections to the Newton's gravitational law. The results of one of most recent Cavendish-type experiments can be found in Ref. [11].

Let us now outline the strongest constraints on the corrections to Newton's gravitational law obtained up to date from the gravitational experiments. The constraints on the parameters of Yukawa-type correction, given by Eq. (1), are presented in Fig. 1. In this figure, the regions of (λ, α_G)-plane above the curves are prohibited by the results of the experiment under consideration, and the regions below the curves are permitted. By the curves 1 and 2 the results of the best Eötvos-type experiments are shown (Refs. [41] and [10], respectively). Curve 4 represents constraints obtained from the Cavendish-type experiment of Ref. [11]. At the intersection of curves 2 and 4 the better constraints are given by curve 3 following from the results of older Cavendish-type experiment of Ref. [47]. As is seen from Fig. 1, rather strong constraints on the Yukawa-type corrections to Newton's gravitational law ($\alpha_G < 10^{-5}$) are obtained only within the interaction range $\lambda > 0.1\,\mathrm{m}$. With decreasing λ the strength of constraints falls off, so that

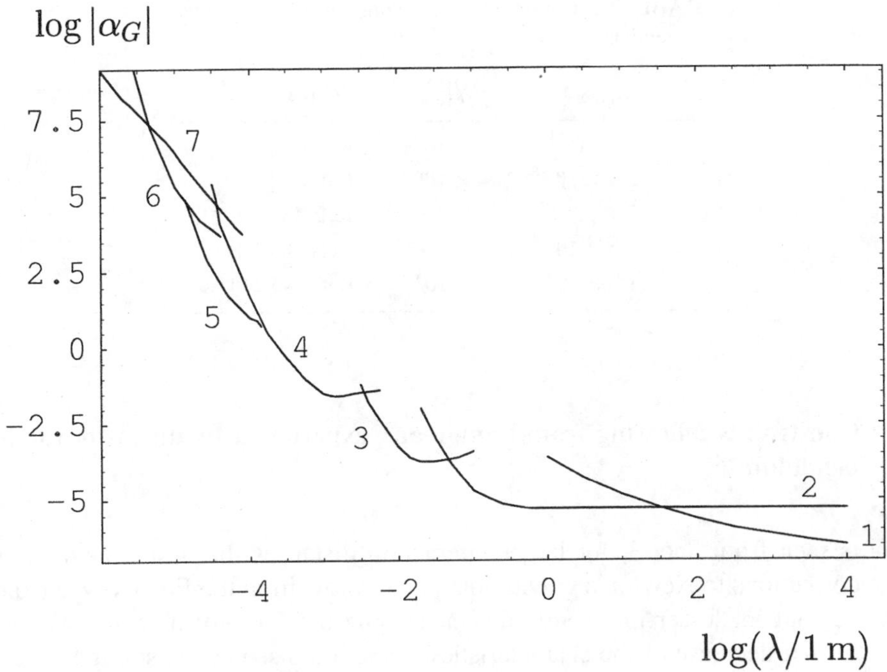

Figure 1. Constraints on the Yukawa-type corrections to Newton's gravitational law. Curves 1, 2 follow from the Eötvos-type experiments, and curves 3–6 follow from the Cavendish-type experiments. The beginning of curve 7 shows constraints from the measurements of the Casimir force. Permitted regions on (λ, α_G)-plane lie beneath the curves.

at $\lambda = 0.1$ mm $\alpha_G < 100$. By the beginning of curve 7 the constraints are shown following from the Casimir force measurements (see Sec. 4).

Recently two more precise Cavendish-type experiments were performed [12, 13] by the use of the micromachined torsional oscillator. They have permitted significantly increase the strength of constraints on α_G in the interaction range around (10–100) μm (see curve 5 [12] and curve 6 [13]).

Now we consider constraints on the power-type corrections to Newton's gravitational law given by Eq. (7). The best of them follow from the Eötvos- and Cavendish-type experiments. They are collected in Table 1.

For $l = 1, 2$ the constraints presented in Table 1 are obtained from the Eötvos-type experiments, and for $l = 3, 4, 5$ from the Cavendish-type ones. It is seen that the strength of constraints falls greatly with the increase of l.

TABLE I. Constraints on the constants of power-type potentials.

l	$\|\Lambda_l\|_{\max}$	$\|\Lambda_l^G\|_{\max}$	Source
1	6×10^{-48}	1×10^{-9}	Ref. [48]
2	2.4×10^{-30}	4×10^{8}	Ref. [10]
3	7×10^{-17}	1.2×10^{22}	Refs. [47, 49, 50]
4	7.5×10^{-4}	1.3×10^{35}	Refs. [45, 50]
5	1.2×19^{9}	2×10^{47}	Refs. [45, 50]

4. Constraints following from Lamoreaux experiment by means of torsion pendulum

As is seen from Sec. 3, for larger interaction distances the best constraints on the corrections to Newton's gravitational law follow from the Eötvos-type experiments and for lesser interaction distances from the Cavendish-type ones. With the further decrease of the characteristic interaction distance the strength of constraints following from the gravitational experiments greatly reduces. Within a micrometer separations, the Casimir and van der Waals force [17,51,52] becomes the dominant force between two macrobodies. As was shown in Ref. [14] for the case of Yukawa-type interactions with a micrometer interaction range and in Ref. [15] for the power-type ones, the measurements of the van der Waals and Casimir forces lead to the strongest constraints on non-Newtonian gravity (see the discussion about the Casimir effect as a test for non-Newtonian gravitation in Ref. [53]).

Currently a lot of precision experiments on the measurement of the Casimir and van der Waals force has been performed (see Ref. [34] for a review). As was mentioned in Introduction, the extensive theoretical study of different corrections to the Casimir force due to surface roughness, finite conductivity of a boundary metal and nonzero temperature gave the possibility to compute the theoretical value of this force with high precision. At the moment the agreement between theory and experiment at a level of 1% is achieved for the smallest experimental separation distances [34]. This permitted to obtain stronger constraints on the corrections to Newton's gravitational law from the results of the Casimir force measurements [26–32,54,55]. Here we briefly present the strongest constraints of this type starting from the first modern experiment performed by Lamoreaux [18].

In Ref. [18] the Casimir force between a spherical lens and a disk made of quartz (with the densities $\rho' = 2.23 \times 10^3\,\mathrm{kg/m^3}$ and $\rho = 2.4 \times 10^3\,\mathrm{kg/m^3}$, respectively) and coated by Cu and Au layers of thickness $\Delta_1 = \Delta_2 = 0.5\,\mu\mathrm{m}$ (with the densities $\rho_1 = 8.96 \times 10^3\,\mathrm{kg/m^3}$, $\rho_2 = 19.32 \times 10^3\,\mathrm{kg/m^3}$) was measured by the use of torsion pendulum. The disk radius was $L = 1.27$ cm and a lens height and curvature radius were $H = 0.18$ cm and $R = 12.5$ cm, respectively.

The absolute error of force measurements in Ref. [18] was about $\Delta F = 10^{-11}$ N for the separation range between a disk and a lens $1\,\mu\text{m} \leq a \leq 6\,\mu\text{m}$. In the limits of this error the theoretical expression for the Casimir force was confirmed

$$F^{(0)}(a) = -\frac{\pi^3}{360}\frac{R}{a^3}. \tag{9}$$

No corrections to Eq. (9) due to surface roughness, finite conductivity of the boundary metal or nonzero temperature were reported. These corrections, however, may not lie within the limits of the absolute error ΔF. By way of example, at $a = 1\,\mu\text{m}$ roughness correction $\Delta_R F(a)$ may be around 12% of $F^{(0)}$ or even larger [26]. The finite conductivity correction $\Delta_{\delta_0} F(a)$ for gold surfaces at $a = 1\,\mu\text{m}$ separation is 10% of $F^{(0)}$ [56]. (Note that ΔF is around 3% of $F^{(0)}$ at $a = 1\,\mu\text{m}$.) As to temperature correction $\Delta_T F(a)$, it achieves 174% of $F^{(0)}$ at the separation $a = 6\,\mu\text{m}$, where, however, ΔF is around 700% of $F^{(0)}$. For this reason, the constraints on the Yukawa-type interaction following from Lamoreaux experiment were found from the inequality [26]

$$|F_{th}(a) - F^{(0)}(a)| \leq \Delta F, \tag{10}$$

where F_{th} is the theoretical force value including $F^{(0)}$, all the corrections to it mentioned above, and also the hypothetical Yukawa-type interaction

$$F_{th}(a) = F^{(0)}(a) + \Delta_R F(a) + \Delta_{\delta_0} F(a) + \Delta_T F(a) + F_{Yu}(a) \tag{11}$$

(we remind that the sign of a finite conductivity correction is opposite to the sign of other corrections).

The hypothetical interaction in a configuration of a spherical lens above a disk was computed in Ref. [26]. For λ smaller or of order of separation a the result is given by

$$\begin{aligned} F_{Yu}(a) &= -4\pi^2 \alpha_G G \lambda^3 e^{-a/\lambda} R \\ &\times \left[\rho_2 - (\rho_2 - \rho_1)\, e^{-\Delta_2/\lambda} - (\rho_1 - \rho')\, e^{-(\Delta_2+\Delta_1)/\lambda}\right] \\ &\times \left[\rho_2 - (\rho_2 - \rho_1)\, e^{-\Delta_2/\lambda} - (\rho_1 - \rho)\, e^{-(\Delta_2+\Delta_1)/\lambda}\right]. \end{aligned} \tag{12}$$

For larger λ, $F_{Yu}(a)$ was computed numerically [26].

The obtained constraints [26] are shown in Fig. 2 (curve 7,a for $\alpha_G > 0$ and curve 7,b for $\alpha_G < 0$). In this figure, the regions of (α_G, λ)-plane above the curves are prohibited, and the regions below the curves are permitted by the results of an experiment under consideration. By the curves 6 the results of the best Cavendish-type experiments are shown (Ref. [13]). Curve 8 represents constraints obtained from the Casimir force measurements between dielectrics

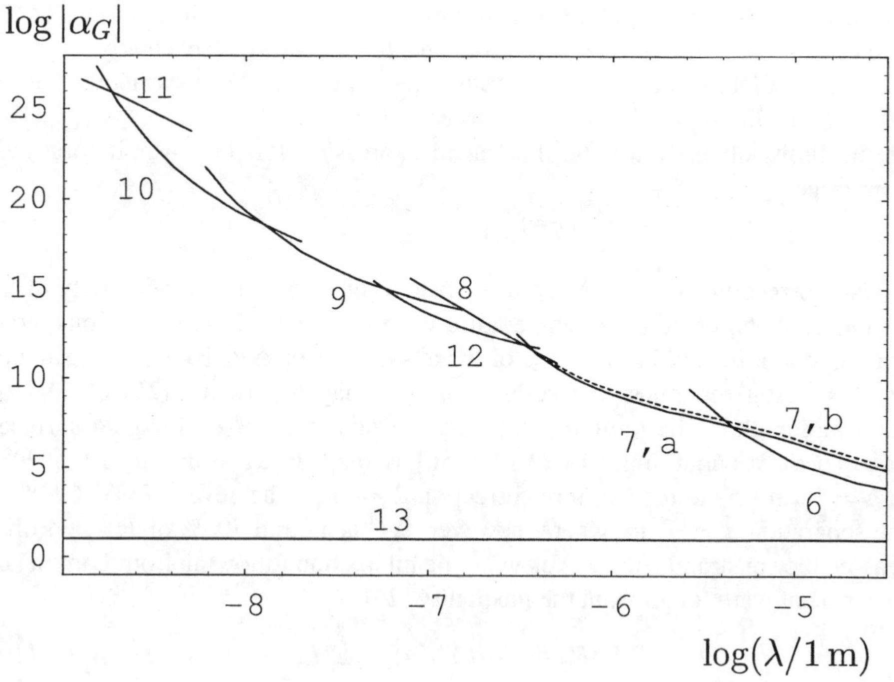

Figure 2. Constraints on the Yukawa-type corrections to Newton's gravitational law. Curves 8–10, 12 follow from the Casimir, and curve 11 from the van der Waals force measurements. The typical prediction of extra dimensional physics is shown by curve 13.

[16,17]. Line 13 demonstrates the typical prediction of extra dimensional theories. The strengthening of constraints given by curves 7,a and 7,b comparing curve 8 is up to a factor of 30 in the interaction range $2.2 \times 10^{-7}\,\mathrm{m} \leq \lambda \leq 5 \times 10^{-6}\,\mathrm{m}$ (a weaker result was obtained later in Ref. [29] where the corrections to the ideal Casimir force of Eq. (9) were not taken into account). This shows that the Casimir force measurements are competitive with the Cavendish-type experiments in a micrometer interaction range.

5. Constraints following from Mohideen et el experiments with *Al* surfaces by means of atomic force microscope

A major progress in obtaining more strong constraints on the Yukawa-type interactions within a nanometer range was achieved due to the measurements of the Casimir force by means of an atomic force microscope [19–21]. In Refs. [19,56] the results of the Casimir force measurement between a flat supphire disk ($L =$

$0.625\,\text{cm}$, $\rho = 4.0 \times 10^3\,\text{kg/m}^3$) and a polystyrene sphere ($R = 98\,\mu\text{m}$, $\rho' = 1.06 \times 10^3\,\text{kg/m}^3$) were presented in comparison with a complete theory taking into account the finite conductivity and roughness corrections. Temperature corrections are not essential in the separation range $0.12\,\mu\text{m} \leq a \leq 0.9\,\mu\text{m}$ used in Refs. [19,56]. The test bodies were coated by the aluminum layer ($\rho_1 = 2.7 \times 10^3\,\text{kg/m}^3$) of $\Delta_1 = 300\,\text{nm}$ thickness and Au/Pd layer ($\rho_2 = 16.2 \times 10^3\,\text{kg/m}^3$) of the thickness $\Delta_2 = 20\,\text{nm}$ (this latter was used to prevent the oxidation processes; it is almost transparent for electromagnetic oscillations of characteristic frequency). The absolute error of force measurements in Refs. [19,56] was $\Delta F = 2 \times 10^{-12}\,\text{N}$. In the limits of this error the theoretical expression for the Casimir force with corrections to it due to the surface roughness and finite conductivity was confirmed.

In the improved version of this experiment [20] the Au/Pd layer was made thiner ($\Delta_2 = 7.9\,\text{nm}$) and the other experimental parameters were as follows: $\Delta_1 = 250\,\text{nm}$, $L = 0.5\,\text{cm}$, $R = 100.85\,\mu\text{m}$, $100\,\text{nm} \leq a \leq 500\,\text{nm}$. Due to experimental improvements like the use of vibration isolation, lower systematic errors, independent measurement of a surface separation and smoother metal surface, the smaller absolute error of force measurement $\Delta F = 1.3 \times 10^{-12}\,\text{N}$ was achieved. In the limits of this error experimental data were in agreement with a complete theory.

To obtain constraints on hypothetical interactions, the hypothetical force was computed [27,28] with account of surface roughness contribution which is especially important at the closest separations

$$F_{Yu}(a) = \sum_i w_i F_{Yu}(a_i). \tag{13}$$

Here w_i are the probabilities for different values of a separation distance between the distorted surfaces, and F_{Yu} is given by Eq. (12). The values of w_i were found [56] on the basis of atomic force microscope measurements of surface roughness. The typical roughness heights were 40 nm and 20 nm (Refs. [19,56]) and 14 nm and 7 nm (Ref. [20]).

The constraints were obtained from the inequality

$$|F_{Yu}(a)| \leq \Delta F, \tag{14}$$

because the theoretical expression for the Casimir force with all corrections to it was confirmed experimentally unlike the case considered in Sec. 4.

Using the above experimental parameters of Refs. [19,56], the strengthening of constraints up to 140 times was obtained [27] as compared with the measurements of the van der Waals and Casimir force between dielectrics. The strengthening holds within the interaction range $5.9\,\text{nm} \leq \lambda \leq 100\,\text{nm}$.

Even stronger constraints were obtained [28] from the experiment of Ref. [20] in a wider interaction range $5.9\,\text{nm} \leq \lambda \leq 115\,\text{nm}$. The new constraints are up to 560 times stronger than the old ones given by curves 8 and 11 in Fig. 2 which follow from old mesurements of the Casimir and van der Waals force between dielectrics (the final constraint curve from the atomic force microscopy measurements will be obtained in Sec. 7).

The above constraints obtained on the basis of Refs. [19, 20, 56] are found for the closest separation distance a (120 nm in Refs. [19, 56] and 100 nm in Ref. [20]). If to decrease the minimal value of a, stronger constraints can be obtained. This is true also if the heavier metal coating is used (see below).

6. Constraints on hypothetical interactions following from Ederth experiment with two crossed cylinders

In Ref. [22] the Casimir force acting between two crossed quartz cylinders of 1 cm radius was measured (quartz density $\rho = \rho' = 2.23 \times 10^3\,\mathrm{kg/m^3}$). Each cylinder was coated by a layer of Au (density $\rho_1 = 18.88 \times 10^3\,\mathrm{kg/m^3}$ and thickness $\Delta_1 = 200\,\mathrm{nm}$) and outer layer of hydrocarbon (density $\rho_2 = 0.85 \times 10^3\,\mathrm{kg/m^3}$, thickness $\Delta_2 = 2.1\,\mathrm{nm}$). The absolute error of force measurements was $\Delta F = 10\,\mathrm{nN}$ which is much larger than in the experiments discussed above. Within the limits of this error the theoretical expression for the Casimir force between cylinders was confirmed. The separation range between the cylinders was in the limits $20\,\mathrm{nm} \le a \le 100\,\mathrm{nm}$, i.e. the more close separations were achieved. The other experimental improvement of Ref. [22] lies in the use of smoother surfaces. The root mean square roughness of the cylindrical surfaces was decreased up to 0.4 nm.

There were also some disadvantages in the experiment of Ref. [22] as compared with the previous experiments. One of them is connected with the presence of hydrocarbon coating which complicates the independent measurement of the residual electrostatic force. The other disadvantage is a substantial deformation of the Au coating caused by the attractive forces in contact and by relatively soft glue used to support the Au layer. As a result, there is no independent and exact determination of surface separation in Ref. [22].

The constraints on the parameters of Yukawa-type interaction, following from the experiment of Ref. [22], were obtained from Eq. (14) in Ref. [30]. It was shown [30] that the hypothetical force between two cylinders, crossed at a right angle, is given once more by Eq. (12). Surface roughness contribution is not essential here as the roughness amplitude was considerably decreased.

The obtained constraints [30] are shown by curve 10 in Fig. 2 (by curve 11 the constraints following from the van der Waals force measurements between dielectrics are demonstrated). They are up to 300 times stronger than the previously known ones within the separation range $1.5\,\mathrm{nm} \le \lambda \le 11\,\mathrm{nm}$. This result was obtained at the closest separation distance $a = 20\,\mathrm{nm}$.

7. Constraints on hypothetical interactions following from Mohideen et el experiments with gold surfaces

The most conclusive measurement of the Casimir force by means of atomic force microscope was performed between a sapphire disk and polystyrene sphere ($R = 95.65\,\mu$m) coated by Au layer of $\Delta_1 = 86.6\,$nm thickness [21]. No additional coating was used which added complexity to interpretation of experimental data of Refs. [19,20,22]. Some other improvements were implemented also in this experiment. Specifically, the root mean square amplitude of surface roughness was decreased up to $1.0 \pm 0.1\,$nm which is comparable with Ref. [22] (see the preceding section) but did not require additional hydrocarbon coating. The electrostatic forces were reduced to a value much smaller of the Casimir force at the shortest separation and used for an independent measurement of surface separation. Also the measurement was performed over smaller separations $62\,\text{nm} \leq a \leq 350\,\text{nm}$ than in previous measurements by means of atomic force microscope.

The absolute error of force measurements in Ref. [21], $\Delta F = 3.5 \times 10^{-12}\,$N, was a bit larger than that in Refs. [19,20]. This was caused by the poor thermal conductivity of the cantilever resulting from the thiner metal coating used. The increase of ΔF is, however, compensated for by the greater increase of the Casimir force at smaller separations.

Constraints on the constants of Yukawa-type interaction were obtained [31,57] from Eq. (14) using the agreement of experimental data with a theoretical Casimir force. Hypothetical force was computed by Eq. (12) having regard to $\Delta_2 = \rho_2 = 0$. The computational results are shown by curve 9 in Fig. 2. The obtained constraints are stronger up to 19 times, comparing the previous experiments using the atomic force microscope, within the interaction range $4.3\,\text{nm} \leq \lambda \leq 150\,\text{nm}$. If to compare with the experiment of Ref. [22], the constraints following from Mohideen et al experiment with Au surfaces prove to be the best ones in the interaction range $11\,\text{nm} \leq \lambda \leq 150\,\text{nm}$. As a consequence, the constraints, which are up to 4500 times more stringent than those from older Casimir and van der Waals force measurements between dielectrics, are obtained from the experiments by means of the atomic force microscope.

8. Constraints from measurements of the lateral Casimir force and from experiment using a microelectromechanical torsional oscillator

In 2002, the new physical phenomenon, the lateral Casimir force, was demonstrated first [24,25] acting between a sinusoidally corrugated gold plate and large sphere. This force acts in a direction tangential to the corrugated surface. The experimental setup was based on the atomic force microscope specially adapted for the measurement of the lateral Casimir force. The measured force oscillates sinusoidally as a function of the phase difference between the two corrugations in

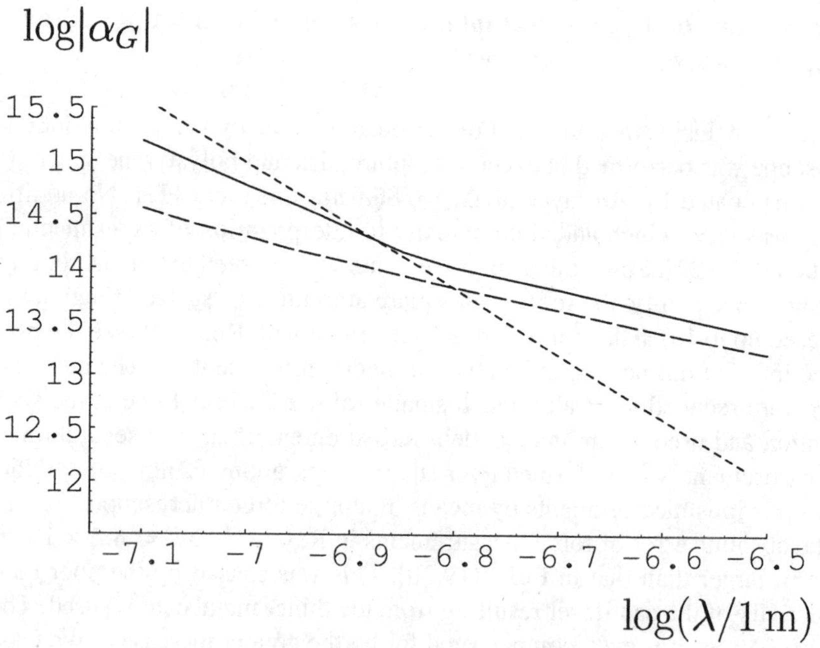

Figure 3. Constraints on the Yukawa-type corrections to Newton's gravitational law from the measurement of the lateral Casimir force between corrugated surfaces (solid curve). For comparison the short-dashed and long-dashed curves reproduce curves 8 and 9 of Fig. 2, respectively, obtained from the meausrements of the normal Casimir force between dielectrics and between gold surfaces.

agreement with theory with an amplitude of 3.2×10^{-13} N at a separation distance 221 nm. So small value of force amplitude measured with a resulting absolute error 0.77×10^{-13} N [25] with a 95% confident probability gives the opportunity to obtain constraints on the respective lateral hypothetical force which may act between corrugated surfaces.

The obtained constraints [25,32] are shown in Fig. 3 as the solid curve. In the same figure, the short-dashed curve indicates constraints obtained from the old Casimir force measurements between dielectrics (curve 8 in Fig. 2), and the long-dashed curve follows from the most precision measurement of the normal Casimir force between gold surfaces [21] (these constraints were already shown by curve 9 in Fig. 2). The constraints obtained by means of the lateral Casimir force measurement are of almost the same strength as the ones known previously in the interaction range 80 nm$< \lambda <$150 nm. However, with the increase of accuracy of the lateral Casimir force measurements more promising constraints are expected.

Recently one more experiment was performed on measuring the normal Casimir force. For this purpose the microelectromechanical torsional oscillator has been used. This permitted to perform measurements of the Casimir force between a sphere and a plate with an absolute error of 0.3 pN and of the Casimir pressure between two parallel plates with an absolute error of about 0.6 mPa [58,59] for separations $0.2 - 1.2\,\mu$m. As a result, the new constraints on the Yukawa-type corrections to Newton's law of gravitation were obtained [59] which are more than one order of magnitude stronger than the previously known ones within a wide interaction range from 56 nm to 330 nm. These constraints are shown by curve 12 in Fig. 2. It is notable that the constraints given by curve 12 almost completely cover the gap between the modern constraints obtained by means of an atomic force microscope and a torsion pendulum. Within this gap the constraints found from the old measurements of the Casimir force between dielectrics (curve 8 in Fig. 2) were the best ones. Now they are changed to the more precise and reliable constraints obtained from the Casimir force measurements between metals by means of a microelectromechanical torsional oscillator.

As is seen from Figs. 2, 3, the present strength of constraints is not sufficient to confirm or to reject the predictions of extra dimensional physics with the compactification scale $R_* < 0.1$ mm (line 13 in Fig. 2). However, Fig. 2 gives the possibility to set constraints on the parameters of light hypothetical particles, moduli, for instance. Such particles are predicted in superstring theories and are characterized by the interaction range from one micrometer to one centimeter [60].

9. Conclusions and discussion

In the above, the modern constraints on the constants of hypothetical interactions are reviewed following from the Casimir force measurements and gravitational experiments. As is evident from Fig. 2, for separations smaller than 10^{-4} m much work is needed to achieve the strength of Yukawa interaction predicted by extra dimensional physics ($\alpha_G \sim 10$). However, one should remember that Casimir force experiments are also sensitive to other non-extra dimensional effects such as exchange by light elementary particles which can lead to the Yukawa-type forces with $\alpha_G \gg 10$ (see Refs. [9,16,60]). Therefore, all experiments which can strengthen the constraints on constants of the Yukawa-type interaction are of immediate interest to both elementary particle physics and gravitation [61].

The following conclusions might be formulated. The idea of hypothetical interactions has gained recognition. New long-range forces additional to the usual gravitation and electromagnetism are predicted by the extra dimensional physics. They may be caused also by the exchange of light elementary particles predicted by the unified gauge theories of fundamental interactions.

The modern measurements of the Casimir force already gave the possibility to strengthen constraints on hypothetical long-range forces up to several thousand times in a wide interaction range from 1 nanometer to 100 micrometers.

Further strengthening of constraints on non-Newtonian gravity and other hypothetical long-range interactions from the Casimir effect is expected in the future. In this way the Casimir force measurements are quite competitive with the modern accelerator and gravitational experiments as a test for predictions of fundamental physical theories.

In near future we may expect to obtain the resolution of the problem are there exist large extra dimensions and the Yukawa-type corrections to Newtonian gravity at small distances.

Acknowledgements

The author is grateful M. Bordag, R. Decca, E. Fischbach, B. Geyer, G. L. Klimchitskaya, D. E. Krause, D. López, U. Mohideen and M. Novello for helpful discussions and collaboration. He thanks V. de Sabbata and the staff of the "Ettore Majorana" Center for Scientific Culture at Erice for kind hospitality. The partial financial support from CNPq (Brazil) is also acknowledged.

10. References

1. Gillies, G.T. (1997) The Newtonian gravitational constant: recent measurements and related studies, *Rep. Prog. Phys.* **60**, 151–225.
2. Kaluza, Th. (1921) On the problem of unity in physics, *Sitzungsber. Preuss. Akad. Wiss. Berlin Math. Phys.* **K1**, 966.
3. Klein, O. (1926) Quantum theory and 5-dimensional theory of relativity, *Z. Phys.* **37**, 895.
4. Arkani-Hamed, N., Dimopoulos, S., and Dvali, G. (1999) Phenomenology, astrophysics, and cosmology of theories with submillimeter dimensions and TeV scale quantum gravity, *Phys. Rev.* **D59**, 086004-1–4.
5. Randall, L. and Sundrum, R. (1999) Large mass hierarchy from a small extra dimension, *Phys. Rev. Lett.* **83**, 3370–3373.
6. Floratos, E.G. and Leontaris, G.K. (1999) Low scale unification, Newton's law and extra dimensions, *Phys. Lett.* **B465**, 95-100.
7. Kehagias, A. and Sfetsos, K. (2000) Deviations from the $1/r^2$ Newton law due to extra dimensions, *Phys. Lett.* **B472**, 39–44.
8. Kim, J. (1987) Light pseudoscalars, particle physics and cosmology, *Phys. Rep.* **150**, 1–177.
9. Fischbach, E. and Talmadge, C.L. (1999) *The Search for Non-Newtonian Gravity*, Springer-Verlag, New York.

10. Smith, G.L., Hoyle, C.D., Gundlach, J.H., Adelberger, E.G., Heckel, B.R., and Swanson, H.E. (2000) Short range tests of the equivalence principle, *Phys. Rev.* **D61**, 022001-1–20.
11. Hoyle, C.D., Schmidt, U., Heckel, B.R., Adelberger, E.G., Gundlach, J.H., Kapner, D.J., and Swanson, H.E. (2001) Submillimeter test of the gravitational inverse-square law: a search for "large" extra dimensions, *Phys. Rev. Lett.* **86**, 1418–1421.
12. Long, J.C., Chan, H.W., Churnside, A.B., Gulbis, E.A., Varney, M.C.M., and Price, J.C. (2003) Upper limits to submillimeter range forces from extra space-time dimensions, *Nature* **421**, 922–925.
13. Chiaverini, J., Smullin, S.J., Geraci, A.A., Weld, D.M., and Kapitulnik, A. (2003) New experimental constraints on non-Newtonian forces below 100 μm, *Phys. Rev. Lett.* **90**, 151101-1–4.
14. Kuz'min, V.A., Tkachev, I.I., and Shaposhnikov, M.E. (1982) Restrictions imposed on light scalar particles by measurements of van der Waals forces, *JETP Lett. (USA)* **36**, 59–62.
15. Mostepanenko, V.M. and Sokolov, I.Yu. (1987) The Casimir effect leads to new restrictions on long-range forces constants, *Phys. Lett.* **A125**, 405–408.
16. Mostepanenko, V.M. and Sokolov, I.Yu. (1993) Hypothetical long-range interactions and restrictions on their parameters from force measurements, *Phys. Rev.* **D47**, 2882–2891.
17. Mostepanenko, V.M. and Trunov, N.N. (1997) *The Casimir Effect and Its Applications*, Clarendon Press, Oxford.
18. Lamoreaux, S.K. (1997) Demonstration of the Casimir force in the 0.6 to 6 μm range, *Phys. Rev. Lett.* **78**, 5–8; (1998) Erratum, **81**, 5475.
19. Mohideen, U. and Roy, A. (1998) Precision measurement of the Casimir force from 0.1 to 0.9 μm, *Phys. Rev. Lett.* **81**, 4549–4552.
20. Roy, A., Lin, C.Y., and Mohideen, U. (1999) Improved precision measurement of the Casimir force, *Phys. Rev.* **D60**, 111101-1–5.
21. Harris, B.W., Chen, F., and Mohideen, U. (2000) Precision measurement of the Casimir force using gold surfaces, *Phys. Rev.* **A62**, 052109-1–5.
22. Ederth, T. (2000) Template-stripped gold surface with 0.4-nm rms roughness suitable for force measurements: Application to the Casimir force in the 20–100 nm range, *Phys. Rev.* **A62**, 062104-1–8.
23. Bressi, G., Carugno, G., Onofrio, R., and Ruoso, G. (2002) Measurement of the Casimir force between parallel metallic surfaces, *Phys. Rev. Lett.* **88**, 041804-1–4.
24. Chen, F., Mohideen, U., Klimchitskaya, G.L., and Mostepanenko, V.M. (2002) Demonstration of the lateral Casimir force, *Phys. Rev. Lett.* **88**, 101801-1–4.
25. Chen, F., Klimchitskaya, G.L., Mohideen, U., and Mostepanenko, V.M. (2002) Experimental and theoretical investigation of the lateral Casimir force between corrugated surfaces, *Phys. Rev.* **A66**, 032113-1–11.

26. Bordag, M., Geyer, B., Klimchitskaya, G.L., and Mostepanenko, V.M. (1998) Constraints for hypothetical interactions from a recent demonstration of the Casimir force and some possible improvements, *Phys. Rev.* **D58**, 075003-1–16.
27. Bordag, M., Geyer, B., Klimchitskaya, G.L., and Mostepanenko, V.M. (1999) Stronger constraints for nanometer scale Yukawa-type hypothetical interactions from the new measurement of the Casimir force, *Phys. Rev.* **D60**, 055004-1–7.
28. Bordag, M., Geyer, B., Klimchitskaya, G.L., and Mostepanenko, V.M. (2000) New constraints for non-Newtonian gravity in nanometer range from the improved precision measurement of the Casimir force, *Phys. Rev.* **D62**, 011701-1–5.
29. Long, J.C., Chan, H.W., and Price, J.C. (1999) Experimental status of gravitational-strength forces in the sub-centimeter range, *Nucl. Phys.* **B539**, 23–34.
30. Mostepanenko, V.M. and Novello, M. (2001) Constraints on non-Newtonian gravity from the Casimir force measurement between two crossed cylinders, *Phys. Rev.* **D63**, 115003-1–5.
31. Fischbach, E., Krause, D.E., Mostepanenko, V.M., and Novello, M. (2001) New constraints on ultrashort-ranged Yukawa interactions from atomic force microscopy, *Phys. Rev.* **D64**, 075010-1–7.
32. Klimchitskaya, G.L. and Mohideen, U. (2002) Constraints on Yukawa-type hypothetical interactions from recent Casimir force measurements, *Int. J. Mod. Phys.* **A17**, 4143–4152.
33. Carugno, G., Fontana, Z., Onofrio, R., and Ruoso, G. (1997) Limits on the existence of scalar interactions in the submillimeter range, *Phys. Rev.* **D55**, 6591–6595.
34. Bordag, M., Mohideen, U., and Mostepanenko, V.M. (2001) New developments in the Casimir effect, *Phys. Rep.* **353**, 1–205.
35. Randall, L., and Sundrum, R. (1999) An alternative to compactification, *Phys. Rev. Lett.* **83**, 4690–4693.
36. Feinberg, G.and Sucher, J. (1979) Is there a strong van der Waals force between hadrons, *Phys. Rev.* **D20**, 1717–1735.
37. Stubbs, C.W., Adelberger, E.G., Raab, F.J., Gundlach, J.H., Heckel, B.R., McMurry, K.D., Swanson, H.E., and Watanabe, R. (1987) Search for an intermediate-range interactions, *Phys. Rev. Lett.* **58**, 1070–1073.
38. Stubbs, C.W., Adelberger, E.G., Heckel, B.R., Rogers, W.F., Swanson, H.E., Watanabe, R., Gundlach, J.H., and Raab, F.J. (1989) Limits on composition-dependent interactions using a laboratory source — is there a 5th force coupled to isospin, *Phys. Rev. Lett.* **62**, 609–612.
39. Heckel, B.R., Adelberger, E.G., Stubbs, C.W., Su, Y., Swanson, H.E., and Smith, G. (1989) Experimental bounds of interactions mediated by ultralow-mass bosons, *Phys. Rev. Lett.* **63**, 2705–2708.

40. Braginskii, V.B. and Panov, V.I. (1972) Verification of equivalence of inertial and gravitational mass, *Sov. Phys. JETP* **34**, 463.
41. Su, Y., Heckel, B.R., Adelberger, E.G., Gundlach, J.H., Harris, M., Smith, G.L., and Swanson, H.E. (1994) New tests of the universality of free fall, *Phys. Rev.* **D50**, 3614–3636.
42. Holding, S.C., Stacey, F.D., and Tuck, G.J. (1986) Gravity in mines — an investigation of Newtonian law, *Phys. Rev.* **D33**, 3487–3497.
43. Stacey, F.D., Tuck, G.J., Moore, G.I., Holding, S.C., Goodwin, B.D., and Zhou, R. (1987) Geophysics and the law of gravity, *Rev. Mod. Phys.* **59**, 157–174.
44. Chen, Y.T., Cook, A.H., and Metherell, A.J.F. (1984) An experimental test of the inverse square law of gravitation at range of 0.1 m, *Proc. R. Soc. London* **A394**, 47–68.
45. Mitrofanov, V.P. and Ponomareva, O.I. (1988) Experimental check of law of gravitation at small distances, *Sov. Phys. JETP* **67**, 1963.
46. Müller, G., Zurn, W., Linder, K., and Rosch, N. (1989) Determination of the gravitational constant by an experiment at a pumped-storage reservoir, *Phys. Rev. Lett.* **63**, 2621–2624.
47. Hoskins, J.K., Newman, R.D., Spero, R., and Schultz, J. (1985) Experimental tests of the gravitational inverse-square law for mass separated from 2 to 105 cm, *Phys. Rev.* **D32**, 3084–3095.
48. Gundlach, J.H., Smith, G.L., Adelberger, E.G., Heckel, B.R., and Swanson, H.E. (1997) Short-range test of the equivalence principle, *Phys. Rev. Lett.* **78**, 2523–2526.
49. Mostepanenko, V.M. and Sokolov, I.Yu (1990) Stronger restrictions on the constants of long-range forces decreasing as r^{-n}, *Phys. Lett.* **A146**, 373–374.
50. Fischbach, E. and Krause, D.E. (1999) Constraints on light pseudoscalars implied by tests of the gravitational inverse-square law, *Phys. Rev. Lett.* **83**, 3593–3596.
51. Milonni, P.W. (1994) *The Quantum Vacuum*, Academic Press, San Diego.
52. Milton, K.A. (2001) *The Casimir Effect*, World Scientific, Singapore.
53. Krause, D.E. and Fischbach, E. (2001) Searching for extra dimensions and new string-inspired forces in the Casimir regime, in C. Lämmerzahl, C.W.F. Everitt, and F.W. Hehl (eds.), *Gyros, Clocks, and Interferometers: Testing Relativistic Gravity in Space*, Springer-Verlag, Berlin, pp. 292–309.
54. Mostepanenko, V.M. (2002) Constraints on forces inspired by extra dimensional physics following from the Casimir effect, *Int. J. Mod. Phys.* **A17**, 722–731.
55. Mostepanenko, V.M. (2002) Experimenatl status of corrections to Newtonian gravitation inspired by extra dimensions, *Int. J. Mod. Phys.* **A17**, 4307–4316.

56. Klimchitskaya, G.L., Roy, A., Mohideen U., and Mostepanenko, V.M. (1999) Complete roughness and conductivity corrections for the recent Casimir force measurement, *Phys. Rev.* **A60**, 3487–3497.
57. Mostepanenko, V.M. and Novello, M. (2001) Weak scale compactification and constraints on non-Newtonian gravity in submillimeter range, in A.A. Bytsenko, A.E. Gonçalves, and B.M. Pimentel (eds.), *Geometric Aspects of Quantum Fields*, World Scientific, Singapore, pp.128–138.
58. Decca, R.S., López, D., Fischbach, E., and Krause, D.E. (2003) Measurement of the Casimir force between dissimilar metals, *Phys. Rev. Lett.* **91**, 050402-1–4.
59. Decca, R.S., Fischbach, E., Klimchitskaya, G.L., Krause, D.E., López, D., and Mostepanenko, V.M. Tests of multydimensional physics and thermal Quantum Field Theory from the Casimir force measurements using a microelectromechanical torsional oscillator, *Phys. Rev. D*, to appear.
60. Dimopoulos, S. and Guidice, G.F. (1996) Macroscopic forces from supersymmetry, *Phys. Lett.* **B379**, 105–114.
61. De Sabbata, V., Melnikov, V.N., and Pronin, P.T. (1992) Theoretical approach to treatment of non-Newtonian forces, *Progress Theor. Phys.* **88**, 623–661.

Searching for Scalar-Tensor Gravity with Lunar Laser Ranging

Kenneth Nordtvedt
Northwest Analysis, 118 Sourdough Ridge, Bozeman MT 59715 USA
kennordtvedt@imt.net

October 15, 2003

Abstract

After more than 30 years of lunar laser ranging (LLR), the Moon's orbit is fit to a general model of relativistic gravity with precision of a few millimeters. This fit provides comprehensive confirmation of the general relativistic equations of motion for multi-body dynamics in the solar system, achieving several key tests of Einstein's tensor theory of gravity, and strongly constraining presence of any additional long range interactions between bodies such as scalar contributions. Earth and Moon are found to fall toward the Sun at equal rates equal to better than a couple parts in 10^{13} precision, confirming both the universal coupling of gravity to matter's stress-energy tensor, and gravity's non-linear coupling to itself. The expected *deSitter* precession (with respect to the distant 'fixed' stars) of the local inertial frame moving with the Earth-Moon system through the Sun's gravity is confirmed to 3.5 parts in 10^3 precision ($\pm$ .07 $mas/year$), and Newton's gravitational coupling parameter G shows no cosmological time variation at the part in 10^{12} *per year* level. Most all of the $1/c^2$ order, post-Newtonian terms in the N-body equations of motion — motional, non-linear, gravito-magnetic, inductive — contribute to the measured details of the lunar orbit; so LLR achieves near-completeness as a gravity experiment and probe of body dynamics. Scalar-tensor gravity, whether metric or non-metric, will generally modify all these mentioned observational features of Earth-Moon dynamics; LLR achieves some of the strongest constraints on such alternatives in physical law. The precision of these measurements, especially those connected with lunar orbit frequencies and rates of change of frequencies, will further improve as upgraded LLR observations continue into the future using latest technologies.

V. de Sabbata et al. (eds.),
The Gravitational Constant: Generalized Gravitational Theories and Experiments, 289–311.
© 2004 *Kluwer Academic Publishers. Printed in the Netherlands.*

1 Outline of Scalar-Tensor Theories.

In scalar-tensor gravitational theories a dynamical tensor field $g_{\mu\nu}(x^\eta)$ is supplemented by a dynamical scalar field $\phi(x^\eta)$ in establishing metrical properties for the space-time arena. These fields may couple to matter and to each other in a special manner which establishes a unique, composite metric field

$$g^*_{\mu\nu} = e^{-2f(\phi)}\, g_{\mu\nu} \tag{1}$$

which molds in a geometrical fashion the nuclear, electromagnetic, and weak physics within rods, clocks, and devices through out the space time arena. Or more generally, the scalar field's coupling to matter could also produce space-time variation of one or more parameters of physical theory such as the fine structure 'constant', $e^2/\hbar c \equiv \alpha\,(x^\eta) = \alpha\,(\phi(x^\eta))$.

The Metric Gravity Case.

An action integral which exemplifies the former *metric gravity* case is

$$\mathbf{A} = \frac{1}{4\kappa}\int [R(g) + 2\,\phi_\mu\phi_\nu g^{\mu\nu} - 2\,\Lambda(\phi)]\,\sqrt{-g}\,d^4x + \int L(M, g^*_{\mu\nu})\,\sqrt{-g^*}\,d^4x \tag{2}$$

with g and $R(g_{\mu\nu})$ being the determinant and the Ricci scalar, respectively, of the tensor field, $\phi_\mu \equiv \partial\phi/\partial x^\mu$. L is the lagrangian for all *matter and fields* other than the scalar field, and the composite metric tensor $g^*_{\mu\nu}$ used therein being previously introduced in Equation (1). The tensor and scalar field equations which result from this action (using the *least action* principle) are

$$R_{\mu\nu} - \frac{1}{2}Rg_{\mu\nu} \;=\; -\,2\kappa\,T_{\mu\nu} - \Lambda(\phi)g_{\mu\nu} - 2\,\phi_\mu\phi_\nu + \phi_\lambda\phi^\lambda g_{\mu\nu} \tag{3}$$

$$g^{\mu\nu}\phi_{,\mu|\nu} + \frac{1}{2}\Lambda'(\phi) \;=\; \kappa\,S \qquad\qquad \kappa = 4\pi\,G_o/c^4 \tag{4}$$

$\phi_{,\mu|\nu}$ indicating the field's second tensor gradients, and matter's energy-momentum tensor given by

$$T_{\mu\nu} \equiv \frac{2}{\sqrt{-g}}\frac{\partial(\sqrt{-g^*}L)}{\partial g^{\mu\nu}} \quad \textit{and} \quad S \equiv \frac{1}{\sqrt{-g}}\frac{\partial(\sqrt{-g^*}L)}{\partial\phi}$$

In this metric theory case the scalar source is proportional to the energy-momentum tensor's trace

$$S = f'\,T_{\mu\nu}g^{\mu\nu} \equiv f'\,T \qquad f' \equiv df(\phi)/d\phi$$

Depending on the detailed form of the scalar field potential function $\Lambda(\phi)$ in the action integral, a scalar source acting like *dark energy* may be present in the cosmological evolution equations for $\phi(t)$ and the expanding universe's scale function $R(t)$ which appears in the background tensor field, and an accelerated expansion rate of the universe recently suggested by observations could result. This issue is explored in other lectures of this school.

Back closer to home in our solar system, for very small scalar potential curvature, $\Lambda'' \cong 0$, Equations (3,4) indicate that both tensor and scalar fields are vehicles for transmitting a long range gravitational force between matter. This field paradigm for interactions is schematically shown in Figure 1. If the scalar and tensor field contributions to the bodies' internal energies can be neglected, and in the static limit of the interaction, the coupling of each object to the tensor field is the total energy content from other forms of matter and fields, while the object's coupling strength to the scalar field is proportional to the scalar trace of these energy sources' total energy tensor. To the degree that the bodies are unstressed by the external world and are in equilibrium internally, these two coupling strengths are equal

$$\int T_{00}\, d^3x \;=\; \int T\, d^3x$$

The two interactions merge in this limit to replicate Newtonian gravity in which each body's gravitational coupling is proportional to its inertial mass, and all bodies are therefore gravitationally accelerated equally by other bodies. An effective Newtonian coupling parameter G results from the sum of the two interactions which when expressed in terms of the proper space-time units which are established by the composite metric field $g^*_{\mu\nu}$ relates to the underlying gravitational parameter G_o as

$$G \;=\; G_o\,\left(1 + (f')^2\right)\, e^{-2f}$$

The strength of Newtonian gravity has essentially become a field quantity; it must therefore generally vary in both space and time as the scalar field so varies

$$\frac{1}{G}\frac{\partial G}{\partial x^\mu} \;=\; 2\,f'\,\left(\frac{f''}{1+(f')^2} - 1\right)\,\frac{\partial \phi}{\partial x^\mu}$$

There are two ways this affects the lunar orbit. If G shows a spatial gradient toward the Sun, Earth and Moon will fall at different rates toward the Sun because of their different fractional levels of gravitational binding energies which "feel" anomalous forces proportional to the gradient of G

$$\delta\vec{a} \;=\; -\,\frac{\partial lnM}{\partial lnG}\,c^2\,\vec{\nabla} lnG \qquad (5)$$

And if G varies in time because of the cosmological time variation in the background value of the scalar field, then the lunar orbit radius and its frequencies of motion will correspondingly change in time. LLR is well-positioned to measure all these effects. These two gradients, the cosmological time variation and the spatial gradient of ϕ (and G) due to a local source of matter, will relate to each other from the common scalar field Equation (4) as solved in the appropriate contexts.

Non-metric Scalar Field Coupling

An example of non-metrical coupling of the scalar field to matter which has interesting experimental consequences results from modification of the electromagnetic field action ($F_{\mu\nu} = \partial_\mu A_\nu - \partial_\nu A_\mu$). Since the standard lagrangian density for the electromagnetic field is invariant under the rescaling $g_{\mu\nu} \rightarrow g^*_{\mu\nu}$ given in Equation (1), explicit inclusion of a scalar field function in this action

$$\mathbf{A_{em}} \;=\; \frac{-1}{16\pi}\int k(\phi) F_{\mu\lambda} F_{\nu\eta}\, g^{\mu\nu} g^{\lambda\eta}\,\sqrt{-g}\, d^4x \;+\; \sum_i e_i \int A_\mu(x^\nu_i) dx^\mu_i$$

has the effect of producing a fine structure parameter $\alpha = e^2/\hbar c$, which varies in space-time as $1/k(\phi)$. The inhomogeneous field equation for the electromagnetic field is now

$$\partial_\mu\left(k(\phi)\,F^{\mu\nu}\right) \;=\; 4\pi\, j^\nu$$

The concept or metaphor of a metric for space-time now becomes limited, as clocks, rulers, and other physical devices whose structure and dynamics depend on $e^2/\hbar c$ will behave in a device-specific, location-dependent manner in a world in which the scalar field ϕ varies throughout the arena, and bodies will accelerate at different rates in response to the spatial gradients of this fundamental parameter. The dynamics and trajectories of light are about the only probes that would directly "see" the space-time geometry defined by the tensor field.

Because various materials are composed of different fractional amounts of electromagnetic binding energy, they will now fall toward an external body at different rates, because the scalar coupling strength of different bodies will no longer be in universal proportion to their mass-energies. These anomalous accelerations will be

$$\delta \vec{a}_i = - \frac{\partial ln\, M_i}{\partial ln\, \alpha}\, c^2\, \vec{\nabla} ln\, \alpha \tag{6}$$

in close analogy with the variable G metric field case in Equation (5). An atom's dominant energy content proportional to α is the Coulomb energy between its nuclear protons and amounts to about

$$\delta \left(M_i c^2 \right)_\alpha \cong .71\, \frac{\alpha}{\alpha_o}\, \frac{Z_i (Z_i - 1)}{A_i^{1/3}} \quad MEV$$

with Z_i and A_i being the element's atomic and baryonic numbers, respectively. This energy represents a fraction of an atom's total mass-energy which varies from about 4.5 10^{-4} for an element like beryllium up to about 3.9 10^{-3} for a high-Z element like lead.

2 The Lunar Laser Ranging Experiment

The precise fit of the lunar laser ranging (LLR) data to theory yields a number of the most exacting tests of Einstein's field theory of gravity, General Relativity, because almost any alternative theory of gravity predicts one change or the other (from that produced by General Relativity) in the Moon's orbit which would be readily detected in the attempt to fit theory to observation. Some of the most interesting and fundamental of the possible variations in theory, and ones which are particularly well-measured by LLR, include 1) a difference in the free fall rate of Earth and Moon toward the Sun due to an anomalous non-linear coupling of the Sun's gravity with the gravitational binding energy within the Earth, 2) a time variation of Newton's gravitational coupling parameter, $G \to G(t)$, related to the expansion rate of the universe, and 3) precession of the local inertial frame (relative to distant inertial frames) because of the Earth-Moon system's motion through the Sun's gravity.

Measurements of the round trip travel times of laser pulses between Earth stations and sites on the lunar surface have been made on a frequent basis ever since the Apollo 11 astronauts placed the first passive laser reflector on the Moon in 1969. By now about 17,000 such range measurements have been made, and are archived and available for use by analysis groups wishing to fit the data to theoretical models for the general relativistic gravitational dynamics of the relevant bodies, the speed of light function in the solar system gravity, plus auxiliary model requirements for the tidal distortions of Earth and Moon, corrections to the light speed through the atmosphere and troposphere, etc. An individual range measurement today has precision of about a centimeter (one-way), but a new generation observing program at Apache Peak, New Mexico plans to improve this range measurement precision down to a millimeter. Because of the large

Scalar-Tensor Gravity

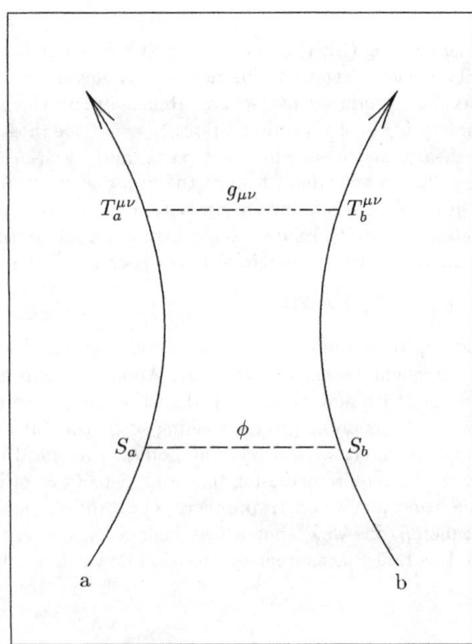

Figure 1: Elements of matter a and b interact by acting as source of either scalar or tensor fields which then *propagate* through space-time to where they then *couple* to the other matter element. The tensor field couples to matter's stress-energy tensor $T^{\mu\nu}$, while the scalar field couples to a scalar density S of the matter. The lowest order linearized interactions are shown; more complicated propagation diagrams would illustrate field interactions between the scalar and tensor fields and non-linear interactions of these fields with themselves.

number of range measurements, some of the key length parameters which describe the lunar orbit are already estimated with precision of a few millimeters (with formal measurement precisions of less than a millimeter), and key lunar motion frequencies are estimated to fractional precisions of a few parts in 10^{12}.

Because both the Earth and Moon have masses which are sufficiently large, the non-gravitational accelerations on these bodies are negligible or adequately modeled at present, and the orbits of these bodies can be fit to the data as single orbital "arcs" extending over more than thirty years in time. The complete model used to fit the many thousands of range measurements contains in excess of a hundred parameters P_m — mostly dynamical initial conditions — which are optimally adjusted from their nominal model values $P_m^{(o)}$ by amounts $\delta P_m = P_m - P_m^{(o)}$ determined in a weighted least squares fit type procedure

$$\textit{Minimize} \qquad \sum_{i,j=1}^{N} W_{ij} \sum_{m,n=1}^{M} [f(m)_i \, \delta P_m \; - \; r_i] \, [f(n)_j \, \delta P_n \; - \; r_j]$$

with the N range measurements being identified by the labels i and j, and the M model parameters being identified by the labels m and n. W_{ij} are the weightings given to each measurement (pair); these weights are usually taken to be diagonal in ij and inversely proportional to the inferred mean squared and independent measurement errors. The *residuals* r_i are the differences between observed range values and those calculated from the nominal model, $r_i = R_{obs}(t_i) - R_{calc}^{(o)}(t_i)$; and the functions $f(m)_i$ are the *partials* associated with each parameter which give the sensitivity of the modeled (calculated) range to change in model parameter values

$$f(m)_i \; = \; \frac{\partial R_{calc}^{(o)}(t_i)}{\partial P_m}$$

evaluated at the times t_i for each of the range measurements.

Although the complete ranging model for fitting LLR data has more than a hundred model parameters, the information needed for testing relativistic gravity theory is concentrated in only a handful of them. The needed orbital parameters are connected with four key oscillatory contributions to the lunar motion, the *eccentric, evective, variational, and parallactic inequality* motions, which are illustrated in Figure 2. The eccentric motion produces an oscillatory range contribution proportional to $\cos(A)$, A being anomalistic (eccentric) phase measured from perigee, and is a natural, undriven perturbation of circular motion. If this eccentric motion perturbed the orbit at the same frequency as the orbital motion, it would simply generate the Keplerian elliptical orbit, but it doesn't; the lunar orbit's major axis revolves in space with 8.9 year period. The Sun's leading order quadrupole and octupole tidal accelerations of the Moon relative to Earth

$$\begin{aligned} \vec{g}\left(\vec{R}+\vec{r}\right)_s \; - \; \vec{g}\left(\vec{R}\right)_s &= \frac{GM}{R^3}\left(3\vec{r}\cdot\hat{R}\hat{R} \, - \, \vec{r}\right) \\ &- \; \frac{3GM}{2R^4}\left((5(\hat{R}\cdot\vec{r})^2 \, - \, r^2)\,\hat{R} \, - \, 2\hat{R}\cdot\vec{r}\vec{r}\right) + \ldots \end{aligned}$$

force the *variation* and *parallactic inequality* perturbations, respectively, with the former range oscillation being proportional to $\cos(2D)$ and the latter proportional to $\cos(D)$, D being synodic phase measured from *new moon*. The *evection* is a hybrid range perturbation with time dependence $\cos(2D-A)$ which accompanies and is proportional to the eccentric motion; it is due to the previously mentioned *variational* modification of the lunar orbit which alters the natural eccentric motion. The eccentric and evective motions which both affect the times of eclipses were discovered by the astronomers of antiquity; the variation and parallactic inequality which do not affect times of eclipses were only found during the era of Newton.

The Earth-Moon range model can be expressed in terms of these primary perturbations illustrated in Figure 2

$$r(t) = L_o - L_{ecc}\cos(A) - L_{evc}\cos(2D - A) - L_{var}\cos(2D) - L_{PI}\cos(D) + ...$$

with the phases advancing as $A = A_o + \dot{A}(t - t_o) + \ddot{A}(t - t_o)^2/2 + ...$ and similarly for synodic phase D. The LLR measurement of L_{PI}, $\dot{A}$, $\dot{D}$, $\ddot{A}$, and $\ddot{D}$ are the foundations for the key gravity theory tests. The amplitude of the parallactic inequality, L_{PI}, is unusually sensitive to any difference in the Sun's acceleration rate of Earth and Moon [2]. The frequency of the eccentric motion, the *anomalistic* frequency $\dot{A}$, when compared to other lunar frequencies determines the precession rate of the Moon's perigee. This rapid 8.9 year precessional rotation is primarily produced by the Sun's tidal acceleration, but there is a leading order relativistic contribution to this precession rate commonly interpreted as an actual rotation of the local inertial frame which accompanies the Earth-Moon system through the Sun's gravity field. Its prediction from General Relativity was made by deSitter in 1917[10]. And from measurement of the time rates of change of the Moon's *anomalistic* and *synodic* frequencies, $\ddot{A}$ and $\ddot{D}$, a rather clean measurement can be made of any time rate of change of Newton's coupling parameter G.

3 Dynamical Equations For Bodies, Light, and Clocks.

LLR comprehensively tests the $1/c^2$ order, gravitational N-body equations of motion which analysis groups computer-integrate to produce orbits for Earth, Moon, and other relevant solar system bodies. The Sun-Earth-Moon system dynamics is symbolically illustrated in Figure 3, with the rest of the solar system bodies sufficiently considered at the Newtonian level of detail. The Earth moves with velocity $\vec{V}$ and acceleration $\vec{A}$ with respect to the solar system center of mass, while the Moon is moving at velocity $\vec{V} + \vec{u}$ and acceleration $\vec{A} + \vec{a}$. (If *preferred frame* effects were to be considered for cases when gravity is not locally Lorentz-invariant, the Sun's cosmic velocity $\vec{W}$ also becomes relevant.) [4] These illustrated motions of Earth and Moon create a variety of post-Newtonian forces by Sun acting on Earth and Moon, and the latter exerting post-Newtonian forces on each other and on themselves (self forces). Among these are *non-linear* gravitational forces for which each mass element of the Earth or Moon experiences forces due to the interactive effect of the Sun's gravity with that of the other mass elements within the same body. The accelerations of individual mass elements within these bodies also induce accelerations on the other mass elements within, thereby altering each body's inertial mass. Acceleration of Earth as a whole induces an acceleration of the Moon, etc. Altogether, these $1/c^2$ order accelerations produce a rich assortment of modifications of the Earth-Moon range which LLR can measure.

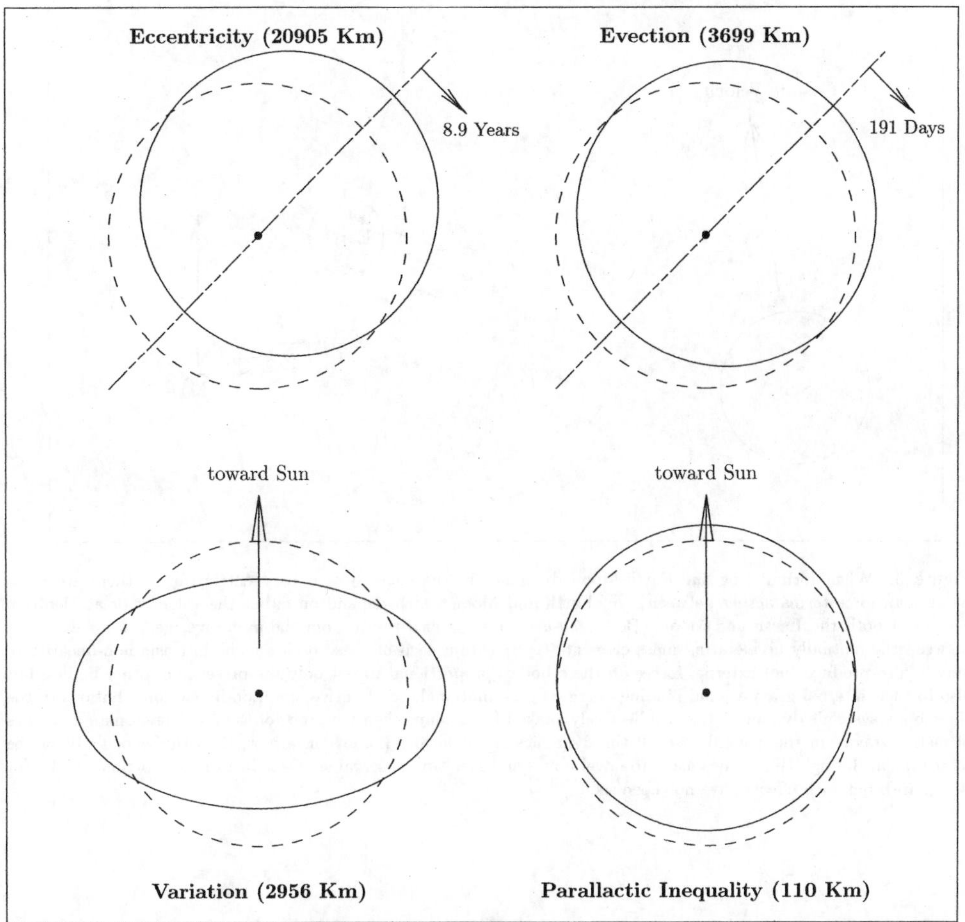

Figure 2: Four lunar orbit perturbations from a nominal circular orbit (dotted) are shown. They produce oscillatory Earth-Moon range terms: the eccentric oscillation $\sim cos(A)$, the variation oscillation $\sim cos(2D)$, the parallactic inequality oscillation $\sim cos(D)$, and the evective oscillation $\sim cos(2D - A)$, with respective amplitudes indicated. Key tests of general relativity are achieved from precise measurements of amplitudes or phase rates of these perturbations. Measurement of the amplitude of the parallactic inequality determines whether Earth and Moon fall toward the Sun at same rate. Measurements of the synodic phase D and anomalistic (eccentric) phase A rates and rate of change of these rates determine the deSitter precession of the lunar orbit and time rate of change of Newton's G.

Velocities and Accelerations of Sun, Earth, and Moon

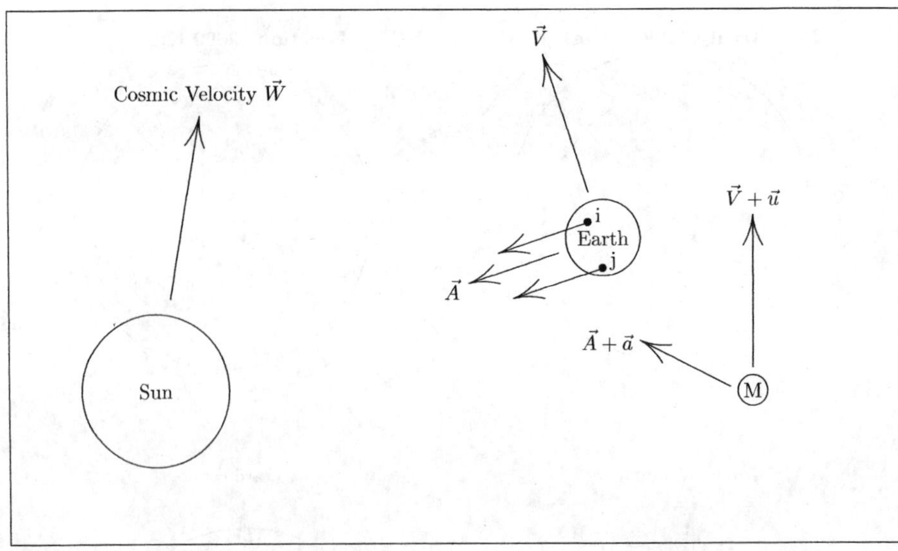

Figure 3: When formulating the Earth-Moon dynamics in the solar system barycentric frame, there are post-Newtonian force terms acting between Sun, Earth and Moon which depend on either the velocity or acceleration vectors of both the Earth and Moon. Body *self-accelerations* also result from the inductive inertial forces acting between the mutually accelerating mass elements (i, j) within each of these bodies. The intrinsic non-linearity of gravity also produces net external forces on these bodies proportional to not only the presence of other bodies, but also to their internal gravitational binding energies. The motional, accelerative, and non-linear contributions to the three body system's dynamics, taken collectively, make LLR a comprehensive probe of the post-Newtonian dynamics of metric gravity in the general case. If the dynamics is not locally Lorentz invariant, then the velocity $\vec{W}$ of the solar system through the cosmos leads to novel forces and resulting observable effects in LLR proportional to $\vec{W}$ (or its square); but such effects have not been seen.

The N-body equation of motion in metric gravity has been formulated in the literature for the completely general case [19]. Since no violations of local Lorentz-invariance or breakdown of conservation laws have been yet found in solar system gravity, and such exotic effects require unusual theoretical foundation to say the least, I here specialize consideration to the fully conservative, locally Lorentz-invariant, lagrangian-based gravitational equation of motion (plus cosmological variation of Newton's G). For N bodies in general motion and configuration, and valid for a broad class of plausible metric theories of gravity, scalar-tensor theories in particular, and stated in a special spatial coordinate system which puts all bodies on an equal footing and preserves the isotropy of light's coordinate speed to this order, the $1/c^2$-order equations of motion for bodies take the form

$$
\begin{aligned}
&A \quad \vec{a}_i = \left(1+\frac{\dot{G}}{G}(t-t_o)\right)\left(\frac{M(G)}{M(I)}\right)_i \vec{g}_i + \qquad (7)\\
&B \quad \frac{1}{c^2}\left\{-\beta^*\sum_{j\neq i}\left(\sum_{k\neq i}\frac{Gm_k}{r_{ik}}+\sum_{k\neq j}\frac{Gm_k}{r_{jk}}\,\vec{g}_{ij}\right)\right.\\
&C \quad +(2\gamma+2)\,\vec{v}_i\times\sum_{j\neq i}\vec{v}_j\times\vec{g}_{ij}\\
&D \quad +(\gamma+1/2)\sum_{j\neq i}\left[v_i^2\,\vec{g}_{ij} - 2\,\vec{v}_i\,(\vec{v}_i-\vec{v}_j)\cdot\vec{g}_{ij}\right]\\
&E \quad +\sum_{j\neq i}\left(((\gamma+1)\,v_j^2 - 3(\vec{v}_j\cdot\hat{r}_{ij})^2/2)\,\vec{g}_{ij} - (2\gamma+1)\,\vec{v}_j\,\vec{v}_j\cdot\vec{g}_{ij}\right)\\
&F \quad \left. +\sum_{j\neq i}\frac{1}{2}\frac{Gm_j}{r_{ij}}\vec{a}_j - \frac{1}{2}v_i^2\,\vec{a}_i + \sum_{j\neq i}\frac{1}{2}\frac{Gm_j}{r_{ij}}\vec{a}_j\cdot\hat{r}_{ij}\,\hat{r}_{ij} - \vec{a}_i\cdot\vec{v}_i\vec{v}_i - (2\gamma+1)\sum_{j\neq i}\frac{Gm_j}{r_{ij}}(\vec{a}_i-\vec{a}_j)\right\}
\end{aligned}
$$

with $\vec{v}_i = d\vec{r}_i/dt$, $\vec{a}_i = d\vec{v}_i/dt$, $r_{ij} = |\vec{r}_i-\vec{r}_j|$, and $i,j,k = 1 \ldots N$. The Newtonian acceleration vectors are indicated

$$
\vec{g}_{ij} = \frac{Gm_j}{r_{ij}^3}\,\vec{r}_{ji} \qquad \textit{and} \qquad \vec{g}_i = \sum_{j\neq i}\vec{g}_{ij} \equiv \vec{\nabla}U(\vec{r},t)
$$

while γ and β^* which appear in lines $B-F$ are two *Eddington* parameters which quantify deviations of metric gravity theory from Einstein's pure tensor theory in which each of these parameters equals one[1]. The several lines of this total equation of motion warrant individual descriptions and brief discussions.

Line A. Newton's law of static, weak field gravity will, in the general case, be modified, itself. When the theory-dependent parameters γ and β^* differ from their general relativistic values, application of the equation of motion relativistic corrections from lines B through F to the internal nature of bodies, specifically their inertia and total coupling strength to external gravity, it is found that the gravitational to inertial mass ratio of celestial bodies differ from each other and from small test bodies in proportion to their gravitational

[1] Eddington used a different but related parameter $\beta = (1+\beta^*)/2$. Detailed investigations of the Equivalence Principle suggests that β^* is a better representation of gravity's non-linearity

binding energies [1]

$$\frac{M(G)}{M(I)} = 1 - (2\beta^* - 1 - \gamma)\,\frac{G}{2Mc^2}\int \frac{\rho(\vec{x})\rho(\vec{y})}{|\vec{x}-\vec{y}|}\,d^3x\,d^3y \;+\; order\;1/c^4 \tag{8}$$

Another way to understand the presence of this ratio is as a consequence of the spatially varying gravitational coupling parameter G which exists in alternative theories

$$G(\vec{r},\,t) \cong G_\infty\left[1 - (2\beta^* - 1 - \gamma)\,\frac{U(\vec{r},\,t)}{c^2}\right]$$

A body for which a significant part of its mass-energy coming from its gravitational binding energy then experiences the additional acceleration

$$\delta\vec{a}_i \cong -\frac{\partial\, lnM_i}{\partial\, lnG}\,c^2\,\vec{\nabla} lnG$$

with the $1/c^2$-order sensitivity of such bodies' masses being

$$\frac{\partial\, M}{\partial\, lnG} = -\frac{G}{2c^2}\int \frac{\rho(\vec{r})\rho(\vec{r}\,')}{|\vec{r}-\vec{r}\,'|}\,d^3r\,d^3r'$$

When the cosmological equations from a metric theory are considered, Newton's coupling parameter G will also generally be found to vary in time in proportion to the Hubble expansion rate of the universe

$$\frac{\dot{G}}{G} \sim (4\beta - 3 - \gamma)\;H$$

The presently most precise way to measure any deviation of β^* from its general relativistic value is presently through measurement of the $M(G)/M(I)$ ratio of Earth using LLR data.

Line B. Gravity couples to itself, thereby producing non-linear gravitational forces inside of, between, and among bodies.

Line C. Just as pairs of moving charges produce magnetic forces between themselves in proportion to the velocities of both charges, pairs of moving masses generate *gravito-magnetic* forces between themselves. This force therefore acts between the mutually moving Earth and Moon and contributes not only to the necessary Lorentz-contraction of the lunar orbit as viewed from the solar system barycenter, but to other features of the orbit as well.

Line D. The motion of a mass alters the gravitational force it experiences from other bodies. These corrections contribute to perihelion precession of orbits, and for the case of the Earth and Moon moving at different velocities through the Sun's gravity, it leads to a *geodetic* precession of the lunar orbit's major axis.

Line E. The motion of gravity sources also alters their gravitational fields; a moving Earth accelerates the Moon slightly differently than an Earth at rest.

The entire package of velocity-dependent accelerations, lines (C-E) for any value of γ, leads to locally Lorentz-invariant gravity which produces no perturbations revealing the motion of the solar system through the cosmos. Any tampering with parts of this package of accelerations

will produce perturbations which violate Lorentz-invariance and can be proportional to one or two powers of the solar system's velocity in the universe. These *preferred frame* effects have not been seen, and LLR provides many of these null tests.

Line F. The inertia of mass elements within a celestial body are altered by their own motion and also by their proximity to other masses and by the inductive effects of the accelerating and proximite mass elements. The first two terms in this line are necessary in order that a body's gravitational self-energy contributes to its total inertial mass in accord with special relativity's prescription $M = E/c^2$.

Except for the modifications on line A, the body equation of motion given by Equation (7) derives from a $1/c^2$ order, Lorentz-invariant, N-body lagrangian

$$\begin{aligned} L &= \sum_i \left(\frac{1}{2} m_i v_i^2 + \frac{1}{8c^2} m_i v_i^4 \right) + \frac{G}{2} \sum_{i \neq j} \frac{m_i m_j}{r_{ij}} \left(1 - \frac{1}{2c^2} (\vec{v}_i \cdot \vec{v}_j + \vec{v}_i \cdot \hat{r}_{ij} \hat{r}_{ij} \cdot \vec{v}_j) \right) \\ &+ (2\gamma + 1) \frac{G}{4c^2} \sum_{i \neq j} \frac{m_i m_j}{r_{ij}} (\vec{v}_i - \vec{v}_j)^2 - \beta^* \frac{G^2}{2c^2} \sum_{i \neq j,\, k} \frac{m_i m_j m_k}{r_{ij} r_{ik}} \end{aligned}$$

with the indices i, j, k each being summed over the N bodies. The two dimensionless strength parameters in this lagrangian create the $1/c^2$-order *theory space* in which lives pure tensor gravity $\gamma = \beta^* = 1$ and the scalar-tensor theory alternatives which are accessible to experimental test and differentiation from LLR and other solar system observations. Each body's equation of motion then results from the principle of least action which requires

$$\frac{d}{dt} \frac{\partial L}{\partial \vec{v}_i} = \frac{\partial L}{\partial \vec{r}_i}$$

In LLR the round trip time of propagation of light between the Earth and Moon trajectories is measured, and these round trip times are recorded by a specific clock moving on a particular trajectory through the solar system. So in the same coordinates used to express the body dynamics, the post-Newtonian $1/c^2$-order modifications to the light speed and to the clock rates are required; they are respectively

$$c(\vec{r}, t) \cong c_\infty \left[1 - (1+\gamma)\, U(\vec{r}, t)/c^2 \right]$$

and

$$d\tau \cong dt \left[1 - v^2/2c^2 - U(\vec{r}, t)/c^2 \right]$$

in which $U(\vec{r}, t)$ is the total Newtonian gravity potential function of all the solar system bodies

$$U(\vec{r}, t) = \sum_j \int \frac{G\rho(\vec{r}\,'(t))_j}{|\vec{r} - \vec{r}\,'(t)|}\, d^3r'$$

Because the Earth moves in the solar system barycentric frame, and it rotates at rate $\vec{\nu}$, two small corrections must be applied to any Earth surface coordinate $\vec{a}$ which might locate a LLR observatory, for instance. There is first the Lorentz-contraction of the extended Earth (or Moon)

$$\delta\vec{a} \cong -\vec{a} \cdot \vec{V}\, \vec{V}/2c^2$$

and because of the non-absolute nature of time simultaneity in special relativity there is a location-dependent displacement of the rotating Earth surface coordinate positions

$$\delta\vec{a} \cong \vec{V} \cdot \vec{a}\, (\vec{\nu} \times \vec{a})/c^2$$

These light and clock equations, and special relativistic body distortion effects play supportive roles necessary for unbiased fitting of the LLR data; but the main science tests emerge from the body equations of motion as given by Equation (7).

4 LLR's Key Science-Related Range Signals.

Assocated with each property of gravitational theory which is tested by LLR, there are specific range signals in the LLR data whose measurements yield the information about theory. Several of these signals are here described.

Violation of Universality of Free-fall.

Because of gravitational self-energies (internal gravitational binding energies) in celestial bodies, their gravitational to inertial mass ratios will generally differ from each other as indicated in line A of Equation (7), and this ratio will have values as given by Equation (8). But there are other ways in which bodies may accelerate at different rates toward other bodies. Within the paradigm that forces between objects are carried or transmitted by fields, an additional long-range interaction in physical law will usually generate a force between bodies i and j with a static limit form

$$\vec{f_i} = \pm K_i \, \vec{\nabla}_i \, \frac{K_j}{R_{ij}} \, e^{-\mu \, R_{ij}}$$

The bodies' coupling strengths K_i and K_j, except in special cases such as metric scalar-tensor gravity in which $K_i \sim M_i$, will be attributes of the bodies which are different than total their total mass-energies; and the dependence on distance of this force will be either inverse square if the field is massless, or it will be Yukawa-like if the underlying field transmitting this force between bodies has mass. A new force not coupling to mass-energy will produce a difference in the Sun's acceleration of Earth and Moon, because the latter two bodies are of different compositions — the Earth has a substantial iron core while the Moon is composed of low-Z mantle-like materials. The fractional difference in acceleration rates of Earth and Moon amounts to

$$\frac{|\delta \vec{a}_{em}|}{|\vec{g}_s|} = \left| \frac{K_s}{GM_S} \left(\frac{K_m}{M_m} - \frac{K_e}{M_e} \right) \right| (1 + \mu R) \, e^{-\mu R}$$

and it will supplement any difference of accelerations resulting from the possible anomalies in the bodies' gravitational to inertial mass ratios due to gravitational self energies. These are basically unrelated possibilities, though somewhere deep in theory they may be linked to common underlying origins. LLR has become a sufficiently precise tool for measuring $|\delta \vec{a}_{em}|$, it now competes favorably with ground-based laboratory measurements which look for composition-dependence of free fall rates; and LLR is the premier probe for measuring a body's $M(G)/M(I)$ ratio as given by Equation (8).

If Earth and Moon fall differently toward the Sun due to either of the mechanisms discussed, then the lunar orbit is polarized along the solar direction[2]. Detailed calculation of this polarization reveals an interesting interactive feedback mechanism which acts between this $\cos(D)$ polarization and the $\cos(2D)$

Newtonian solar tide perturbation of the lunar orbit, the *variation*. The result is an amplification of the synodic perturbation[15, 18]

$$\delta r(t)_{me} = \frac{3}{2}\frac{\Omega}{\omega} R \left(\frac{1 - 4\Omega/3\omega + ...}{1 - 7\Omega/\omega + ...}\right) \delta_{me} \cos D$$
$$\cong 3\ 10^{12}\ \delta_{me} \cos D \quad cm \tag{9}$$

with $\delta_{em} = |(\vec{a}_e - \vec{a}_m)/\vec{g}_s|$, R is distance to the Sun, Ω and ω are the sidereal frequencies of solar and lunar motion, and D is the lunar phase measured from new moon. The exhibited feedback amplification factor for the response of the lunar orbit results as follows: if an initial perturbation polarizes the lunar orbit toward the Sun, the Moon feels a stronger tidal force toward the Sun when it is on half the orbit toward the Sun and a weaker tidal force away from the Sun when the Moon is on the other half of the orbit, producing an average force which further amplifies the perturbation. (This feedback grows stronger for orbits beyond the Moon with an interesting resonant divergence for an orbit with about a two months period.) Computer integration of the complete Equation (7) for the Sun-Earth-Moon system dynamics confirms these analytically determined polarization sensitivities.

The most recent fits of the LLR data find no anomalies in the $\cos(D)$ amplitude to precision of 4 *millimeters*, so from Equation (9) δ_{me} is constrained to be less than $1.3\ 10^{-13}$. Neglecting a composition dependence, and using Equation (8) with an estimate for the fractional gravitational self energy of the Earth being $4.5\ 10^{-10}$, a constraint on a combination of the *Eddington* parameters is achieved

$$|2\beta^* - 1 - \gamma| \equiv |4\beta - 3 - \gamma| \leq 4\ 10^{-4}$$

If metric gravity is a combination of scalar and tensor interaction, the smallness of this constraint is an approximate measure of the scalar interaction strength compared to the dominant tensor interaction. A scenario which could explain today's weakness of the scalar interaction is illustrated in Figure (4). The first derivative (squared) of scalar-tensor metric gravity's single coupling function $f(\phi)$ gives the strength of the scalar interaction, and value of $1 - \gamma$, and in combination with that function's second derivative (curvature) determines the other $1/c^2$-order non-linearity PPN coefficient β^*. Near an extremum of $f(\phi)$ the Eddington parameters are given by

$$1 - \gamma \cong \frac{1}{2}\left(\frac{df(\phi)}{d\phi}\right)^2$$
$$\beta^* - 1 \cong \frac{1-\gamma}{4}\frac{d^2 f(\phi)}{d\phi^2}$$

As the universe expands, the dynamical equations for the background scalar field will drive the scalar toward a minimum of its coupling function to matter, if it exists, at which location γ and β approach their general relativistic values. But although scalar gravity may tend to turn itself off naturally if an "attractor" exists in its coupling function, that process, being dynamical, should not be entirely complete today, and a small remnant of scalar interaction will still exist and be detectable by sufficiently precise testing of relativistic gravity using LLR and other experiments [8, 9].

The LLR result also places a limit on the spatial gradient of the *fine structure constant*, $\alpha = e^2/\hbar c$, in the vicinity of the Sun. If α is a function of a scalar field whose source includes ordinary matter, a spatial gradient of α near bodies should exist, and composition-dependent accelerations of other objects toward this body should occur

$$\delta\vec{a}_i = -\frac{\partial ln M_i}{\partial ln\alpha} c^2 \vec{\nabla} ln\alpha$$

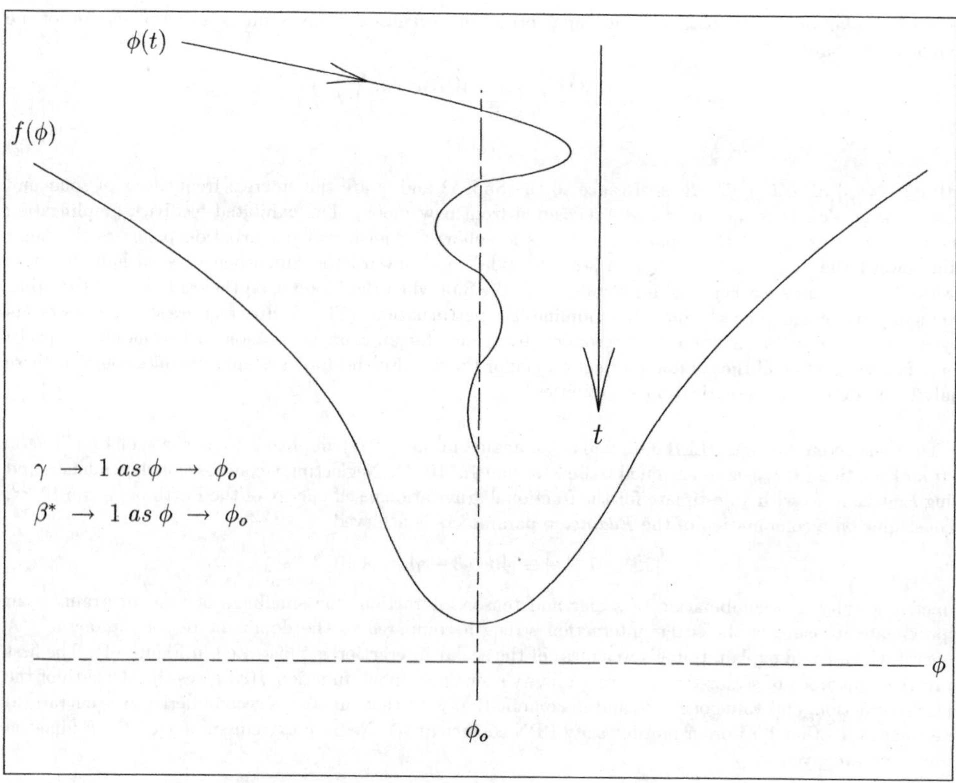

Figure 4: Typical cosmological dynamics of a background scalar field is shown if that field's coupling function $f(\phi)$ has an attracting point ϕ_o. The strength of the scalar interaction's coupling to matter, proportional to the derivative of the coupling function, weakens as the attracting point is approached; so in a scalar-tensor metric theory, for example, the Eddington parameters γ and β^* both approach the pure tensor gravity values of one.

The dominant electromagnetic contribution to the mass-energy of different elements is due to the electrostatic energy among the Z nuclear protons. This energy fractionally varies by an order of magnitude (from a few parts in 10^4 to a few parts in 10^3) as one proceeds through the periodic table from low-Z to high-Z elements. For the Earth with its iron-core and Moon composed almost entirely of mantle-like materials, one can conclude from the LLR constraint on δ_{me} that any gradient of α produced by the Sun is quite small compared to the Sun's gravitational field $\vec{g}_s$

$$\frac{c^2\,|\vec{\nabla}ln\alpha|}{|\vec{g}_s|} \leq 4\,10^{-10}$$

This should be compared with the best constraints on time variation of α, which in units of the Hubble expansion rate are substantially weaker

$$H\frac{\dot{\alpha}}{\alpha} \leq 10^{-5}\ or\ 10^{-6}$$

This suggests that unless there are obscure sources for the scalar field which controls the value of α, e.g., sources which are present in a cosmological context but which do not concentrate within ordinary matter, or other special situations, then today's LLR constraint on the spatial gradient of α is the most restrictive present measure of the constancy of α in space and time.

Cosmic Test of Freefall Universality

Several years ago the LLR data was also tested for an anomalous polarization of the lunar orbit in any fixed cosmic direction within the ecliptic plane; no polarization was found[16]. If Earth and Moon were differentially accelerated by some fixed source in the cosmos and in amount δa, the amplitude of the resulting polarization would be[17]

$$\delta r(t) \;=\; -\frac{3}{2}\,\frac{\delta a}{\omega(\omega-\dot{A})}\;\cos(\omega t-\theta_c)$$

If this test were to be repeated today with inclusion of the range measurements more recently made, a measurement of δa with precision of about $4\,10^{-15}\ cm/s^2$ can result. This converts into a measurement of a cosmological gradient of the fine structure 'constant' to a level below

$$|\vec{\nabla}ln\alpha| \;\sim\; 10^{-32}\ cm^{-1}$$

Does this have implications for cosmological models? It is equivalent to a variation $\delta\alpha/\alpha \sim 10^{-4}$ across the scale size of the entire universe; and on the scale of our galaxy, $L_{gal} \sim 10^{22}\ cm$, this acceleration difference equals amounts to the fraction 10^{-3} of the galactic acceleration of our solar system which is presently assumed to be produced primarily by *dark matter*. LLR therefore provides an upper limit on the source strength of dark matter for generating space-time variation in α — a dark matter constraint which is weaker than the corresponding constraint for ordinary matter by several orders of magnitude, suggesting that dark matter could be the dominant source for a scalar field's background value.

Cosmological Time Evolution of Gravity's Coupling Strength G.

Time evolution of Newton's coupling parameter G results in corresponding evolutions in both the radial size and frequencies of the lunar motion. But slightly different orbital changes occur when a torque (indicated by $\dot{L}$) acts on the orbit

$$\begin{aligned} \frac{\dot{r}}{r} &= -\frac{\dot{G}}{G} + 2\frac{\dot{L}}{L} \\ \frac{\dot{\omega}_n}{\omega_n} &= 2\frac{\dot{G}}{G} - 3\frac{\dot{L}}{L} \end{aligned}$$

and these two mechanisms complicate analysis of the LLR data. During the earlier years of the LLR experiment it was the signal from a linear-in-time orbital radius evolution

$$\delta r(t)_{me} = \left(2\frac{\dot{L}}{L} - \frac{\dot{G}}{G}\right) r\,(t - t_o)$$

which was used to measure $\dot{G}$. But this involved estimating and subtracting the contribution to $\dot{r}$ resulting from the orbital torque exerted on the Moon by the Earth's ocean tidal bulges which because of friction lag in orientation from the direction toward the Moon. The inclination and 18.6 year precession of the lunar orbit's plane result in a modulation of the tidal contribution to $\dot{r}$ which helped separate the two perturbations after accumulation of sufficient years of data. In the last few years the data set produced by LLR has become sufficiently extended in time so that the range signals associated with frequency shifts, which grow quadratically in time, have become dominant in the fit for $\dot{G}$. Recall that the two key lunar phases can be expanded in terms of initial phase, rate, and *acceleration*

$$\begin{aligned} D(t) &= D + \dot{D}\,(t - t_o) + \frac{1}{2}\ddot{D}\,(t - t_o)^2 + ... \\ A(t) &= A + \dot{A}\,(t - t_o) + \frac{1}{2}\ddot{A}\,(t - t_o)^2 + ... \end{aligned}$$

with all the coefficients in these expansions being measurable parameters. While the synodic frequency is simply the difference of lunar sidereal rate and the Sun's sidereal rate around the Earth

$$\dot{D} = \omega - \Omega$$

the Moon's *anomalistic* rate results from the underlying Newtonian plus post-Newtonian equations of motion and has expression in infinite series

$$\dot{A} = \omega - \frac{3}{4}\frac{\Omega^2}{\omega} - \frac{225}{32}\frac{\Omega^3}{\omega^2} - ... - (\gamma + 1/2)\frac{GM}{c^2 R}\Omega + ...$$

which consists of the slowly converging, classical Newtonian series driven by the solar tidal acceleration, plus relativistic modifications of which the dominant geodetic precession contribution is exhibited. From these two expressions the solar sidereal rate can be isolated as a function of the two lunar rates measured by LLR

$$\Omega = \dot{A} - \dot{D} + \frac{3}{4}\frac{(\dot{A} - \dot{D})^2}{\dot{A}} + ...$$

and time derivative of this expression gives

$$\dot{\Omega} = \ddot{A} - \ddot{D} + ...$$

While the lunar phases A and D each suffer accelerations due to any tidal torques acting between Earth and Moon, the combination of lunar phase accelerations which equals solar rate acceleration, $\dot{\Omega}$, is not affected by these tidal torques. This latter is therefore a rather untainted measure of a time variation of G. Noting that the partials for $\ddot{A}$ and $\ddot{D}$ grow in amplitude quadratically with time

$$\begin{aligned} \frac{\partial R_{calc}}{\partial \ddot{A}} &= \frac{1}{2}t^2 \, [L_{ecc} \cos(A) \; - \; L_{evc} \cos(2D - A)] \\ \frac{\partial R_{calc}}{\partial \ddot{D}} &= t^2 \, [L_{var} \cos(2D) \; + \; L_{evc} \cos(2D - A)] \end{aligned}$$

the formal error in measuring $\dot{G}$ consequently decreases as the inverse square of the time span T of LLR observations even before consideration of the statistical $1/\sqrt{N}$ improvements of a longer experiment. For a uniform time distribution of observations one estimates this measurement's variance

$$with \qquad \frac{\dot{G}}{G} = \frac{1}{2}\frac{\dot{\Omega}}{\Omega} \qquad \left(\frac{\delta \dot{G}}{G}\right)_{rms} = \sqrt{\frac{360}{N}} \, \frac{1}{\Omega \, T^2} \, \frac{\sigma}{\sqrt{4L_{var}^2 + 3L_{evc}^2}}$$

with σ being the rms size of individual range measurement errors, and N being the total number of measurements spread over the time T. A recent fit of about 30 years of LLR data yields the excellent measurement constraint [11]

$$\frac{\dot{G}}{G} \cong (0 \, \pm \, 1.1) \, 10^{-12} \, y^{-1}$$

which is a small fraction, about 1/60, of the observed Hubble expansion rate of the universe. With this measurement's formal precision expected to continue to improve quadratically in time, LLR should remain at the cutting edge in supplying the premier measurement of $\dot{G}$.

Geodetic Precession of the Local Inertial Frame.

Because Earth and Moon are moving at different velocities through the Sun's gravitational field, the terms from line D of Equation (7) accelerate the Moon relative to Earth. A particularly interesting part of this differential acceleration is proportionalal to both the velocity of Earth-Moon system relative to Sun, $\vec{V}$, and velocity of Moon relative to Moon, $\vec{u}$, as shown in Figure 3. This produces a coriollis acceleration signaling an effective rotation rate of the local inertial frame which accompanies the Earth-Moon system through the solar gravity

$$\delta \vec{a}_m = 2 \, \vec{\Omega}_{dS} \times \vec{u} \qquad with \; \vec{\Omega}_{dS} = \frac{2\gamma + 1}{2} \, \frac{GM_s}{c^2 R^3} \, \vec{R} \times \vec{V}$$

This *deSitter* rotation has magnitude of about 19.2 mas/y. The effect of this perturbing acceleration on the orbit is primarily an additional rate of perigee precession with respect to distant inertial space. It is observed by comparing measurements of the Moon's anomalistic frequency $\dot{A}$ (rate of eccentric motion) with its synodic frequency $\dot{D}$ (rate of monthly phase). The latter converts into lunar sidereal frequency ω (orbital rate) by adding to $\dot{D}$ the annual rate around the Sun, Ω, which is provided by results from other solar system experiments. Sidereal minus anomalistic frequency of lunar motion includes deSitter's precession rate as a supplement to the Newtonian tidal contributions to perigee precession. These lunar frequencies are measured from range signal perturbations whose size grows linearly in time. From the three dominant oscillatory contributions to Earth-Moon range

$$\delta r_{me} = -L_{ecc} \, \cos(A) \; - \; L_{var} \, \cos(2D) \; - \; L_{evc} \, \cos(2D - A) \; + \; ...$$

with L_{ecc}, L_{var} and L_{evc} being the amplitudes of eccentric motion, the amplitude of solar tidal perturbation called *variation*, and the amplitude of the hybrid *evection* perturbation due to both the solar tidal force and the eccentric motion of the Moon. The least-squares-fit of the LLR data, which yields best estimates for the two key lunar frequencies, will employ the parameter 'partials'

$$\begin{aligned}\frac{\partial}{\partial \dot{A}}(\delta r_{me}) &= t\ [L_{ecc}\ \sin(A)\ -\ L_{evc}\ \sin(2D-A)]\\ \frac{\partial}{\partial \dot{D}}(\delta r_{me}) &= 2t\ [L_{var}\ \sin(2D)\ +\ L_{evc}\ \sin(2D-A)]\end{aligned}$$

Measurement precision of the deSitter precession, being based on frequency measurements, grows especially well with total time of the LLR experiment, not only because of the growing quantity and quality of the accumulated range measurements, but also because of the linear growth in the pertinent partials. A most recent fit of the LLR data confirms presence of the geodetic precession with precision of 0.07 mas/y [11]. Although it does not yield a measurement of PPN γ with precision equal to that obtained by other observations involving light propagation, confirmation of the deSitter precession has been of interest to the astrophysical community.

5 Constraints on a Yukawa Interaction in the Sun-Earth-Moon System

If a supplementary interaction acting between Earth and Moon is of a Yukawa nature rather than inverse square, it contributes to precession of perigee for the near-circular lunar orbit of radius r in amount

$$\frac{\delta(\omega-\dot{A})}{\omega} = \frac{1}{2}\frac{K_i K_j}{GM_i M_j}\,(\mu r)^2\, e^{-\mu r}$$

with ω and $\dot{A}$ being the lunar orbit's sidereal and eccentric (anomalistic) frequencies, respectively. This perturbation occurs for both the metric and non-metric cases for the K_i/M_i ratio. With the Moon's perigee precession rate measured to precision .07 mas/y and showing no anomaly, then for Yukawa ranges in the vicinity of that for maximum sensitivity of lunar perturbation, $\mu r \cong 2$, the strength of the Yukawa force is decisively constrained

$$\frac{|K_e K_m|}{GM_e M_m} \leq 5\ 10^{-12}\left(\frac{4}{(\mu r)^2}\,e^{(2-\mu r)}\right)$$

6 Gravito-magnetism.

Line C of the N-body gravitational equation of motion given by Equation (7) expresses a *gravito-magnetic* post-Newtonian gravitational force proportional to the velocities of both Earth and Moon as present in the center of mass frame of the solar system

$$\delta\vec{a}_i = (2+2\gamma)\sum_{j\neq i}\frac{Gm_j}{c^2 r_{ij}^3}\vec{v}_i\times(\vec{r}_{ij}\times\vec{v}_j)$$

This gravitomagnetic acceleration plays a significant role in producing the shape of the calculated lunar orbit, albeit in conjunction with the rest of the total equation of motion. The precision fit of the LLR data can only occur with inclusion of the gravitomagnetic force and with the strength specified in the equation of motion. With the Earth having velocity $\vec{V}(t)$ and the Moon $\vec{V}(t) + \vec{u}(t)$, perturbations to the Earth-Moon range from the gravitomagnetic acceleration result which are proportional to both V^2 and Vu

$$\begin{aligned}\delta r(t) \quad &\cong \quad \frac{Gm_e}{r^2}\left(-\frac{4}{3\omega^2}\frac{V^2}{c^2}\cos(2D) + \frac{2}{\omega\Omega}\frac{Vu}{c^2}\left(\frac{1-4\Omega/3\omega+....}{1-7\Omega/\omega+....}\right)\cos(D)\right)\\ &\cong \quad -\,530\,\cos(2D) \;+\; 525\,\cos(D) \quad cm\end{aligned}$$

And when account is taken of the very small motion of the Sun relative to the solar system barycenter (due primarily to acceleration by Jupiter), a further perturbation of the lunar orbit is produced by the gravitomagnetic interaction acting slightly differently between the Sun and Earth, and Sun and Moon

$$\delta r(t) \cong 2\,r\,\frac{M_j}{M_s}\frac{u\,V_j}{c^2}\frac{\omega}{\omega - \dot{A} - \Omega_j}\cos(\omega t - L_j) \cong 5.5\,\cos(\omega t - L_j) \quad cm$$

$\omega - \dot{A}$ being the Moon's perigee precession rate (8.9 *year* period), Ω_j being Jupiter's sidereal rate (11.9 *year* period) and sidereal orbital longitude of Jupiter L_j indicated.

The amplitudes of the lunar motion at both the monthly and semi-monthly periods are determined to better than half a centimeter precision in the total orbital fit to the LLR data. It would be impossible to understand this fit of the LLR data without the participation of the gravitomagnetic interaction in the underlying model, and with strength very close to that provided by general relativity, $\gamma = 1$. As in electromagnetic theory, the velocity-dependent force terms on lines $C - E$ of Equation (7) can individually be changed by formulating the dynamics in different frames of reference, but the very ability to reformulate the equations of motion in different frames without introducing new frame-dependent terms depends on the local Lorentz invariance (LLI) of gravity. It is the entire package of velocity-dependent, post-Newtonian terms which includes the gravitomagnetic terms, lines $C - E$ of Equation (7), that produces the LLI; the *Eddington* parameter γ represents the only freedom in the structure of this LLI package. Our confidence in the exhibited structure of this total collection of velocity-dependent terms is established in proportion to the precision with which the various *preferred frame*, LLI-violating effects in the solar system proportional to W^2, WV, and Wu have been found to be absent [5]. LLR has been one of the main contributors in establishing this invariance of gravity through the null measurements of several W-dependent effects [6, 4, 21, 7].

7 Inductive Inertial Forces.

Among the forces shown on line F of Equation (7) are *inductive forces* in which the acceleration of one mass element induces an acceleration of another proximite mass element (e.g., i and j in Figure 3)

$$\delta\vec{a}_i = \sum_{j\neq i}\frac{Gm_j}{2c^2 r_{ij}}\left((4\gamma+3)\vec{a}_j + \vec{a}_j\cdot\hat{r}_{ij}\hat{r}_{ij}\right)$$

These inductive accelerations play a necessary part in altering the inertial masses of the Earth and Moon so as to account for their internal gravitational binding energies in the shorthand $E = mc^2$ relationship of special relativity. Either the absence or an anomalous strength of these inductive forces would translate directly into differences between the acceleration rates of these whole bodies toward the Sun. A polarization of the Moon's orbit in the solar direction, as previously discussed, would result. The inductive forces by themselves acting between the mass elements of Earth, for example, would lead to an anomalous polarization of the lunar orbit of very large magnitude

$$\delta r(t) \cong 130 \cos(D) \quad meters$$

Only when these inductive forces are combined with the other post-Newtonian inertial forces shown on line F of Equation (7) does the total inertial self force of a body become

$$\begin{aligned} \delta\vec{f} = & -\frac{1}{c^2}\left(\frac{1}{2}\sum_i m_i v_i^2 - \frac{G}{2}\sum_{i,j}\frac{m_i m_j}{r_{ij}}\right)\vec{a} \\ & -\frac{1}{c^2}\left[\sum_i m_i \vec{v}_i\, \vec{v}_i - \frac{G}{2}\sum_{i,j}\frac{m_i m_j}{r_{ij}^3}\vec{r}_{ij}\,\vec{r}_{ij}\right]\cdot\vec{a} \end{aligned}$$

with $\vec{a}$ being the common acceleration of the body. The first line of this self force is the expected inertial force due to the internal kinetic energy and gravitational binding energy within the body. The second line represents contributions to the body's internal *virial* which, when totaled over all internal force fields, vanishes for a body in internal equilibrium and experiencing negligible external tidal-like forces (This virial vanishes as it stands for a gas model of the celestial body). These body self forces are an integral part of the determination of the total gravitational to inertial mass ratio of bodies discussed previously, and in general relativity they are compensated by equal contributions of internal energies to a body's gravitational mass. The acceleration-dependent terms in the equation of motion also play other smaller roles in perturbing the final lunar orbit. They were explicitly discussed here in order to show the large size of such inductive force contributions which must necessarily be taken into account in the fit of a relativistic theory of gravity to the LLR data.

This work has been supported by the National Aeronautics and Space Administration through contract NASW-97008 and NASW-98006.

References

[1] Nordtvedt K 1968 *Phys. Rev.* **169** 1017

[2] Nordtvedt K 1968 *Phys. Rev.* **170** 1186

[3] deSitter W 1916 *Mon. Not. R. Astron. Soc.* **77** 155

[4] Nordtvedt K 1973 *Phys. Rev.* **D 7(8)** 2347

[5] Nordtvedt, K, 1987. Probing gravity to the second post-Newtonian order and to one part in 10^7 using the spin axis of the Sun. *Astrophys. J.* **320**, 871-874

[6] Nordtvedt, K and Will, C, 1972. Conservation laws and preferred frames in relativistic gravity. II. Experimental evidence to rule out preferred-frame theories of gravity. *Astrophys. J.* **177**, 775-792.

[7] Nordtvedt K (1996) The isotropy of gravity from lunar laser ranging *Class. Quantum Grav.* **13** 1309-1316

[8] Damour T and Nordtvedt K (1993) General relativity as a cosmological attractor of tensor-scalar theories. *Phys. Rev. Lett.* **70**, 2217-2219

[9] Damour T and Nordtvedt K (1993) Tensor-scalar cosmological models and their relaxation toward general relativity *Phys. Rev.* **D48** 3436

[10] deSitter W 1916 *Mon. Not. R. Astron. Soc.* **77** 155-184

[11] Williams J G, Boggs D H, Dickey J O and Folkner W M (2001)Lunar laser tests of gravitational physics, in *Proc. of the Marcel Grossmann Meeting IX, 2-8 July 2000, Rome, Italy (World Scientific)*, eds. Ruffini R,

[12] Shapiro I I *et al* 1988 *Phys. Rev. Lett.* **61** 2643

[13] Bertotti B, Ciufolini I and Bender P L 1987 *Phys. Rev. Lett.* **58** 1062

[14] Williams J G, Newhall X X and Dickey J O 1996 *Phys. Rev.* **D 53** 6730

[15] Nordtvedt K 1995 *Icarus* **114** 51

[16] Nordtvedt K (1994) Cosmic Acceleration of the Earth and Moon by dark matter *Astrophys. J* **437(1)** 529-531

[17] Nordtvedt K L, Müller J and Soffel M (1995) Cosmic acceleration of the Earth and Moon by dark matter. *Astron. Astrophys.* **293**, L73-L74.

[18] Damour T and Vokrouhlicky D 1996 *Phys. Rev.* **D 53** 4177

[19] Will C and Nordtvedt K 1972 *Astrophys. J.* **177** 757

[20] Nordtvedt K 1996 *Class. Quant. Grav.* **13(6)** 1317

[21] Müller J, Nordtvedt K, and Vokrouhlicky D 1996 *Phys. Rev.* **D54** 5927

[22] Müller J and Nordtvedt K 1998 *Phys. Rev.* **D58(6)** 2001

IS A HYPOTHETICAL LONG RANGE SPIN INTERACTION OBSERVABLE WITH A LABORATORY DETECTOR?

R.C. RITTER and G.T. GILLIES
Department of Physics
University of Virginia
Charlottesville, VA 22904-4714, U.S.A.
E-mail: rcr8r@virginia.edu, gtg@virginia.edu

ABSTRACT

A quadrupole arrangement on a torsion pendulum of two polarized masses each having intrinsic spin ~ 10^{21}h acts as a detector of hypothetical anomalous spin interactions. Unlike earlier experiments in our laboratories using local source masses acting on these detectors to investigate anomalous spin interactions, e.g. existence of the axion, this experiment seeks a possible anomalous spin sensing of matter on a scale as large as our galaxy. Rotation of the Earth provides a scan of the sky by the detector, and pendulum position variation is time-correlated with a predicted daily pattern. Our original motivation was the possibility of detecting an exotic dark matter cloud roughly centered in our galaxy, although other sources are conceivable. After eight years of essentially continuous operation, a long-term pattern has developed in the correlations of pendulum torque with predicted pattern. This is analyzed as an unspecified signal, and is referenced to a sidereal frame to separate it from from local noise and systematic causes. The expected high noise-to-signal level requires unusual analytical methods. A 1997 report at this School discussed results of the first two years of this experiment, which could not anticipate the 8-year pattern we now observe.

1. Introduction and Motivation

Intrinsic spin has occupied the interest of general relativists since the acceptance of Einstein's principle of equivalence. A significant step in understanding the role of spin in fundamental interactions was taken when Leitner and Okubo [1] discussed it in the context of P- and T- symmetry breaking. Others, e.g. [2] extended this to the larger

V. de Sabbata et al. (eds.),
The Gravitational Constant: Generalized Gravitational Theories and Experiments, 313–329.
© 2004 *Kluwer Academic Publishers. Printed in the Netherlands.*

perspective of CPT symmetry considerations. Lammerzahl [3] has reviewed the status of these and of other aspects of spin interactions extensively.

In the Leitner-Okubo manner the symmetry-breaking terms in the CPT potential are written

$$U_{CPT}(\mathbf{r}) = U_o(\mathbf{r}) \{1 + A_1 \boldsymbol{\sigma} \cdot \mathbf{r} + A_2 \boldsymbol{\sigma} \cdot (\mathbf{v}/c) + A_3 \mathbf{r} \cdot [(\mathbf{v}/c) \times \boldsymbol{\sigma}]\} \quad (1)$$

Here the terms A_1, A_2 and A_3 deal with (P- and T-), (P- and C-), and (C- and T-) symmetry breaking, respectively.

A more specific look at spin interactions involving the coupling of fermions by the exchange of light pseudoscalar bosons between fermions was given by Moody and Wilczek [4]. This work has provided significant guidance for experimenters. In practical experiments these (single exchange) couplings involve two types of vertices: the scalar and pseudoscalar, where g_s and g_p are the respective coupling constants.

1.1 Background: spin potentials and macroscopic forces

1.1.1 Possible experimental categories involving single exchange of these bosons

Moody and Wilczek labeled the vertices in several ways corresponding to the experimental components of an interaction: the scalar vertex is also a “mass” or a “monopole” vertex, and the pseudoscalar vertex is a “spin” or a “dipole” one. Then, exchange between unpolarized masses is a mass-mass or monopole-monopole interaction, exchange between a polarized and an unpolarized mass is a spin-mass or dipole-monopole interaction and, lastly, that between two polarized masses is a spin-spin or dipole-dipole interaction.

A listing [5] of the status of recent experimental limits on the second two of these types of interactions was presented in 1997 at the XVth Course of this School, “Spin in Gravity: Is It Possible to Give an Experimental Basis to Torsion?” [6]. It has been noted by Fischbach [7] that the unnatural parity at each pseudoscalar vertex results in p-wave coupling of the exchange particle, with the repulsion weakening the interaction relative to that of a scalar vertex by a factor of $\sim 10^{-12}$ or 10^{-13}, as is apparent in the experiments reviewed in [5]. It is for this reason that comparable experiments without spin, such as those seeking a composition-dependent fifth force (see, e.g. [8]) are $\sim 10^{25}$ times more sensitive than the spin-spin experiments [5]. Some numerical relations put the experimental conditions into context.

It is useful to consider the macroscopic scale of interactions in relation to the scalar gravitational coupling strength of two nucleons:

$$G_s^2 \sim (M_n/M_{Pl})^2 \sim 10^{-38} \quad (2)$$

For two 1 gram masses the centers of which are separated by 2 cm, the strength of the gravitational force between them is

$$F \sim (6 \times 10^{23})^2 (M_n/M_{Pl})^2 \, hc/1\text{ cm} = 6.7 \times 10^{-8} \text{ dyne} \quad (3)$$

Forces for the three possible first-order fermion couplings can then be considered relative to each other: (a) the mass-mass interaction, with both masses unpolarized (eg., composition dependence or 5th force experiments) yields ~ 6.7 x 10^{-8} dyne, (b) the spin-mass interaction, with one mass polarized and the other unpolarized (eg., the Axion experiment) yields ~ 10^{-20} dyne, and (c) the spin-spin interaction, with both masses polarized yields ~10^{-32} – 10^{-33} dyne. Clearly, all of these experiments require a high degree of shielding against electromagnetic interactions and other ambient disturbances.

1.1.2 Laboratory measurements: test for non-magnetic spin coupling between masses

A major problem addressed in these experiments is how to measure these weak couplings in the presence of magnetic moments of the masses, and in the presence of non-zero local magnetic fields. One solution to this problem was proposed by Wei-Tou Ni, who helped initiate its use in our laboratory [9]. It involves what are termed compensated masses, which can act like macroscopic electrons having a net intrinsic spin of ~ 10^{21} to 10^{23} h/2. The material used in our case is one of four phases of the DyFe compound. The particular stoichiometry for it is Dy_6Fe_{23}. Slightly below room temperature the magnetic moment of the aligned electron intrinsic spins in this material is just compensated by that of the nucleons and electron orbital spin in it.

To take advantage of this, an unpolarized cylinder of sintered Dy_6Fe_{23} is put in a 210 kilogauss field which essentially polarizes the ~2000 electrons in each molecule of this compound in one direction, and the remaining (e.g. orbital and nuclear) spins in the other. The single crystals of such materials have been studied and are well described by two Brillouin functions, having opposite polarities and shifted peaks and slopes in their temperature dependence, as well as scaling differences appropriate to the different spin mechanisms. Thus, a crossing is achieved, and its slope provides a measure of the degree of polarization of the electrons. In our tests, operation at the exact compensating temperature showed a 74% reduction in strength relative to the well-studied pure single crystals of the material [9]. The exact compensation was found at –24 celsius. For experimental convenience we used light shielding to operate these spin masses at room temperature instead. This combination provided a factor of 10^{-16} reduction in the magnetic interaction between our masses relative to the same number of polarized electrons not compensated or shielded, but only when the masses were maintained in a one milligauss environment.

There is no overall magnetic shielding of this experiment. However, large Helmhotz coils are set up, which reduce the ambient magnetic background at the detector chamber that houses the masses to ~ 1 milligauss, except for occasional magnetic storms or unusual local activity during which temporary changes up to about 3 milligauss have been seen using a flux-gate magnetometer to monitor at the chamber. Figure 1 illustrates one of the earlier pendulum experiments carried with these masses, the spin-spin experiment at the University of Virginia [9].

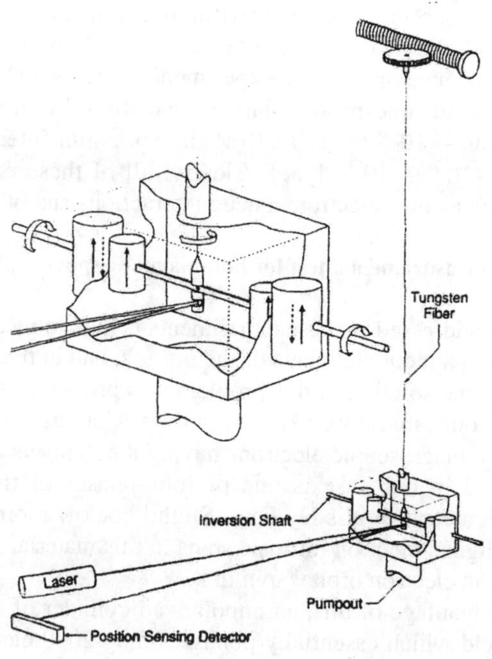

Figure 1. Torsion pendulum used in the **σ•σ** experiment at the University of Virginia. An evacuated chamber contains small spin masses with ~ 10^{21}h aligned electron spins, while the larger external masses have ~5 x 10^{22}h aligned spins. The external masses were inverted every 3 hours.

The most interesting of these types of experiments have been of the spin-mass category, seeking axions or other possible P- and T- violating exchanges, e.g. the massless arions of An'selm [10]. The status of the limits from a variety of these experiments is included in our 1997 review [6]. In 1998 an experiment was carried out at the University of Washington [11] that improved the limit significantly but did not traverse the restrictive attribute mentioned above for the relative strength of spin experiments. Figure 2 shows their alternative way of achieving spin-polarized electrons in their detecting torsion pendulum. The results from that study are shown in Figure 3, along with those from our laboratory [12] and an ion-trap experiment at Amherst University [13]. Two sections of the Washington experimental limit are shown, associated with the use of different source masses: laboratory and terrestrial.

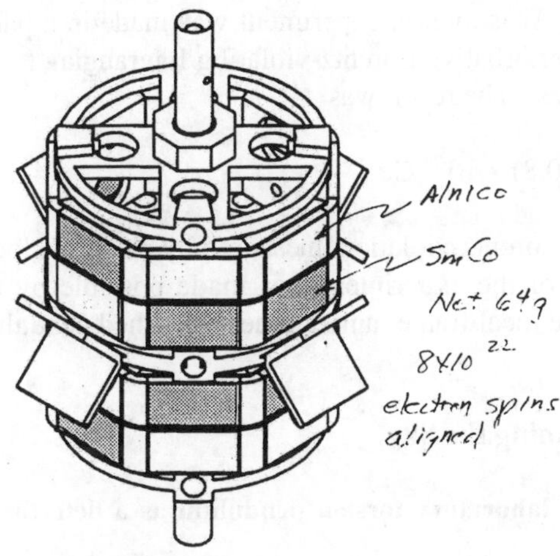

Figure 2. The spin mass on the rotating U.Washington ("Eöt-Wash") pendulum. This arrangement contained $\sim 8 \times 10^{22}$ h aligned electron spins.

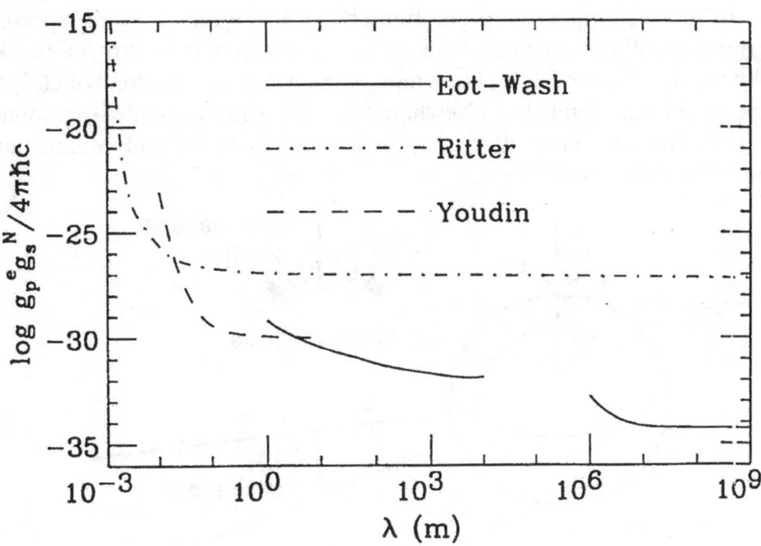

Figure 3. Limits on the dimensionless coupling of the spin-mass interaction from three experiments of relatively different types (as discussed in the text). Dual limits are shown for the Eöt-Wash experiment as a result of the use of different source masses.

An interesting interpretation of the Washington experiment was made in a paper by Bluhm and Kostelecky [14] , using a perturbative Lorentz-violating Lagrangian to put a new limit on Lorentz violating interactions. The result was

$$b_z \sim (1.4 +/- 0.8) \times 10^{-28} \text{ GeV} \quad (4)$$

This surpasses the best previous bound on Lorentz-violation, measured by electron clocks to be 10^{-27} GeV. Such an interpretation of the experiment was made possible by the rotation of the Washington pendulum in the local frame, unlike other published pendulum spin experiments.

2. Quadrupole Spin Pendulum Scanning the Sky

2.1 The concept and brief history of a laboratory torsion pendulum as a detector of astronomical spins

Following the completion of the spin-mass experiment at Virginia [9], which set a limit on the axion existence in part of the mass range of the Turner Window [15], a radical revision was made in the use of this experimental arrangement. Instead of using local source masses to act on the pendulum's detector masses (then unpolarized) for the purpose of studying an interaction, the spin masses were put on the pendulum, and the rotation of the Earth swept the sky for interactions from remote sources. The spin masses are arranged in the quadrupole configuration shown in Figure 4. This represents an unnatural arrangement for intrinsic spin and it is not found in any of the known fundamental particles. Figure 5 is a photograph of the apparatus. Reports of early results with this experiment were published elsewhere [16 - 19]. Subsequent observations have caused us to modify our view of the experimental analysis, and of the potential interpretation of the experiment.

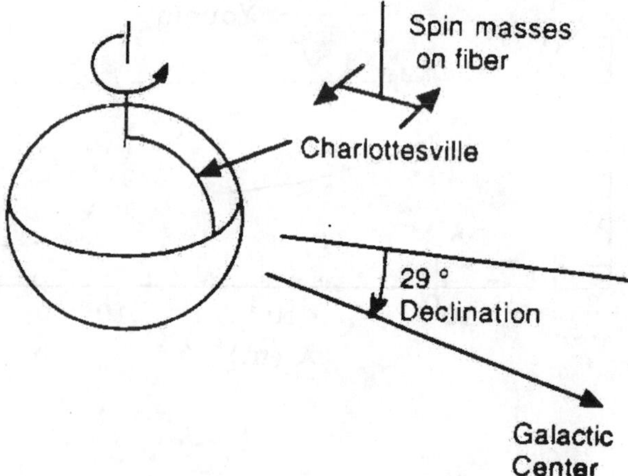

Figure 4. Quadrupole spin mass detector, rotating with the Earth.

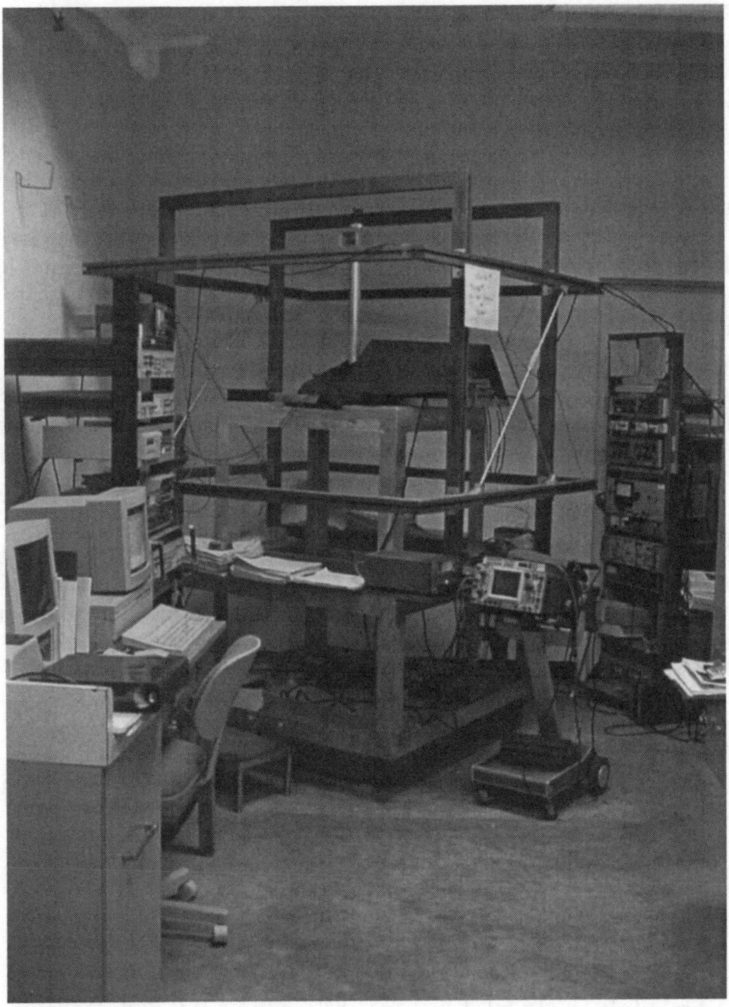

Figure 5.

Photograph of the quadrupole experiment including the external Helmholtz coils. The masses and laser position-sensing beam are shown separately in close-up view.

2.2 The University of Virginia Method

Two previous experiments, by Phillips [20] and Ni [21], had used dipole spin masses to similarly scan the sky, each with roughly a week's running of the experiment. In their studies, spectral analyses were used to search for a spatial anisotropy, and none was observed. In a radically different manner, the present experiment aims at pattern matching with a predicted $\boldsymbol{\sigma} \bullet \mathbf{r}$ interaction (the P- and T- symmetry violating form) for each mass with a hypothetical dark matter halo centered on our galactic center. The predicted pattern based on this hypothesis is shown for a five-day run in Figure 6, along with a typical series of data, correlated in time.

In our data analysis, pendulum torque, i.e., angular position departure from equilibrium in the local frame, provides the raw signal. This signal, five 87-point days (and the corresponding time, temperature and background magnetic field) is subsequently conditioned by first order correction for fiber drift, temperature and rate of change of temperature. Five days was chosen as a suitable compromise between diurnal averaging benefits and occasional interfering noise and disruptions in the laboratory. Given the small cross section and the hypothetical source distances, a very weak response was expected (and is observed). In our experiment each five-day run, when appropriately compared with the predicted pattern, becomes a single data point, to be seen below. Thus our very long term observations constitute a search for an effect that might manifest itself over a period of years.

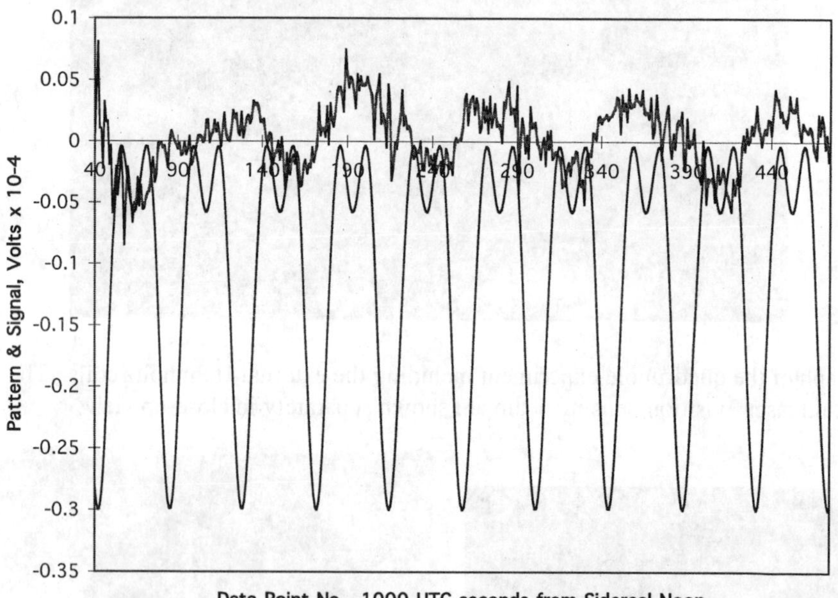

Figure 6. Pattern predicted from the $\boldsymbol{\sigma} \bullet \mathbf{r}$ interaction of both masses with a spherical source centered in our galaxy, and one 5-day raw data set from the torsion pendulum.

As Phillips and Ni would likely also have expected, such a noise-prone experiment presents significant difficulty in finding a desired signal. To cope with this, we have developed techniques to take advantage of the properties of the quadrupole arrangement [17 - 18] in addition to the averaging effects of compressing each five days of run time into a single data point. The experiment is run by computer, thus facilitating nearly continuous runs for many years. The starting times of each run are converted to sidereal time, and all of the analyses are carried out in sidereal time.

Because of the long (now 9 years) sequence of observations, the pendulum position, temperature, and magnetic field background data are taken only each 1000 s in UTC time, and then converted to sidereal time. Even so, the management of such long data trains is limiting. There are no direct or compelling physical reasons for sampling data more often because our pendulum time constant is 530 s. In addition we use an analog filter with a pass band of 1000 s to avoid aliasing effects in our effectively 1000 s digital filter. But one negative consequence of these temporally sparser data trains is the limited resolution in the spectral analyses which are used for ancillary tests.

2.2.1 Signal Charaterization

In a daily scan, the experiment has a 12-hour (sidereal time) repetition period. The summation of two data runs, termed “S+”, separated by 12 sidereal hours would constitute a summation of any true signal, as well as noise. The subtraction of data runs, or “S-“, also separated by 12 hours, would yield a signal train that could not contain a quadrupole signal, and thus would constitute a pure background. To avoid overlap and summations of temporally close noise in S+, the signal train summations and subtractions were actually made 183 sidereal days apart, i.e., over one-half sidereal year plus a 12 sidereal hour shift for each day. We have called this operation on the data a “sidereal filter.” [More appropriately it might be called a “quadrupole filter.”]

The signal conditioning prior to operating this filter is not highly effective, but does remove a significant part of unusual temperature changes. For the temperature corrections, a calibrated thermister reads the temperature on the thick-walled aluminum box that houses the masses and is thermally contiguous with the 90 cm long tube surrounding the fiber. This system was evacuated during experiments with local source magnets. Now it is ported to atmosphere to provide some damping on the pendulum.

Typically the conditioning corrections reduced the rms noise level by 30 to 50 percent. A large number of other corrections were attempted, including nonlinear thermal equation fitting and correction for the ambient magnetic field variation. None of these provided a statistically significant improvement, and are not used in the data presented here.

During the period of these runs, the pendulum fiber had to be replaced four times due to accidents of several kinds. The “nicotine” fiber (primarily of nickel, cobalt and tin) is of material used in Swiss mechanical clock making. It is loaded to about one fourth of its breaking strength, and is not annealed. It has been found to have much less drift than is commonly reported for tungsten fibers, even when they are annealed.

To test the effectiveness of the sidereal filter, a spectral analysis was performed on approximately 175 days’ averaged raw data. Figure 7 exhibits the spectral density

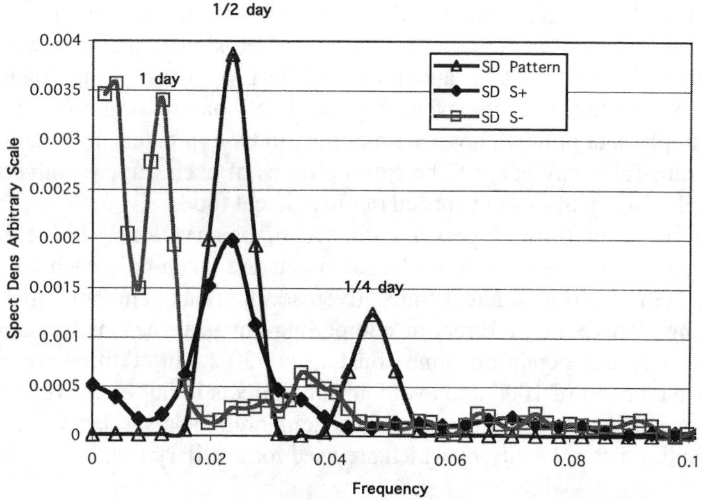

Figure 7. Spectral density from averaged runs of 175 sidereal days, along with that of the predicted pattern.

of runs over this period, along with that of the pattern of Figure 6. It is seen that the S+ signal correlates well with the 1/2-day peak of the predicted pattern, and is nearly absent at the one-day position. Conversely, the S- curve has a strong peak at one day, but has little amplitude at 1/2 day. The 1/4-day peak in the predicted-pattern spectrum is not considered significant, as its appearance only depends on an arbitrary cutoff in the halo integration. In summary, the spectra are consistent with a significant separation between the S+ quadrupole-plus-noise data components and S- components that cannot contain a quadrupole signal. They also indicate that the ambient diurnal effects that change from night to day over one-half year are providing significant filtering.

2.2.2 Six Year Behavior of the Signal

Records of the torsion-pattern statistical correlation value are retained for S+ and for S- data trains, analyzed and plotted as individual points corresponding to each continuous 5-day segment. It is found that a direct plot of the point sequences is a noisy figure. To see the tendencies more clearly, boxcar smoothing was employed on these plots, so that only very long-term effects could remain. Figures 8A and 8B, plotted on the same scale but separated for clarity, show the smoothed S+ and the background S- data, respectively. Smoothing of these data was over each 40 sequential 5-day points, thus retaining effects with a pass band of roughly 200 sidereal days, or about one-half year. In these plots, and others herein, we show the analysis beginning with runs after the first 900 sidereal days. At that time, the spin masses were inverted in the pendulum, in the hope of seeing an inversion in the signal as a test of reality of the effect [18]. However, subsequent longer runs have shown us that the original 2.5 years running was too brief to establish a significant observation of that change.

A broad hump is shown for the S+ plot, while that for S- presents an approximately level background around 0. The vertical scale is the Pearson correlation value in each case. Since the pattern itself is un-scaled, the physical angular shift of the pendulum, along with its properties, is used to calculate the pendulum torque and hence the acceleration of each mass, for a sensitivity reference. From this any observed signal can be compared with the accepted gravitational acceleration a_g of any mass towards the center of the galaxy: 1.8 x 10^{-8} cm/s^2. From the torsional sensitivity of the fiber, 0.09 dyne/radian, the sensor-mass weights and their separations, 6 cm, an observed correlation of unity would be equal to 5.8 times a_g. Thus, the observed pattern in S+ peaks at a value of about 0.4 a_g.

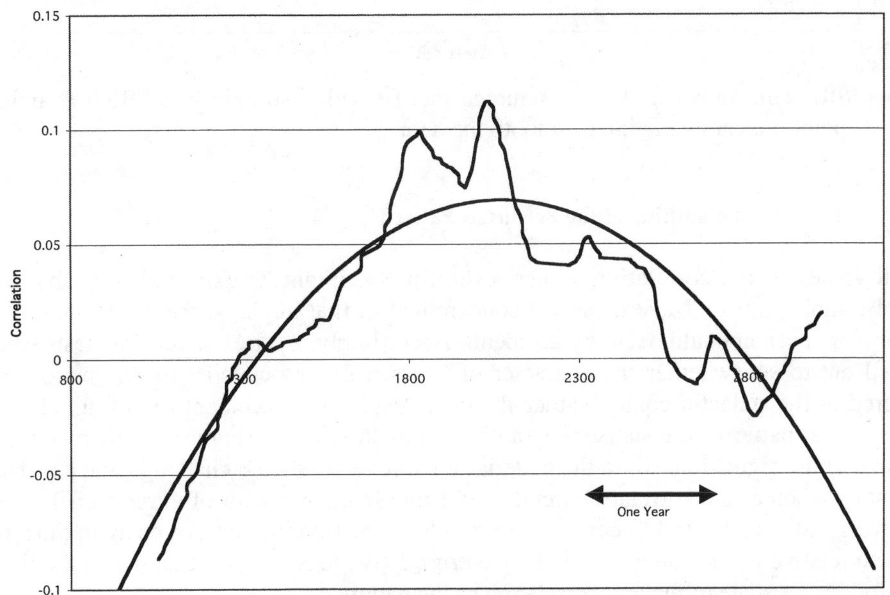

Figure 8A is the plot of S+ for about 2200 sidereal days, fit with a parabola.

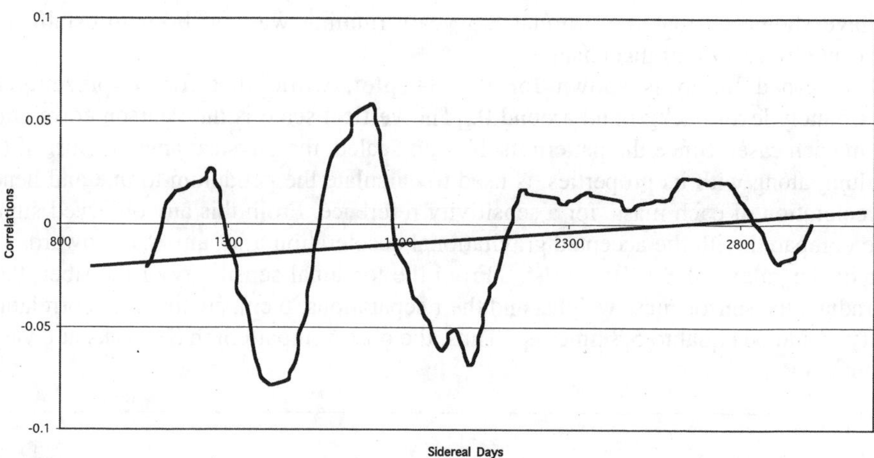

Figure 8B. This shows S- for the same period, fit with a straight line. Both A and B have 40-point boxcar averaging to smooth the data.

2.2.3 Tests by Time Shifting of the Predicted Pattern

Small values of the correlation values, even for S+, might be expected from the low signal-to-noise ratio. Nevertheless, a concomitant of that might be the expectation that the observed effect could occur by accident. Accordingly, a series of separate tests were carried out to see whether the character of S+ is truly associated with the pattern as centered at the galactic center, rather than the result of an accident of fitting. If the experimental pattern were statistically able to show the trend of Figure 8A, then shifting the time of the signal relative to the pattern, or the reverse, should challenge that accident by testing whether the correlation trend would follow the behavior of a true signal. The following test was devised to compare the trends of rotation in space, or delay in time, of the data relative to the pattern, with the pattern relative to itself. That is, would the time shift follow the scale of the autocorrelation of the pattern?

In four separate correlation series the pattern was shifted by certain time values referenced to the assumption of S+ as patterned with 0 hours at the time when the bar supporting the two masses is perpendicular to a line to the center of the galaxy (our initial reference condition). These shifts are series T+ (+6.1hours), U+ (+3 hours), V+ (+1.4 hours), and W+ (-1.9 hours). Figure 9 shows the prediction of correlation value changes plotted versus hours of delay, based on a perfect signal-pattern correlation, i.e. on an autocorrelation of the pattern. First, to see this directly, T+, which is expected to show the most significant change, is plotted in Figure 10 with S+ over the 2400-day period, to illustrate the significant change in the 8-year trends occurring for an approximately 6-hour shift (one-half pattern repeat time). Here smaller (55 day) boxcar smoothing was employed, but the pattern trends are shown by optimal quadratic fits to each pattern.

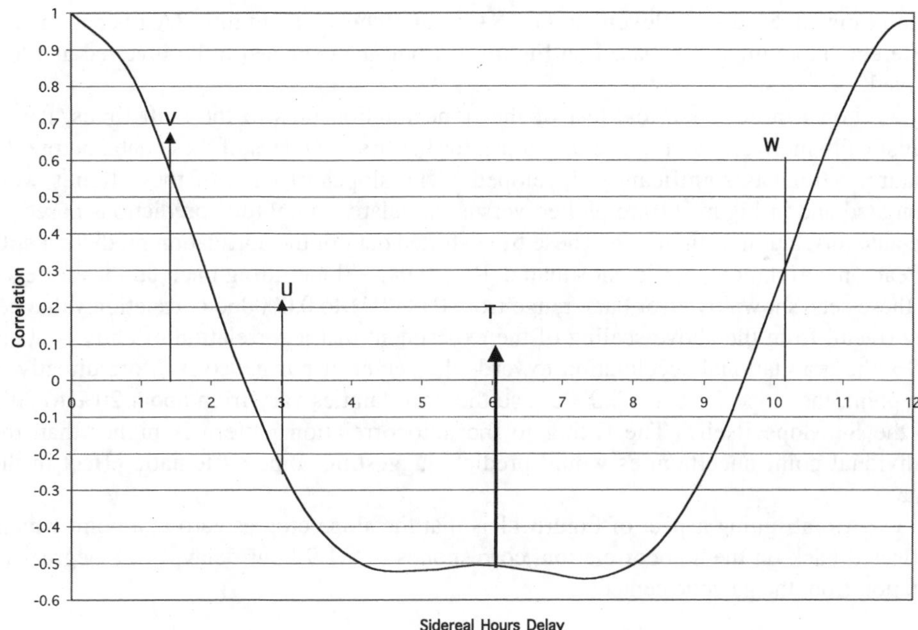

Figure 9. Autocorrelation of the predicted pattern, showing the time shifts for the four data-pattern correlations, series T, U, V, and W.

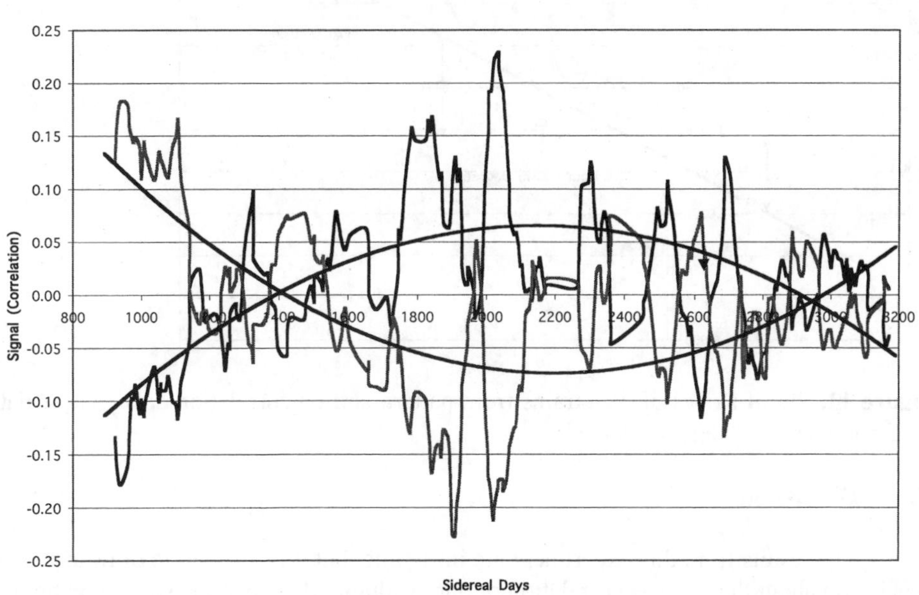

Figure 10. S+ Trend (0 hours shift) and T+ Trend (6.1 hours shift), having boxcar

smoothing of 55 days (eleven runs). S+ is that shown in Figure 8A except for the different smoothing. T+, based on Figure 9, would be expected to be inverted and less curved.

For a more numerical test of the trend relationships to the shift times, it was statistically more efficient to use the slopes for the first 700 days of these data, before the quadratic trend is significantly developed. The slopes of each of these trends were tabulated and in Figure 11 are plotted versus the relative amplitude predictions based on the autocorrelation of Figure 9. These time-shifted data fit the correlation prediction with a Pearson correlation coefficient squared, $R^2 = 0.94$. Slope fitting uncertainties for each of these sets, shown by error bars, range from 0.000041 to 0.000066 correlation value per day (recall from the above scaling of the experiment that a correlation of unity is about 5.8 x the gravitational acceleration towards the center of our galaxy.) More directly to the point, the slope for S+ is 0.00021, so the uncertainties vary from about 20% to 30% of the S+ slope itself. The fitting to the autocorrelation pattern is higher than the individual point uncertainties would predict, suggesting some systematic effect in the data.

An intriguing aspect of Figure 11 is that the slope crosses zero at a point which, reflected back on the autocorrelation, corresponds to a 1.9 hour delay, or 28 degrees of rotation from the galactic center.

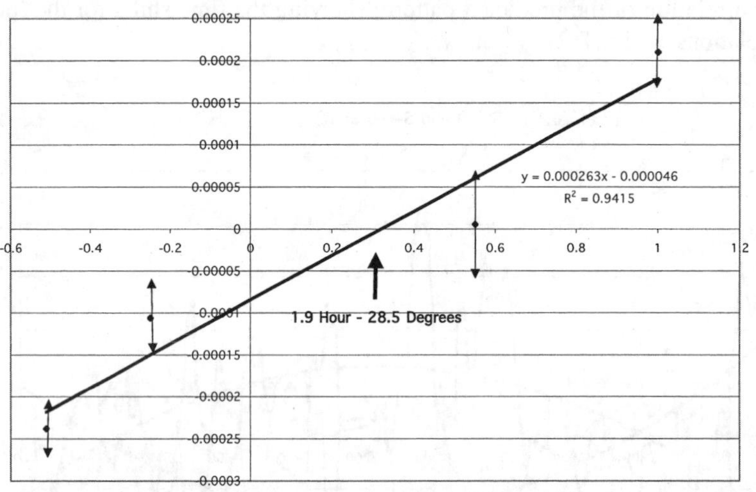

Figure 11. Fit of slopes of data trains from pattern-shifted correlation data. These data are from over 700 sidereal days.

3. Discussion

The results from this experiment are intriguing, but it is difficult to draw any clear conclusion about them. The correlation or our pendulum data with the sidereal pattern is significant by standard statistical reasoning. And, a trend in the 6.5-year sequence of points shows a substantial difference in the character of the quadrupole-filtered data train and the background data train, but with an appearance that is not easily related to known astronomical features. The extremely low signal-to-noise ratio has required filtering with a very low frequency cutoff to provide a simple appearance capable of clearly distinguishing S+ from the background S-. Nevertheless, fitted values, e.g. slopes in Figure 11 have been tested, and change almost imperceptibly between boxcar filtered and unfiltered data trains, so that numerical aspects of the analysis are not affected by the filtering. The large positive rise in this filtered S+ data train has crossed from negative to positive a few years ago, and seems to have turned downward again. Given the time scale for this feature, a physical source for this would seem to be solar in scale rather than galactic. sd

From early in the experiment, local artifactual sources of the features of the data have been suspected and studied. For example, potential magnetic interactions with the pendulum have been evaluated, and the interplanetary magnetic field seems too weak and of unlikely character to cause these features. The possibility of coupling of the experiment to the Earth's magnetic core is interesting [22], but a component of alignment in sidereal space is in question.

Within an order of magnitude the observed strength in S+ is that of the gravitational acceleration of mass towards the galactic center. A mass imbalance of our pendulum, attracted to the galactic center cannot, however, explain this possibility. That effect would lead to a 24-sidereal hour pattern, rather than the 12-hour quadrupolar pattern that characterizes our filtered S+ data.

We mention a few other conjectures of plausible physical mechanisms leading to these experimental results. The appearance of our S+ data trend is somewhat similar to that associated with the peak in sunspot number distribution [23]. It fits that pattern for the first part of the cycle, but falls off much too early for the profile of that eleven year cycle. Solar planetary system features that we have been made aware of were potentially interesting from their magnitudes, but seem not to fit in the time scale implied by S+.

One might postulate moving through a time varying field or clouds of particles providing a pseudoscalar interaction with our masses. The spatial scale is seen from the fact that six years of our path around the galaxy would correspond to a distance on the scale of one light year. The A_2 term of equation (1) has been considered briefly for fitting these data. A roughly similar pattern to the A_1 term might be expected since $\mathbf{v}$ is orthogonal to $\mathbf{r}$. But the component of $\mathbf{v}$ on $\boldsymbol{\sigma}$ would be expected to peak plus or minus six sidereal hours away from the $\boldsymbol{\sigma} \cdot \mathbf{r}$ peak.

V. de Sabbata, in these Proceedings, discusses the lack of necessity for dark matter to explain the flattened galactic rotation curves, which originally motivated this experiment. He cites several alternatives, the most interesting from the point of view of this experiment being the interaction of spin with torsion in space. From Einstein-Cartan theory and appropriate cosmological parameters, he finds that torsion can impose on spin a minimal acceleration $a_{min} \sim 10^{-8}$ cm/s^2, which is similar to the gravitational acceration toward the galactic center and to what we observe. This raises the question: could torsion in Earth's passage through space have patterns on a scale and manner to produce such a data trend?

Acknowledgments

The authors thank the Ettore Majorana Centre for Scientific Culture for its hospitality in hosting the XVIIIth Course of the International School of Cosmology and Gravitation held in Erice, Sicily, Italy in May 2003.

References

1. Leitner, J. and Okubo, S., (1964) *Phys. Rev.* **136**, B1542.
2. Klein, J.R. and Thorsett, S.E., (1990), *Phys. Lett.* **A 145**, 79.
3. Lammerzahl, C., (1997), *Spin in Gravity: Is it Possible to Give an Experimental Basis to Torsion?,* ed. Bergmann, P.G., de Sabbata, V., Gillies, G.T., Pronin, P.I.,World Scientific, Singapore. pp 91-117.
4. Moody, J.E., and Wilczek, F., (1984), *Phys. Rev.* D **30**, 130.
5. Ritter, R.C. and Gillies, G.T., (1997), *Spin in Gravity: Is it Possible to Give an Experimental Basis to Torsion?,* ed. Bergmann, P.G., de Sabbata, V., Gillies, G.T., Pronin, P.I., World Scientific, Singapore. pp 199-212.
6. *Spin in Gravity: Is it Possible to Give an Experimental Basis to Torsion?,* ed. Bergmann, P.G., de Sabbata, V., Gillies, G.T., Pronin, P.I., World Scientific, Singapore. pp 1-255.
7. Krause, D.E., Fischbach, E. and Talmadge, C. (1993) *Perspectives in Neutrinos, Atomic Physics, and Gravitation,* ed. Tran Thanh Van, J., Damour, T., Hinds, E. and Wilkerson, J., Editions Frontieres, Gif-sur-Yvette Cedes – France. pp 455-463.
8. Adelberger, E.G., 2001 *Classical and Quantum Gravity,* **18**, 2397-2405.
9. Ritter, R.C. *et al,* (1990), *Phys. Rev.* D **42**, 977-991.
10. Ansel'm, A. (1982), *Pis'ma Zh. Eksp. Teor. Fiz.* **35**, 266
11. Heckel, B.R. *et al,* (2000) *Advances in Space Research,* **25**, pp. 1225-1230; Adelberger, E. *et al* (1999) *Proc. Of the Fifth International Wein Symposium: Physics Beyond the Standard Model,* ed. Herczeg, P., Hoggman, C.M., and Klapdor-Kleingrothaus, H.G., World Scientific, Singapore, pp 717-737.
12, Ritter, R.C., Winkler, L.I. and Gillies, G.T. (1993) *Phys. Rev. Lett.* **70**, 701-704.
13. Youdin, A.N. *et al,* (1996), *Phys. Rev. Lett.* **77**, 2170-2173.
14. Bluhm, and Kostelecky, (2000) *Phys. Rev. Lett.* 84, 138-141.
15. Turner, M.S., (1990) *Phys. Rep.* **197**, 67
16. Ritter, R.C., Winkler, L.I., and Gillies, G.T. (1994), *Particle Astrophysics, Atomic Physics, and Gravitation,* ed. . Tran Thanh Van, J., Fontaine, G., and Hinds, E., Editions Frontieres, Gif-sur-Yvette Cedes – France. pp 441-444.
17. Ritter, R.C., Winkler, L.I., and Gillies, G.T., (1996) *Dark Matter in Cosmology, Quantum Measurements, and Experimental Gravitation,* Ed. Ansari, R., Giraud-Heraud, Y. and Tran Thanh Van, J., E., Editions Frontieres, Gif-sur-Yvette Cedes – France. pp 417-422.
18. Ritter, R.C., Winkler, L.I., and Gillies, G.T., (1997) *Very High Energy Phenomena in the Universe,* , Ed. Giraud-Heraud, Y. and Tran Thanh Van, J., Editions Frontieres, Gif-sur-Yvette Cedes – France. pp 349-352.

19. Ritter, R.C. and Gillies, G.T.,), *Spin in Gravity: Is it Possible to Give an Experimental Basis to Torsion?,* ed. Bergmann, P.G., de Sabbata, V., Gillies, G.T., Pronin, P.I., World Scientific, Singapore. pp 213-224.
20. Phillips, P.R., (1987) *Phys. Rev. Lett.* **59**, 1784-1787.
21. Wang, S-L., Ni, W-T, and Pan, S-S, (1993) *Mod. Phys. Lett.* A **8**, 3715.
22. K. Nordtvedt has also suggested that the Earth's core might itself be a potential large mass with some intrinsic spin alignment for detecting exotic coupling to regions in space.
23. We thank Andrew Hall for the suggestion of that possibility and for many other interesting discussions of this experiment.

PROSPECTS FOR A SPACE-BASED DETERMINATION OF *G* WITH AN ERROR BELOW 1 PPM

Alvin J. Sanders
Department of Physics & Astronomy
401 Nielsen Building
University of Tennessee
Knoxville, TN 37996-1200
and
Engineering Science and Technology Division
Oak Ridge National Laboratory
PO Box 2008
Building 5800, Mail Stop 6054
Oak Ridge, TN 37831-6054
ASanders@utk.edu

George T. Gillies
Department of Mechanical and Aerospace Engineering
University of Virginia
Charlottesville, VA 22904
gtg@virginia.edu

Abstract

It is not clear that ground-based laboratory determinations of Newton's gravitational constant *G* will eventually converge on a common value that has an accepted uncertainty appreciably below 100 ppm. For many years *G* has been one of the least well known of the fundamental physical constants, and the uncertainty in the CODATA value of *G* was raised recently from 128 ppm to 1500 ppm, in part because of the very puzzling results from the Physikalisch-Technische Bundesanstalt (PTB) in 1995. Intense efforts by a number of groups during the past 8 years to resolve this difficulty are now beginning to bear fruit, but unfortunately the basic situation is unchanged. New results from three groups doing precision measurements of *G* disagree by over 200 ppm, although the reported errors of the individual experiments are below 50 ppm. The seeming intractability of this situation suggests that a useful alternative may be a space-based determination of *G*. We present and discuss a projected error budget for one such proposed measurement of *G* using the SEE (Satellite Energy Exchange) observatory. A SEE mission is also foreseen as being capable of measuring the time variation of *G* (*G*-dot) to about 1 part in 10^{14} per year. A finding of a non-zero value of *G*-dot would have immediate and profound cosmological significance. The authors have previously shown how a measurement of the required accuracy might be accomplished, entailing a synergism between SEE and a geopotential mission similar to the current GRACE mission of NASA.

Key words: Gravity, gravitational constant, *G-dot*, inverse-square law, unification theory.

Introduction

At the 1995 Erice school in Gravitation and Cosmology the authors compared and analyzed a number of proposals for determining *G* in space, including SEE (Satellite Energy Exchange) [Sanders & Gillies, 1996; Sanders & Gillies, 1998]. The general conclusion was threefold: (1)

V. de Sabbata et al. (eds.),
The Gravitational Constant: Generalized Gravitational Theories and Experiments, 331–340.
© 2004 *Kluwer Academic Publishers. Printed in the Netherlands.*

The space environment offers the possibility of a significant improvement in the uncertainty of *G*, (2) great care in the identification and treatment of perturbations is required to realize the promise of space, and (3) most extant *G*-in-space proposals only explored basic principles, giving very inadequate attention to perturbations.

We conjectured that terrestrial determinations are unlikely to be able to reduce the uncertainty in *G* significantly below existing levels. This is based on the long history of the efforts in this field [Gillies, 1997]. In 1995 and 1997 the publication of four very careful *G* determinations [Fitzgerald & Armstrong, 1995; Michaelis *et al.*, 1995; Walesch *et al.*, 1995; and Bagley & Luther, 1997] abruptly undermined the until-then relative complacency about our knowledge of *G*: Although all four experiments claimed accuracies of ~100 ppm, the scatter of the results was much larger. The results of Michaelis *et al.* and of Fitzgerald & Armstrong were in substantial disagreement with the accepted CODATA value. As is well known, the result of Michaelis *et al.* from the PTB was 0.6% (6000 ppm) above the CODATA value. Largely because of this, there was an unprecedented burst of new activity in *G* determination. A conference in London in November, 1998, on the occasion of the 200th anniversary of Cavendish's experiment, was attended by virtually every active research group (see special section of the June, 1999 issue of *Measurement Science and Technology*). Three of these groups now have new results that claim errors less than 100 ppm (units are $m^3/(kg\ s^2)$):

Gundlach & Merkowitz (U. Wash):

$$G = 6.674215\text{x}10^{-11} \pm 14 \text{ ppm} \qquad (1)$$

Quinn *et al.* (BIPM):

$$G = 6.67559\text{x}10^{-11} \pm 41 \text{ ppm} \qquad (2)$$

Schlamminger *et al.* (Zürich):

$$G = 6.67407\text{x}10^{-11} \pm 33 \text{ ppm} \qquad (3)$$

Additional presently-active groups are those led by Faller (U. Colo/JILA), Karagioz (Moscow), Luo Jun (China), Newman (UC, Irvine) and Paik (U. Maryland), among others.

The results of the University of Washington and the BIPM differ by 206 ppm, while those of the University of Zurich and the BIPM differ by 228 ppm. Thus, the results of these recent experiments, while all done with great care and precision, still allow for a significant overall uncertainty in the value of *G*, about 100 ppm. Unless substantially new technologies are brought to bear on the problem, it seems likely that the limits of *G* determination by terrestrial methods (*i.e.*, torsion fiber instruments) are close to being reached. The question is then open as to whether or not a space-based determination might be able to help the situation.

The concept of SEE, a novel approach to space-based determination of *G* and other gravitational parameters, was proposed in considerable detail in 1992 [Sanders & Deeds, 1992a; Sanders & Deeds, 1992b; Sanders & Deeds, 1993; Sanders, Deeds & Gillies, 1993] and was subsequently analyzed by a team from the Russian Gravitational Society based at Gosstandard [Melnikov *et al.*, 1993; Bronikov *et al.*, 1193a; Bronikov *et al.*, 1993b; Alekseev *et al.*, 1993a; Alekseev *et al.*, 1993b; Alekseev *et al.*, 1994; Alekseev *et al.*, 1994]. Further analysis by the SEE teams during the past ten years has led to a number of enhancements of the SEE concept [Harris, 1994; Schunk, 1996; Nordtvedt, 1998; Sanders & Gilies, 1998a; Sanders & Gilies, 1998b; Sanders *et al.*, 1998; Corcovilos & Gatford, 1998; Antonov, 1999a; Antonov, 1999b; Sanders *et al.*, 1999; Sanders *et al.*, 2000; Sanders & Gillies, 2002; Irick & Hornback; 2001]. A SEE mission entails essentially observing the mutual orbital perturbation of test bodies in a restricted three-body situation. The distinguishing interaction is an *exchange of energy* between two co-orbiting satellites, as first envisaged and described by Sir George Darwin a century ago [Darwin, 1897]. A number of papers have appeared in the recent literature regarding experimental observations of SEE-type interactions, including papers on a pair of co-orbiting satellites of Saturn [Dermott & Murray, 1981a; Dermott & Murray, 1981b; Yoder *et al.*, 1983] and the discovery paper for a companion asteroid of the Earth itself, with which it engages in a SEE-type interaction [Wiegert *et al.*, 1997].

Objectives of a SEE Mission

Since the data from SEE observations reflect the net result of all forces acting on the test bodies, they are naturally adapted not only to determining *G*, but also to detecting and measuring various deviations from Newtonian gravity. The capability for simultaneous determination of a number of different gravitational parameters is very likely possible only in space. Even under the most favorable scenarios for ground-based determination of *G*, it is doubtful that there could be enough intrinsic sensitivity to permit multiple-parameter determinations.

Specifically, a SEE mission can in principle make the following measurements and tests:

1. Determination of the gravitational constant *G*
2. Search for non-zero *G*-dot (time variation of *G)*
3. Test of WEP by ISL at test-body separations ~few meters
4. Test of WEP by ISL at test-body separations ~R_E (radius of Earth)
5. Test of WEP by CD at test-body separations ~few meters
6. Test of WEP by CD at test-body separations ~R_E
7. Possible search for other violations of General Relativity

The estimated accuracies of the various tests and measurements are listed in Table 1. We note that the results expected for *G*, *G*-dot, and the two ISL tests are anticipated to be significantly better than those of any other existing or proposed experiment. Existing WEP/CD limits from ground-based experiments are already tighter than those expected from SEE at intermediate ranges (~meters), and STEP is expected to do much better than SEE on the long-range WEP/CD test (although the SEE result would be a significant improvement over the *existing* limit).

Table 1
Expected Accuracy of SEE Tests and Measurements

Test/Measurement	Expected Accuracy
G	0.33 ppm (330 ppb)
(*G*-dot)/*G*	$<10^{-13}$/yr in one year
WEP/ISL at ~few meters	2×10^{-7}
WEP/ISL at ~R_E	1×10^{-10}
WEP/CD at ~few meters	$<10^{-7}$ ($\therefore \alpha<10^{-4}$)
WEP/CD at ~R_E	$<10^{-16}$ ($\therefore \alpha<10^{-13}$)
GR violations	Undetermined

Implications for Unification Theories

The results of precise tests of gravitational parameters have considerable implications for unification theories, in at least three respects:

A. Gravity remains essentially the "missing link" in many aspects of unification theories.

B. Nearly all modified theories of gravity and unified theories predict some violations of the Weak Equivalence Principle (WEP), either by deviations from the Newtonian law (inverse-square-law, ISL) or by composition-dependent (CD) gravity accelerations, due to the appearance of new possible massive particles (partners) [De Sabbata *et al.*, 1992; Melnikov, 1988; Melnikov, 1994; Ivashchuk & Melnikov, 1998].

C. Most such theories also predict non-zero values of *G*-dot.

These areas are very attractive for tests by orbiting satellites such as SEE.

More broadly, the current generation of existing and proposed gravitation experiments—including LIGO, LLR, LISA, MICROSCOPE, STEP, GP-B, and SEE—promises great increases in our knowledge of Nature. We note that this is, to a great extent, the fruit of recent advances in superconductivity and high-precision technology. Until about 20 years ago the state of technology was simply insufficient for the empirical exploration of gravity at the degree of precision and accuracy that is required for comparative tests of theory. As Prof. Zichichi has pointed out, general relativity has only recently entered the realm of experimental investigation. Thus, we are now on the brink of the prospect that our knowledge of gravity can become comparable to our knowledge of the other fundamental forces. We owe our understanding of these forces chiefly to the investment that society has made in accelerator-based physics during the past half century. It has paid very rich dividends. If funding for experimental gravity now expands commensurately, further great leaps in our understanding of Nature may be expected.

Projected Error Budget for Determination of the Gravitational Constant *G*

The various perturbing effects which are thought to have the potential to contribute to uncertainty in the realization of the gravitational constant *G* on a SEE mission are now being evaluated. The status of this evaluation is shown in Table 2 [after Sanders *et al.*, 1999]. We estimate that the value of *G* can be determined with an uncertainty of about 0.33 ppm by a SEE mission. Note that the total uncertainty is dominated by that of the mass of the Shepherd [Debler, 1991]. The total remaining uncertainty is under 0.15 ppm, with significant contributions coming from distance resolution, electrostatic effects, and mass defects in the Shepherd and in the SEE observatory. Careful design is required to achieve the relatively small size of the errors shown in Table 2. For example, the small value for blackbody radiation results from a combination of the choice of a sun-synchronous, continuous-sunlight orbit, several open cylinders surrounding the experimental chamber to act as radiative shields, and the slow rotation of the observatory (Sanders & Deeds, 1992; Harris, 1994; Schunk, 1996; Sanders et al., 1997; Corcovilos & Gatford, 1998; Antonov, 2000a). Similarly, the uncertainty due to the gravitational field of the SEE observatory *per se* may be limited by designing its mass distribution so that its internal field is zero in principle and then mapping (calibrating) the field in the manner of satellite geodesy (Sanders & Deeds, 1992; Sanders & Gillies, 1996; Corcovilos & Gatford, 1998; Antonov, 2000b).

Table 2. Error Budget for Determination of *G*

Error Source	Average or RMS Force	$\delta G/G$ ($\times 10^{-9}$)	Comments
Distance resolution I	NA	49	Assuming ISL
Distance resolution II	NA	54	Testing ISL
Time resolution	NA	negligible	
Blackbody radiation	$<0.6\times10^{-18}$ N	<4	$\Delta T=0.0001$ K
Electrostatic, Shepherd-Particle	2.9×10^{-18} N @s=3m	<50	$q_s=0.24$ pC, $V_s=12$ mV $q_p=0.24$ pC, $V_p=103$ mV
Electrostatic, Particle Image	$<2.4\times10^{-17}$ N ($\perp$)	<<170	$\Delta z=1$m from end wall
Lorentz forces	$<<5\times10^{-14}$ N ($\perp$)	negligible	Shielded/compensated
Earth's gravity gradient	NA	zero	SEE is an orbital-perturbation experiment
Earth's non-spherical field	calculable	negligible	
Fields of moon, Jupiter, etc.	calculable	negligible	
Shepherd mass uncertainty	Shepherd's gravity x 300×10^{-9}	300	Debler, PTB (1991)
Shepherd's mass defects	small	<50	
Mass defects in observatory	$<13\times10^{-18}$ N	<90	Many defects ~10 mg
Outgassing	small	small	Obviate by baking
Total I, assuming ISL		<330	
Total II, testing ISL		<331	

For computing the total error shown in Table 2, the individual errors are added in quadrature. The bounded errors are evaluated at the upper bound, while the gross overestimate for the Particle image is evaluated—very conservatively—as one-third of the given overestimate. To wit,

$$\sqrt{[49^2 + 4^2 + 50^2 + (170/3)^2 + 300^2 + 50^2 + 90^2]} = 330$$

We note two significant differences between the error budgets for G and for G-dot (discussed below): Uncertainties in the Earth's gravitational field cause no appreciable uncertainty in G, although they are very important for G-dot. On the other hand, the time-averaged internal field of the SEE observatory can be very effectively reduced by cancellation techniques for purposes of the G-dot measurement, but cancellation techniques do not work for G.

Theoretical Calculations of *G*

Until recently the gravitational constant G has often been regarded as simply a constant whose value has only an empirical meaning and is unrelated to any other fundamental constant. The view of G may now be changing among theorists, as serious attempts are being made to derive G from theoretical considerations [see for example, Damour, 1999]. As further progress is made in this area, the importance of the value of G as a test of theories is likely to grow. Thus, a measurement of G at the level of <1 ppm could in the near future be an important discriminator among theories.

Time Variation of *G*

Perhaps the most important contribution that SEE might make to physics would be the provision of a very accurate measurement/test for (*G*-dot)/*G*. We expect an accuracy of $\sim 10^{-14}$/yr, which if achieved would be about two orders of magnitude beyond the present capability of Lunar Laser Ranging (LLR) [Williams *et al.*, 2001] and is sufficiently precise to provide significant discrimination among theoretical models [Sanders *et al.*, 2000]. The basic approach is to use the orbital period of the SEE Shepherd as a clock, inferring a possible change in *G* from the dependence of the orbital period on *G*.

The various perturbing effects that are thought to have the potential to contribute to error in the measurement of *G*–dot on a SEE mission are being evaluated. The status of this evaluation is shown in Table 3 [after Sanders *et al.*, 2000].

Table 3. Error Budget for *G*-dot (One-Year and Four-Year Observation Periods)

Error Source	Average Force ($\times 10^{-17}$ N)	δ(*G*-dot/*G*) ($\times 10^{-15}$)		Brief Comments (details below)
		1 yr	4 yrs	
Tracking error	NA	15.6	2.0	GPS/SLR accuracy = 1 cm
Timekeeping error	NA	~1	~1	Next generation clocks
Blackbody radiation	10.0	8.6	4.3	$\Delta\Theta < 0.1$ mK
Electrostatic forces	<15	<10	<5	Surface potential < 6.4 mV
Lorentz forces	small	zero	zero	Perpendicular to velocity
Earth's field	<1.4	<0.9	<0.5	With GRACE or equivalent
Capsule mass defects	22.2	15	7.4	Many defects ~10 mg
Gravity of particle	<0.22	<0.15	<0.08	Newton's 3rd law
Shepherd's moments	small	small?	small?	Not evaluated yet
Outgassing jets	small	small	small	Obviate by baking
Total	**NA**	**25**	**10**	

At the 2000 Erice School in Gravitation and Cosmology, the authors presented a paper on the error budget and expected accuracy of a (*G*-dot)/*G* measurement by SEE [Sanders & Gillies, 2002]. The focus of this paper was on the time variation of the largest harmonic coefficient of the Earth's field, $J_{2,0}$, which corresponds to the Earth's equatorial bulge. Uncertainties in the Earth's field are potentially very important in the error budget for *G*-dot (unlike *G per se*, as indicated above). This is a major issue because the field of the Earth varies on a time scale of weeks or months by amounts that, if unknown, would spoil our *G*-dot accuracy goal (Ritter, 1997). Fortunately, the new NASA geopotential mission GRACE can provide most of the information needed by SEE to attain (*G*-dot)/*G* to $\sim 10^{-14}$/yr accuracy. However, GRACE cannot obtain a sufficiently accurate value of $J_{2,0}$, because spurious heating/cooling effects at the period of its orbit become mixed with the real gravitational signals from $J_{2,0}$. Fortuitously, we believe that SEE can extract $J_{2,0}$ from the precession rates of the ascending node and/or the argument of perigee. It is well known that this could be done if $J_{2,0}$ were the *only* non-spherical harmonic (See, for example, Sanders & Gillies, 1996). It follows that, if the other harmonics are known—*i.e.*, if their values are supplied by GRACE or an equivalent geopotential mission—then the effects of the higher harmonics can be subtracted out, thus isolating the impacts of $J_{2,0}$ alone on the precession rates [Sanders & Gillies, 2002]. Hence, combining the capabilities of SEE and GRACE (or an equivalent mission) may result in a synergism that substantially exceeds what either could accomplish without the other.

For a *G*-dot measurement, it is also critically important to control the internal gravitational field of the SEE observatory. This is done in three stages. First, the capsule is designed and manufactured to make the internal field zero in principle [Sanders & Deeds, 1992]. Second, it is mapped (calibrated) [op. cit.; Sanders & Gillies, 1996]. Finally, the residual effects, which are too small to measure in the calibration, may be largely *cancelled* by flying the observatory and the Shepherd in "Four Flight Configurations"; this results in "almost-zero time-averaged drag" (AZTAD) on the Shepherd [Sanders et al., 2000]. Taking all identified effects into account, we estimate that the time-averaged drag on the SEE Shepherd will be about 10^{-18} *g*. This is expected to be sufficient to determine *G*-dot to ~10^{-14}/yr within a few years of observation. We note that this would make the *time-averaged* drag on the SEE Shepherd less than the *instantaneous* drag on the Moon.

To avoid misunderstanding, we stress that the drag on the SEE observatory and the instantaneous drag on the Shepherd will be much larger than 10^{-18} *g*. The expectation of a very small drag applies to the Shepherd and only in a time-averaged sense—which is what is important to the experiment.

Conclusions

It is reasonable to expect to attain an uncertainty of less than 1 ppm from a space-based determination of the gravitational constant *G*. We estimate the uncertainty from a SEE mission will be about 0.33 ppm. The degree of consistency among the best terrestrial determinations indicates that their actual uncertainties are about 100 ppm, and some argue that this represents something approaching the ultimate accuracy for terrestrial determinations, barring the development at some future time of an entirely new approach, now completely unforeseen, with very low systematic errors. The time variation of *G* is foreseen as being measurable to about 1 part in 10^{14} within a few years of observation by using the orbital motion of the SEE Shepherd as a clock. Thus, a space-based approach may hold promise of providing at least two orders of magnitude improvement in both *G* and *G*-dot. An unambiguously non-zero experimental value of *G*-dot would have an immediate and profound impact on our understanding of Nature. An accurate experimental value of *G* is likely to have cosmological significance in the near future, as the process of testing unified models continues and as the capability of the models to predict the value of *G* improves.

Acknowledgements

We are pleased to acknowledge the encouragement of Venzo de Sabbata, the hospitality of the E. Majorana Center, the support by and interest of NATO Division des Afffaires scientifiques et de l'Environnement, ESTEC, NASA Fundamental Physics in Microgravity Program, NASA Marshall Space Flight Center, and NASA Jet Propulsion Laboratory.

References

A.D. Alekseev, Bronnikov KA, Kolosnitsyn NI, Konstantinov MYu, Mel'nikov VN, and Radynov AG, 1993a. "Analysis of the project for a space experiment to measure the gravitational constant and possibly the ones of non-Newtonian-type interactions." Paper presented to Eighth Russian Gravitational Conference (Moscow, May, 1993).

A.D. Alekseev, Bronnikov KA, Kolosnitsyn NI, Mel'nikov VN, and Radynov AG, 1993b. "Error sources in Earth satellite measurements of gravitational interaction parameters." *Meas. Tech.* **36** (No. 10), 1070-1077.

A.D. Alekseev, Bronnikov KA, Kolosnitsyn NI, Mel'nikov VN, and Radynov AG, 1994. "Simulation of the procedure for measuring the gravitational constant on an Earth satellite." *Meas. Tech.* **37** (No. 1), 1-5.

A.D. Alekseev, Bronnikov KA, Kolosnitsyn NI, Konstantinov MYu, Melnikov VN, and Radynov AG, 1994. "On two-body space experiments for measuring gravitational interaction parameters." *Int. J. Mod. Phys. D* **3** (No. 4), 773.

V. Antonov, 1999a. "Simulation of the azimuthal temperature variation in the experimental chamber in Project SEE." Internal UT-ORNL Science Alliance Center of Excellence report (July 28, 1999).

V. Antonov, 1999b. "Fourier coefficients of the gravitational potential and perturbation forces due to capsule mass defects in Project SEE." Internal UT-ORNL Science Alliance Center of Excellence report (July 28, 1999).

C.H. Bagley & G.G. Luther, 1997. *Phys. Rev. Lett.* **78** 3047-50 (1997).

K.A. Bronnikov, Kolosnitsyn NI, Konstantinov MYu, Mel'nikov VN, and Radynov AG, 1993b. "Numerical modeling of the trajectories of particles for measuring the gravitational constant on an artificial satellite." *Meas. Tech.* **36** (No. 9), 951-957. .

K.A. Bronnikov, Kolosnitsyn NI, Konstantinov MYu, Mel'nikov VN, and Radynov AG, 1993a. "Measurement of the gravitational interaction. parameters of an Earth-orbiting satellite." *Meas. Tech.* **36** (No. 8), 845-852.

Minkang Cheng & B.D. TAPLEY, 1999. "Seasonal variation in low-degree zonal harmonics of the Earth's gravity field from satellite laser ranging observations"; *J. Geophys. Res.* **104** (No. B2), 2667-2681.

T.A. Corcovilos & Gadfort T, 1998. "Issue-scoping for Project SEE"; University of Tennessee Technical Report (August 1998).

T. Damour, 1999. "The theoretical significance of G,"*Meas. Sci. Technol.*, 10(No. 6) 567–569.

G.H. Darwin GH, 1897. "Periodic Orbits." *Acta Math.* **21**, 99.

E. Debler, 1991. "Set-up of mass scales above 1 kg illustrated by the example of a 5 t mass scale." *Metrologia* **28**, 85-94.

S.F. Dermott & Murray CD, 1981a. "The dynamics of tadpole and horseshoe orbits: Theory." *Icarus* **48**, 1-11.

S.F. Dermott & Murray CD, 1981b. "The dynamics of tadpole and horseshoe orbits II: Experiment." *Icarus* **48**, 12-22.

V. de Sabbata, Melnikov VN, and Pronin PI, 1992. "Theoretical Approach to Treatment of Nonnewtonian interactions." *Progr.Theor.Phys.* **88**, 623.

M.P. Fitzgerald & T.R. Armstrong, 1995. *IEEE TRANS. INSTRUM. MEAS.* **44**, 494-497.

G.T. Gillies, 1997. "The Newtonian gravitational constant: Recent measurements and related studies." *Reports Progr. Phys.* **60**, 151-225.

M.J. Harris, 1994. "Using Radiation Shields to Achieve Cryogenic Temperatures in the SEE Capsule"; Univ. of Tennessee Technical Report (August 1994).

B.C. Irick & D.E. Hornback, 2001. "Analysis of solar flare proton flux on Satellite Energy Exchange (SEE) orbiter"; Univ. of Tenn. internal report (August 9, 2001).

V.D. Ivashchuk and Melnikov VN, 1998. "Multidimensional and Spherically Symmetric Solutions with Intersecting p-branes." In *Proc. 2nd Samos Meeting on Cosmology, Geometry and Relativity, September 1998* (Springer); Los Alamos gr-qc/9901001.

V.N. Melnikov, 1988. "Gravitational Relativistic Metrology." In: *Gravitational Measurements, Fundamental Metrology and Constants*, eds. V. de Sabbata and V.N. Melnikov (Kluwer Acad. Publishers, Dordtrecht), p. 283.

V.N. Melnikov, 1994. "Fundamental physical constants and their stability." *Int. J. Theor. Phys.* **33** (No. 7), 1569.

V.N. Melnikov, Konstantinov MYu, Kolosnitsyn NI, Bronnikov KA, Radynov AG, Alekseev AD, and Antonkuk PN, 1993. "Report for Project SEE," Moscow (Feb. 1993).

W. Michaelis, 1995. "A new determination of the gravitational constant." *Bull. Am. Phys. Soc.* **40** (No. 2), 976.

Kenneth Nordtvedt, private communication, 1998.

R.G. Ritter, priviate communication, 1997. One of the authors (AJS) is grateful to Rogers Ritter for pointing out the hitherto oversight of the need for precision time-varying geopotential for a *G*-dot experiment based on using the orbital period of an earth satellite as a clock.

A.J. Sanders & Deeds WE, 1992a. "Proposed new determination of the gravitational constant *G* and tests of Newtonian gravitation." *Phys. Rev. D* **46** (No. 2), 489-504 .

A.J. Sanders & Deeds WE, 1992b. "Project SEE: Proposed New Determination of the Gravitational Constant *G* and Tests of Newtonian Gravitation." *Bull. Am. Phys. Soc.* **37** (No. 7), 1675 .

A.J. Sanders & Deeds WE, 1993. "Reply to 'Perturbative Forces in the Proposed Satellite Energy Exchange Experiment'." *Phys. Rev. D* **47** (No. 8), 3660-3661.

A.J. Sanders, Deeds WE, and Gillies GT, 1993. "Proposed new space-based method for more accurate gravitational measurements." In *The Earth and the Universe: Festschrift in honour of Hans-Jürgen Treder*, edited by WILFRIED SCHRÖDER (International Association of Geomagnetism and Aeronomy, Bremen-Rönnebeck, Germany, 1993), pp. 360-365.

A.J. Sanders & Gillies GT, 1996. "A comparative survey of proposals for space-based determination of the gravitational constant *G*." *Rivista Nuovo Cimento* **19** (No. 2), 1-54

A.J. Sanders, Allison SW, Campbell JW, Cates MR, Earl DD, Harris MJ, Newby RJ, and Schunk GR, 1997. "Precision control and measurement issues in the SEE (Satellite Energy Exchange) satellite." *Grav. & Cosmol.* **3** (No. 4) 287-292 (in English and Russian).

A.J. Sanders & Gillies GT, 1998. "Project SEE (Satellite Energy Exchange): Proposed Space-Based Method for More Accurate Gravitational Measurements." In Bergmann, P. G., de Sabbata, V., Gillies, G. T., and Pronin, P. I., eds., *Spin in Gravity: Is it Possible to Give an Experimental Basis to Torsion?* (International School of Cosmology and Gravitation XVth Course), The Science and Culture Series - Physics, No. 16 (World Scientific, Singapore, 1998), pp. 225-230.

A.J. Sanders & G.T. Gillies, 1998. "A Comparative Survey of Proposals for Space-Based Determination of the Gravitational Constant *G*," In Bergmann, P. G., de Sabbata, V., Gillies, G. T., and Pronin, P. I., eds., *Spin in Gravity: Is it Possible to Give an Experimental Basis to Torsion?*

(International School of Cosmology and Gravitation XVth Course), The Science and Culture Series - Physics, No. 16 (World Scientific, Singapore, 1998), pp. 231–234.

A. J. Sanders, A. D. Alexeev, S. W. Allison, K. A. Bronnikov, J. W. Campbell, M. R. Cates, T. A. Corcovilos, D. D. Earl, T. Gadfort, G. T. Gillies, M. J. Harris, N. I. Kolosnitsyn, M. Yu. Konstantinov, V. N. Melnikov, R. J. Newby, R. G. Schunk, and L. L. Smalley, 1999. "Project SEE (Satellite Energy Exchange): Proposal for Space-based Gravitational Measurements," *Meas. Sci. Technol.*, 10(No. 6) 514–524.

A.J. Sanders, A.D. Alexeev, S.W. Allison, V. Antonov, K.A. Bronnikov, J.W. Campbell, M.R. Cates, T.A. Corcovilos, D.D. Earl, T. Gadfort, G.T. Gillies, M.J. Harris, N.I. Kolosnitsyn, M.Yu. Konstantinov, V.N. Melnikov, R.J. Newby, R.G. Schunk, and L.L. Smalley, 2000. "Project SEE (Satellite Energy Exchange): An International Effort to Develop a Space-Based Mission for Precise Measurements of Gravitation"; *Class. Quant. Grav.* **17** (No. 12), 2331-2346.

A.J. Sanders & G.T. Gilies, 2002. "Prospects for a Test or measurement of *(G-dot)/G* by a SEE (Satellite Energy Exchange) Mission"; Proceedings of the XVIth Course of the International School of Cosmology and Gravitation: Advances in the Interplay between Quantum and Gravity Physics (Kluwer Academic, Dordrecht, 2002), pp. 353-363.

R.G. Schunk, 1996, Internal NASA memorandum.

H. Walesch, H. Meyer, H. Piel, and J. Schurr, 1995. *IEEE Trans. Instrum. Meas.* 44, 491-493.

P.A. Wiegert, Innanen KA, and Mikkola S, 1997. "An asteroidal companion to the Earth." *Nature* **387**, 685-686 (12 June 1997).

J.G. Williams, Boggs, D.H., Dickey, J.O., and Folkner, W.M., 2001. "Lunar Laser Tests of Gravitational Physics"; Proceedings of Ninth Marcel Grossmann Meeting, World Scientific Publ., Ed. R. Jantzen, web version posted 2001, paper version in press.

C.F. Yoder, Colombo G, Synnott SP and Yoder KA, 1983. "Theory of motion of Saturn's coorbiting satellites." *Icarus* **53**, 431-443.

PROJECTIVE UNIFIED FIELD THEORY REVISITED AND ADAPTED TO THE NEW MEASURED VALUES (WMAP). NEW RESULTS REFER TO APPROXIMATE STATIC INTERIOR AND EXTERIOR SOLUTION OF A SPHERICALLY SYMMETRIC PERFECT FLUID SPHERE WITH APPLICATIONS TO CELESTIAL BODIES, EINSTEIN EFFECTS WITH PARTICULAR TREATMENT OF THE PERIHELION SHIFT OF MERCURY INCLUDING THE QUADRUPOLE MOMENT OF THE SUN

E. SCHMUTZER
University of Jena, Germany

Abstract. The contents of this paper refers to following main subjects: numerical adaptation of our cosmological parameters recently used within the framework of the Projective Unified Field Theory [1] to the new measured values of the Wilkinson Microwave Anisotropy Probe (WMAP) published by Ch.L. Bennett et al. [2]; study of the resulting consequences for our cosmological model of a closed homogeneous isotropic cosmos; approximate treatment of a quasi-static spherically symmetric perfect fluid sphere (interior and exterior solution) being under the adiabatic influence of the expanding cosmos; application of these results to sun-like and earth-like cosmic bodies; time dependence of the "empirical gravitational constant"; treatment of the Einstein effects, particularly of the periastron (perihelion) motion of an orbiting body (mercury) about a central body (sun) with a mass quadrupole moment und calculation of its numerical value for the sun. Further cosmological and astrophysical applications. An appendix points to a different set of parameters.

1. Revisiting PUFT cosmology on the basis of the new empirical results by WMAP

Two years ago we presented in [1] our 5-dimensional Projective Unified Field Theory (PUFT) in detail and treated on this basis a closed homogeneous isotropic cosmological 2-component model consisting of a uniform gas of mechanical particles (mechanical gas) and of a uniform photon gas. In the meantime following

V. de Sabbata et al. (eds.),
The Gravitational Constant: Generalized Gravitational Theories and Experiments, 341–380.
© 2004 *Kluwer Academic Publishers. Printed in the Netherlands.*

interesting empirical cosmological parameters were published [2] which we now use as important points of reference for the further numerical fixing of our closed cosmological model (p present, y year, K Kelvin):

$$\begin{aligned} &\text{a)}\quad T_{ph|p} = 2.725\,\text{K}\,, &&\text{(present temperature of the cosmic microwave background / photon gas),}\\ &\text{b)}\quad t_{AW} = 13.7\cdot 10^{9}\,\text{y}\,, &&\text{(age of the world / cosmos),}\\ &\text{c)}\quad H_p = 71\,\frac{\text{km}}{\text{s Mpc}}\,, &&\text{(present Hubble parameter).} \end{aligned} \tag{1}$$

When we wrote our article mentioned [1], on the basis of different methods (nucleogenesis methods of age determination of elements, age of globular clusters of stars etc.) astrophysicists tended to an age of the world being in the interval (14 to 22) billions of years, i.e. much bigger than the age of the world (1b) recently determined by WMAP.

By the publication [2] the basis of PUFT will not be altered, but we have to revisit the numerical cosmological values. In order to avoid repetition of the rather voluminous set of formulas in [1], we introduce the following notation explained at the example: ((44) in [1]) means formula (44) in the quoted publication [1]. Further we remember the relationship between the temporal parameter η and the physical time $t : \eta = ct/A_0$, where $A_0 = 10^{27}$ cm is the cosmological rescaling factor.

The recently improved value of the Einstein gravitational constant (γ_N Newton gravitational constant)

$$\varkappa_0 = \frac{8\pi\gamma_N}{c^4} = 2.07038\cdot 10^{-48}\,\text{g}^{-1}\text{cm}^{-1}\text{s}^2 \tag{2}$$

instead of ((46b) in [1]) gives the slightly altered numerical value

$$\hat{\varkappa} = \frac{\varkappa_0 A_0^2}{6} = 3.45063\cdot 10^{5}\,\text{g}^{-1}\text{cm}\,\text{s}^2\,. \tag{3}$$

For practical reasons it is convenient to go over from the world radius K to the rescaled world L by means of the relation $K = A_0 L$. Our essential changes start with section 4.2 of [1].

Then instead of ((47) in [1]) stands (1a). Let us now remember the dimensionless cosmological electromagnetic radiation constant ((46a) in [1])

$$\Delta = \frac{\varkappa_0 \Lambda_0}{A_0^2}\,, \tag{4}$$

where the cosmological electromagnetic radiation constant ((39d) in [1])

$$\Lambda_0 = \frac{1}{3}\,w_{(r)}K^4 = \frac{1}{3}\,w_{(r)|p}\,A_0^4 L_p^4 \qquad (p \text{ present}) \tag{5}$$

with the radiation energy density according to the Stefan–Boltzmann electromagnetic radiation law

$$w_{(r)} = a_{SB}T^4 \qquad (a_{SB} \text{ Stefan–Boltzmann electromagnetic constant}) \tag{6}$$

occurs. Hence results

$$\Delta = 2.8791 \cdot 10^{-7}\, L_p^4\,. \tag{7}$$

As in paper [1] we have to solve numerically the system of 3 non-linear differential equations ((45) in [1]) for the rescaled world radius L, the scalaric world function σ and the cosmological scalaric energy density (scalerg density) ϑ, but in this case with the following numerically changed initial conditions (dot means derivative with respect to η)

$$\begin{aligned} &\text{a)} \quad L_0 = L(\eta = 0) = 9.95 \cdot 10^{-5}\,, \qquad \dot{L}(\eta = 0) = \sqrt{\frac{\Delta}{L_0^2} - 1}\,, \\ &\text{b)} \quad \sigma(\eta = 0) = 0\,, \qquad \dot{\sigma}(\eta = 0) = 0\,, \\ &\text{c)} \quad \vartheta(\eta = 0) = 3.1 \cdot 10^9\,\mathrm{g\,cm^{-1}s^{-2}} \end{aligned} \tag{8}$$

and the changed free parameters

$$\begin{aligned} &\text{a)} \quad C_0 = 2.73 \cdot 10^{-4}\,, \\ &\text{b)} \quad L_p = 2.10508 \qquad \text{(present rescaled world radius).} \end{aligned} \tag{9}$$

As explained in our recent publications, it was a rather complicated problem to find such values for the free parameters leading to the empirical values (1).

Results of the computer calculations being important for application are besides (9b) the following present numerical values:

$$\begin{aligned} &\text{a)} \quad \dot{L}_p = 0.16\,, \quad \text{b)} \quad \sigma_p = 22.4626 \quad \text{(scalaric world function),} \\ &\text{c)} \quad \dot{\sigma}_p = 0.392\,. \end{aligned} \tag{10}$$

According to our cosmology the cosmos starts with the "big start" (Urstart), introduced by us in our previous publications, instead of the usual big bang. In the interval (0, σ, 1) lies the cosmological repulsion era (antigravitational era) followed by the cosmological attraction era (gravitational era) for $\sigma > 1$. The cosmological transition point lies at $\sigma = 1$. In both eras act gravitational and scalaric forces simultaneously, where in the repulsion era the scalaric forces and in the attraction era the gravitational (metric) forces dominate.

The following 4 figures demonstrate the temporal course of the 3 integration functions wanted (ordinate values). The abscissa denotes the rescaled time parameter η from the big start at $\eta = 0$ to the present time at $\eta = \eta_p = 13$, where the unit $\Delta\eta = 1$ approximately corresponds to 1 billion years. In Figure 1 the abscissa refers to about 100 years. In Figure 2 the range of the abscissa is circa 14 billions of years. Similar is the situation in Figures 3 and 4.

After knowing the 3 integration functions mentioned above, we are now able to calculate the time-inverse temporal course of the relative cosmological frequency shift ($\bar{\eta} = \eta_p - \eta$ denotes the inverse time parameter, where the unit $\Delta\bar{\eta} = 1$ corresponds to the distance of circa 1 billion light years)

$$z(\bar{\eta}) = \frac{L_p}{L(\bar{\eta})} - 1 \tag{11}$$

shown in Figure 5.

The results presented in the 5 preceding figures reflect, apart from the problem of the waviness of some curves, the empirical situation described by the 3 equations (1). As mentioned above, for the present time the value $\eta_p = 13$ is valid. On this basis we get following further results for the present time:

$$t_{AW} = 13.75 \cdot 10^9 \text{ y} \qquad \text{(age of the world)}, \tag{12a}$$

$$K_p = 2.11 \cdot 10^{27} \text{ cm} \qquad \text{(world radius)} \tag{12b}$$

and

$$H_p = 70.6 \, \frac{\text{km}}{\text{s Mpc}} \qquad \text{(Hubble parameter)}, \tag{13a}$$

$$q_p = 21.8 \qquad \text{(deceleration parameter)}. \tag{13b}$$

The cosmological transition point defined above corresponds to the value

$$\text{a)} \quad \sigma = 1, \quad \text{i.e.} \quad \text{b)} \quad \eta = 2.675 \cdot 10^{-8}. \tag{14}$$

Hence follows for the duration of the repulsion era

$$t_{RE} = 28.3 \text{ y}. \tag{15}$$

Of particular interest of the last two diagrams are the intervals with negative-valued quantities H and q. According to our cosmology at the present time both quantities are positive. With respect to recent negative-valued observational results of the deceleration parameter one has to take into account the time era of the emission of electromagnetic waves of the sources.

The recently published figures and particularly the above figures show that our cosmology is able to explain a waviness (non-monotony) of the temporal curves, i.e. equal redshift values for different distances if such a waviness would be observed at all. This prediction may be of basic interest for the observational research. If only a very weak waviness or no waviness would empirically be observed, then such a situation were not a disproof of PUFT, but would primarily raise for us the question after the finality of the cosmological WMAP-parameters chosen as basis for our numerical calculations here. In the appendix to this paper we once more will shortly treat this subject.

2. Numerical values for further important cosmological quantities

2.1. MECHANICAL QUANTITIES:

For further application we need the following formulas:
Cosmological mass density ((28) in [1])

$$\mu = \mu^{(\sigma)} \sigma \tag{16}$$

defined by means of the scalaric world function and the scalaric mass density (scalmass density)

$$\mu^{(\sigma)} = \frac{1}{c^2} \vartheta ; \tag{17}$$

kinetic temperature of the mechanical gas ((42) in [1]) (k Boltzmann constant)

$$T = \frac{m_0^{(\sigma)} c^2}{3 \, \mathrm{k} L \sqrt{C_0} \sqrt{1 + C_0 \sigma^2 L^2}} ; \tag{18}$$

mechanical pressure of the mechanical gas ((44) in [1])

$$p = \frac{\mu c^2}{3(1 + C_0 \sigma^2 L^2)} = \frac{m_0^{(\sigma)} c^2 n}{3 \sqrt{C_0} L \sqrt{1 + C_0 \sigma^2 L^2}} ; \tag{19}$$

particle number density of the mechanical gas

$$n = \frac{\sigma L \vartheta \sqrt{C_0}}{m_0^{(\sigma)} c^2} \cdot \frac{1}{\sqrt{1 + C_0 \sigma^2 L^2}} ; \tag{20}$$

particle velocity (mean value) of the mechanical gas

$$v = \frac{c}{\sqrt{1 + C_0 \sigma^2 L^2}} . \tag{21}$$

In the formulas (18) and (19) appears the hypothetically introduced prescalmass (Urmasse) of the particles of the mechanical gas:

$$m_0^{(\sigma)} = \frac{3\sqrt{C_0}}{A_0} \left(\frac{45 \Lambda_0 h^3}{8 \pi^5 c^5} \right)^{1/4} = 4.36 \cdot 10^{-38} \, \mathrm{g} . \tag{22}$$

Later it will be shown how we came to this formula.

Numerical values of the mechanical quantities (16) to (21) at the big start:

$$\begin{aligned} &\text{a)} \quad \mu = 0\,, \quad \text{b)} \quad \mu^{(\sigma)} = 3.44 \cdot 10^{-12}\,\mathrm{g\,cm^{-3}}\,, \quad \text{c)} \quad T = 5.77 \cdot 10^{4}\,\mathrm{K}\,, \\ &\text{d)} \quad p = 0\,, \quad \text{e)} \quad n = 0\,, \quad \text{f)} \quad v_{rel} = 1\,. \end{aligned} \tag{23}$$

Numerical values of the mechanical quantities (16) to (21) at the present time:

$$\begin{aligned} &\text{a)} \quad \mu = 4.54 \cdot 10^{-28}\,\mathrm{g\,cm^{-3}}\,, && \text{b)} \quad \mu^{(\sigma)} = 2.02 \cdot 10^{-29}\,\mathrm{g\,cm^{-3}}\,, \\ &\text{c)} \quad T = 2.15\,\mathrm{K}\,, && \text{d)} \quad p = 8.45 \cdot 10^{-9}\,\mathrm{Pa}\,, \\ &\text{e)} \quad n = 2.85 \cdot 10^{8}\,\mathrm{cm^{-3}}\,, && \text{f)} \quad v_{rel} = 0.79\,. \end{aligned} \tag{24}$$

2.2. ELECTROMAGNETIC RADIATION QUANTITIES:

Similar to the above treatment of the mechanical quantities we now present the radiation (photon) quantities:

First we note that with the aim of comparing the radiative and mechanical quantities it is convenient to pass over from the above used radiation energy density to the corresponding radiation mass density by means of the relationship

$$\mu_{(r)} = \frac{1}{c^2}\, w_{(r)}\,. \tag{25}$$

Further we remember that the temperature of the electromagnetic radiation (photon gas) can be derived from the Stefan-Boltzmann radiation equation (6):

$$T_{ph} = \frac{1}{A_0 L}\left(\frac{3\Lambda_0}{a_{SB}}\right)^{1/4}\,. \tag{26}$$

As it is well known, the electromagnetic radiation pressure is related to the electromagnetic energy density by the formula

$$p_{(r)} = \frac{1}{3}\, w_{(r)}\,. \tag{27}$$

Further by using the following formula for the energy of one photon (ν frequency)

$$\varepsilon_{ph} = h\nu \tag{28}$$

from (6) results the electromagnetic energy of the whole closed cosmos

$$E_{(r)|cos} = \frac{6\pi^2\Lambda_0}{A_0 L}\,. \tag{29}$$

In recent publications we could show that the number of photons of the whole cosmos is according to our cosmology a conservation constant, here of following numerical value

$$N_{ph|cos} = \frac{E_{(r)|cos}}{\varepsilon_{ph}} = \frac{2\pi^2}{\mathrm{k}}\left(\frac{\Lambda_0^3 a_{SB}}{3}\right)^{1/4} = 6.81 \cdot 10^{85} \,. \tag{30}$$

From (27) follows the formula for the photon number density

$$n_{(r)} = \frac{1}{\mathrm{k} A_0^3 L^3}\left(\frac{a_{SB}\Lambda_0^3}{3}\right)^{1/4} \,. \tag{31}$$

Numerical values of the radiation quantities (25), (26), (27), (31) at the big start:

$$\begin{aligned} &\text{a)}\quad \mu_{(r)} = 9.29 \cdot 10^{-17}\,\mathrm{g\,cm^{-3}}\,, &&\text{b)}\quad T_{ph} = 5.77 \cdot 10^{4}\,\mathrm{K}\,, \\ &\text{c)}\quad p_{(r)} = 2.79 \cdot 10^{3}\,\mathrm{Pa}\,, &&\text{d)}\quad n_{(r)} = 3.5 \cdot 10^{15}\,\mathrm{cm^{-3}}\,. \end{aligned} \tag{32}$$

Numerical values of the radiation quantities (25), (26), (27), (31) at the present time:

$$\begin{aligned} &\text{a)}\quad \mu_{(r)} = 4.64 \cdot 10^{-34}\,\mathrm{g\,cm^{-3}}\,, &&\text{b)}\quad T_{ph} = 2.725\,\mathrm{K}\,, \\ &\text{c)}\quad p_{(r)} = 1.39 \cdot 10^{-14}\,\mathrm{Pa}\,, &&\text{d)}\quad n_{(r)} = 369.6\,\mathrm{cm^{-3}}\,. \end{aligned} \tag{33}$$

Annotation 1:

From the present Hubble parameter (13a) and the age of the cosmos (12a) we find for the product the numerical result

$$Ht = H_p\, t_{AW} = 0.993 \,. \tag{34}$$

It is interesting that new empirical results lie rather near to this value [3]:

$$\begin{aligned} &\text{a)}\quad Ht = 0.93 \quad \text{(S. Perlmutter et al. 1997)}, \\ &\text{b)}\quad Ht = 0.95 \quad \text{(J. L. Tonry et al. 2003)}. \end{aligned} \tag{35}$$

Annotation 2:

For empirical purposes the following two quantities are of basic nature:

$$\text{a)}\quad h(\eta) = \frac{1}{L(\eta)}\frac{\mathrm{d}L(\eta)}{\mathrm{d}\eta} \quad \text{with} \quad \text{b)}\quad h_p = 7.62 \cdot 10^{-2} \tag{36}$$

and

$$\text{a)} \quad \sigma_{rel}(\eta) = \frac{1}{\sigma(\eta)} \frac{\mathrm{d}\sigma(\eta)}{\mathrm{d}\eta} \quad \text{with} \quad \text{b)} \quad \sigma_{rel|p} = 1.75 \cdot 10^{-2} \,. \tag{37}$$

The first quantity plays an important role with respect to the Hubble parameter and more general with respect to our predicted influence of the cosmological expansion on the frequency of periodically acting physical systems:

$$\begin{aligned} &\text{a)} \quad H(\eta) = 927\, h(\eta)\, \mathrm{s}^{-1} \,, \quad \text{b)} \quad \frac{1}{\omega(\eta)} \frac{\mathrm{d}\omega(\eta)}{\mathrm{d}\eta} = -h(\eta) \,, \\ &\text{c)} \quad \left(\frac{1}{\omega} \frac{\mathrm{d}\omega}{\mathrm{d}t} \right)_p = -h_p \frac{\mathrm{d}\eta}{\mathrm{d}t} = -7.21 \cdot 10^{-11}\, \mathrm{y}^{-1} \,. \end{aligned} \tag{38}$$

The second quantity is of considerable importance for the deviation of the motion of bodies, orbiting about a gravitating center, from Newtonian physics. As pointed out in the appendix, with respect to the numerical values we are reserved.

3. Hypothetical cosmological scalon and dark matter

Our 2-component model of the cosmos is based on two different sorts of gases: photon gas and mechanical particle gas. In the papers [5] we were able to find a way to the determination of the concrete mass of such a cosmological particle by the hypothesis:

The temperature of the mechanical particle gas (18) and of the photon gas (26) may coincide at the big start of the cosmological expansion.

This assumption even allowed us to derive the formula and to calculate the numerical value of the mass of such a particle (22). Because of the basic character of this cosmological gas particle we name it "scalon" pointing to our term "scalarism". The word scalaron could not be chosen, since for a long time it already has been used in a different context. For brevity we analogously take the corresponding obvious word "scalon gas" for the mechanical particle gas.

Working theoretically with the mass of the scalon we are confronted with the distinction between different mass notions of such a particle. The reason for this necessity lies in the specific nature of the expanding cosmos.

For a better understanding of this situation let us remember the different notions of mass for a body in PUFT and in the Einstein theory:

As explained in our publications on PUFT since many years, we attribute to a usual body the scalmass (scalaric mass) $\mathcal{M}$ as the primary constant basic quantity. Study of the orbital motion of such a body leads to the introduction of the inertial mass caused by the scalaric field

$$M = \mathcal{M}\sigma \tag{39}$$

and further in application of this concept to the formula for the inertial mass of the central body

$$M_c = \mathcal{M}_c \sigma. \tag{40}$$

One should remember that in PUFT the "equivalence principle of motion" (independence of the motion on the mass of the moving test body) is fulfilled.

In the Einstein theory the "equivalence principle of mass" (equality/identity of the inertial mass and the gravitational mass) is used. Since with respect to this equivalence principle both the Einstein theory and the Newton theory coincide, we get the identification of the gravitational mass in both theories:

$$M_{cE} = M_{cN} \,. \tag{41}$$

The essential difference between PUFT and the Einstein theory is so that in PUFT the inertial mass M_c (40) is because of the cosmological adiabatic time dependence also time dependent, whereas in the Einstein theory the gravitational mass M_{cE} is (usually) a time independent constant, i.e. for basic reasons the following inequality yields:

$$M_c \neq M_{cE} \,. \tag{42}$$

Furthermore, we know from the Einstein theory that, starting from the rest mass of a body, we have to take into account the relativistic velocity effect leading to the dynamical mass (momentum mass) and the metric inducing effect leading to a metrically changed mass of a body being in a metric (gravitational) field.

As mentioned above, with respect to a cosmological scalon we have to begin with the constant scalmass (in this case urmass) and then to go over to the inertial mass according to (39) caused by the time dependent scalaric world function σ. Now we have to take into account the diminishing effect of the velocity of the scalon by the cosmological expansion which according to the relativistic velocity effect leads to a mass which in the framework of the Einstein theory is called rest mass of a particle being the starting point in this theory.

Our further cosmological considerations are based on an approximately uniform gas with a mean velocity of the scalons. The peculiar motions of the scalons are neglected.

Applying this concept to the cosmological scalon, we are led to following results:

urmass of the scalon according to (22):

$$m_0^{(\sigma)} = 4.36 \cdot 10^{-38}\ \mathrm{g}\,; \tag{43}$$

scalaric-inertial mass of the scalon (scalaric effect):

$$m_{s|scalon} = m_0^{(\sigma)} \sigma = 9.8 \cdot 10^{-37}\ \mathrm{g}\,; \tag{44}$$

scalaric-dynamical mass of the scalon (short: scalaric rest mass identical with the usual rest mass of an elementary particle) caused by the scalaric and the velocity effect (cosmological expansion)

$$m_{scalon} = m_0^{(\sigma)} \frac{\sqrt{1 + C_0 \sigma^2 L^2}}{L\sqrt{C_0}} = 1.59 \cdot 10^{-36}\,\mathrm{g}$$
$$= 8.95 \cdot 10^{-4}\,\frac{\mathrm{eV}}{\mathrm{c}^2}\,. \qquad (45)$$

Reflecting on this rather small rest mass of the cosmological scalon, one is tempted to remember the parton concept of R.P. Feynman (1969). Here the correspondence reads: 10^{12} scalons $\longrightarrow$ 1 proton instead of 10^8 partons $\longrightarrow$ 1 proton.

But one should note that this is only a correspondence, since the physical situation is quite different, as we shortly point out in the following.

According to our basic cosmological concept of a 2-component system immediately after the big start (scalon gas and photon gas) the scalon gas comprises per definition all matter ("materia" without electromagnetism). Now we divide the matter into main-baryonic matter (baryons and electrons, perhaps mostly identical with the visible matter) and the "rest matter" (dark matter and the various sorts of massive (with rest mass) and non-massive (without rest mass) elementary particles). As empirical experience shows, in the second category dark matter overwhelmingly dominates the other sorts of matter. Therefore we name this category shortly dark matter without knowing in detail the exact physical property of this amount of matter.

In our paper [1] and in other publications we tentatively proposed to interpret the big value of the cosmological mass density $\mu_p = 4.54 \cdot 10^{-28}\,\mathrm{g\,cm^{-3}}$ (24a) (here obtained as a result of the used WMAP parameters (1)), compared to the visible mass density (stars, globular clusters, galaxies etc.) of about $\mu_{visible} \approx 5 \cdot 10^{-31}\,\mathrm{g\,cm^{-3}}$ as follows:

The ratio $\mu_{visible}/\mu_p \approx 10^{-3}$ shows that only a "marginal" part of the cosmological matter is "normal matter", i.e. the overwhelming part of the matter is the so-called "dark matter". This scalon gas penetrates all matter, particularly also the existing compact objects of normal matter which look like buoys in the dark matter sea.

This concept corresponds to the empirical estimates on dark matter being under discussion, particularly within the framework of the 4-dimensional scalar-tensor theory leading on a fully different theoretical basis to the concepts of dark matter and "dark energy".

In this context we point to our results presented in the appendix of this paper, where this above ratio of only about 1/10 is perhaps physically better understandable.

4. Treatment of a static spherically symmetric perfect fluid sphere

Assumption:

For practical reasons we restrict our further calculations for the whole article to the second order Taylor expansion in the expansion parameter γ which will be defined later. Numerical computations are restricted to the first order (linear approximation).

4.1. EXTERIOR SOLUTION

In 1995 we succeeded in finding an exact static spherically symmetric vacuum exterior solution [4], [1]. Using the metric

$$\mathrm{d}s^2 = e^{\alpha(r)}\mathrm{d}r^2 + r^2(\mathrm{d}\theta^2 + \sin^2\theta\mathrm{d}\varphi^2) - e^{\beta(r)}(\mathrm{d}x^4)^2 \tag{46}$$

this solution takes the following form if we consider as source a perfect fluid sphere (central body with radius R):

$$\begin{gathered} \text{a)}\quad e^{\alpha(r)} = 1 + \frac{r_g}{r} + \left(\frac{r_g}{r}\right)^2\left(1 + \frac{\tau}{4}\right), \qquad \text{b)}\quad e^{\beta(r)} = 1 - \frac{r_g}{r}, \\ \text{c)}\quad \sigma(r) = \sigma_\infty + \frac{r_g}{R}\sigma_1 + \left(\frac{r_g}{R}\right)^2\sigma_2 + \frac{1}{2}\left(\frac{r_g}{r}\right)\sqrt{\tau} + \frac{1}{4}\left(\frac{r_g}{r}\right)^2\sqrt{\tau}\,. \end{gathered} \tag{47}$$

Here σ_∞, σ_1, σ_2, τ, r_g are constants of integration whose physical meaning will become clear later. The Einstein theory corresponds to the limiting case $\tau \to 0$. The last mentioned constant of integration r_g will be interpreted as the gravitational radius of the central body with the inertial mass M_c:

$$r_g = \frac{\varkappa_0 M_c c^2}{4\pi} = \frac{2\gamma_N M_c}{c^2}\,. \tag{48}$$

In this context one should pay attention to the following situation: The solution (47) refers to pseudo-euclidicity at infinity. It is our aim to link the time dependent cosmology with the cosmogony in the cosmos by admitting a weak (adiabatic) time dependence of the scalaric world function $\sigma(t)$.

For the following Taylor series expansions it is convenient to use the above mentioned expansion parameter

$$\gamma = \frac{r_g}{R} \tag{49}$$

which in the astrophysical cases treated is extremely small.

Recently A. Gorbatsievich could clarify the relationship between our exact exterior solution and the Heckmann-Jordan-Fricke solution in the Jordan theory. Further he found the interesting fact that in the case of a point-like source our solution allows the interpretation of the source as a naked singularity, i.e. a singularity not having a horizon [6]. Of course, this fact raises the question after the existence of black holes in PUFT.

Now we expand the solution (47) and find the result

$$\text{a)}\quad \alpha(r) = \frac{R\gamma}{r} + \frac{R^2\gamma^2\left(1+\frac{\tau}{2}\right)}{2r^2}\,, \quad \text{b)}\quad \beta(r) = -\frac{R\gamma}{r} - \frac{R^2\gamma^2}{2r^2}\,,$$
$$\text{c)}\quad \sigma(r) = \sigma_\infty + \gamma\sigma_1 + \gamma^2\sigma_2 + \frac{R\gamma\sqrt{\tau}}{2r} + \frac{R^2\gamma^2\sqrt{\tau}}{4r^2}\,. \tag{50}$$

Concluding this section we introduce the gravitational potential corresponding to the metric coefficient g_{44} by the definition

$$\phi = c^2 \ln \sqrt{-g_{44}}\,. \tag{51}$$

From the above metric we obtain for the exterior potential

$$\phi_e(r) = -\frac{R\gamma c^2}{3r} - \frac{R^2\gamma^2 c^2}{4r^2} \tag{52}$$

if we apply the usual normalization: vanishing gravitational potential at infinity.

4.2. SKETCH OF THE APPROXIMATE INTERIOR SOLUTION

4.2.1. *Introduction to the computer approach*

We consider a perfect fluid sphere (R radius) with the physical properties of mass density $\mu(r)$ and pressure $p(r)$. The metric used corresponds to the exterior metric (46) with exchanged metric functions ($\alpha \to a$, $\beta \to b$, $\sigma \to s$):

$$\mathrm{d}s^2 = e^{a(r)}\mathrm{d}r^2 + r^2(\mathrm{d}\theta^2 + \sin^2\theta\,\mathrm{d}\varphi^2) - e^{b(r)}(\mathrm{d}x^4)^2\,. \tag{53}$$

Similar to the external case we have to exploit both the field equations ((17)[1]) and ((21)[1]):

$$\text{a)}\quad R^{ij} - \tfrac{1}{2}\,Rg^{ij} = \varkappa_0 T^{ij}\,, \quad \text{b)}\quad \sigma^{,j}{}_{;j} = -\frac{\varkappa_0\vartheta}{2} \tag{54}$$

with the energy tensors (u^i four-velocity, ϑ scalerg density)

$$\text{a)}\quad T^{ij} = \Theta^{ij} + S^{ij}\,, \quad \text{b)}\quad \Theta^{ij} = -\left(\mu + \frac{p}{c^2}\right)u^i u^j - pg^{ij}\,,$$
$$\text{c)}\quad S^{ij} = \frac{2}{\varkappa_0}\left(\sigma^{,i}\sigma^{,j} - \tfrac{1}{2}\,g^{ij}\sigma_{,k}\sigma^{,k}\right)\,. \tag{55}$$

Applying this set of equation to the sphere considered we are left with the following coupled system of 4 non-linear differential equations (prime means derivative with respect to r):

$$\begin{aligned}
&\text{a)}\quad b'' + \frac{2b'}{r} - \frac{1}{2}\, b'(a' - b') = \varkappa_0 e^{a}(\vartheta s + 3p)\,,\\
&\text{b)}\quad s'' + \frac{2s'}{r} + \frac{1}{2}\, s'(b' - a') = -\frac{1}{2}\,\varkappa_0 e^{a}\vartheta\,,\\
&\text{c)}\quad a' + rs'^{\,2} + \frac{1}{r}\left(e^{a} - 1\right) = \varkappa_0 e^{a}\vartheta s\,,\\
&\text{d)}\quad p - s\vartheta = \frac{2}{\varkappa_0 r^2}\left(e^{-a} - 1\right) + \frac{1}{\varkappa_0 r}\, e^{-a}(b' - a')\,.
\end{aligned} \tag{56}$$

We spent some time to find an exact solution but without success. Finally we decided to use the method of Taylor expansion: first with respect to the radial coordinate r, then with respect to the parameter τ appearing in our cosmology as rather small, and at last with respect to the expansion parameter γ defined above. Already in rather low order expansion it turned out that, apart from structural insights and qualitative output, a considerable effort is necessary to find good results for concrete physical prediction. We gave up and started with computer methods, strongly supported by A. Gorbatsievich. The numerical results are even up to the second order in γ so voluminous that here we are only able to sketch the way of calculation and to report the set of formulas in linear approximation.

Our ansatzes read:

$$\begin{aligned}
&\text{a)}\quad a(r) = A_0 + \gamma a_1(r) + \gamma^2 a_2(r)\,,\\
&\text{b)}\quad b(r) = B_0 + \gamma b_1(r) + \gamma^2 b_2(r)\,,\\
&\text{c)}\quad s(r) = S_0 + \gamma s_1(r) + \gamma^2 s_2(r)\,,\\
&\text{d)}\quad \mu(r) = \gamma(m_1 + m_2 r^2 + m_3 r^4) + \gamma^2(M_1 + M_2 r^2 + M_3 r^4)\,,\\
&\text{e)}\quad p(r) = \gamma p_1(r) + \gamma^2 p_2(r)\,.
\end{aligned} \tag{57}$$

The expansion coefficients are constants. Further we have to remember the contents of the relations (16) and (17) in the form

$$\vartheta = \frac{\mu c^2}{\sigma}\,. \tag{58}$$

During the process of calculating, including the integration of the occurring differential equations, we must take into account the regularity of the appearing integration functions. Further we have to apply the comparison of coefficients of the resulting expressions of the same radial order.

Since there is not publication space enough, we have to refrain from reproducing the 2-order results for the field functions $a(r)$, $b(r)$, $s(r)$ and the physical quantities $\mu(r)$, $p(r)$.

4.2.2. *Boundary conditions*

Exploiting the field equations (56) on the surface of the sphere ($r \to R$), we arrive at the following results (combining the interior and corresponding exterior solutions):

$$\begin{aligned} &\text{a)} \quad \text{continuity of the } a\,\alpha\text{-function, } b\,\beta\text{-function, } s\,\sigma\text{-function and} \\ &\qquad \text{of the first derivatives of the } b\,\beta\text{-function, } s\,\sigma\text{-function,} \\ &\text{b)} \quad \text{jump of the derivative of the } a\,\alpha\text{-function,} \\ &\text{c)} \quad \text{vanishing pressure at the surface: } p(R) = 0. \end{aligned} \tag{59}$$

4.2.3. *Results in linear approximation*

The jump (59b) is given by

$$\Delta = a'(R) - \alpha'(R) = \Delta_1 \gamma = \frac{7}{R} - \frac{2\varkappa_0}{15}\, c^2 R(10m_1 + 3m_2 R^2)\,. \tag{60}$$

The field functions read:

$$a(r) = \frac{\gamma}{15R^6}\Big[15r^6 + c^2R^2\varkappa_0 r^2\Big(-5m_1r^4 - 3m_2R^2r^4 + 5m_1R^4 + 3m_2R^4r^2\Big)\Big], \tag{61a}$$

$$b(r) = \frac{\gamma}{180R^6}\Big[30(r^6 - 7R^6) - c^2R^2\varkappa_0(r^2 - R^2)^2 \times \times\Big\{3m_2R^2(2r^2 + R^2) + 10m_1(r^2 + 2R^2)\Big\}\Big], \tag{61b}$$

$$s(r) = \sigma_\infty + \frac{\gamma}{16216200R^6\sigma_\infty}\Big[\text{order } r^6\text{, very long}\ldots\Big], \tag{61c}$$

$$\phi_i(r) = \frac{\gamma c^2}{360R^6}\Big[30(r^6 - 7R^6) - c^2R^2(r^2 - R^2)^2\Big\{3m_2R^2(2r^2 + R^2) + 10m_1(r^2 + 2R^2)\Big\}\varkappa_0\Big], \tag{61d}$$

whereas the physical quantities acquire the form:

$$\mu(r) = \gamma\Big[m_1 + m_2r^2 - \frac{7r^4(-15 + 5m_1c^2R^2\varkappa_0 + 3m_2c^2R^4\varkappa_0)}{15c^2R^6\varkappa_0}\Big], \tag{62a}$$

$$p(r) = \frac{\gamma^2}{13500\sigma_\infty^2 R^{12}\varkappa_0}\Big[\text{order } r^{10}\text{, very long}\ldots\Big]. \tag{62b}$$

One should realize that the Taylor expansion of the pressure begins with a second-order term.

Further one can prove that the boundary conditions (59) are fulfilled, where one has to specialize the interior solution (61) and (62b) as well as the exterior solution (50) to the values at the surface of the sphere.

Let us furthermore mention that during this rather tedious recent calculation process we arrived at the following results for the parameters:

$$\tau = \frac{1}{\sigma_\infty}, \tag{63a}$$

$$m_3 = -\frac{7}{15c^2R^6\varkappa_0}\left(-15 + 5c^2m_1R^2\varkappa_0 + 3c^2m_2R^4\varkappa_0\right), \tag{63b}$$

$$\begin{aligned}\sigma_1 = &-\frac{(1+\sigma_\infty^2)}{4054050\sigma_\infty}\Big[1091475 + 210c^2R^2(110m_1 + 27m_2R^2)\varkappa_0 \\ &+ 4c^4R^4(1540m_1^2 + 600m_1m_2R^2 + 63m_2^2R^4)\varkappa_0^2\Big].\end{aligned} \tag{63c}$$

Both the last relations were used for eliminations during the calculations.

The relation (63a) has for astrophysics a significant importance and therefore it deserves a concrete discussion:

Let us first mention that the Schwarzschild solution turns out to be the special case of the external solution (50) for $\tau \to 0$. This knowledge opens insight into the relationship between PUFT and the Einstein theory.

Furthermore, by a view on (50c) we learn that the constant of integration σ_∞ represents the scalaric world function at infinity. Thus this parameter gets its astrophysical interpretation. But one has to realize that according to (50c) the condition $\left|\frac{\gamma\sigma_1}{\sigma_\infty}\right| \ll 1$ has to be fulfilled. This is indeed the case, since the parameter γ is for the examples numerically considered extremely small.

Another problem has also to be discussed: Our exterior solution (50) is valid for the case of pseudo-euclidicity at infinity. But reality for the present time is that our cosmological model shows a rather small curvature and a small (adiabatic) time dependence, as it can be seen from (37b), compared to the rather "quick" temporal motion of astrophysical bodies. Therefore we think that our recently described decisive step of linking cosmology and astrophysics, done by us in order to get grasped the cosmological influence on the astrophysical events and even on planetary processes.

Further parameters introduced above either vanish or do not occur in the linear approximation. Resuming this situation we notice: our linear interior solution of such a perfect fluid sphere exhibits the two free parameters m_1, m_2. When later applying our theory to astrophysical objects, we will go into details.

5. Application of the theory to spherically symmetric astrophysical bodies

5.1. TRANSITION TO RELATIVE RADIAL COORDINATE

For applications it is convenient to pass over from the radial coordinate r to the dimensionless relative radial coordinate x by means of the transformation $x = r/R$. Then from both the above sets of formulas (61) and (50) result both the transformed sets of formulas:

$$a(x) = -\frac{\gamma}{15}\Big[-15x^6 + c^2R^2x^2(x^2-1)\Big\{3m_2R^2x^2 + 5m_1(1+x^2)\Big\}\varkappa_0\Big], \tag{64a}$$

$$b(x) = -\frac{\gamma}{180}\Big[-30(x^6-7) + c^2R^2(1-x^2)^2\Big\{10m_1(2+x^2) + 3m_2R^2(1+2x^2)\Big\}\varkappa_0\Big], \tag{64b}$$

$$s(x) = \sigma_\infty + \frac{\gamma}{16216200R^6\sigma_\infty}\Big[\text{order } x^6\text{, very long} \ldots\Big], \tag{64c}$$

$$\phi_i(x) = -\frac{\gamma c^2}{360}\Big[-30(x^6-7) + c^2R^2(x^2-1)^2\Big\{10m_1(2+x^2) + 3m_2R^2(1+2x^2)\Big\}\varkappa_0\Big]; \tag{64d}$$

$$\text{a)}\quad \alpha(x) = \frac{\gamma}{x}, \quad \text{b)}\quad \beta(x) = -\frac{\gamma}{x}, \quad \text{c)}\quad \sigma(x) = \sigma_\infty + \gamma_1\sigma_1 + \frac{\gamma}{2x\sigma_\infty}; \tag{65}$$

Similarly we obtain from (62a) for the mass density the set:

$$\mu(x) = \gamma\Big[m_1 + m_2R^2x^2 - \frac{7x^4}{15c^2R^2\varkappa_0}(-15 + 5c^2m_1R^2\varkappa_0 + 3c^2m_2R^4\varkappa_0)\Big], \tag{66a}$$

$$\mu_{CE} = m_1\gamma \qquad (CE\text{ center}), \tag{66b}$$

$$\mu_{SF} = \gamma\Big[m_1 + m_2R^2 - \frac{7}{15c^2R^2\varkappa_0}(-15 + 5c^2m_1R^2\varkappa_0 + 3c^2m_2R^4\varkappa_0)\Big] \qquad (SF\text{ surface}) \tag{66c}$$

and from (62b) for the pressure the set:

$$p(x) = \frac{\gamma^2}{13500 R^{12} \sigma_\infty^2 \varkappa_0} \left[\text{order } x^{10}, \text{ very long} \ldots \right], \tag{67a}$$

$$\begin{aligned} p_{CE} = \frac{\gamma^2(\sigma_\infty^2 - 1)}{13500 R^2 \sigma_\infty^2 \varkappa_0} \Big[& 4725 + 15 c^2 R^2 (40 m_1 + 9 m_2 R^2) \varkappa_0 \\ & + c^4 R^4 (400 m_1^2 + 105 m_1 m_2 R^2 + 9 m_2^2 R^4) \varkappa_0^2 \Big], \end{aligned} \tag{67b}$$

$$p_{SF} = 0\,. \tag{67c}$$

5.2. SPECIALIZATION TO MODELS

5.2.1. *Central mass, parameters and models*

Integrating the mass density (up to the second order in γ) over the whole sphere (R radius) leads to a formula of following structure:

$$M_{c|2} = \frac{4\pi r_g}{c^2 \varkappa_0} + \frac{4\pi r_g^2}{R^2} \Big[\quad \Big]. \tag{68}$$

In first order we obtain

$$M_c = \frac{4\pi r_g}{c^2 \varkappa_0}\,. \tag{69}$$

This result coincides with (48). By means of (49) hence we find

$$M_c = \frac{4\pi R \gamma}{c^2 \varkappa_0}\,. \tag{70}$$

In application of our theory to astrophysical bodies let us consider both the parameters M_c (mass) and R (Radius) as astrophysically determined quantities. Hence by means of (70) the parameter γ is determined. Remembering the relation (63b), we see that our treatment of the sphere in linear approximation leads to a theory of the sphere with the two free parameters m_1 and m_2, i.e. to a 2-parametric family of curves which has been investigated further by us.

5.2.2. *Parabola model*

Here we restrict our treatment to the 1-parametric parabola model defined by

$$m_3 = 0 \tag{71}$$

i.e.

$$m_2 = \frac{5}{c^2 R^4 \varkappa_0} - \frac{5 m_1}{3 R^2}\,. \tag{72}$$

In this case we obtain simpler formulas: from (66) for the mass density

$$\begin{aligned} &\text{a)} \quad \mu(x) = \gamma\Big[m_1\Big(1 - \frac{5x^2}{3}\Big) + \frac{5x^2}{c^2R^2\varkappa_0}\Big]\,, \\ &\text{b)} \quad \mu_{CE} = \gamma m_1\,, \quad \text{c)} \quad \mu_{SF} = -\frac{2m_1\gamma}{3} + \frac{5\gamma}{c^2R^2\varkappa_0} \end{aligned} \tag{73}$$

and from (67) for the pressure

$$\begin{aligned} p(x) = {} & \frac{\gamma^2(\sigma_\infty^2 - 1)(1 - x^2)}{108R^2\sigma_\infty^2\varkappa_0}\Big[45 + 6c^2m_1R^2\varkappa_0 + 2c^4m_1^2R^4\varkappa_0^2 \\ & - 9c^4m_1^2R^4\varkappa_0^2x^2 + (-36c^2m_1R^2\varkappa_0 + 12c^4m_1^2R^4\varkappa_0^2)x^4 \\ & + (-45 + 30c^2m_1R^2\varkappa_0 - 5c^4m_1^2R^4\varkappa_0^2)x^6\Big]\,, \end{aligned} \tag{74a}$$

$$p_{CE} = \frac{\gamma^2(\sigma_\infty^2 - 1)}{108R^2\sigma_\infty^2\varkappa_0}\Big(45 + 6c^2m_1R^2\varkappa_0 + 2c^4m_1^2R^4\varkappa_0^2\Big)\,, \tag{74b}$$

$$p_{SF} = 0\,. \tag{74c}$$

6. Kinetic temperature in the perfect fluid sphere

As performed above, we calculated via the field equations without explicit use of the equation of state the radial course of the field quantities but also of the mass density and pressure. By integration we obtained some constants of integration. This way we were able to treat the problem posed to a result to be handled physically.

Further one should remember that the calculation of the radial courses of the mass density and the pressure did not need any input on the kinetic temperature (here short: temperature) of the fluid. The problem of the radial course of the temperature will be treated now.

In general for a fluid material without specific properties the equation of state reads (suppressing further parameters):

$$\text{a) } F(p\,,\mu\,,T) = 0 \text{ (implicit form) or b) } T = T(p\,,\mu) \text{ (explicit form).} \tag{75}$$

From the latter form we can calculate the radial course of the temperature if we know the radial courses of the mass density and the pressure.

Similar to the situation of a perfect (ideal) gas with the equation of state in the form

$$p = n\,\mathrm{k}T \tag{76}$$

(n particle number density, k Boltzmann constant) we now for simplicity model a perfect fluid with the rearranged equation of state (76) in the form

$$\text{a)}\quad T = \frac{m_0}{\mathrm{k}}\frac{p}{\mu} \quad \text{with} \quad \text{b)}\quad \mu = n m_0 \tag{77}$$

(m_0 mass of the particles of the perfect fluid). Of course in this idealization of a fluid the interaction between the particles and irreversible properties are neglected.

If we restrict our further treatment of the temperature to the parabola model, by means of (74a) for the pressure and (73a) for the mass density, we find from (77a) the following formula for the temperature:

$$\begin{aligned} T(x) = {} & \frac{\gamma m_0 c^2 (1-\sigma_\infty^2)(x^2-1)^2}{36\,\mathrm{k}\sigma_\infty^2} \times \\ & \times \frac{1}{-15x^2 + m_1 c^2 R^2(-3+5x^2)\varkappa_0} \Big[45(1+x^2+x^4) \\ & + 6 m_1 c^2 R^2 (1+x^2-5x^4)\varkappa_0 + m_1^2 c^4 R^4 (2-7x^2+5x^4)\varkappa_0^2\Big], \end{aligned} \tag{78a}$$

$$T_{CE} = \frac{\gamma(\sigma_\infty^2-1) m_0 (45 + 6 m_1 c^2 R^2 \varkappa_0 + 2 m_1^2 c^4 R^4 \varkappa_0^2)}{108\,\mathrm{k}\, m_1 R^2 \sigma_\infty^2 \varkappa_0}, \tag{78b}$$

$$T_{SF} = 0\,. \tag{78c}$$

Because of the vanishing pressure at the surface of the sphere this model implies vanishing temperature at the surface, too. In spite of the simplicity of such an approach, we nevertheless in the following apply this model to two typical celestial bodies with the parameters of the sun and the earth, without any claim of identification.

7. Application of the parabola model to the interior of a sun-like body

7.1. GENERAL DATA

We consider the interior of the sun, i.e. we will not take into account the hydrogen convection zone (28.7 % of the radius outside the interior), because in this zone turbulence prevails and therefore the equation of state for a perfect gas is not valid. On the basis of this geometrical restriction the rough parameters read:

$$M_{sun|in} = 1.97 \cdot 10^{33}\,\mathrm{g} \qquad \text{(mass of the interior of the sun),} \tag{79a}$$

$$R_{sun|in} = 4.96 \cdot 10^{10}\,\mathrm{cm} \qquad \text{(radius of the interior of the sun),} \tag{79b}$$

$$r_{g|sun|in} = 2.93 \cdot 10^{5}\,\mathrm{cm} \qquad \text{(gravitational radius)} \tag{79c}$$

$$\gamma_{sun|in} = 5.91 \cdot 10^{-6} \qquad \text{(expansion parameter),} \tag{79d}$$

$$m_{1|sun|in} = 1.52 \cdot 10^{6} \qquad \text{(free choice of this parameter).} \tag{79e}$$

7.2. MASS DENSITY

The radial course (Figure 6) is described by the equation (73a), whereas the mass density in the center (73b) and at the surface (73c) read:

$$\text{a)} \quad \mu_{CE} = 9\,\mathrm{g\,cm}^{-3}\,, \qquad \text{b)} \quad \mu_{SF} = 0.42\,\mathrm{g\,cm}^{-3}\,. \tag{80}$$

In the corresponding literature on traditional sun models one finds for the center the considerably higher value $\mu_{CE} \approx 130\,\mathrm{g\,cm}^{-3}$.

7.3. PRESSURE

For the radial course (Figure 7) we have to use formula (74a). The values of the pressure in the center (74b) and at the surface (74c) are given by

$$\text{a)} \quad p_{CE} = 1.17 \cdot 10^{15}\,\mathrm{Pa}\,, \quad \text{b)} \quad p_{SF} = 0\,. \tag{81}$$

Traditional sun models give for the center the value $2 \cdot 10^{16}$ Pa which is one order of magnitude higher than the value (81a).

7.4. TEMPERATURE

The formula for the course of the temperature (78a) exhibits the mass of the particles of the gas considered (Figure 8). It is convenient to relate this mass to the mass of the proton (as a representative of the 1-atomic hydrogen) $m_{0|proton} = 1.675 \cdot 10^{-24}$ g by the formula

$$m_0 = S_m \cdot m_{0|proton}\,, \tag{82}$$

where we name S_m "scalaric material parameter". In the case of the non-ionized hydrogen gas of the sun this parameter takes the value $S_m = 2$. The radial course of the temperature (figure 8) is given by the formula (78a), whereas the values at the center (78b) and at the surface (78c) read:

$$\text{a)} \quad T_{CE} = 3.15 \cdot 10^7\,\mathrm{K}\,, \quad \text{b)} \quad T_{SF} = 0\,. \tag{83}$$

Traditional sun models find the value $T_{CE} \approx 1.5 \cdot 10^7$ K which is very near to the value (83a). Because of the vanishing pressure at the surface the value (83b) is understandable.

8. Application of the parabola model to the interior of an earth-like body

8.1. GENERAL DATA

In the case of the sun the object of application was a gaseous sphere. In the case of the interior of the earth we are confronted with a plastically deformable material of extremely high viscosity, i.e. far away from a perfect fluid. But since our calculations are oriented to quasi-statics (nearly vanishing velocities after the long cosmological time of evolution), we have some hope for a certain applicability. As a matter of interest we nevertheless apply our theory to such a material, taking over the parameters of the earth (neglecting the crust). On this basis the rough parameters read:

$$M_{earth} = 5.974 \cdot 10^{27}\,\text{g} \qquad \text{(mass)}, \tag{84a}$$
$$R_{earth} = 6.37 \cdot 10^{8}\,\text{cm} \qquad \text{(radius)}, \tag{84b}$$
$$r_{g|earth} = 0.889\,\text{cm} \qquad \text{(gravitational radius)}, \tag{84c}$$
$$\gamma_{earth} = 1.4 \cdot 10^{-9} \qquad \text{(expansion parameter)}, \tag{84d}$$
$$m_{1|earth} = 7.74 \cdot 10^{9} \qquad \text{(free choice of this parameter)}. \tag{84e}$$

8.2. MASS DENSITY

As in the preceding case the radial course is described by the equation (73a), whereas the mass density in the center (73b) and at the surface (73c) read (Figure 9):

$$\text{a)} \quad \mu_{CE} = 10.8\,\text{g}\,\text{cm}^{-3}\,, \quad \text{b)} \quad \mu_{SF} = 2\,\text{g}\,\text{cm}^{-3}\,. \tag{85}$$

In the corresponding literature on traditional earth models one finds for the center the near-by value $\mu_{CE} \approx 13\,\text{g}\,\text{cm}^{-3}$.

8.3. PRESSURE

For the radial course (Figure 10) we have to use formula (74a). The values of the pressure in the center (74b) and at the surface (74c) are given by

$$\text{a)} \quad p_{CE} = 3.18 \cdot 10^{11}\,\text{Pa}\,, \quad \text{b)} \quad p_{SF} = 0\,. \tag{86}$$

Traditional earth models give for the center the value $p_{CE} \approx 3.6 \cdot 10^{11}$ Pa being near-by to the value (86a).

8.4. TEMPERATURE

The formula for the radial course of the temperature (78a) exhibits the mass of the particles of the gas considered (figure 10). We are aware that in the case of the interior of the earth we are confronted with a material far away from a perfect fluid. Nevertheless let us test the situation.

In the section on the temperature of the sun model we related the mass of the particles of the fluid to the mass of the proton by introducing the "scalaric material parameter" S_m.

The radial course of the temperature is given by the formula (78a), whereas the values at the center (78b) and at the surface (78c) read (Figure 11):

$$\text{a)}\quad T_{CE} = 7151\,\mathrm{K}\,, \quad \text{b)}\quad T_{SF} = 0\,. \tag{87}$$

Traditional models of the earth orient towards the value 7000 K which is very near to the value (87a). Because of the vanishing pressure at the surface the value (87b) is understandable. Compared to the big value of the temperature in the center a small value has to be expected at the surface. It is an odd situation that the rather reasonable values found above were the result of the choice $S_m = 2$ as we did it in the case of the interior of the sun.

Annotation 3:

The above application of our theory on celestial bodies (here we chose two extreme cases, further cases were treated) is based on a 1-component homogeneous isotropic material with an extremely simple equation of state. Nevertheless we obtained surprisingly practical results. We think that after refinements our fully new approach to this subject could be of use for the research in cosmogony.

9. Relativistic motion of a test body in a gravitational-scalaric field and the Einstein effects

In this section and in further following parts we resume recent results published in a series of papers. For simplification we mostly refer to the Erice School (2001) [1].

9.1. EQUATIONS OF MOTION OF A MECHANICAL CONTINUUM AND A TEST BODY

Neglecting the inner interaction of the matter considered as well as the thereby resulting irreversible phenomena, we successfully used the model of a perfect fluid described by mass density, pressure and the four-velocity ((25a) in [1]). In the Gauss system of units we obtained on this basis for a continuum the equation of motion ((26) in [1])

$$\begin{aligned}\left(\mu + \frac{p}{c^2}\right) u^k{}_{;l}u^l = \frac{1}{c}\,\varrho_0 B^k{}_l u^l - \left(p^{,k} + \frac{1}{c^2}\,\frac{\mathrm{d}p}{\mathrm{d}\tau}\,u^k\right)\\ -\,\vartheta\left(\sigma^{,k} + \frac{1}{c^2}\,\frac{\mathrm{d}\sigma}{\mathrm{d}\tau}\,u^k\right)\end{aligned} \tag{88}$$

and for a point-like test body the equation of motion ((30) in [1])

$$Mu^k{}_{;l} = \frac{Q}{c}\,B^k{}_l u^l - Mc^2\left[(\ln\sigma)^{,k} + \frac{1}{c^2}\,\frac{\mathrm{d}\ln\sigma}{\mathrm{d}\tau}\,u^k\right]. \tag{89}$$

The symbols used mean: τ proper time, ϱ_0 electric rest charge density, $B^k{}_l$ electromagnetic field strength tensor, M inertial mass according to (39), Q electric charge.

9.2. EINSTEIN EFFECTS

The three Einstein effects refer to the periastron (perihelion) shift of an electrically neutral body ($Q = 0$) orbiting about a central body, to the deflection of electromagnetic waves by a central body and to the frequency shift of electromagnetic waves in the field of a central body.

9.2.1. *Periastron (perihelion) shift*

In this case the equation of motion (89) takes the form

$$u^k{}_{;l}u^l = -c^2\left[(\ln\sigma)^{,k} + \frac{1}{c^2}\frac{d\ln\sigma}{d\tau}\,u^k\right] . \tag{90}$$

We remember that in this equation the gravitational (metric) field occurs in the covariant derivative, whereas the scalaric field appears explicitly. The further treatment goes on as usual. Here the solution (50) for the metric and the scalaric field has to be inserted, where the constant of integration τ (no mix-up with the proper time) has to be eliminated by (63a).

A rather long calculation, where the numerically well fulfilled approximation assumption

$$\frac{r_g}{12a} \ll \frac{1}{\sigma_\infty^2} \tag{91}$$

is used (a semi-axis), leads by means of (10b) to the relative result (compared to the Einstein theory)

$$(\Delta\varphi)_{rel} = \frac{(\Delta\varphi)_{1PUFT}}{(\Delta\varphi)_{1Einstein}} = \frac{1 - \dfrac{2}{3\sigma_\infty^2} - \dfrac{1}{6\sigma_\infty^4}}{\left(1 - \dfrac{1}{\sigma_\infty^2}\right)^2} \approx 1 + \frac{4}{3\sigma_\infty^2} + \frac{3}{2\sigma_\infty^4} \tag{92}$$

(φ polar angle of motion) valid up to the 4th order in $\dfrac{1}{\sigma_\infty}$. This result was already published in recent papers ((68) in [1]).

We remember that the Einstein result for 1 revolution of the orbiting body with the angular momentum L_0 is

$$\text{a)}\quad (\Delta\varphi)_{1Einstein} = \frac{3\pi r_{gE}^2 M_{0E}^2}{2L_0^2}\,, \qquad \text{b)}\quad (\Delta\varphi)_{Einstein|cent} = 42.98''\,, \tag{93}$$

where in formula (93a) the Einstein-Newton concept of mass has to be applied to the orbiting body ($M_{0E} = M_{0N}$). By exchanging in formula (48) $M_c \to M_{cE}$ we are led to the Einstein gravitational radius r_{gE} .

Using here the recent numerical value of the scalaric world function at the present time (10b)

$$\sigma_\infty = \sigma_p = 22.4626\,, \tag{94}$$

we find from (92) the value

$$(\Delta\varphi)_{rel} = 1 + 2.6484 \cdot 10^{-3}\,. \tag{95}$$

Applying this result, we obtain from (92) for the perihelion shift per century within the framework of PUFT

$$(\Delta\varphi)_{PUFT|cent} = 43.094''\,. \tag{96}$$

A rather reliable observational result for the planet mercury orbiting about the sun is

$$(\Delta\varphi)_{obs|cent} = 43.13''\,. \tag{97}$$

Though the difference between (93b) and (97) is very small, this question is of basic interest. The main proposal discussed during the last decades aims at taking into account the additional contribution caused by the mass quadrupole moment of the sun whose empirical value has been rather exactly measured [7],[8]:

$$J_2^\odot \approx 2.3 \cdot 10^{-7}\,. \tag{98}$$

It seems that the Einstein theory is not able to remove this discrepancy between the values (93b) and (97) by adding the quadrupole contribution. Let us therefore investigate the analogous situation for the PUFT result (96).

Transcribing the quadrupole moment correction term given in [9] for PUFT, we find (ε eccentricity, N_{cent} number of revolutions per century)

$$(\Delta\varphi)_{total|cent} = (\Delta\varphi)_{PUFT|cent} + \frac{3\pi R_\odot^2 J_2^\odot N_{cent}}{a^2(1-\varepsilon^2)^2}\,. \tag{99}$$

Equating this result with the observational value of the perihelion shift (97) leads to the theoretical value of the quadrupole moment of the sun

$$J_2^\odot = 2.84 \cdot 10^{-7} \tag{100}$$

being near to the observational value (98).

9.2.2. *Deflection of electromagnetic waves*
Not taking into account the mass quadrupole moment of celestial bodies we some years ago arrived at the formula [1]

$$\frac{\Delta\chi}{(\Delta\chi)_{Einstein}} = \frac{1}{1 - \dfrac{1}{\sigma_p^2}} . \tag{101}$$

If we use the present value (94) of the scalaric world function, we obtain numerically

$$\frac{\Delta\chi}{(\Delta\chi)_{Einstein}} = 1.002 . \tag{102}$$

9.2.3. *Frequency shift of electromagnetic waves*
On the same basis as before we find

$$\frac{\Delta\omega}{(\Delta\omega)_{Einstein}} = \frac{1}{1 - \dfrac{1}{\sigma_p^2}} , \tag{103}$$

$$\frac{\Delta\omega}{(\Delta\omega)_{Einstein}} = 1.002 . \tag{104}$$

10. Non-relativistic motion of a test body in a gravitational-scalaric field and resulting predictions

10.1. EQUATION OF MOTION OF A TEST BODY

We begin our research with the non-relativistic approximation of (89) and arrive in the case of spherically symmetric exterior fields around a central body at

$$M\Big(\frac{\mathrm{d}\boldsymbol{v}}{\mathrm{d}t} + \operatorname{grad}\phi + c^2 \operatorname{grad}\ln\sigma + \frac{\mathrm{d}\ln\sigma}{\mathrm{d}t}\boldsymbol{v}\Big) = Q\Big(\boldsymbol{E} + \frac{\boldsymbol{v}\times\boldsymbol{B}}{c}\Big) . \tag{105}$$

If we restrict our further considerations to an electrically neutral body ($Q = 0$) but moving in a gravitational-scalaric field depending on the radial coordinate r and on time (not only time dependence as in the cosmological case of the scalaric world function), then from (105) results

$$\frac{\mathrm{d}\boldsymbol{v}}{\mathrm{d}t} + \boldsymbol{e}_r a_r + \frac{\mathrm{d}\ln\sigma}{\mathrm{d}t}\boldsymbol{v} = 0 , \tag{106}$$

where

$$a_r = \frac{\partial\phi}{\partial r} + c^2\frac{\partial\ln\sigma}{\partial r} \tag{107}$$

is the radial component of the acceleration of the moving body.

Now we restrict our further calculations to the approximations

$$\begin{aligned} &\text{a)} \quad \frac{r_g}{r} \ll 1\,, \\ &\text{b)} \quad \frac{\sigma - \sigma_\infty}{\sigma_\infty} \ll 1 \quad \text{(neglecting the scalaric self-interaction)}, \\ &\text{c)} \quad \left|\frac{r_g \sigma_1}{R}\right| \ll \sigma_\infty\,. \end{aligned} \tag{108}$$

Then from the exterior solutions (52) and (50c) we obtain

$$\text{a)} \quad \phi_e(r) = -\frac{r_g c^2}{2r}\,, \qquad \text{b)} \quad \sigma(r) = \sigma_\infty + \frac{r_g}{2\sigma_\infty r}\,. \tag{109}$$

Hence the radial acceleration (107) takes the form

$$a_r = \frac{r_g c^2}{2r^2}\left(1 - \frac{1}{\sigma_\infty^2}\right) \tag{110}$$

if we neglect the cosmological adiabatic time dependence in σ_∞.

Now rearranging (48) by means of (40) gives

$$r_g = \frac{2\gamma_N \mathcal{M}_c \sigma_\infty}{c^2}\,. \tag{111}$$

Inserting into (110) leads to

$$a_r = \frac{\gamma_N \mathcal{M}_c \sigma_\infty}{r^2}\left(1 - \frac{1}{\sigma_\infty^2}\right)\,. \tag{112}$$

For the physical interpretation of this expression we split the radial acceleration into the two parts

$$a_r = a_N + a_s \tag{113}$$

with

$$\begin{aligned} &\text{a)} \quad a_N = \frac{\gamma_N \mathcal{M}_c \sigma_\infty}{r^2} && \text{(Newtonian acceleration)}, \\ &\text{b)} \quad a_s = -\frac{\gamma_N \mathcal{M}_c}{\sigma_\infty r^2} = -\frac{a_N}{\sigma_\infty^2} && \text{(scalaric acceleration)}. \end{aligned} \tag{114}$$

Since both these accelerations have the same reciprocal radial dependence, they cannot be separated radially, but this could be done temporally.

10.2. PROBLEM OF TEMPORAL DEPENDENCE OF THE GRAVITATIONAL CONSTANT FROM THE PUFT POINT OF VIEW

This problem is primarily a problem of the time dependence of the radial acceleration (112). Here our investigation does not point to the time dependence through the motion of the test body ($r = r(t)$) but to the adiabatic time dependence via $\sigma_\infty(t)$. Since per definition in our theory γ_N and $\mathcal{M}_c$ are true constants, the questionable time dependence is caused by the modified scalaric world function

$$\sigma^{(S)} = \sigma_\infty \left(1 - \frac{1}{\sigma_\infty^2}\right) . \tag{115}$$

Using this expression, the radial acceleration reads

$$a_r = \frac{\gamma_N \mathcal{M}_c \sigma^{(S)}}{r^2} . \tag{116}$$

This equation suggests to define within the framework of PUFT the gravitational mass of the central body as

$$M_c^{(S)} = \mathcal{M}_c \sigma^{(S)} . \tag{117}$$

By this step formally the Newtonian shape of the radial acceleration (116) is reached:

$$a_r = \frac{\gamma_N M_c^{(S)}}{r^2} . \tag{118}$$

We showed in our recent papers that in PUFT the product "gravitational parameter (instead of gravitational constant) $\times$ mass" can be factorized in different ways. Which one seems to be physically mostly acceptable?

Following our line, we use as gravitational parameter the quantity

$$G_S = \gamma_N \sigma^{(S)} . \tag{119}$$

Then (116) reads

$$a_r = \frac{G_S \mathcal{M}_c}{r^2} \qquad \text{(universal effect).} \tag{120}$$

Here one should realize that G_S numerically differs considerably from γ_N but without affecting the empirical situation.

From (119) results the relative temporal change of the scalaric gravitational parameter ((80) in [1]):

$$\frac{1}{G_S} \frac{\mathrm{d}G_S}{\mathrm{d}t} = \frac{1 + \dfrac{1}{\sigma_\infty^2}}{1 - \dfrac{1}{\sigma_\infty^2}} \left(\frac{\mathrm{d} \ln \sigma}{\mathrm{d}t} \right)_\infty . \tag{121}$$

Using (37b) we find for the present time the value

$$\left(\frac{1}{G_S}\frac{\mathrm{d}G_S}{\mathrm{d}t}\right)_p \approx 1.66 \cdot 10^{-11}\,\mathrm{y}^{-1}\,. \tag{122}$$

Of course, it is of interest how the experiments on this important subject are usually theoretically based: One takes the Newton mechanics with the concept of a time independent constant central mass M_c. Then instead of (118) stands ($G(t)$ time dependent "empirical gravitational constant")

$$a_r = \frac{G(t)M_c}{r^2}\,. \tag{123}$$

Observing the motion of an orbiting satellite gives the new result [10]

$$\left|\frac{1}{G}\frac{\mathrm{d}G}{\mathrm{d}t}\right|_p \lessapprox 1.8 \cdot 10^{-12}\,\mathrm{y}^{-1} \tag{124}$$

being one order of magnitude smaller than (122). If one expects a rough agreement between (122) and (124) on the basis of PUFT, one arrives at doubts about some WMAP parameters.

11. Miscellaneous

In our paper [1] but also in various other publications we treated astrophysical subjects which should also be considered on the basis of PUFT, since the occurring scalaric field leads to new aspects of such basic subjects of research. Let us list some of them partly treated in the paper quoted above:

Adiabatic scalaric approximation of the non-relativistic equation of motion of a test body orbiting about a central body and hence by the cosmological expansion resulting cosmological astrophysical universal effects: positive value of the secular angular acceleration, negative values of the time derivatives of the orbital radius (decrease), revolution period (decrease), excentricity (transition from elliptic to cyclic orbits).

Considering the non-relativistic equation of motion of a test body (105), we realize (in the language of mechanics) a linear "cosmological friction term" with a "cosmological friction coefficient" caused by the temporal dependence of the scalaric world function. This kind of "cosmological induction" leads to a "cosmological heat production" in the moving body. Here the question after the frame of reference for measuring the velocity occurring in the heat power production term appears. The obviously preferred fame of reference is of course the cosmological frame of reference defined by the cosmological metric, but frames of reference of galaxies or stars with their overwhelming masses compared to the masses in the surrounding space are under discussion. The heat production within a heated body leads of course to cosmologically caused expansion of the moving body.

Dark matter accretion through the attractive gravitational-scalaric force around a celestial body yields to an additional negative radial force (towards the central body) which could be of interest in explaining the so-called "Pioneer effect" of satellites.

Further subjects of research refer to the rotation curves of stars in galaxies and to the spiral motion of stars in spiral galaxies.

These listed subjects were shortly treated in [1].

12. Appendix

My two lectures at the Erice School in May 2003 tested my Projective Unified Field Theory with respect to the then new empirical cosmological WMAP values just published three months ago. This publication comprises the results of this comparison. If one takes these basic values (1) as a doubtless input, my theory leads to some results considered as partially problematic by astrophysicists, e.g. the typical waviness effect of cosmological quantities or the time dependence of the gravitational parameter leading to the numerical value (122) compared to (124). As I emphasized above, this possible discrepancy is not a disproof of my theory, since it can be avoided by another choice of the initial values (8) for the system of differential equations. Some months ago Ch.S. Kochanek and P.L. Schechter [11] published an interesting paper on their new measuring result of the Hubble parameter:

$$H_p = (48 \pm 3)\,\frac{\mathrm{km}}{\mathrm{s\,Mpc}} \tag{125}$$

obtained by the method of gravitational lens time delay. This provoking small number is a challenge for PUFT. Therefore I also tested this result (125). In the following I shortly report on my comparison of PUFT with this empirical outcome. For brevity I do not repeat the previous text explaining the formulas, but communicate the changed input and changed outcome, mostly by numerical data.

Instead of (8a), (8c), (9), (10), (12), (13), (36), (37), (38), (14), (15), (22), (43), (44), (45): (32), (33), (36b), (37b), (43), (44), (45):

$$\begin{aligned} &\text{a)}\quad L_0 = 10^{-4}\,, && \text{b)}\quad \vartheta_0 = 3.2\cdot 10^{13}\,\mathrm{g\,cm^{-1}s^{-2}}\,, \\ &\text{c)}\quad C_0 = 1\,, && \text{d)}\quad L_p = 273.41\,; \end{aligned} \tag{126}$$

$$\text{a)}\quad \sigma_p = 5.83\,, \qquad \text{b)}\quad \vartheta_p = 6.75\cdot 10^{-10}\,\mathrm{g\,cm^{-1}s^{-2}}\,; \tag{127}$$

$$\text{a)}\quad \dot{L}_p = 14.18\,, \qquad \text{b)}\quad \dot{\sigma}_p = 9.11\cdot 10^{-3}\,; \tag{128}$$

$$
\begin{array}{lll}
\text{a)} & t_{AW} = 13.75\,\mathrm{y} & \text{(age of the world),} \\
\text{b)} & K_p = 2.73 \cdot 10^{29}\,\mathrm{cm} & \text{(world radius),} \\
\text{c)} & H_p = 48 & \text{(Hubble parameter),} \\
\text{d)} & q_p = 0.487 & \text{(deceleration parameter);}
\end{array}
\tag{129}
$$

$$\left(\frac{\mathrm{d}\ln L}{\mathrm{d}\eta}\right)_p = 5.18 \cdot 10^{-2}\,, \tag{130a}$$

$$\left(\frac{\mathrm{d}\ln \sigma}{\mathrm{d}\eta}\right)_p = 1.56 \cdot 10^{-3}\,, \tag{130b}$$

$$\left(\frac{\mathrm{d}\ln L}{\mathrm{d}t}\right)_p = 4.9 \cdot 10^{-11}\,\mathrm{y}^{-1}\,, \tag{130c}$$

$$\left(\frac{\mathrm{d}\ln \sigma}{\mathrm{d}t}\right)_p = 1.48 \cdot 10^{-12}\,\mathrm{y}^{-1}\,, \tag{130d}$$

$$\left(\frac{1}{\omega}\frac{\mathrm{d}\omega}{\mathrm{d}t}\right)_p = -4.9 \cdot 10^{-11}\,\mathrm{y}^{-1}\,; \tag{130e}$$

$$\text{a)}\quad \sigma = 1\,, \qquad \text{i.e.} \qquad \text{b)}\quad \eta_{RE} = 1.37 \cdot 10^{-9} \tag{131}$$

$$t_{RE} = 1.45\,\mathrm{y} \qquad \text{(duration of the repulsion era).} \tag{132}$$

$$m_0^{(\sigma)} = 3.44 \cdot 10^{-34}\,\mathrm{g} \qquad \text{(urmass),} \tag{133a}$$

$$m_{s|scalon} \approx 2 \cdot 10^{-33}\,\mathrm{g} \qquad \text{(scalaric-inertial mass),} \tag{133b}$$

$$m_{scalon} \approx 2 \cdot 10^{-33}\,\mathrm{g} = 1.13\,\frac{\mathrm{eV}}{\mathrm{c}^2} \qquad \text{(scalaric rest mass).} \tag{133c}$$

Some mechanical quantities at the big start (23):

$$
\begin{array}{l}
\text{a)}\quad \mu = 0\,, \quad \text{b)}\quad T = 7.46 \cdot 10^{6}\,\mathrm{K}\,, \quad \text{c)}\quad p = 0\,, \\
\text{d)}\quad n = 0\,, \quad \text{e)}\quad v_{rel} = 1\,;
\end{array}
\tag{134}
$$

some mechanical quantities at the present time (24):

$$
\begin{array}{ll}
\text{a)}\quad \mu = 4.37 \cdot 10^{-30}\,\mathrm{g\,cm^{-3}}\,, & \text{b)}\quad T = 1.71 \cdot 10^{-3}\,\mathrm{K}\,, \\
\text{c)}\quad p = 5.15 \cdot 10^{-17}\,\mathrm{Pa}\,, & \text{d)}\quad n = 2.18 \cdot 10^{3}\,\mathrm{cm^{-3}}\,, \\
\text{e)}\quad v_{rel} = 6.27 \cdot 10^{-4}\,; &
\end{array}
\tag{135}
$$

$$N_{ph|cos} = 1.5 \cdot 10^{92} = const \quad \text{(number of the photons in the cosmos);} \tag{136}$$

Some radiation quantities at the big start (32):

$$\begin{array}{ll} \text{a)} \;\; \mu_{(r)} = 2.61 \cdot 10^{-8}\,\mathrm{g\,cm}^{-3}\,, & \text{b)} \;\; T_{ph} = 7.46 \cdot 10^{6}\,\mathrm{K}\,, \\ \text{c)} \;\; p_{(r)} = 7.82 \cdot 10^{11}\,\mathrm{Pa}\,, & \text{d)} \;\; n_{(r)} = 7.59 \cdot 10^{21}\,\mathrm{cm}^{-3}\,; \end{array} \tag{137}$$

some radiation quantities at the present time (33):

$$\begin{array}{ll} \text{a)} \;\; \mu_{(r)} = 4.64 \cdot 10^{-34}\,\mathrm{g\,cm}^{-3}\,, & \text{b)} \;\; T_{ph} = 2.725\,\mathrm{K}\,, \\ \text{c)} \;\; p_{(r)} = 1.39 \cdot 10^{-14}\,\mathrm{Pa}\,, & \text{d)} \;\; n_{(r)} = 369.62 \cdot 10^{21}\,\mathrm{cm}^{-3}\,. \end{array} \tag{138}$$

Physically remarkable data stand for following quantities at the present time: age of the world (129a), world radius (129b), Hubble parameter (129c), deceleration parameter (129d), relative temporal change of the frequency (130e), duration of the repulsive era (132), predicted rest mass of the hypothetical scalon (133c).

Further results for the ratio of the present mass density (135a) and the above used estimated value of the visible mass density:

$$\frac{\mu_{visible}}{\mu_p} \approx \frac{5 \cdot 10^{-31}}{4.37 \cdot 10^{-30}} \approx \frac{1}{10}\,. \tag{139}$$

This outcome means that according to this particular PUFT cosmology dark matter (scalon gas or already transmutations happened) overweighs normal matter by a factor of about 10.

With respect to (133c) let us point to the ratio

$$\frac{m_{proton}}{m_{scalon}} = 8.38 \cdot 10^{8} \tag{140}$$

which reminds the reader of the Feynman hypothesis mentioned above concerning the proton/parton ratio.

Furthermore we mention that instead of (122) by means of (130d) follows

$$\left(\frac{1}{G_s}\,\frac{\mathrm{d}G_s}{\mathrm{d}t}\right)_p \approx 1.48 \cdot 10^{-12}\,\mathrm{y}^{-1}\,. \tag{141}$$

This value fits well to the empirical result (124).

Let us finally emphasize that the new parameters chosen in this appendix do not change our picture of the expanding cosmos without the big bang phenomenon.

The following figures: 12 (instead 2), 13 (instead 4) and 14 (instead 5) show the intended disappearance of the above waviness.

I hope that during the next years the various methods of the cosmological distance determination will converge. If one compares, of course on the basis of

PUFT, both the world radii of the cosmos (12b) and (129b), an enormous discrepancy becomes obvious.

If the reader wants to get more information on our 5-dimensional Projective Unified Field Theory, he may look into our new monograph [12].

I am very grateful to the E. Majorana Center Erice for splendid hospitality as well as to the professors Venzo de Sabbata and Alexander Gorbatsievich (University of Minsk) for scientific discussions and help. Furthermore I have to thank Dr. J.G. Williams (JPL) for a series of discussions on empirical astrophysics.

13. Figures

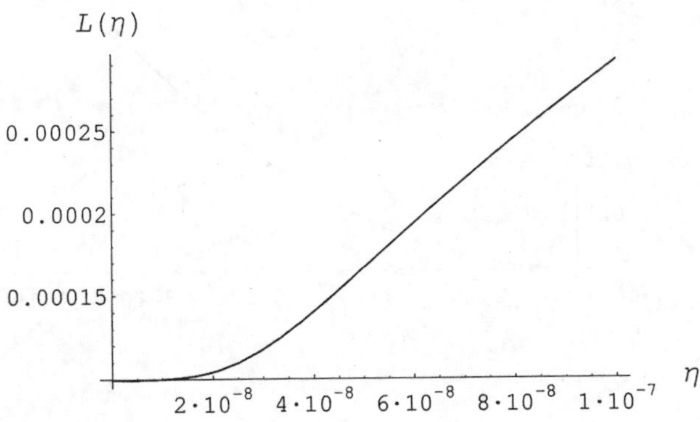

Figure 1. Temporal course of the rescaled world radius immediately after the big start

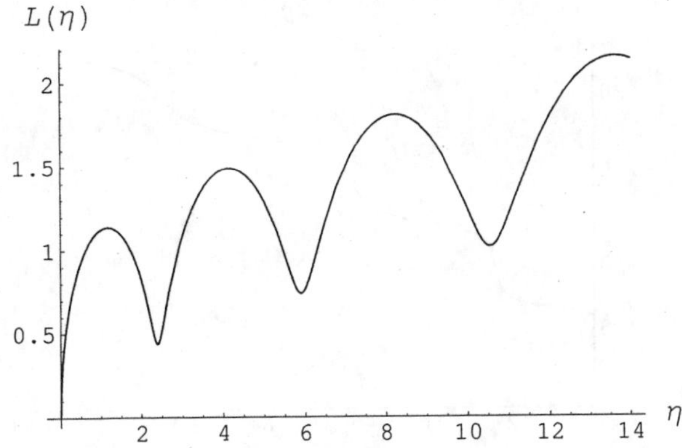

Figure 2. Temporal course of the rescaled world radius for the whole time

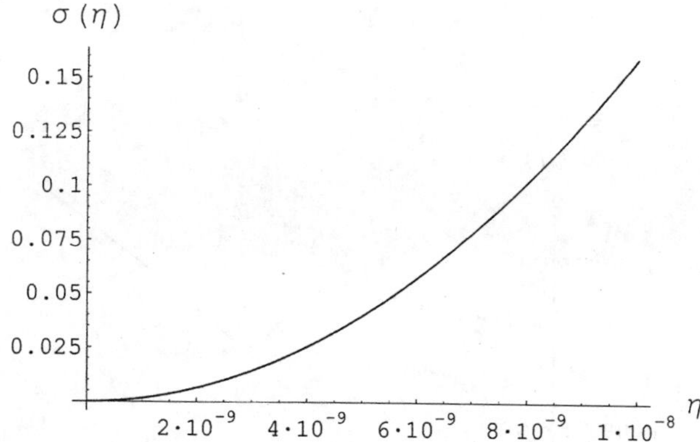

Figure 3. Temporal course of the scalaric world function immediately after the big start

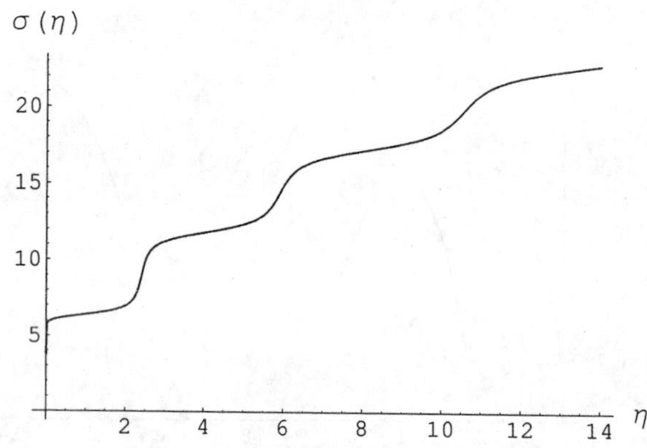

Figure 4. Temporal course of the scalaric world function for the whole time

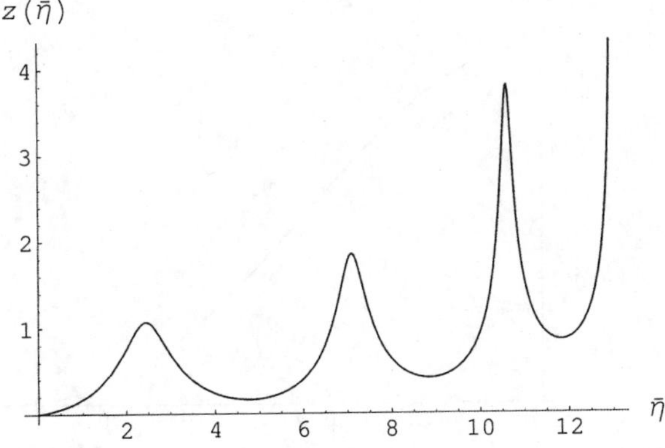

Figure 5. Temporal course of the relative cosmological frequency shift backwards from the present time as a function of the distance of the emitting sources

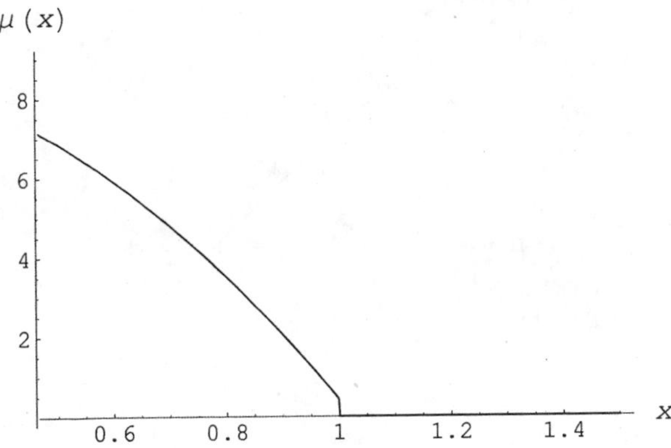

Figure 6. Radial course of the mass density for the interior of the sun-like model

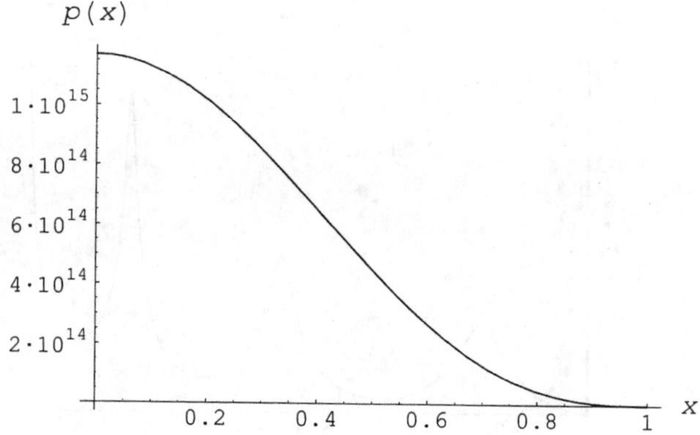

Figure 7. Radial course of the pressure for the interior of the sun-like model

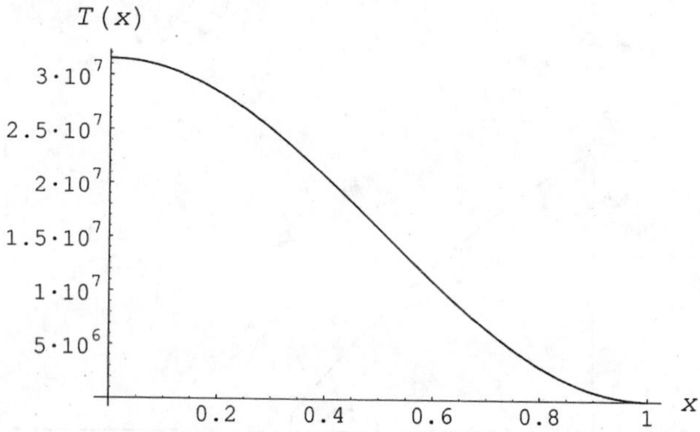

Figure 8. Radial course of the temperature for the interior of the sun-like model

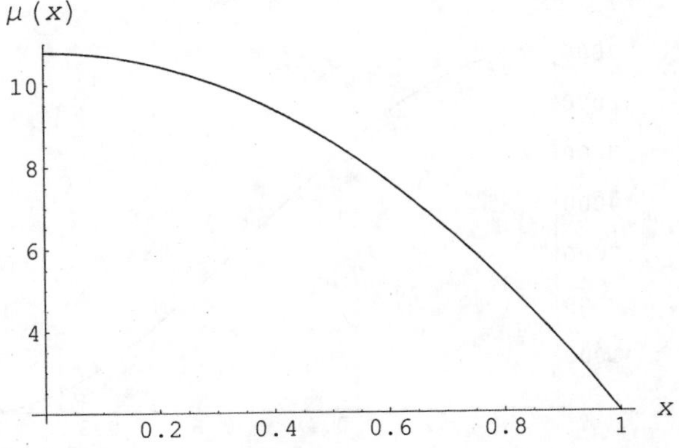

Figure 9. Radial course of the mass density for the interior of the earth-like model

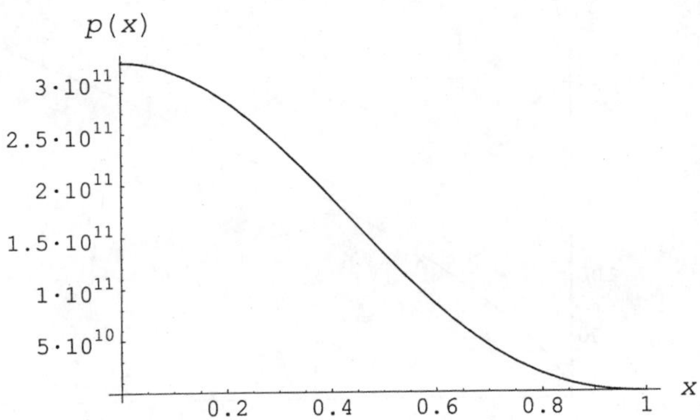

Figure 10. Radial course of the pressure for the interior of the earth-like model

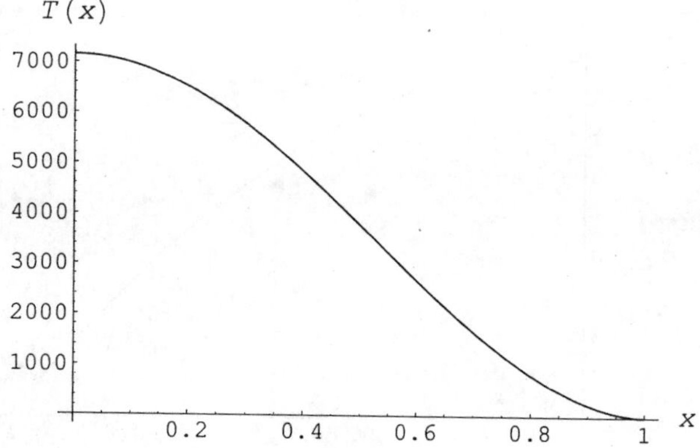

Figure 11. Radial course of the temperature for the interior of the earth-like model

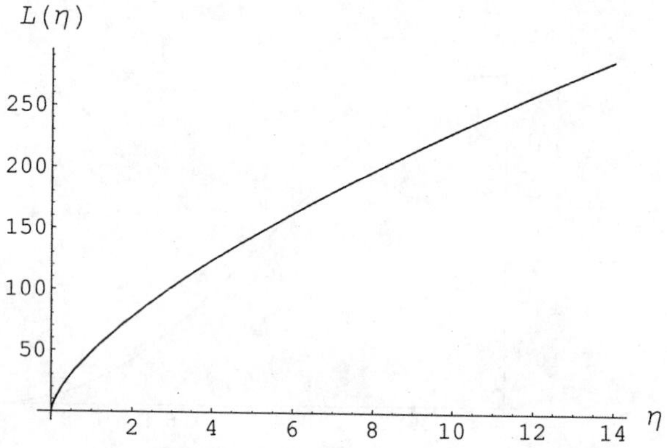

Figure 12. Temporal course of the rescaled world radius for the whole time

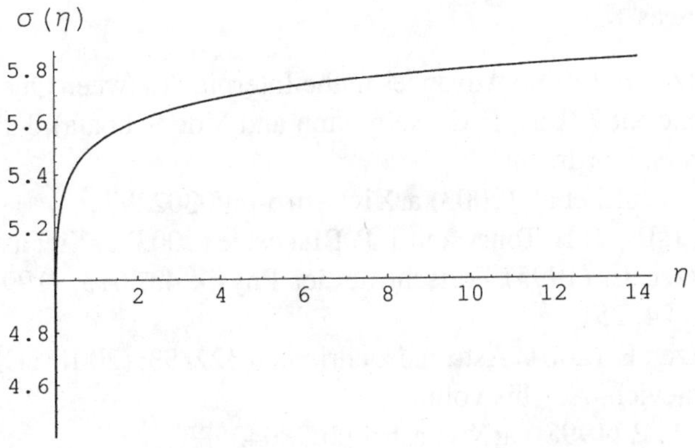

Figure 13. Temporal course of the scalaric world function for the whole time

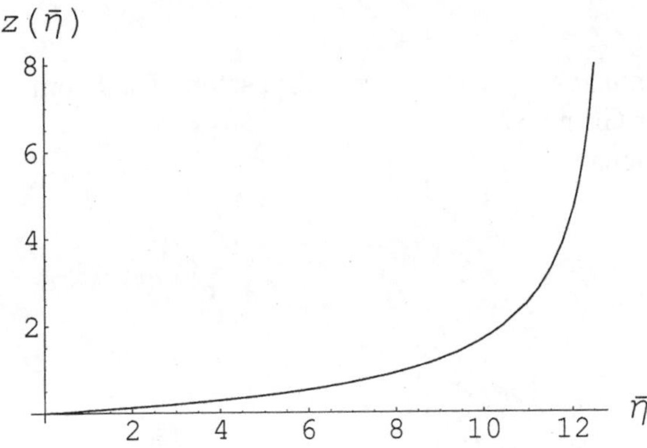

Figure 14. Temporal course of the relative cosmological frequency shift backwards from the present time as a function of the distance of the emitting sources

14. References

1. Schmutzer, E. (2002) Advances in the Interplay Between Quantum and Gravity Physics 387 (Eds. P. G. Bergmann and V.de Sabbata), Kluwer Academic Publishers, Dordrecht
2. Bennett, Ch.L. et al. (2003) arXiv: astro-ph/0302207
3. Jensen, J. B. , J. L. Tonry and J. P. Blakeslee (2003) arXiv: astro-ph/0304427
4. Schmutzer, E. (1995) Fortschritte der Physik **43**, 613; (1995) Ann. Physik (Leipzig) **4**, 251
5. Schmutzer, E. (2001) Astron. Nachrichten **322**, 93; (2001) arXiv: gr-qc/010604
6. Gorbatsievich, A. (this volume)
7. Pijpers, F. P. (1998) arXiv: astro-ph/9804258
8. Anderson, J. D., E. L. Lau, S. Turyshev, J. G. Williams and M. M. Nieto (2000) Bull. Amer. Astron. Society **34**, 660
9. Misner, Ch. W., K. S. Thorne and J. A. Wheeler (1973) Gravitation, W. H. Freeman and Company, San Francisco
10. Williams, J. G. (private communication 2003)
11. Kochanek, Ch. S. and P. L. Schechter (2003) arXiv: astro-ph/0306040
12. Schmutzer, E. (2004) Projektive Einheitliche Feldtheorie mit Anwendungen in Kosmologie und Astrophysik, Harri Deutsch Verlag Frankfurt (in preparation)

Ernst Schmutzer
Cospedaer Grund 57
D-07743 Jena
Germany

e-mail: eschmu@aol.com

THE INTERFACE OF QUANTUM MECHANICS AND GRAVITY

C. S. UNNIKRISHNAN
Tata Institute of Fundamental Research,
Homi Bhabha Road, Mumbai - 400 005, INDIA

1. INTRODUCTION

Quantum theories of gravity have been discussed for more than half a century, and the progress has been slow. Apart from theoretical technical reasons, lack of guidance by experimental or observational clues has been a major problem. While this situation will not be changed very much if the scale of quantum gravity is the Planck scale, there are two areas of study one should pursue where there is hope of arriving at some helpful indications. One is obviously cosmology, and astrophysical observations on the cosmological scale where Planck scale signatures could be visible in principle. Small quantum effects can get accumulated over cosmological scales of propagation and become observable. Another avenue is the interface of gravity and quantum mechanics as studied through the dynamics of quantum systems and fundamental particles in gravitational fields. Even classical experiments, like test of Universality of Free-Fall (UFF) using classical mass elements come under this category, and certain limited questions concerning gravity-quantum mechanics interface can be addressed and possibly answered. In this article I review some of these possibilities. Some of the issues we address are, 'Is the equivalence principle valid for a general quantum system?', 'Do Casimir energy (energy of the quantum vacuum) obey the equivalence principle?', 'Can observations on cosmological scales give clues about quantum gravity?', 'What is the influence of our specific Universe on local quantum systems like atoms?', and 'Are there signatures of quantum processes in the original of the Universe observable today?' Some of the answers have to be speculative since it is often difficult to have independent confirmations since even arriving at one clue is difficult. Yet, the hope is that some of this discussion will initiate new thoughts that can

V. de Sabbata et al. (eds.),
The Gravitational Constant: Generalized Gravitational Theories and Experiments, 381–393.
© 2004 *Kluwer Academic Publishers. Printed in the Netherlands.*

make progress regarding the issue of a general theory of gravity, including aspects of quantum gravity.

2. THE EQUIVALENCE PRINCIPLE AND QUANTUM MECHANICS

It is often mentioned that the equivalence principle (EP) would be violated at some level when quantum effects are dominant. While there is no specific empirical or theoretical clue regarding the exact nature and amount of such a violation, it would be surprising if such a violation is indicated in experiments that use quantum systems dynamically evolving in a gravitational field that could be considered classical, like the field of the earth or Sun or even the exterior field of a stellar black hole. However, one often sees remarks that the equivalence principle is not obeyed in experiments employing quantum particles, as in matter wave interferometry experiments. I will analyze some of these remarks and claims and establish the firm result that the equivalence principle is valid for quantum evolution at the same level as established using classical detectors like a torsion balance[1].

As examples, I quote the following representative statements;

1) D. Greenberger [2], Ann. Phys, 47, 116 (1968): "A direct application of quantum mechanics to the problem of a particle in an external gravitational field leads to results which seem to defy the equivalence principle". He goes on to state that the results indicate violation of the classical conception of the principle of weak equivalence.

2) J. J. Sakurai [3], Modern Quantum Mechanics, Chapter 2: Since the quantum phase depends on the mass of the particle, it is stated, "gravity is not geometrical in nature at the quantum level".

3) D. V. Ahluwalia [4], Nature 398, 199 (1999): "In many situations, phase measurements give an enormous advantage and can also highlight the differences between the quantum and classical realms of gravity".

There are several other examples as well, which I do not list here, including the much discussed experiment by Page and Geilker, misinterpreted by the authors themselves[5]. In contrast, there is also a complete and correct analysis of the main issues in ref. [6].

The reason for the statements quoted above is the observation that the expression for the quantum phase change depends explicitly on the gravitational mass, the inertial mass, and the Planck's constant, and not just on the ratio of the inertial mass and the gravitational mass. This is usually contrasted with the expression for a classical trajectory, which depends only on the ratio of the inertial and gravitational mass. Since this ratio is supposed to be universal, the mass dependence drops out. Therefore many people think that there is some violation of the equivalence principle in the quantum case.

This is just a misinterpretation of the principle of equivalence in any of its forms. The mass independence of the classical trajectory in a gravitational field is the consequence of the EP, applied to the classical case, and not a statement of the EP itself. If the gravitational field is locally equivalent to an accelerating frame, then the mass independence of a classical trajectory follows. The same EP demands that the quantum phase change depends quadratically on either the gravitational mass or the inertial mass, exactly as noticed in the theoretical expressions and in experiments. We now derive these relations to establish the validity of the EP for the quantum case. Since these points are discussed in detail in ref. [1], I briefly state the results here, and only outline the proof.

It is important to start with a correct statement of the equivalence principle before deciding whether the principle is violated or not in an experiment. The equivalence principle asserts that there is no possibility of distinguishing locally the effect of gravity and that of acceleration. (It is this statement that is formally used in constructing the general theory of relativity.) Another precise statement would be about the equivalence of the gravitational charge (gravitational mass) and the inertial mass. It is easy to check that these statements of the equivalence principle are never violated even in the slightest way in any known quantum situations.

The basic idea in showing that standard quantum dynamics will obey the EP to the same level as shown to be valid by classical detectors like a torsion balance is simple. All one needs to notice is that the full quantum propagator without any approximation is derived from the classical action, since the Lagrangian is at most quadratic in coordinates and velocities. Therefore if the EP is verified with a classical object, as the mass element of a torsion balance, then it is valid for all those fundamental constituents, like electrons, protons and neutrons that make up the macroscopic classical detector. Since the EP is verified even for various forms of binding energies, it does not make any difference whether one is experimenting with individual particles or bound systems.

We consider a matter interferometer consisting of coherently split paths $x1$ and $x2$, recombined at a detector. The relative phase accumulated by the quantum wavefunction propagating in a gravitational potential is

$$\delta\varphi = \frac{m_g}{\hbar}\left(\int_{x1} \phi_1(x)dt_1 - \int_{x2} \phi_2(x)dt_2\right) \tag{1}$$

Here $\phi(x)$ is the gravitational potential. The integrals are over the two relevant paths. This phase shift is the same as the gravitational phase shift measured in neutron interferometry experiments[7]. The phase shift in a gravitational field for this interferometer is well known [3, 7] and it is given by

$$\delta\varphi = \alpha \frac{m_g m_i g A \lambda}{\hbar^2} \tag{2}$$

g is the gravitational acceleration, A is the area of the interferometer, and λ is the de Broglie wavelength. α is a geometrical factor that depends on the orientation of the interferometer. Since this expression depends explicitly on the gravitational mass, there is a mistaken notion that a separate quantum version of the EP is demanded. The presence of the Planck's constant is sometimes interpreted as 'some quantum signature in gravity'[3]. Neither statement is correct. To see this, we now calculate the expression for the phase shift, without gravity, in an accelerating frame. This can be easily done (both for an accelerating frame and for the case in which there is a gravitational field) using the perturbation technique in path integral formalism. The Lagrangians in both cases are at most quadratic and the gravitational field or the acceleration is treated as a perturbation (applicable in the relevant experiments with neutrons, atoms and electrons). The key result is that the phase shift is given by the time integral of the perturbation of the Lagrangian over the unperturbed classical path [8]. For the gravitational case, the perturbation to the Lagrangian is $m_g g z$, for a uniform gravitational field in the vertical direction, z. Then the phase shift is

$$\delta\varphi_G = \frac{1}{\hbar} \oint m_g g z(t) dt = \frac{m_g g A}{\hbar v_0} = \frac{m_g m_i g A \lambda}{\hbar^2} \tag{3}$$

Now we calculate the phase shift in the same interferometer in an accelerating frame, with uniform acceleration a. The perturbation to the free Lagrangian, to first order, is $\delta L = \frac{1}{2} m_i v^2(t) - \frac{1}{2} m_i v_0^2 = m_i v_0 a t$. The phase shift then is

$$\delta\varphi_a = \frac{v_0}{\hbar} \oint m_i a t dt = \frac{v_0 m_i a A}{\hbar v_0^2} = \frac{m_i a A}{\hbar v_0} = \frac{m_i m_i a A \lambda}{\hbar^2} \tag{4}$$

The quantum phase shifts calculated in a gravitational field and in accelerated frame are identical, if the classical equivalence principle is valid for the particles in the interferometer. The correct statement of the EP is the equivalence of $\delta\varphi_G$ and $\delta\varphi_a$. The fact that the phase shift depend on the gravitational mass is no indication of any violation of the equivalence principle, contrary to repeated claims. In particular, if the classical equivalence principle is verified for neutrons and atoms using a torsion balance or by any other technique, then quantum interferometry using these particles cannot violate the equivalence principle. The issue whether the atoms or neutrons are bound in the form of a macroscopic object or not is not relevant, since the equivalence principle is also verified to great accuracy for various forms of binding energies [9].

Now we outline the general proof of the statement of validity of the EP for nonrelativistic quantum systems, assuming the validity of EP tested with a classical system(while this means systems obeying the Schrodinger equation, the presence of spin does not alter the result).
Statement: The time evolution (dynamics) from time t_a to t_b of the most general quantum state consisting of an arbitrary superposition of mass eigenstates will give exactly the same result in a uniform gravitational field g and in an accelerating frame with acceleration $a = -g$ if the classical equivalence principle is valid.

We start with the quantum propagator,

$$K(t_b, t_a) \equiv \langle \psi_b | U(t_b, t_a) | \psi_b \rangle = F(t_b, t_a) \exp(iS_{cl}/\hbar) \tag{5}$$

The classical action that determines the classical trajectory completely decides the quantum evolution as well. $F(t_b, t_a)$ does not depend on the properties of the potential; it is identical for the cases of an external gravitational field and that of an accelerating frame, for evolution between fixed times t_a and t_b. The classical equivalence principle dictates that the classical path in a gravitational field g is exactly same as the classical path in an accelerated frame with acceleration $-a$. Then the classical actions are identical for the two situations. Therefore, the path integrals, and the phase shifts are identical in the two situations, without any approximations.

We get

$$K(t_b, t_a)|_g = K(t_b, t_a)|_{a=-g} \tag{6}$$

where the subscripts identify the gravitational field and the accelerating frame. Since the quantum propagators are identical, the quantum evolutions in the two situations are identical for arbitrary quantum states.

3. TESTS OF EP: 'MOTHER OF ALL GRAVITATION EXPERIMENTS'

Tests of equivalence principle is the most important of all gravitation experiments since EP in conjunction with special relativity implies all that can be tested about gravity. Of course, experiments aiming to detect and study gravitational waves constitute another important set of experiments, since they go beyond just testing gravity theories. The importance of the test of EP results from its multiple sensitivity – apart from providing the test of Universality of free fall, these experiments are also sensitive to existence of new forces, spatial and temporal variations of fundamental constants, exotic couplings, gravitomagnetism etc.

Every classical test of the EP compares the trajectories of two test masses which differ in various internal properties. Typically, this is done by comparing the accelerations experienced by the test masses in approximately uniform gravitational field. Consider two masses, 1 and 2, with their gravitational masses M_{g1} and M_{g2}. The total inertial mass can be written as

$$M_{i1} = \sum_k m_{k1} \rightarrow m_{01} + \sum_k \frac{E_{(b)k1}}{c^2}$$

and

$$M_{i2} = \sum_k m_{k2} \rightarrow m_{02} + \sum_k \frac{E_{(b)k2}}{c^2} \tag{7}$$

where $E_{(b)}$ represents the various binding energies, like electrostatic, nuclear, gravitational etc. The acceleration a of each test mass in a gravitational field then depends on the ratio M_g/M_i, and if any form of energy within the system does not couple to gravity normally, then the acceleration will not be universal and the EP is violated. The difference

$$\left(\frac{M_{g1}}{M_{i1}}\right) - \left(\frac{M_{g2}}{M_{i2}}\right) = \frac{\Delta a}{a} \equiv \eta \tag{8}$$

defines the sensitivity of the test. Of course, the acceleration will not be universal if there are new long range forces that couple selectively to certain generalized charges, like baryon number. Thus, tests of EP are versatile and important class of experiments and continued effort to improve their sensitivity is always justified well. At present, only Lunar Laser Ranging experiments [10] are sensitive to the issue of free fall of the gravitational binding energy itself.

3.1. DOES THE CASIMIR ENERGY FREE FALL?

This is a very important question that one would like to answer with a direct experiment. There is no obvious answer to this question since the gravitational properties of quantum vacuum are not known. The Casimir energy is a derivative of the concept of quantum vacuum, and hence only direct or indirect experiments are qualified to answer this question.

It turns out to be nearly impossible to test whether the Casimir energy obeys the equivalence principle because the Casimir energy contribution in any conceivable test mass happens to be far smaller than the relative accuracy of even future tests of the EP. To see this let us estimate the amount of Casimir energy is some standard configurations and compare it with the rest mass of the test object. For the standard 'parallel plates' configuration considered in the calculations of Casimir energy, the energy per unit area [11] is

$$\frac{E_c}{A} = \frac{\pi^2 \hbar c}{720 d^3} \tag{9}$$

and the Casimir energy density is

$$\frac{E_c}{V} = \frac{\pi^2 \hbar c}{720 d^4} \simeq 10^{-4} \quad J/m^3 \tag{10}$$

for $d \simeq 1$ micron. This energy density of about 10^9 eV/cm^3 should be compared with the typical rest mass energy of $10^{33} - 10^{34}$ eV/cm^3 for metallic materials. Thus in this case, the relative contribution of the Casimir energy is at most only 10^{-24}, and a test of EP for Casimir energy then requires a sensitivity better than 10^{-24}!

The situation does not improve with other configurations of the test masses. For a hollow sphere of radius R, the Casimir energy density (for $R \simeq 1\mu$) is

$$\frac{E_c}{V} \simeq \frac{0.01 \hbar c}{R^4} \simeq 3 \times 10^{-4} \quad J/m^3 \tag{11}$$

Again the relative contribution is only about 10^{-25}.

There has been an interesting proposal by D. K. Ross [12] that it should be possible to test the free fall of the Casimir energy by considering the amount of vacuum energy expelled by a solid test mass, rather than considering the Casimir energy contained within the test mass. If one considers the amount of Casimir energy expelled by a solid conducting sphere, since the wavelength longer than those corresponding to an energy of 15-20 electron Volts will not be able to penetrate through, we get (written symbolically)

$$E_{Expelled}/V \simeq \frac{4\pi}{(\hbar c)^3} \int_0^{20eV} E^3 dE \tag{12}$$

to be integrated over all the frequencies upto $\hbar\omega \simeq 20$ eV. This is about 10^{19} eV/cm^3, which is about 10^{-14} of rest mass energy. This is an interesting range of energy contribution for EP tests since future tests can in principle reach this sensitivity. In particular, space based experiments are expected to perform at a sensitivity of 10^{-15} to 10^{-17}. A recent paper by Moffat and Gillies discusses in detail the proposal for a satellite based experiment incorporating Ross's idea, for testing whether Casimir energy obeys the EP [13].

I think that there is basic flaw in the argument that the EP for Casimir energy can be tested by considering the fraction of the *expelled energy*, rather than the fraction of *included energy*. Since I have found that the detailed counter-argument [14] is not transparent enough, here I discuss a short argument invoking cosmology that prohibits the test of the EP for Casimir energy by considering the expelled energy density. This simple argument also demonstrates the power of cosmological constraints.

Any calculation of expelled energy first of all has to assume that there is so much energy available to be expelled. In particular if vacuum energy of 10^{19} eV/cm^3 has to be expelled from a volume, then that much energy has to be there to start with. But this is not allowed by even the crudest of cosmological observations! The maximum amount of smooth energy density that can couple to gravity that can be present in the Universe cannot exceed 10^{-29} g/cm^3, which is only about 10^4 eV/cm^3. There are direct observations of the amount of matter in the Universe and also there has to be consistency with the age of the Universe. Both constrain the amount of matter to be of the order of the critical density. So, the Casimir energy that can be expelled even in principle, while living in this particular Universe, is a mere 10^{-30} of typical rest mass energies, and there is no way such a concept can be used for a test of the EP.

Does that mean that we have no hope of settling this issue using experimental means? Fortunately, not. I will outline a simple and powerful general argument assuming only conservation of energy (absence of perpetual motion) to show that there is experimental evidence already validating EP for Casimir energy.

First point to note is that the Casimir energy is convertible to kinetic energy. This is evident in several experiment in the past, where the Casimir force was measured by looking at the 'fall' of a test body when brought closer to another plate. In general one can assume that experiment have confirmed with an accuracy of at least 10% that the Casimir energy can be converted to kinetic energy. The second point to note is that kinetic energy obeys the equivalence principle. The kinetic energy contribution, $v^2/2c^2$, for the earth in the Lunar Laser Ranging test of the EP, for example, is about 2×10^{-9}. Since the test itself has a sensitivity of 10^{-13}, it is assured with good accuracy that kinetic energy obeys the EP. Clearly, if out of two forms of interconvertible energies, one does obey the EP and the other does not, it will be possible to build a perpetual motion machine. All one needs to do is to start with a configuration that has some Casimir energy, and no kinetic energy, transport it in a gravitational field, convert the Casimir energy to kinetic energy and then transport the system back to the starting spatial point. This cycle will have an extra energy loss or gain if the Casimir energy did not couple to gravity 'minimally'. This is a very significant result considering that direct experiments are out of reach. This general argument is applicable to any form of energy that can be converted to a form for which the EP is tested adequately. We combine the result of a known EP test (kinetic energy in this case), and an experiment that confirms the convertiility and together they provide a direct empirical test of the validity of EP for even those forms of energy where single direct test seems impossible.

4. ATOMS AND THE UNIVERSE

For most of laboratory experiments, the properties of the Universe 'outside the laboratory' are considered not relevant. In fact we do not bother about the density of the Universe or its expansion rate while analyzing results from an atomic physics experiment. Therefore, it may be surprising to notice that some of the purely quantum mechanical observables related to atoms might be intimately connected to the properties of the Universe we live in. Next I briefly explore this connection between the gravity of the Universe and the quantum mechanical atom.

In Einstein's theory, and in metric theories in general, the metric components $g_{\mu\nu}$ are like 'potentials'. In particular, g_{00} is related to the Newtonian gravitational potential ($\phi = c^2(g_{00}-1)/2$), and the g_{0i} are like vector potentials (velocity dependent). In the relativistic framework, a scalar potential like ϕ implies an induced vector potential in a moving frame, given by $\frac{\overrightarrow{v}}{c}\phi$. If this induced vector potential is time dependent, it implies an induction force. This simple physical picture assumes great importance when one realizes that there is a lot of matter in this Universe, each bit of it contributing to our local gravitational potential (ϕ_U). Moving through such a Universe induces the vector potential proportional to the velocity (therefore, no forces for uniform motion), and accelerated body will feel the 'force of the Universe'. Following Sciama [15],

$$F = -\frac{m_g}{c}\frac{\partial \overrightarrow{A}}{\partial t} = -\frac{m_g}{c}\left(\sum \phi\right)\frac{d\overrightarrow{v}}{dt} = -\frac{m_g}{c^2}\phi_U \overrightarrow{a} \tag{13}$$

There is a reaction force proportional to the acceleration, and that is what we call inertia. Thus, we can identify the inertial mass with the quantity $\frac{m_g}{c^2}\phi_U$, and the ratio $m_g/m_i = 1$, if the total local 'potential' due to all the masses in the Universe, $\phi_U = c^2$. This is the case for a critical Universe. It may also be noted that this calculation is a formal quantitative statement of the Mach's principle [16].

Going one step further gives an important new insight regarding atomic spectra. For an electron in a state with angular momentum $l \neq 0$, the induced vector potential in its frame is such that $\overrightarrow{\nabla} \times \overrightarrow{A} \neq 0$. This gravitomagnetic field has a quantized interaction with the two possible spin projections, creating an energy level splitting. This is precisely the contribution to the fine structure splitting, usually derived as the Thomas precession contribution! When consistently done, this part of the fine structure is a general relativistic effect of the gravity of Universe on the atom. It is indeed startling to see the signature of gravity of the critical Universe in the quantum mechanical atomic fine structure. It is important to note that this is not a mere reinterpretation; it is a physical effect that is a consequence of the present gravitational properties of the Universe, and usual interpretations like the Thomas precession is the geometric reinterpretation of this concrete gravitational effect.

5. ARE THERE QUANTUM GRAVITY SIGNALS?

Finally we discuss the question whether there are direct or indirect signals that can be taken as evidence for quantum gravity, or as constraints on the scale of quantum gravity. The possibility of such observations will be of great importance to any theory of gravity that goes beyond the classical general relativity. We consider two possibilities (though there are more proposals in the literature) – one is based on the expectation within any quantum gravity theory that at the quantum gravity scale, space-time is not smooth, and there are quantum fluctuations that make space-time granular, or foamy. This is related to a lowest scale for distance in a typical quantum gravity theory. Another possibility we briefly consider is based on the expectation that the origin of the Universe is determined by quantum gravitational effects. If this has determined the initial conditions in a unique way, then the present observed parameters of the Universe would be quantitatively connected in a verifiable way to these initial parameters.

5.1. ASTROPHYSICAL SIGNAL FOR QUANTUM GRAVITY

The basic idea in this interesting proposal is that the space-time form or granularity can induce a energy (frequency) dependent dispersion for light [17]. This will then result in a frequency dependent time delay in a pulsed signal reaching us from a cosmological distance. The large distance is necessary since the effect is small and there should be sufficient time to accumulate the small dispersion into an observable time delay. The frequency independent free-space dispersion relation for light is $c^2p^2 = E^2$. Phenomenologically, this is modified as

$$c^2p^2 = E^2(1 + f(E/E_{QG})) \tag{14}$$

E_{QG} is the quantum gravity scale. The functional form of the correction depends on quantum gravity physics, and in the absence of anything known, a parametrized expansion is used.

$$c^2p^2 = E^2(1 + \xi\frac{E}{E_{QG}} + O(\frac{E}{E_{QG}})^2 + ...) \tag{15}$$

The (frequency dependent) propagation velocity is

$$v = \frac{\partial E}{\partial p} = c(1 - \xi E/E_{QG}) \tag{16}$$

For a propagation distance of L, a time delay

$$\Delta t \simeq \xi \frac{E}{E_{QG}} \cdot L/c \tag{17}$$

could be expected. L is cosmological, greater than 10^{26} cm. For photon energy of $E \simeq 1$ MeV and a pulse timing resolution of 10^{-2} s, E_{QG} around $10^{15} - 10^{16}$ GeV can be probed. In fact, observations have done marginally better. As an example, measurements on gamma ray bursts from GRB 930229 and GRB 930131 in the 30-200 keV and 30 keV to 80 MeV respectively enabled to establish that the differential propagation velocity $\frac{\Delta c}{c} < 6 \times 10^{21}$, and (within the specific phenomenological model with the linear correction term) that $E_{QG} > 8 \times 10^{16}$ GeV, or $L_{QG} < 10^{-31}$ cm [18].

There are other proposals involving noise in gravity wave detectors, which we have not discussed here [19].

5.2. ARE THERE SIGNATURES OF A QUANTUM BIG-BANG?

Big-bang theory is now the majority view on the evolution of the Universe and it is therefore the standard model of the Universe. Almost all the observations concerning the Universe can be fitted within this model, provided that the original standard big-bang model is extended with unknown and unestablished particle physics at very high energies. It is interesting to see whether quantum mechanics and relativistic gravity together decided the initial conditions (density, temperature etc. at origin) of the Universe in a unique way. Here I discuss briefly one such proposal [20]. If one assumes that the 'realization' of the Universe was over a relativistic gravitational barrier and that the uncertainty principle was operative (these are the only two fundamental inputs for a quantum gravity scenario without a proper theory), then we have

$$E\Delta t \simeq \hbar/2 \frac{2G(E/c^2)}{c^2 r} \simeq \frac{2G(E/c^2)}{c^3 \Delta t} \simeq 1 \tag{18}$$

E is the average energy content in a elementary volume causally connected during time scale Δt. From these, we get the time scale of the quantum birth as

$$\Delta t \simeq \left(\frac{\hbar G}{c^5}\right)^{1/2} \equiv t_P \tag{19}$$

as expected from dimensional considerations. t_P is the Planck time. This allows us to estimate the initial energy density, and temperature E/k_B of the new-born Universe. Since we have assumed that the origin was over a gravitational barrier, naturally this Universe is critical, evolving at a critical density. The initial temperature, predicted by purely quantum considerations coupled with relativistic gravity is about $7 \times 10^{31}\ ^oK$. Without further assumptions, these initial values can be extrapolated to present time within the standard big bang theory without inflation etc. and the result of about 2.4 ^oK is remarkably close to the observed present temperature[20]. One may be tempted to dismiss this result as 'obvious' pointing out that it is well known that when the present temperature is extrapolated back to the Planck time one gets the Planck temperature (close to $10^{32}\ ^oK$), and there is nothing surprising. That is also the result one expects from dimensional considerations. But the issue is not as trivial as it may appear. Note that both dimensional analysis as well as our first principles analysis assumes that quantum mechanics and relativistic gravity together determined the initial conditions of the Universe. Otherwise there is no physical reason to have the temperature at Planck time of the order of $10^{32}\ ^0K$. If the present temperature matches with the expectations from such a scenario, as it does, then it is substantial evidence for a quantum gravitational origin of the Universe.

6. SOME REMARKS ABOUT QUANTUM MECHANICS

Many particle theorists are of the view that at very high energies the Einstein's theory of gravity will break down, and that it is only a 'low energy approximation' to a more complete theory of quantum gravity. While this view is very reasonable, there is a remarkable difference between Einstein's theory of gravity and Quantum mechanics as far the interpretation and understanding of the theory is concerned. Einstein's theory is what could be called a 'beautiful theory', solid in its fundamental structure, definite in its detailed predictions and without contradictions with our general notions on causality, locality etc. Quantum mechanics on the other hand does not seem to be a complete theory. It has only probabilistic predictions, though single experiments and measurements can be performed on single quantum systems. At present quantum mechanics does not address the description of such single measurements. Quantum mechanics also violates the spirit of special relativity since it explicitly violates Einstein locality in its theoretical structure. So much are the interpretational problems with quantum mechanics that one may wonder whether it is the quantum theory that may require a change at its core to be reconciled with the relativistic theory of gravity. These issues are open, and more and more explorations into possible experimental clues at the interface are essential before a definite direction in building the quantum theory of gravity can be clearly seen.

Acknowledgements:

I thank Prof. Venzo de Sabbata and the Ettore Majorana Centre for Scientific Culture for the generous hospitality and excellent opportunities for discussions offered during the VIIIth Course of the School of Cosmology and Gravitation. Several useful inputs from George Gillies, Rogers Ritter and Kenneth Nordtvedt have sharpened and improved my understanding of topics discussed in these notes.

References

1. Unnikrishnan, C. S. *Mod. Phys. Lett.*, **16**, 429, (2001).
2. Greenberger, D. *Ann. Phys.* **47**, 116, (1968).
3. Sakurai, J. J. Modern Quantum Mechanics (Addison Wesley, 1994), see chapter 2.
4. Ahluwalia, D. V. *Nature* **398**, 199 (1999).
5. Page, D. N. and Geilker. C. D. *Phys. Rev. Lett. 47, 979,* (1981).
6. Borzeszkowski, H.-H.v and Treder, H.-J. *Found. Phys.* **12**, 1113, (1981).
7. Colella, P. A., Overhauser, A. W. and Werner, S. A. *Phys. Rev. Lett.*, **34,** 1472, (1975).
8. Storey, P. and Cohen-Tannoudji, C. *J. Phys. II France*, **4**, 1999 (1994).
9. Will, C. M. *Theory and Experiment in Gravitational Physics*, (Cambridge University Press 1983).
10. For a review see Dickey, J. O *et al, Science* **265**, 482 (1994).
11. Milonni, P. W. *The Quantum Vacuum*, (Academic Press, 1994).
12. Ross, D. K. *Nuovo Cimento* **B114**, 1074 (1999).
13. Moffat J. W. and Gillies, G. T. *New Jl. Phys.* **4**, 92.1-6 (2002).
14. Unnikrishnan C. S. unpublished notes.
15. Sciama, D. *MNRAS*, **113**, 34, (1953).
16. Unnikrishnan, C. S. *Experiments motivated by the Mach's principle: A review with comments*, in the Proceedings of the Workshop on Mach's Principle and the Origin of Inertia, Kharagpur, India, 2002, Eds. M. Sachs and A. R. Roy, (Apeiron, Montreal).
17. Amelino-Camelia, G. *et al, Nature* **393,** 763 (1998).
18. Schaefer, B. E. *Phys. Rev. Lett. 82, 4964* (1999).
19. Amelino-Camelia, G. *Nature* **398**, 216 (1999), *Nature* **410**, 1065 (2001).
20. Unnikrishnan, C. S, Gillies, G. T. and Ritter. R. C., *Pramana-Jl. Phys.* **59**, 369 (2002).

QUATERNIONIC PROGRAM

ALEXANDER P. YEFREMOV
Peoples Friendship University of Russia
Russia, Moscow, Miklukho-Maklaya str., 6

1. Introduction

Discovery of quaternionic (Q) numbers is dated as 1843 and commonly attributed to Hamilton [1,2] although in previous century Euler and Gauss investigated the Q-type mathematical objects, and Rodriguez even designed the multiplication rule for elements of similar algebra [3-5]. Active struggle of Gibbs and Heavyside against Hamilton's disciples brought to life contemporary vector algebra and later analysis while quaternions stayed aside of main stream of mathematical physics, despite striking exclusive character of Q-algebra proved by Frobenius. At the beginning of 20 century the last bastion of Q-numbers' amateurs, "Association for the Promotion of the Study of Quaternions", was ruined, and the only reminiscence of once famous hypercomplex numbers appeared in the Pauli matrices. Later on a fragmentary interest was noticed to quaternions mostly utilized as a tool for alternative description of already known physical models [6,7] or otherwise due to enigmatic and fascinating simplicity of their multiplication they were used in rigid body kinematics [8]. Attention and interest to Q-numbers was renewed a couple of decades ago when the time came, and younger generation of theoreticians started feeling deep potential of quaternions yet unrevealed (e.g. [9-11]). Herewith a sketch of a study is given which represents an attempt of more systematic analysis and further investigation of hypercomplex Q-numbers. As well it offers some amazing mathematical observations and physical models claimed original, sometimes surprising and doubtlessly worth paying attention to.

The paper is organized as follows. Sect.1 briefly describes axioms and generic correlations of quatrnionic algebra in "traditional" Hamiltonian symbols and in tensor-like approach. Sect.2 is devoted to structure of three "imaginary" quaternionic units. In Sect.3 elements of differential Q-geometry are presented with examples of mathematical applications. Sect.4 offers Newtonian mechanics in rotating systems of reference presented by Q-frames. In Sect.5 the Theory of Quaternionic Relativity is introduced with a number of newly offered effects calculated for non-inertial relativistic frames. Sect.6 contains observation of 'Great Quaternionic Coincidences' and concluding discussion.

2. Quaternionic algebra

TRADITIONAL APPROACH

According to Hamilton, quaternion is a mathematical object

$Q \equiv a + b\mathbf{i} + c\mathbf{j} + d\mathbf{k}$

where *a, b, c, d* are real numbers, a is multiplier of the real unit 1, and **i**, **j**, **k** are three different imaginary quaternionic units. The multiplication rule for these units was written by Hamilton and still is commonly repeated as

V. de Sabbata et al. (eds.),
The Gravitational Constant: Generalized Gravitational Theories and Experiments, 395–409.
© 2004 *Kluwer Academic Publishers. Printed in the Netherlands.*

$$\mathbf{1i} = \mathbf{i1} \equiv \mathbf{i}, \qquad \mathbf{1j} = \mathbf{j1} \equiv \mathbf{j}, \qquad \mathbf{1k} = \mathbf{k1} \equiv \mathbf{k},$$

$$\mathbf{i}^2 = \mathbf{j}^2 = \mathbf{k}^2 = -1.$$

$$\mathbf{ij} = -\mathbf{ji} = \mathbf{k}, \qquad \mathbf{jk} = -\mathbf{kj} = \mathbf{i}, \qquad \mathbf{ki} = -\mathbf{ik} = \mathbf{j}.$$

This rather bulky set of equations implies that the Q-multiplication is no more commutative

$$Q_1 Q_2 \neq Q_2 Q_1,$$

so that there are left and right Q-products, but still associative

$$(Q_1 Q_2) Q_3 = Q_1 (Q_2 Q_3).$$

One can distinguish in a quaternion two algebraically very different parts and define them as a scalar part

$$scal\, Q = a,$$

and vector part

$$vect\, Q = b\mathbf{i} + c\mathbf{j} + d\mathbf{k}$$

The following step is quaternionic conjugation introduced analogously to that of complex numbers

$$\overline{Q} \equiv scal\, Q - vect\, Q = a - b\mathbf{i} - c\mathbf{j} - d\mathbf{k},$$

and to define a Q-number modulus

$$|Q| \equiv \sqrt{Q\overline{Q}} = \sqrt{a^2 + b^2 + c^2 + d^2}.$$

All this helps to properly formulate quaternionic division, which expectedly is two-sided: right and left

$$Q_L = \frac{Q_1 \overline{Q}_2}{|Q_2|^2}, \qquad Q_R = \frac{\overline{Q}_2 Q_1}{|Q_2|^2}.$$

Having Q-modulus definition at hand, one instantly gets famous four squares identity

$$|Q_1 Q_2|^2 = |Q_1|^2 |Q_2|^2.$$

All these means that quaternions form an algebra that was proved to belong to elite of 4 exclusive "very good" algebras: of real, complex, quaternionic and octanionic numbers (Frobenius-Gourvic theorems of 1878-1898).

Special attention must be paid to Q-units representation. In Hamilton's notation real unit was just 1, while imaginary units (in analogy with complex numbers) were denoted as (**i**, **j**, **k**). Later on a simple constant 2×2-matrix representation was found

$$\mathbf{i} = -i\begin{pmatrix} 0 & 1 \\ 1 & 0 \end{pmatrix}, \qquad \mathbf{j} = -i\begin{pmatrix} 0 & -i \\ i & 0 \end{pmatrix}, \qquad \mathbf{k} = -i\begin{pmatrix} 1 & 0 \\ 0 & -1 \end{pmatrix},$$

but of course not unique. If one expresses here familiar imaginary unit of complex numbers i with the help of 2×2-matrix containing only real unit

$$i = \begin{pmatrix} 0 & 1 \\ -1 & 0 \end{pmatrix},$$

then three vector Q-units are represented by real 4×4-matrices. It is clear that the procedure of duplication of representation matrix dimensionality can be continued infinitely.

“TENSOR” FORM AND REPRESENTATIONS

If each Q-unit vector is given its number (like component of a tensor)

$$(\mathbf{i},\mathbf{j},\mathbf{k}) \to (\mathbf{q}_1,\mathbf{q}_2,\mathbf{q}_3) = \mathbf{q}_k, \qquad j, k, l, m, n, \ldots = 1, 2, 3,$$

then quaternionic multiplication rule acquires compact form

$$1\mathbf{q}_k = \mathbf{q}_k 1 = \mathbf{q}_k, \qquad \mathbf{q}_j\mathbf{q}_k = -\delta_{jk} + \varepsilon_{jkn}\mathbf{q}_n,$$

where δ_{kn}, ε_{knj} are 3-dimensional (3D) Kroneker and Levi-Civita symbols.

Further on, for any 2×2-matrices

$$A = \begin{pmatrix} a & b \\ c & -a \end{pmatrix},\; B = \begin{pmatrix} d & e \\ f & -d \end{pmatrix}, \qquad Tr\,A = Tr\,B = 0,$$

one can represent first two Q-units as

$$\mathbf{q}_1 = \frac{A}{\sqrt{\det A}}, \qquad \mathbf{q}_2 = \frac{B}{\sqrt{\det B}},$$

while the third one is

$$\mathbf{q}_3 \equiv \mathbf{q}_1\mathbf{q}_2 = \frac{AB}{\sqrt{\det A \det B}} \quad \text{if} \quad Tr(AB) = 0.$$

Scalar unit is always the same $1 = \begin{pmatrix} 1 & 0 \\ 0 & 1 \end{pmatrix}$.

TRANSFORMATION OF Q-UNITS AND INVARIANCE OF MULTIPLICATION RULE

Spinor-type transformations

If U is an operator changing all units and $\exists\; U^{-1}$: $UU^{-1} = E$ then under transformations

$$\mathbf{q}_{k'} \equiv U\mathbf{q}_k U^{-1} \qquad \text{and} \qquad 1' \equiv U1U^{-1} = E1 = 1$$

the multiplication rule

$$1\mathbf{q}_k = \mathbf{q}_k 1 = \mathbf{q}_k, \; \mathbf{q}_j\mathbf{q}_k = -\delta_{jk} + \varepsilon_{jkn}\mathbf{q}_n$$

is form invariant

$$\mathbf{q}_{k'}\mathbf{q}_{n'} = U\mathbf{q}_k U^{-1}U\mathbf{q}_n U = U\delta_{kn}U^{-1} + \varepsilon_{knj}U\mathbf{q}_j U^{-1} = \delta_{kn} + \varepsilon_{knj}\mathbf{q}_{j'}.$$

The transformation can be represented either as 2×2-matrix

$$U = \begin{pmatrix} a & b \\ c & d \end{pmatrix}, \; \det U = 1,$$

or as unimodular quaternion $\qquad U = \dfrac{a+d}{2} + \sqrt{1 - \left(\dfrac{a+d}{2}\right)^2}\,\mathbf{q},$

with $$\mathbf{q} \equiv \left(\sqrt{1-\left(\frac{a+d}{2}\right)^2}\right)^{-1} \begin{pmatrix} \frac{a-d}{2} & b \\ c & -\frac{a-d}{2} \end{pmatrix}.$$

In general case there are 3 independent complex parameters (6 real), then $U \in SL(2,C)$. In special case there are 3 real parameters, then $U \in SU(2)$.

Vector-type transformations

The vector Q-units can be transformed as

$$\mathbf{q}_{k'} = O_{k'n}\mathbf{q}_n$$

with the help of 3×3-matrix $O_{k'n}$.

If invariance of the Q-multiplication rule is demanded then

$$O_{k'n}O_{j'n} = \delta_{kn}, \quad \text{hence} \quad O^{-1}_{nk'} = O_{k'n} \quad \text{and} \quad \det O = 1.$$

In general case the transformation has 6 parameters, and $O \in SO(3,C)$.

In special case there are 3 parameters, and $O \in SO(3,R)$. Below a variant of representation of matrix O is given with x, y, z being arbitrary real or complex functions

$$O=\begin{pmatrix} \sqrt{1-x^2-z^2} & -\frac{x\sqrt{1-y^2-z^2}+yz\sqrt{1-x^2-z^2}}{1-z^2} & \frac{xy-z\sqrt{1-x^2-z^2}\sqrt{1-y^2-z^2}}{1-z^2} \\ x & \frac{\sqrt{1-x^2-z^2}\sqrt{1-y^2-z^2}-xyz}{1-z^2} & \frac{-y\sqrt{1-x^2-z^2}-xz\sqrt{1-y^2-z^2}}{1-z^2} \\ z & y & \sqrt{1-y^2-z^2} \end{pmatrix}.$$

This can be rewritten in the form of product of three irreducible factors

$$O=\begin{pmatrix} \sqrt{\frac{1-x^2-z^2}{1-z^2}} & -\frac{x}{\sqrt{1-z^2}} & 0 \\ \frac{x}{\sqrt{1-z^2}} & \sqrt{\frac{1-x^2-z^2}{1-z^2}} & 0 \\ 0 & 0 & 1 \end{pmatrix} \begin{pmatrix} \sqrt{1-z^2} & 0 & -z \\ 0 & 1 & 0 \\ z & 0 & \sqrt{1-z^2} \end{pmatrix} \begin{pmatrix} 1 & 0 & 0 \\ 0 & \sqrt{\frac{1-y^2-z^2}{1-z^2}} & -\frac{y}{\sqrt{1-z^2}} \\ 0 & \frac{y}{\sqrt{1-z^2}} & \sqrt{\frac{1-y^2-z^2}{1-z^2}} \end{pmatrix}$$

that after substitution $z \equiv \sin \mathrm{B}$, $x \equiv -\sin \mathrm{A} \cos \mathrm{B}$, $y \equiv -\sin \Gamma \cos \mathrm{B}$, where A, B, Γ are complex “angles”, takes the form

$$O=\begin{pmatrix} \cos\mathrm{A} & \sin\mathrm{A} & 0 \\ -\sin\mathrm{A} & \cos\mathrm{A} & 0 \\ 0 & 0 & 1 \end{pmatrix} \begin{pmatrix} \cos\mathrm{B} & 0 & -\sin\mathrm{B} \\ 0 & 1 & 0 \\ \sin\mathrm{B} & 0 & \cos\mathrm{B} \end{pmatrix} \begin{pmatrix} 1 & 0 & 0 \\ 0 & \cos\Gamma & \sin\Gamma \\ 0 & -\sin\Gamma & \cos\Gamma \end{pmatrix} = O_3^{\mathrm{A}} O_2^{\mathrm{B}} O_1^{\Gamma}.$$

If the angles are real: $\mathrm{A} = \alpha$, $\mathrm{B} = \beta$, $\Gamma = \gamma$, then the transformation is a real rotation made up of tree simple rotations about numbered axes $O \Rightarrow R$, $R = R_3^{\alpha} R_2^{\beta} R_1^{\gamma}$.

"Spinor" and "vector" transformation relations are easily derived:

$$O_{k'n} = -\frac{1}{2} Tr\,(U\mathbf{q}_k U^{-1}\mathbf{q}_n), \qquad U = \frac{1 - O_{k'n}\mathbf{q}_k\mathbf{q}_n}{2\sqrt{1 + O_{mm'}}}.$$

GEOMETRY IN 3-DIMENSIONAL SPACE

Hamilton himself noticed that the triad of Q-units behaves as three interconnected unit vectors (with length *i*) and initiates a constant Cartesian frame, exotic somehow due to its "imaginary" character. In 3D space the triad $(\mathbf{q}_1, \mathbf{q}_2, \mathbf{q}_3)$ will be called the quaternionic basis (Q-basis).

Transformations of Q-units have now geometrical sense of different rotations of Q-basis. An example: a simple rotation at real angle α about axis № 3

$$\mathbf{q}' = R_3^{\alpha}\,\mathbf{q}.$$

Notion of Q-basis helps to introduce 3D quaternionic vectors (Q-vectors), defined as

$$\mathbf{a} = a_k \mathbf{q}_k$$

where all components a_k are real. One of the most important properties of a Q-vector is its invariance with respect to vector transformations SO(3,R): if

$$a_{k'} R_{k'j} \mathbf{q}_j = a_j \mathbf{q}_j,$$

then

$$\mathbf{a}' = a_{k'}\mathbf{q}_{k'} = a_k \mathbf{q}_k = \mathbf{a}.$$

Projections of a Q-vector onto arbitrary coordinate axes can be found in two different ways. First, if at least one set of values of a Q-vector a_n and rotation matrix $R_{nk'}$ are given, then the projection of the rotated vector are found straightforwardly as

$$a_{k'} = a_n R_{nk'}.$$

The second way needs considering the following study of interior structure of Q-units.

3. Structure of quaternionic "imaginary" units

Q-UNITS' EIGENFUNCTIONS [12]

Arbitrary vector Q-unit can be thought of as operator, with resulting formulation for it the eigen-function (EF) problem and equations:

$$\mathbf{q}\psi = \lambda\psi\,,\ \varphi\mathbf{q} = \mu\varphi.$$

The solutions are: eigen-values being "imaginary" lengths of vectors

$$\lambda = \mu = \pm i,$$

and EF being columns $\psi^{\pm}$ and rows $\varphi^{\pm}$, functions of components of $\mathbf{q}$.

An example of explicit form of EF: for the Q-unit represented by matrix

$$\mathbf{q} = -\frac{i}{T}\begin{pmatrix} a & b \\ c & -a \end{pmatrix}$$

where $T \equiv a^2 + bc \neq 0$, $b \neq 0$, $c \neq 0$, the EF are found as

$$\varphi^{\pm} = x\begin{pmatrix} 1 & \pm\frac{b}{T \pm a} \end{pmatrix}, \qquad \psi^{\pm} = y\begin{pmatrix} 1 \\ \mp\frac{c}{T \pm a} \end{pmatrix}$$

with *x*, *y* arbitrary complex factors.

Arising freedom in components in EF is eliminated by normalization condition

$$\varphi^{\pm}\psi^{\pm} = 1,$$

while orthogonality is an inherited property of EF

$$\varphi^{\mp}\psi^{\pm} = 0.$$

One can build tensor products of EF and obtain 2×2-matrices

$$C^{\pm} \equiv \psi^{\pm}\varphi^{\pm}$$

with properties reciprocal to those of the vector $\mathbf{q}$:

$$\det C = 0, \qquad Tr\, C = 1,$$

while

$$\det \mathbf{q} = 1, \qquad Tr\, \mathbf{q} = 0.$$

Moreover, matrices C are idempotent

$$C^n = C,$$

and they can be algebraically expressed through its Q-vector

$$C^{\pm} = \frac{1 \pm i\mathbf{q}}{2}.$$

Being inversed the last relation represents the inner structure of a unit Q-vector

$$\mathbf{q} = \pm i\,(2C^{\pm} - 1) = \pm i\,(2\psi^{\pm}\varphi^{\pm} - 1),$$

that turns out to be constructed out of its own EF and scalar unity. Naturally, each vector Q-unit of the triad has its own EF, so a set of EF belonging to any triad selected exists $\{\varphi^{\pm}_{(k)}, \psi^{\pm}_{(k)}\}$. There is an interesting observation about the sets. Q-units are interconnected by a non-linear combination (multiplication), e.g.:

$$\mathbf{q}_3 = \mathbf{q}_1\mathbf{q}_2,$$

But it is easy to show that their EF are interconnected lineary:

$$\varphi^{\pm}_{(3)} = \sqrt{\mp i}\,\varphi^{\pm}_{(1)} \pm \sqrt{i}\,\varphi^{\pm}_{(2)}, \qquad \psi^{\pm}_{(3)} = \sqrt{\pm i}\,\psi^{\pm}_{(1)} \pm \sqrt{-i}\,\psi^{\pm}_{(2)}.$$

Having the quaternionic EF at hand one can express the spinor type transformations of Q-units leaving the Q-multiplication rule invariant in the familiar form

$$\psi^{\pm}_{(k')} = U\psi^{\pm}_{(k)}, \qquad \varphi^{\pm}_{(k')} = \varphi^{\pm}_{(k)}U^{-1},$$

so that the EF can be thought of as set of specific spinor functions subject in general to SL(2C) transformations. The following mathematical observation may be worth mentioning: a set of scalar SL(2C)-invariants is built as contractions of couples of EF, belonging to different Q-units and having different parity. It is interesting that all the invariants are just real or complex numbers, e.g.

$$\sigma^{+}_{12} \equiv \varphi^{+}_{1}\psi^{+}_{2} = \sqrt{-\frac{i}{2}} = \frac{1-i}{2}.$$

Q-EIGEN FUNCTIONS AS PROJECTORS

EF act on their own Q-basis as following

$$\varphi^{\pm}_{(1)}\mathbf{q}_1\psi^{\pm}_{(1)} = \pm i, \; \varphi^{\pm}_{1}\mathbf{q}_2\psi^{\pm}_{1} = 0, \; \varphi^{\pm}_{1}\mathbf{q}_3\psi^{\pm}_{1} = 0,$$

or in general

$$\varphi^{\pm}_{(k)}\mathbf{q}_n\psi^{\pm}_{(k)} = \pm i\,\delta_{kn} \text{ (no summation for } k\text{)}.$$

It looks as though EF 'choose' projection of their 'native' Q-unit. The idea is confirmed when the action of initial Q-units onto rotated Q-basis is considered

$$\varphi^{\pm}_{(k)}\mathbf{q}_{n'}\psi^{\pm}_{(k)} = \varphi^{\pm}_{(k)}R_{n'm}\mathbf{q}_{m}\psi^{\pm}_{(k)} = \pm i\, R_{n'k} = \pm i\cos\angle(\mathbf{q}_{n'},\mathbf{q}_{k}),$$

the result appears to be nearly projection of Q-basis $\mathbf{q}'$ onto $\mathbf{q}$. Convenient notation for real projection is

$$\langle \mathbf{q}_{n'}\rangle_{k} \equiv \mp i\varphi^{\pm}_{(k)}\mathbf{q}_{n'}\psi^{\pm}_{(k)} = \cos\angle(\mathbf{q}_{n'},\mathbf{q}_{k}).$$

Thus projection of any Q-vector **a** onto arbitrary direction $\mathbf{q}_j$ is readily computed

$$\langle \mathbf{a}\rangle^{+}_{j} \equiv -i\, a_{k'}\varphi^{+}_{(j)}\mathbf{q}_{k'}\psi^{+}_{(j)} = a_{k'}R_{k'j} = a_{j}.$$

Thus quaternionic EF being objects more fundamental than Q-units and having their own interesting mathematics are also useful tool for practical purposes like computing projections of Q-vectors.

4. Differential Q-Geometry

QUATERNIONIC CONNECTION

If vectors of Q-basis are smooth functions of parameters $\mathbf{q}_k(\Phi_\xi)$ with index ξ enumerating parameters, then

$$d\mathbf{q}_k(\Phi) = \omega_{\xi\, kj}\mathbf{q}_j d\Phi_\xi$$

where object $\omega_{\xi\, kj}$ is called quaternionic connection. It is skew in vector indices

$$\omega_{\xi\, kj} + \omega_{\xi\, jk} = 0,$$

and number of its components is

$$N = G\, p(p-1)/2,$$

where G is number of parameters and $p = 3$ is number of dimensions. If G = 6 [*SO(3,C)* case] then $N = 18$; If G = 3 [*SO(3,R)* case] then $N = 9$.

Q-connection may be computed by at least three ways:

using vectors of Q-basis $\omega_{\xi kn} = \left\langle \dfrac{\partial \mathbf{q}_k}{\partial \Phi_\xi}\right\rangle^{+}_{n}$,

using matrices U from SL(2C) and special representation of constant Q-units $\mathbf{q}_{\tilde{k}} = -i\sigma_k$ where σ_k are the Pauli matrices

$$\omega_{\xi kn} = \left\langle U^{-1}\frac{\partial U}{\partial \Phi_\xi}\mathbf{q}_{\tilde{k}} - \mathbf{q}_{\tilde{k}}U\frac{\partial U^{-1}}{\partial \Phi_\xi}\right\rangle^{+}_{n},$$

and finely using matrices O from SO(3,C)

$$\omega_{\xi kn} = \frac{\partial O_{\tilde{k}\tilde{j}}}{\partial \Phi_\xi}O_{\tilde{n}\tilde{j}}.$$

All formulae, naturally, give similar results.

From the viewpoint of vector transformation Q-connection is not a tensor. If $\mathbf{q}_k = O_{kp'}\mathbf{q}_{p'}$, then transformed object is expressed through initial one with additional non-homogeneous term

$$\omega_{\xi kj} = O_{kp'}O_{jn'}\omega_{\xi p'n'} + O_{jp'}\frac{\partial O_{kp'}}{\partial \Phi_\xi}.$$

The Q-connection has clear geometrical and physical meaning in 3D case when variable Q-basis behaves as a Cartan frame. Parameters of its R-rotations may depend on space coordinates $\Phi_\xi = \Phi_\xi(x_k)$ then $\partial_n \mathbf{q}_k = \Omega_{nkj}\mathbf{q}_j$, and modified Q-connection

$$\Omega_{nkj} \equiv \omega_{\xi kj}\, \partial_n \Phi_\xi$$

plays the role of Ricci rotation coefficients. The parameters may depend on line's length or time of observer, then $\Phi_\xi = \Phi_\xi(t)$, $\partial_t \mathbf{q}_k = \Omega_{kj}\mathbf{q}_j$, and the Q-connection

$$\Omega_{kj} \equiv \omega_{\xi kj}\, \partial_t \Phi_\xi$$

is a set of the frame's generalized angular velocities.

Characteristic examples of the Q-frame and Q-connection use are:

a) Frenet frame. For a curve given in constant frame $x_{\tilde{k}}(s)$, the Frenet Q-basis is $\mathbf{q}_k$, satisfying equations

$$\frac{d}{ds}\mathbf{q}_1 = R_I(s)\mathbf{q}_2,\ \frac{d}{ds}\mathbf{q}_2 = -R_I(s)\mathbf{q}_1 + R_{II}(s)\mathbf{q}_3,\ \frac{d}{ds}\mathbf{q}_3 = -R_{II}(s)\mathbf{q}_2,$$

where first and second linear curvatures are

$$R_I = \Omega_{12},\ R_{II} = \Omega_{23};$$

b) Twisted straight line. For a given line $x_{\tilde{1}} = u$, $x_{\tilde{2}} = x_{\tilde{3}} = 0$, associated Q-basis is built with tangent and orthogonal vectors while Q-connection is represented by the only torsion (twist) term

$$\mathbf{q}_1 = -i\begin{pmatrix} 1 & 0 \\ 0 & -1 \end{pmatrix},\ \mathbf{q}_2 = -i\begin{pmatrix} 0 & -ie^{-i\gamma(u)} \\ ie^{i\gamma(u)} & 0 \end{pmatrix},\ \Omega_{23} = \frac{d\gamma}{du},$$

$\gamma(u)$ being an arbitrary angle smoothly depending on the line's length.

QUATERNIONIC SPACES

Tangent Q-space [13]. It is known that over any N-dimensional differentiable manifold U_N with coordinates $\{y^A\}$ a tangent space T_N with coordinates $\{X^{(A)}\}$ can be constructed so that $dX^{(A)} = g_B^{(A)}dy^B$, with $g_B^{(A)}$ being Lamet coefficients.

One more transformation leads to building of a tangent Q-space can $T(U,\mathbf{q})$ with coordinates $\{x_k\}$, k = 1,2,3;

$$dx_k = h_{k(A)}dX^{(A)} = h_{k(A)}g_B^{(A)}dy^B,$$

where $h_{k(A)}$ are non-square (in general) matrices, normalized with the help of space projectors.

Quaternionic space (itself) $\mathbf{U}_3$ is defined as 3D space locally identical to that of $T(\mathbf{U}_3,\mathbf{q})$. Main characteristics of the Q-space are: Q-metric is represented by vector part of Q-multiplication rule $\mathbf{q}_j\mathbf{q}_k = -\delta_{jk} + \varepsilon_{jkn}\mathbf{q}_n$, it is not symmetric, its skew part is a Q-operator (matrix), and so each point of $\mathbf{U}_3$ has interior Q-structure. The Q-connection of $\mathbf{U}_3$ may be (i) proper (metric) $\Omega_{nkj} \equiv \omega_{\xi kj}\,\partial_n \Phi_\xi$, for variable Q-basis it always differs from zero, and (ii) affine (non-metric), independent of Q-basis. Q-torsion for both cases does not vanish while quaternionic curfature $r_{knab} = \partial_a \Omega_{bkn} - \partial_b \Omega_{akn} + \Omega_{ajn}\Omega_{bjk} - \Omega_{bjk}\Omega_{ajn}$ for metric Q-connections identically equals to zero, but exists for affine connection. Notion of Q-space evokes more profound study of differential manifolds and spaces with extra interior properties. Preliminary classification of Q-spaces counting curvatures, torsions and non-metrisities, distinguishes at least 10 different families. As well the Q-spaces can serve as helpful background for classical and quantum physical theories and problems.

5. Newtonian mechanics in Q-basis

EQUATIONS OF DYNAMICS IN ROTATING FRAME [14]

Endowed with a clock the Q-basis becomes classical (non-relativistic) frame of reference. For an inertial observer the dynamic equation of classical mechanics can be written in a constant frame

$$m\frac{d^2}{dt^2} x_{\tilde{k}} \mathbf{q}_{\tilde{k}} = F_{\tilde{k}} \mathbf{q}_{\tilde{k}} .$$

SO(3,R)-invariance of two Q-vectors, radius-vector $\mathbf{r} \equiv x_k \mathbf{q}_k$ and force $\mathbf{F} \equiv F_k \mathbf{q}_k$, allows to represent the equation in Q-vector quaternionic form

$$m\frac{d^2}{dt^2}(x_k \mathbf{q}_k) = F_k \mathbf{q}_k , \qquad \text{or} \qquad m\ddot{\mathbf{r}} = \mathbf{F} .$$

Explicitly the last equation is

$$m\left(\frac{d^2}{dt^2} x_n + 2\frac{d}{dt} x_k \Omega_{kn} + x_k \frac{d}{dt}\Omega_{kn} + x_k \Omega_{kj}\Omega_{jn}\right) = F_n .$$

Due to skew-symmetry of the connection (generalized angular speed)

$$\Omega_j \equiv \Omega_{kn}\frac{1}{2}\varepsilon_{knj} , \qquad \Omega_{kn} = \Omega_j \varepsilon_{knj} ,$$

the dynamic equation can be rewritten in vector components

$$m\left(a_n + 2v_k \Omega_j \varepsilon_{knj} + x_k \frac{d}{dt}\Omega_j \varepsilon_{knj} + x_k \Omega_j \Omega_m \varepsilon_{jkp}\varepsilon_{mpn}\right) = F_n$$

or in more familiar traditional vector form

$$m(\vec{a} + 2\vec{\Omega}\times\vec{v} + \dot{\vec{\Omega}}\times\vec{r} + \vec{\Omega}\times(\vec{\Omega}\times\vec{r})) = \vec{F} .$$

One instantly reveals here four classic accelerations: linear, Coriolis, rotational and centripetal; but in the classical sense it is the case only for a simple rotation, otherwise the angular velocity describe combination of the frame of reference (here Q-basis) many rotations, sometimes quite sophisticated. Note that derivation of the equation, rather prolonged in manuals for the simple case, with the help of Q-math even for much more complicated rotations of the frame is in fact one-line-ready expression.

Examples of Q-formulated mechanical problems

a) 'Chasing' Q-basis is the frame whose one vector, let $\mathbf{q}_1$, always is pointed onto observed particle. Explicit form of the dynamic equations for general case is

$$\ddot{r} - r(\Omega_2^2 + \Omega_3^2) = F_1 / m,$$

$$2\dot{r}\Omega_3 + r\dot{\Omega}_3 + r\Omega_2\Omega_1 = F_2 / m,$$

$$2\dot{r}\Omega_2 + r\dot{\Omega}_2 + r\Omega_1\Omega_3 = -F_3 / m.$$

The connection components are given in terms of angles of two rotations, first (angle α) about vector $\mathbf{q}_3$, second (β) about $\mathbf{q}_2$

$$\Omega_1 = \dot{\alpha}\sin\beta, \qquad \Omega_2 = -\dot{\beta}, \qquad \Omega_3 = \dot{\alpha}\cos\beta.$$

The 'chasing' Q-basis approach helps to solve many mechanical problems involving objects' rotations, complicated too. An illustration is the following.

b) Rotating oscillator. Sought for is motion law $r(t)$ of an oscillator (mass m, spring elasticity k) having freedom to move along smooth rode rotating in a plane about one its end with angular velocity ω; equilibrium point is at distance l from the rotation center; no gravity.

Using 'chasing' Q-basis radial and tangent equations (F is unknown rod-mass reaction force)

$$\ddot{r} - r\omega^2 = -\frac{k}{m}(r - l), \qquad 2\dot{r}\omega = \frac{1}{m}F,$$

one readily finds families of solutions:

(i) $r(t) = r_0 + vt$,

describing linear run of the mass from rotation center,

(ii) $r(t) = const + Ae^{iwt} + Be^{-iwt}$, $\qquad w \equiv \sqrt{k/m - \omega^2}$.

describing three different situations:

- $r = const$,
- harmonic oscillations,
- exponential run.

Amusing is formal similarity of the solutions to four Einstein-DeSitter-Freedman models of the Universe in General Relativity.

6. Quaternionic Relativity

HYPERBOLIC ROTATIONS AND BIQUATERNIONS [15]

As was said the SO(3,C)-transformation of Q-units admit pure imaginary parameters. In this case rotations become hyperbolic (H), e.g. a simple H-rotation of the form $\mathbf{q}' = H_3^{\psi}\mathbf{q}$, matrices of the frame being no more anti-hermitian:

$$H_3^{\psi} = \begin{pmatrix} \cosh\psi & -i\sin\psi & 0 \\ i\sin\psi & \cosh\psi & 0 \\ 0 & 0 & 1 \end{pmatrix}, \qquad \mathbf{q}_{1'} = -i\begin{pmatrix} 0 & e^{\psi} \\ e^{-\psi} & 0 \end{pmatrix},$$

Now it is the time to recall notion of biquaternion (BQ) vectors. A BQ-vector is defined as a Q-vector with complex components $\mathbf{u} = (a_k + ib_k)\mathbf{q}_k$. The vectors of the type obviously have in general problems with definition of their norm. But there is a set of 'good' BQ-vectors with well definable norm $\mathbf{u}^2 = b^2 - a^2$; they are form-invariant under subgroup $SO(2,1) \subset SO(3,C)$, and, in particular under simple H-rotations,

$\mathbf{q}' = H\mathbf{q}$, $\mathbf{u} = u_k \mathbf{q}_k = u_{k'} \mathbf{q}_{k'}$ but only if orthogonality condition of the vector-components is fulfilled $a_k b_k = 0$.

QUATERNIONIC RELATIVITY

The given above observation allows to assume a space-time interval to be a BQ-vector

$$d\mathbf{z} = (dx_k + idt_k)\mathbf{q}_k .$$

with specific features: (i) the change-of-time magnitude is an imaginary vector, (ii) space-time of the model turns out 6-dimensional (6D), (iii) particle space displacement and respective change-of-time vectors must be always orthogonal $dx_k \, dt_k = 0$. The 6D interval then is itself $SO(2,1) \subset SO(3,C)$-invariant as well as, of course, its square (differing from norm only by sign) $d\mathbf{z}^2 = dt^2 - dr^2$, the latter looking precisely as space-time interval of Einstein's Special Relativity (SR). So described 6D model was initially called Quaternionic Relativity (QR). Time variable here as well as space one intrinsically enters the Q-interval, and the set of Q-units can be naturally treated as relativistic frame of reference $\Sigma \equiv (\mathbf{q}_1, \mathbf{q}_2 \, \mathbf{q}_3)$. Transit to another Q-frame is performed by rotational equation (RE) of the type $\Sigma' = O\Sigma$ with O being product of R- and H-rotations from $SO(2,1)$; that is why QR may be also called (and more correctly) Rotational Relativity (RR). Meaning of simple H-rotations is revealed straightforwardly from RE of the type $\Sigma' = H_3^{\psi}\Sigma$, or in explicit form

$$i\mathbf{q}_{1'} = i \cosh\psi \, (\mathbf{q}_1 + \tanh\psi \, \mathbf{q}_2)$$

and with $\cosh\psi = dt / dt'$

$$idt'\mathbf{q}_{1'} = idt(\mathbf{q}_1 + V\mathbf{q}_2)$$

implying that Σ' moves relative to Σ with velocity V along $\mathbf{q}_2$. It is verified that from H-rotations of Q-frames Lorents coordinate transformations follow, hence all known motion effects of SR.

But here it is time to note that nothing forbids to R- and H-rotational parameters to be variable, e.g. to depend on time of observers. This implies possibility to describe in the framework of QR non-inertial motions. The idea has been strongly confirmed. Well-known hyperbolic motion problem considered many times (with additional to SR assumptions) in QR is solved naturally and fast not only from the inertial observer viewpoint, but from viewpoint of accelerated frame too [16].

With the help of RE of the type $\Sigma' = H_2^{\psi(t)} R_1^{\alpha(t)} \Sigma$ cinematic problem was completely solved also for a frame Σ' subject to relativistic circular motion around immobile frame Σ. Resulting correlations are given here for an immobile observer

$$t = \int dt' \cosh\psi(t'), \quad \alpha(t) = \frac{1}{R}\int dt' \tanh\psi(t'),$$

$$a_{\tan}(t) = \frac{1}{\cosh^2\psi}\frac{d\psi}{dt}, \qquad a_{norm}(t) = R\left(\frac{d\alpha(t)}{dt}\right)^2,$$

but analogous relations is easily found for observer in arbitrary rotating frame.

«Classical» Thomas precession relations, normally also demanding assumptions extra to SR, in QR are readily derived from the firs row of the matrix RE $\Sigma'' = R_1^{-\alpha(t)} H_2^{\psi} R_1^{\alpha(t)} \Sigma$ giving correct value of the precession frequency

$$\omega_T = (1 - \cosh\psi) \approx -\frac{1}{2}\omega V^2 .$$

Moreover, the QR approach helps to treat the Thomas-like precession for Q-frame relative motions of general kind. The basic is as always a RE, naturally generalized $\Sigma'' = R^{-\theta(t)} H^{\psi(t)} R^{\theta(t)} \Sigma$, where $\theta(t)$ is the non-uniformly changing angle of instant rotation. Important is that H-rotation axis should be normal to the plane, formed by radius-vector of the observed frame and vector of its velocity, and in the same time orthogonal to the latter. Changing with time generalized Thomas frequency is found as

$$\Omega_T = \frac{d}{dt}(\theta - \theta') .$$

A computed example is the planet Mercury appearing perihelion shift due to Thomas precession $\Delta\varepsilon = 2{,}7''/100$ years.

General character of the frames' motion (non-inertial included) admitted by QR hints to look for other relativistic effects. One is found in Solar System planets' satellites motion. Relative velocity of the Earth and a planet changes and sometimes achieves a considerable magnitude comparable (to some extent) with fundamental speed. This must result in cinematic differences between calculated and observed from the Earth cyclic processes at the planet. In particular there must be observed differences of the planet's satellites positions. Such an angular difference is surprisingly found to be linearly dependent upon the time of observation

$$\Delta\varphi \approx \frac{\omega V_E V_P}{c^2} t ,$$

with ω being satellite angular velocity, V-s being the Earth's and the planet's Solar speeds. For closest Jupiter satellite $\Delta\varphi \cong 12'$ for 100 Earth years, for closest Mars satellite $\Delta\varphi \cong 20'$ for 100 Earth years [17]. Both values seem big enough to risk performing precise observations.

The naturally arising model of QR and its kinematics based on SO(2,1)-invariant RE are elaborated and work. But there is neither QR-dynamics yet nor theory of field; quaternionic gravity, electromagnetism, weak and strong interactions are remote projects. But no doubt, the baby-theory grew up in time. The assurance is supported by many surprising facts and yet non-explainable coincidences that form an irritating quaternion mosaic upon the 'beautiful' picture of modern physics.

7. Great Quaternionic Coincidences

There at least five such coincidences (named here) revealed by different authors in different time.

1) Maxwell equations as 'differentiability conditions' for Q-variable functions. In 1937 Fueter [18] noticed that Cauchy-Riemann equations $\partial f / \partial z^* = 0$ for complex variable functions physically modeling plane motion of fluid without sources and curls has its quaternionic analogue

$$\left(i\frac{\partial}{\partial t}-\mathbf{q}_{\tilde{k}}\frac{\partial}{\partial x_{\tilde{k}}}\right)\mathbf{H}=0\ ,\ \mathbf{H}=(B_{\tilde{n}}+iE_{\tilde{n}})\mathbf{q}_{\tilde{n}}\ .$$

Great was the surprise of the mathematician when he found that respective physical model turns out to be Maxwell equations in vacuum

$$div\,\vec{E}=0,\quad div\,\vec{B}=0,\quad rot\,\vec{E}-\frac{\partial\vec{B}}{dt}=0,\quad rot\,\vec{B}+\frac{\partial\vec{E}}{dt}=0\,.$$

2) Classical Mechanics in rotating frames. This study was given above. It remains only to stress that the primitive one-line form of the equations

$$m\ddot{\mathbf{r}}=\mathbf{F}$$

may hide inside numerous combinations of simple rotations of frame of reference or of the observed body. With the help of naturally emerging differential Q-geometric objects this equation rapidly and compactly is represented in understandable explicit form.

3) Quaternionic Relativity. One-to-one isomorphism of SO(3,C), group of Q-multiplication rule invariance, and Lorents group, associated with SR, brings to life an exotic kind of quaternionic relativity theory with 6D space-time model absolutely different from SR, but with all its cinematic effects. Specific BQ-vector-interval $d\mathbf{z}=(dx_k+idt_k)\mathbf{q}_k$ invariance under SO(3,C) subgroup SO(2,1) and allowed variability of transformations parameters make possible to compute non-inertial relativistic effects.

4) Pauli equation [19]. If one considers a quantum particle with charge e, mass m, and generalized momentum

$$P_k\equiv-i\hbar\frac{\partial}{\partial x_k}-\frac{e}{c}A_k$$

in the most simple quaternionic space (all parameters are constant), then the particle's Hamiltonian calculated with the help of Q-metric

$$H\equiv-\frac{1}{2m}P_kP_m\mathbf{q}_k\mathbf{q}_m\,,$$

appears to precisely coincide with that of Pauli spin equation

$$H=\frac{1}{2m}\left(\vec{p}-\frac{e}{c}\vec{A}\right)^2-\frac{e\hbar}{2mc}\vec{B}\cdot\vec{\sigma}$$

and Bohr magneton as ready coefficient at the spin-magnetic field interaction term.

5. Yang-Mills field intensity. If out of connection Ω_{amn}, where indices a,b,c count coordinates in a base Q-space, while indices j,k,m,n are related to the vector triad in Q-tangent space, one builds a 'potential' vector (in any Q-space)

$$A_{ka}\equiv\frac{1}{2}\varepsilon_{kmn}\Omega_{amn}\,,$$

and also out of Q-curvature $r_{knab}=\partial_a\Omega_{bkn}-\partial_b\Omega_{akn}+\Omega_{ajn}\Omega_{bjk}-\Omega_{bjk}\Omega_{ajn}$ one analogously construct respective 'intensity' tensor

$$F_{kab}\equiv\frac{1}{2}\varepsilon_{kmn}r_{mnab}\,,$$

then these newly defined geometrical objects are related with each other exactly as intensity and vector-potential of Yang-Mills field

$$F_{kab} \equiv \partial_b A_{ka} - \partial_a A_{kb} + \varepsilon_{kmn} A_{ma} A_{nb}.$$

It is necessary to stress that for Q-spaces with metric connection curvature (and 'intensity') of the Q-space vanishes.

DISCUSSION

Quaternionic numbers of course are first of all mathematical objects, and development of their algebra, analyses and geometry is a self-consistent and fascinating task. But since one starts speaking of geometry, especially differential, there is no way to avoid physics. Widespread opinion exists that Einstein was the first to geometrize physics in his General Relativity. But it is also a known fact that Maxwell formulated his electrodynamics in terms of quaternions, natural to geometrically describe tensions, the language that was given up decades later.

Aspects of quaternions presented in this review confirm the ideas of close relationship between geometry and physics: from Q-frame rotations in mechanics and relativity up to manifestation of Q-space structure in Pauli Hamiltonian and Q-curvature in Yang-Mills force.

Non-conventional physical models emerging in the Q-approach context, like 6D space-time or amazing coincidences can force one to feel that quaternion numbers form a sort of mathematical toy, a 'lego', allowing construction of exotic buildings. But there are two following observations.

1) Despite originality of the models the Q-approach helps to calculate effects for very physical problems, so it is a useful practical tool. A noticeable example is arising due to exponential character of a simple rotation the easy description of rotations summation, including of course relative motion of frames found to be an 'imaginary' rotation. It is worth recalling that summation of frame rotations is quite a task even in classical mechanics.

2) All physical Q-theories appear 'naturally' as the Nature reflection of the Q-mathematics correlations, sometimes just beautiful. An amateur of Pythagorean 'number = world' philosophy would see in the fact an extra argument. Indeed the Q-algebra, the last associative algebra, perfectly fits to physical objects, all of them up to contemporary understanding being associative magnitudes from observable cinematic and dynamic to theoretic tensor and spinor ones. This gives a hope that further efforts in 'Q-numbers – Physics' relation study one day will grow into a wide program. A modest but insistent step to it is a recent exact solution of relativistic oscillator problem found by the author of this review in the framework of Quatrnionic Relativity. The solutions' details will be published elsewhere.

Acknowledgments. The author wishes to thank Prof. V.De Sabbata and Prof. V.N.Melnikov for opportunity to give a lecture on the related subject at the Gravitation and Cosmology School, Italy, Erice, May, 2003.

References:

1. Hamilton W.R. (1853) *Lectures on Quaternions*, Dublin, Hodges & Smith.
2. Hamilton W.R. (1969), *Elements of Quaternions*, Chelsey Publ. Co. N.Y.
3. Stroik D.Y. (1969) *Short history of matemattics*, Moscow, Nauka, (in Russian).
4. Burbaki N. (1963) *Essays on history of matemattics*, Moscow, Nauka, (Russian translation).

5. Bogoliubov A.N. (1983) *Mathematicians, Mechanists*, Kiev.
6. Klien F., (1924) *Arithmetic, Algebra, Analyses*, N.Y., Dover Publ. (Translation from 3-d German edition).
7. Rastall P. (1964) Quaternions in Relativity, *Review of Modern Physics*, July, 820-832.
8. Branets V.N., Shmyglievski I.P. (1973) *Quaternions in the Problems of Rigid Body Orientation*, Moscow, Nauka (in Russian).
9. Horwitz L.P., Biedenharn L.C. (1984) *Quaternionic Quantum Mechanics: Second Quantization and Gauge Fields*, Ann. Phys., **157**, 432-488.
10. Berezin A.V., Kurochkin Yu.A., Tolkachev E.A. (1989) *Quaternions in Relativistic Physycs*, Minsk, Nauka (in Russian).
11. Bisht.P.S, Negi O.P., Rajput B.S., (1991) *Quaternionic Gauge Theory of Dyonic Fields*, Progr. Theor. Phys., **85**, №1, 157—168.
12. Yefremov A.P. (1985) Q-Field, Variable Quatrnionic Basis, Fizika, Izvestiya Vuzov, **12**, 14-18.
13. Yefremov A.P. (2001) *Tangent Quaternionic Space*, Gravitation & Cosmology, **7**, № 4 273-275.
14. Yefremov A.P. (1995) Newtonian Mechanics in Quaternionic Basis, Moscow, Peoples Friendship Univ. Publ. (in Russian).
15. Yefremov A.P. (1996) *Quaternionic Relativity .I .Inertial Motion*, Gravitation & Cosmology, **2**, № 1, 77-83.
16. Yefremov A.P. (1996) *Quaternionic Relativity .II. Non-Inertial Motion*, Gravitation & Cosmology, **2**, № 4, 335-341.
17. Yefremov A.P. (2000) *Rotational Relativity*, Acta Phys. Hungarica, New Series – Heavy Ion Physics **11**, № 1-2.147-153.
18. Fueter R. (1934-1935) Comm. Math. Hel., **B7S**, 307-330.
19. Yefremov A.P. (1983) *Quaternionic Multiplication Rule as a local Q-Metric*, Lett. Nuovo Cim. **37**, № 8, 315-316

SUBJECT INDEX